A. Rushton, A. S. Ward, R. G. Holdich

Solid-Liquid Filtration and Separation Technology

Second, Completely Revised Edition

A. Rushton, A. S. Ward, R. G. Holdich

Solid-Liquid Filtration and Separation Technology

Second, Completely Revised Edition

WILEY-VCH

Weinheim · New York · Chichester · Brisbane · Singapore · Toronto

Dr. Albert Rushton
'Colynwood'
Claremont Drive
West Timberley
Cheshire, WA 14 5NE
Great Britain

Dr. Anthony S. Ward
Dr. Richard G. Holdich
Department of
Chemical Engineering
Loughborough University
of Technology
Loughborough LE11 3TU
Great Britain

This book was carefully produced. Nevertheless authors and publishers do not warrant the information contained therein to be free of errors. Readers are advised to keep mind that statements data, illustrations, procedural details or other items may inadvertently be inaccurate.

1st edition 1996
2nd, completely revised edition, 2000

Library of Congress Card No. applied for

A CIP catalogue record for this book is available from the British Library

Die Deutsche Bibliothek – CIP Cataloging-in-Publication-Data
A catalogue record for this publication is available from Die Deutsche Bibliothek

© WILEY-VCH Verlag GmbH, D-69469 Weinheim (Federal Republic of Germany), 2000

Printed on acid-free and chlorine-free paper

All rights reserved (including those of translation into other languages). No part of this book may be reproduced in any form – by photoprinting, microfilm, or any other means – nor transmitted or translated into a machine language without written permission from the publishers. Registered names, trademarks, etc. used in this book, even when not specifically marked as such, are not to be considered unprotected by law.

Composition: Graphik & Text Studio, D-93092 Regensburg-Barbing
Printing: Betzdruck, D-63291 Darmstadt
Bookbinding: Osswald & Co., D-67433 Neustadt (Weinstraße)

Printed in the Federal Republic of Germany

Preface to the Second Edition

The pace of technological development continues and no aspect is free from change. The period since the first edition went to press has seen much new work published in the field of filtration and separation so it was felt important to introduce a revised and up to date version of this text. The first edition was warmly received and kindly reviewed, for which comfort the authors are particularly grateful, but inevitably there were some imperfections and shortcomings evident. So the authors are doubly grateful for this opportunity to present a revised and refreshed edition.

Highlights of this second edition include a major revision and updating of the chapters dealing with the fundamental processes of filtration and sedimentation. Full details are provided on how to simulate the formation of filter compacts and sediments, including compressible compacts and sludges in consolidation tanks. New pictures, illustrations, descriptions and applications of the process equipment are included. Information to create the simulation models in a computer spreadsheet package is contained within the relevant chapters, and the World Wide Web address is provided to allow the relevant files to be downloaded at no further cost.

Extensive revision of the section on crossflow membrane microfiltration includes discussions of emerging applications such as the removal of cryptosporidium oocysts from drinking water. A new section concerns the design of microfilter membranes to minimise fouling and information on the use of critical flux strategy to avoid permeate flux decay.

The role of surfactants in coagulation and flocculation is included and the extended DLVO theory, which usefully explains some anomalies, is introduced. Applications of surfactants and other surface effects such as that of zeta potential are discussed in relation to sedimentation and flotation.

The chapter on process equipment and calculations has been revised and extended to include further analysis of filtration economics.

There are many new diagrams, tables and photographs throughout the book.

A. Rushton
A.S. Ward
R.G. Holdich

January 2000

Preface to the First Edition

The separation of particulate solids from Liquids by filtration and associated techniques constitutes an important and often controlling stage in many industrial processes. The latter generate a somewhat bewildering array of particle-fluid separation problems. Separation by filtration is achieved by placing a permeable filter in the path of the flowing suspension. The barrier, i.e. a filter screen, medium or membrane in some cases is selected with a view to retaining the suspended solids on the filter surface, whilst permitting passage of the clarified Liquid. Other systems, e.g. deep-bed or candle filters, operate in a different mode, in promoting deposition of the particles within the interstices of the medium. Further purification of the clarified liquid may proceed by the use of adsorbents to remove dissolved solutes. Alternatively, the two phases may be separated by sedimentation processes, in the presence of gravitational or centrifugal force fields.

Serious operational problems centre on the interaction between the particles and the filter medium. Plugging of the latter, or collapse of the collected solids under the stress caused by flow through the filter, can result in low productivity. Such effects are often related to the size of particles being processed; enhanced effective particle size can be accomplished by pretreatment with coagulants or flocculants. These techniques are discussed in detail in the text, which also reports recent improvements in the rnachinery of separation, e.g. the variable chamber presses, the cross-flow processes, ceramic dewatering filters, etc.

Several of these newer modifications in filtration plant have followed trends in the developing science of solid–fluid separation and the growing understanding of the processes involved. Fortunately, filtration processes have attracted the attention of increasing numbers of scientists and engineers. A large output of literature has resulted in a copious flow of design and operational information sufficient to place filtration on a much sounder scientific basis.

Nevertheless, the random nature of most particulate dispersions has resulted in a wide range of machines in tlis unit operation. Selection of the best available separation technique is, therefore, a difficult process problem. lt is the authors' viewpoint that many existing separation problems would have been avoided by the application of available scientific data. This text is aimed at the provision of theoretical and practical information which can be used to improve the possibility of selecting the best equipment for a particular separation. lt is relevant to record the recent increased commercial awareness of the need for this information in the selection of plant used in environmental control.

The material presented in the text has been used by the authors in short-course presentations over several years. These courses are illustrated by a large number of practical problems in the SLS field; some of these problems have been used to illustrate the book.

Basic theoretical relationships are repeated in those chapters dealing with process calculations. Tlüs feature minimises the need for back-referencing when using the book.

A. Rushton
A.S. Ward
R.G. Holdich

January 1996

Contents

9 Post-Treatment Processes

Appendix A

Particle Size, Shape and Size Distributions

Appendix B

Slurry Rheology

Appendix C

Computer Spreadsheet Files

Index

1 Solid Liquid Separation Technology

1.1 Introduction

It is difficult to identify a large-scale industrial process which does not involve some form of solid-fluid separation. In its entirety, the latter activity involves a vast array of techniques and machines. This book is concerned only with those parts of this technological diversity which relate to solid-liquid separation (SLS).

Attempts have been made [Svarovsky, 1981] to catalogue the variety of processes and machines used in SLS systems; these are usually based on two principal modes of separation:

1. Filtration, in which the solid-liquid mixture is directed towards a "medium" (screen, paper, woven cloth, membrane, etc.). The liquid phase or filtrate flows through the latter whilst solids are retained, either on the surface, or within the medium.

2. Separation by sedimentation or settling in a force field (gravitational, centrifugal) wherein advantage is taken of differences in phase densities between the solid and the liquid. The solid is allowed to sink in the fluid, under controlled conditions. In the reverse process of flotation, the particles rise through the liquid, by virtue of a natural or induced low "solids" density due to attached air bubbles.

The large range of machinery shown in Figure 1.1 reflects the uncertainty which attaches to the processing of solids, particularly those in small particle size ranges.

The filterability and sedimentation velocity of such mixtures depend on the state of dispersion of the suspension; in turn, the latter is strongly influenced by solid-liquid surface conditions which govern the stability of the mixture and the overall result of particle-particle contact. The properties of such systems may also be time dependent, with filterability and settling rate being a function of the history of the suspension [Tiller, 1974].

The dispersive and agglomerative forces present in these systems are functions of pH, temperature, agitation, pumping conditions, etc. all of which complicate the situation and produce the result that suspension properties cannot be explained in hydrodynamic terms alone. Despite these formidable problems, modern filtration and separation technology continues to produce separations in seemingly intractable situations, and to eliminate the "bottle neck" characteristic of the SLS stage in many processes.

A first step in the rationalisation of such problems is to choose the most appropriate technology from filtration, sedimentation or a combination of these two operations. In general, sedimentation techniques are cheaper than those involving filtration; the use of gravity settling would be considered first, particularly where large, continuous liquid flows are involved [Pierson, 1981].

A small density difference between the solid and fluid phases would probably eliminate sedimentation as a possibility, unless the density difference can be enhanced, or the force field of gravity increased by centrifugal action. Such techniques for enhancing sedimentation would be retained as a possibility in those circumstances where gravity separation proves to be impossible, and the nature of the particulates was such as to make filtration "difficult". The latter condition would ensue when dealing with small, sub-micron material, or soft, compressible solids of the type encountered in waste water and other effluents. Some separations require combinations of the processes of sedimentation and filtration; preconcentration of the solids will reduce the quantity of liquid to be filtered and, therefore, the size of filter needed for the separation.

Having decided upon the general separation method, the next stage is to consider the various separational techniques available within the two fields. These operational modes may be listed as:

A Sedimentation: gravity; centrifugal; electrostatic; magnetic
B Filtration: gravity; vacuum; pressure; centrifugal

Another serious consideration, also indicated in Figure 1.1, is whether the separation is to be effected continuously or discontinuously; the latter method is known as "batch" processing. In this case, the separator acts intermittently between filling and discharge stages. The concentration of solids in the feed mixture and the quantities to be separated per unit time are also factors which affect the selection procedure.

This activity is made more complicated by the fact that the separation stage rarely stands alone. Figure 1.2 [Tiller, 1974] includes various pre- and post-treatment stages which may be required in the overall SLS process. Thus, the settling rate of a suspension, or its filterability, may require improvement by pretreatment using chemical or physical methods. After filtration, wet solids are produced, and these may require further processing to deliquor, i.e. reduce the liquid content in the filter cake; in some cases, the latter, being the principal product, requires purification by washing with clean liquid.

It will be apparent that in the development of a typical process for: (a) increasing the solids concentration of a dilute feed, (b) pretreatment to enhance separation characteristics, (c) solids separation, (d) deliquoring and washing, many combinations of machine and technique are possible. Some of these combinations may result in an adequate, if not optimal, solution to the problem. Full optimisation would inevitably be time consuming and expensive, if not impossible in an industrial situation. Certain aspects of filter selection are considered at the end of this chapter and in Chapter 11, on pressure filter process calculations.

1.2 The Filtration Process

As stated above, a typical medium for the filtration of coarse materials is a woven wire mesh which will retain certain particulates on the surface of the screen. As the size of the

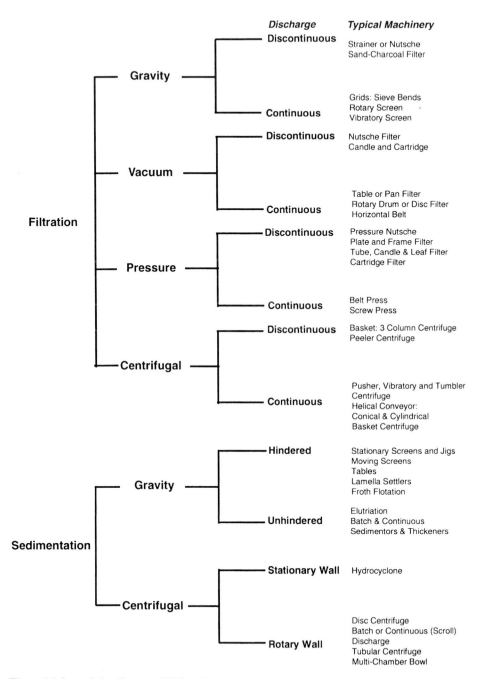

Figure 1.1 General classification of SLS equipment

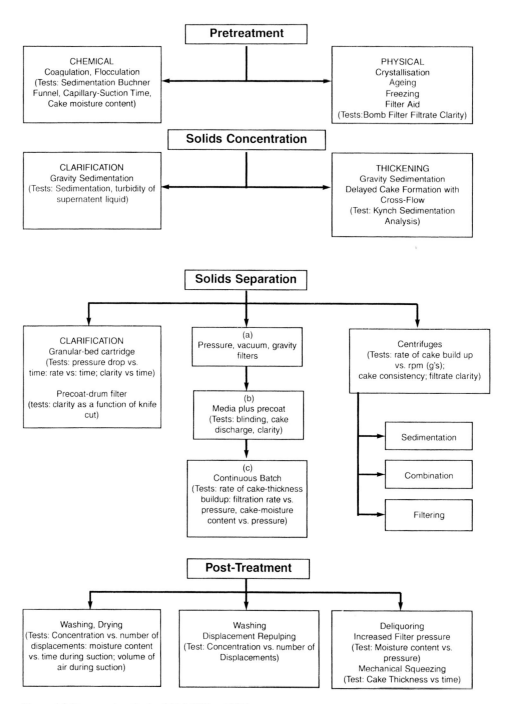

Figure 1.2 Stages and methods of SLS [Tiller, 1974]

particulates decreases, other "screens" are required, e.g. woven cloths, membranes, etc.; these are constructed with smaller and smaller openings or pores. Flow through such a system is shown in Figure 1.3. Where the particulates are extremely small and in low concentration, deposition may occur in the depths of the medium, such as in water clarification by sand filters.

The filtration medium may be fitted to various forms of equipment which, in turn, can be operated in several modes. Thus, plant is available which creates flow by raising the fluid pressure by pumping, or some equivalent device. Such "pressure" filters operate at pressure levels above atmospheric; the pressure differential created across the medium causes flow of fluid through the equipment. Plant of this type can be operated at constant pressure differential or at constant flow rate. In the latter case, the pressure differential will increase with time whilst at constant pressure, liquid flow rate decreases with time.

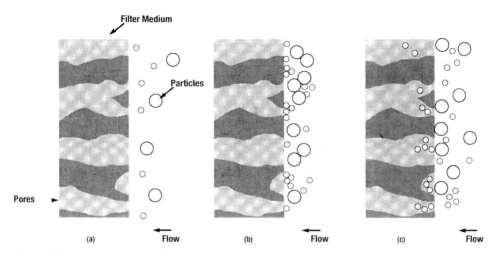

Figure 1.3 Particle deposition in filtration

Subatmospheric, vacuum operation is used in a wide number of applications. In these filters, the absolute pressure downstream of the filter medium is maintained at a low, controlled level by vacuum pumps. The suspension supplied to the filter is delivered at essentially atmospheric pressure; this process is an example of "constant-pressure differential" filtration. Vacuum operation is, of course, a special case of pressure filtration. Here a relatively small pressure differential is available, and attention should be given to pressure losses, e.g. in the frictional effects caused by filtration flow in associated pipes and fittings. Low-pressure conditions in a flowing fluid may lead to the phenomenon of "cavitation" in which dissolved gases, or vapour bubbles, are released into the liquid. Passage of such mixtures through a downstream higher pressure zone, e.g. at the outlet of a pump, causes bubble collapse and material damage to the pump.

Vacuum filters find applications in many areas of industry and are widely used in laboratory tests; the latter are required in order to assess the filterability of the suspension and the suitability of a filter medium. Whilst a complete quantitative description of the SLS process will be described later, at this stage it is sufficient to record that the rate of flow of fluid, at a particular pressure differential, will depend on the resistance to fluid flow of the particles and the filter medium.

Flow can also be created by spinning the suspension, thereby creating a centrifugal force in the system. Thus, centrifugal filters, fitted with suitable filter media are found in many applications in the food, beverage, and pharmaceutical industries.

1.3 Filtration Fundamentals

An understanding of the physical mechanisms involved in SLS is essential in resolving problems developed in existing machinery, or in avoiding future difficulties when selecting new plant. Many years of consultative practice [Purchas, 1987] in this area leads to a realisation of the value of sound quantitative relationships between variables such as liquor flow rate and viscosity, particle size and concentration, filter medium pore rating and their effects on the filtration process.

As mentioned in the previous section, the separation stage is rarely required in isolation and is often followed by drying or dewatering of the porous deposits and/or purification of the recovered solids by washing. Pre- and post-treatment processes such as flocculation, coagulation and liquid expression, may have equivalent process importance in determining, sometimes controlling, the overall separation process time. It is vital to identify process time requirements of the various phases involved in a separation, in order to identify possible bottlenecks. Many examples could be quoted of installed filters with which it is impossible to meet process specifications, and this particularly applies to systems with stringent dewatering and washing requirements. This points to the need for well-designed pilot filter trials, preferably on a small-scale version of the machine of interest, before plant selection. It is, perhaps, unfortunate that such information is not always available, and selection has to proceed with relatively meagre data.

A successful selection procedure is closely linked to the proper choice of the medium to be used in the separation. A large proportion of industrial-scale process difficulties relate to the interaction between the impinging particles and the pores in the filter medium, as depicted in Figure 1.3. The ideal circumstance, where all separated particles are retained on the surface of a medium is often not realised; particle penetration into cloth or membrane pores leads to an increase in the resistance of the medium to the flow of filtrate. This process can ensue to the level of total blockage of the system. Such difficulties can be avoided, if the pores in the medium are all smaller than the smallest particulate in the mixture processed, as discussed below.

The theoretical and practical considerations required for effective SLS are expounded in detail in the various chapters of this book. Of particular importance are the fundamental aspects presented in Chapter 2. Here, the "surface deposition" mode depicted in Figure 1.3 B is described by two series resistances to fluid flow:

a) The resistance of the filter medium R_m
b) The resistance of the particulate layer or "cake" R_c

As obtained in the flow of fluids through pipes and conduits, 'laminar' conditions will be produced by media containing small pores. In these cases, the filtrate velocity v_o through the clean filter medium is proportional to the pressure differential ΔP imposed over the medium; the velocity is inversely proportional to the viscosity of the flowing fluid μ and the resistance of the medium. These relationships may be expressed mathematically as:

$$v_o = \Delta P / \mu R_m \qquad (1.1)$$

Under the same overall pressure differential, the filtrate velocity, after the deposition of particles, decreases to v_f where:

$$v_f = \Delta P / \mu \, (R_c + R_m) \qquad (1.2)$$

These simple expressions are developed further in Chapter 2, to include, inter alia, particle concentration effects, filter cake compression, etc.

The equations above are based on the assumption that the medium resistance does not change during the process. This assumption is considered in depth in Chapter 4, where the influence of the filter medium is explained. Generally, R_m takes low values for open, coarse media, e.g. woven screens; the largest media resistances are found in membranes, used in microporous filtration, ultra filtration and reverse osmosis. The filter media resistances in the latter may be one million times higher than those characterising open screens. Those inherent features of these media which can be used to remove very fine particulates, are reported in detail in Chapters 6 and 10 on membrane technology.

Filter cake resistances vary over a wide range, from free filtering sand-like particulates to high resistance sewage sludges. Generally, the smaller the particle, the higher will be the cake resistance. The latter is sensitive to process changes in slurry concentration, fluid velocity, fluid pressure, temperature, etc. These effects have received much attention [Shirato and Tiller, 1987] in the development of physical and mathematical models of the SLS process.

As mentioned above, post-treatment deliquoring and washing of filter cakes are subjects of great importance in SLS operations. These subjects are fully discussed in Chapter 9. A principal interest in deliquoring wet cakes lies in the economic difference between solids drying by thermal and mechanical methods. Thermal drying costs can be much higher (20-30 times) than costs incurred by mechanical dewatering. Dewatered solids are more easily handled than wet sludges; this is of particular importance in waste water treatment processes. A high solidity in a dewatered filter cake can reduce handling costs and improve the possibility of continuous processing.

In filter cake washing, an important aspect is the time (or wash volume) required to remove residual impurities. Quite often, washing may be the controlling step in overall filtration cycles, as discussed in Chapter 9.

1.4 Sedimentation Processes

In certain circumstances, process conditions preclude the possibility of using direct filtration as a means of separating a solid-liquid mixture. High dilution or extreme fineness of particle lead to uneconomic sizes of filter or, in some cases, make normal filtration impossible without pretreatment or concentration of the feed.

Obviously, any device which, at relatively small cost, reduces the absolute amount of liquid in the feed finds application in processes involving large tonnages. Thus the ubiquitous thickener, used for increasing the solids concentration in dilute feeds, is found in almost every section of process industry.

Here, gravity sedimentation, involving a difference in density between solids and liquids, is used to produce an essentially clear overflow of liquid and a concentrated underflow of the mixture. The latter may now be more readily filterable, or be in a condition suitable for disposal, e.g. in sewage handling. Thickeners are used extensively in hydrometallurgical applications, singly or in series, e.g. in counter-current decantation (washing) plants. Where gravity forces lead to inordinately long settling times, the latter may be reduced by chemical treatment, i.e. flocculation and coagulation, as discussed below.

Mathematical analysis of the sedimentation process starts with the well-known Stokes relationship for the setting velocity of a single particle in an infinite expanse of fluid:

$$u_t = x^2 g (\rho_s - \rho) / 18\mu \tag{1.3}$$

where u_t is the Stokesian gravitational settling velocity, m/s; x is the particle size, m; g is the acceleration of gravity, m/s^2; ρ_s, ρ are the densities of the solid and fluid, respectively, kg/m^3; μ is the viscosity, Pa s. This fundamental relationship must be modified in applications to practical designs of equipment, as shown in Figure 1.4, to allow for the effects of particle concentration on the settling velocity of the suspension U_o.

In general, increases in sludge concentration C lead to decreases in U_o; it follows that the calculation of thickening processes, involving large increases in solids concentration, requires information on the U_o - C relationship. The overall intention in these processes is to produce an overflow of clarified water and an underflow of concentrated sludge.

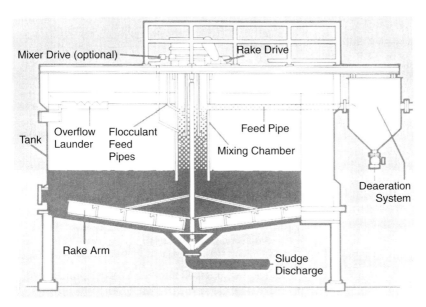

Figure 1.4 High-capacity thickener (Eimco Process Equipment Co. Ltd., United Kingdom)

Suspensions which possess unique U_o-C relationships are termed "ideal"; other mixtures, e.g., biosuspensions, may exhibit "non ideal" settling characteristics where settling rate may be affected by suspensions height, sedimentation column diameter, intensity of mixing before settling, etc. Such suspensions are often described by equations of the type:

$$U_0 = k\,C^{-m}$$ (1.4)

where k relates to the settling velocity at low concentrations. Both k and m vary widely from suspension to suspension. Equation 1.4 indicates that the settling velocity of a suspension is inversely proportional to the solids concentration.

Fundamental aspects of the above processes are considered in detail in Chapter 3. Appropriate tests are required to measure the effect of changes in concentration on U_o. It may be observed that since the downward flux of solids in a settling suspension equates to the product of U_o and C, the possibility of a minimum flux presents itself. Identification of such minima is required in the specification of process plant used for sedimentation and thickening.

Modern sedimentors are sometimes fitted with inclined plates, spaced at intervals, as shown in Figure 1.5. Theoretical aspects of settlement under inclined surfaces are presented in Chapter 3; practical details of the design of such equipment are dealt with in Chapter 7. Gravity sedimentors compete with devices such as sedimenting centrifuges, hydrocyclones, flotation cells, in the process area of fluid clarification and solids concentration.

Along with the selection of filtration machinery, attempts have been made to provide a basis for the selection of sedimentation equipment [Wakeman, 1994]. Table 1.1 contains some of this information, in abridged form which points to the principal factors influencing such selections.

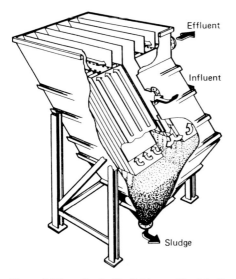

Figure 1.5 Lamella clarifier/thickener (Svedala, Pumps & Process AB, Sala, Sweden)

Table 1.1 Selection of sedimentation machinery [Wakeman, 1994]

	Gravity sedimentation	Sedimentation centrifuges			Hydrocyclones	Flotation
		Tubular bowl	Disc	Scroll		
Scale of process m³/h	1–(>100)	1–10	1–(>100)	1–(>100)	1–(>100)	10–(>100)
Solids settling rate, cm/s	0.1–(>5)	(<0.1)–5	(<0.1)–5	(<0.1)–(>0.5)	0.1–(>0.5)	(>0.2)
Operation C: Continuous B: Batch	B or C	B	B or C	B or C	C	C
Process * Objectives	a,b,c	a,b	a,b	a,b	a,b,c	a,b

* a: clarified liquid; b: concentrated solids; c: washed solids

1.5 Filter Media

The importance of the filter medium in SLS processes cannot be overstated. Whilst any of the filters reported in this text would deal with most solid–liquid suspensions, albeit with low efficiency in some cases, attempts to use inadequate filter media will incur certain failure. It follows that much attention has been given, in the relevant literature, to the role of the medium in SLS processes.

A wide variety of media is available to the filter user; the medium of particular interest will, of course, be of a type which is readily installed in the filter to be used in the process. Thus woven and nonwoven fabrics, (Figure 1.6) constructed from natural or synthetic fibres, are often used in pressure, vacuum and centrifugal filters. Again, these units can be fitted with woven metallic cloths, particularly in those circumstances where filter aids will be used in the process. The same materials, and also rigid porous media (porous ceramics, sintered metals, woven wires, etc.), can be incorporated into cartridge and candle filters. In these applications, the rigid medium will usually be fashioned into a cylinder, although other shapes exist. Random porous media (sand, anthracite, filter aids) will be used in clarification processes. These generalisations can also be extended to flexible and rigid membranes described in Chapter 10.

The filtration mechanisms involved in separations using such media will depend mainly on the mode of separation. Thus in "cake" filtration, ideally, impinging particles should be larger than the pores in the medium. Experience shows that less processing difficulty is experienced in circumstances where media pores are much smaller than the particles. Despite the obviously higher fluid flow resistance of clean tighter media, in practice, the eventual, used medium resistance will be more acceptable.

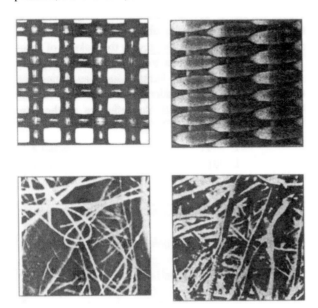

Figure 1.6 Woven and nonwoven filter media

In clarification systems involving depth filtration, loose media are used which are often associated with pores thousands of times larger than the particles requiring filtration. However, deposition of the moving solids onto the medium does occur, and clarified liquids are obtained. Such separations depend on the surface condition and area of the media used in deep-bed systems.

Chapter 4 explores some of the features of filter media, particularly those of the woven fabric variety. The chapter is aimed at developing an understanding of the steps required in media applications to attain process features such as:

(a) clear filtrates
(b) easily discharged filter cakes
(c) economic filtration times
(d) absence of media "blinding"
(e) adequate cloth lifetime

Filter media behaviour is also reported in other sections of the text, e.g. Chapter 10 deals with the crucially important area of membrane separations. Again, Chapter 6 deals with depth filtration systems such as deep sand filters and cartridges, and describes the media used in such equipment.

Some aspects of media selection are also covered in the sections of Chapter 2 which highlight laboratory test procedures. Certainly much can be gained from well-designed laboratory tests in filter media selection. It will also be realised that a vast reservoir of experience and information is available from filter media manufacturers who, fortunately, continue to report their knowledge in the filtration literature.

In this respect, it is interesting to note the newer developments in this subject. Thus in systems where the SLS process calls for filtration and deliquoring of the filtered solids, modern media are available (in woven and ceramic form) which prevent the leakage of gases through the system. Thus Figure 1.7 shows the different behaviour of a modern ceramic "capillary control" medium, which allows the free flow of liquid during filtration and dewatering, but prevents the flow of air used in the last step. This leads to considerable economies in vacuum production [Anlauf and Müller, 1990].

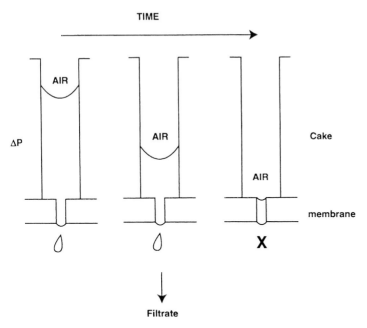

Figure 1.7 Capillary control filter media

Ceramic filter media are being used in a wide range of applications, in both liquid and gaseous filtration systems. This follows from the unique combination of physical characteristics possessed by ceramics, which can be manufactured in tubular, disc or plate forms and with tightly controlled pore structures. These media also exhibit high corrosion resistance and material stability in severe process conditions.

Ceramic membranes are used increasingly in industrial applications, e.g. waste water treatment; these media are discussed in Chapter 4 and Chapter 11, together with the computer spreadsheets described in Appendix C, deals with the influence of the filter medium on process productivities.

1.6 Pretreatment Techniques

In sedimentation systems dealing with very small particles, the action of gravity alone may lead to inordinately long settling times, as reported (Table 1.2).

Table 1.2 Effect of particle size in gravity settling in water at 20^0C

Type of suspension	Particle size range µm	Time required to settle 5 cm *
Coarse	1000–100	1 s–13 s
Fine	100–1	13 s–20 h
Colloidal	1–0.001	20 h–20 years

* Assuming quiescent conditions and a solids density of 2650 kg/m^3

The permanent suspension of colloidal materials follows from the presence of electrically charged ions on the surface of the particles. These ions may be adsorbed from solution, or produced by part-solubilisation of the particle surface. Colloids, with particles less than 1 µm in size, possess large surface area to mass (or volume) ratios. This property enhances the mutual repulsion of charged particles and prevents sedimentation. In these systems, forces of attraction also exist on the surface of the solids. These are of an electrostatic nature, which act over very short distances from the surface of the particle. Stabilisation follows from the action of the repulsive forces in preventing the particles coming close enough for the attractive forces to be operable, despite their random Brownian motion resulting from bombardment by water molecules.

Elimination of the repulsive forces, and a "destabilisation" of the colloidal suspension may be achieved by the addition of certain chemicals. Such processes are of enormous importance in municipal and industrial water clarification plants which contain sedimentation equipment of the type shown in Figure 1.4.

In addition to faster settling, a destabilised system will generally possess enhanced filtration characteristics. These improvements follow from the increased possibility of particulate collisions, with the production of "clusters" or "flocs" of particles. The clusters will have a greater effective diameter. Thus from the Stokes relationship:

u_t proportional to x^2

where x is the cluster diameter, a doubling of the effective diameter will quadruple the sedimentation velocity.

The process of destabilisation is known as coagulation [La Mer and Healey, 1963]; the chemicals used to produce destabilisation are termed primary coagulants, which act to reduce the repulsive potential of the surface charges. The overall result is a relatively small cluster possessing poor filtration characteristics. The process of coagulation may be followed by the action of synthetic long-chained, high molecular weight polyelectrolytes, of the type shown in Figure 1.8. These act to increase the size of the coagulated clusters, thereby increasing sedimentation velocities and filtration rates of the suspension. Similar action may be effected by the use of naturally occurring polyelectrolytes, such as starch, gums, etc.

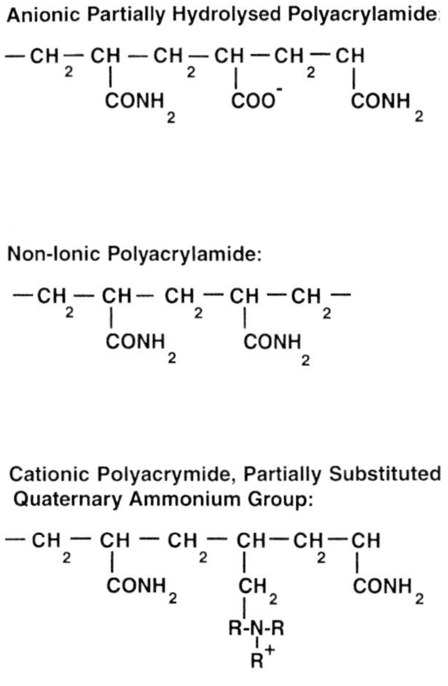

Anionic Partially Hydrolysed Polyacrylamide:

$$-CH_2-CH-CH_2-CH-CH_2-CH-$$
$$\quad\quad\; |\quad\quad\quad\quad |\quad\quad\quad\quad\; |$$
$$\quad\quad CONH_2\quad\; COO^-\quad\quad CONH_2$$

Non-Ionic Polyacrylamide:

$$-CH_2-CH-CH_2-CH-CH_2-$$
$$\quad\quad\; |\quad\quad\quad\quad\; |$$
$$\quad\quad CONH_2\quad\; CONH_2$$

Cationic Polyacrymide, Partially Substituted Quaternary Ammonium Group:

$$-CH_2-CH-CH_2-CH-CH_2-CH-$$
$$\quad\quad\; |\quad\quad\quad\quad\; |\quad\quad\quad\; |$$
$$\quad\quad CONH_2\quad\; CH_2\quad\quad CONH_2$$
$$\quad\quad\quad\quad\quad\;\; |$$
$$\quad\quad\quad\quad\quad R\text{-}N\text{-}R$$
$$\quad\quad\quad\quad\quad\; |\; +$$
$$\quad\quad\quad\quad\quad R^+$$

Figure 1.8 Synthetic organic polyelectrolytes

Cluster growth proceeds firstly by "perikinetic" flocculation. This involves quite small species, submicrometre in diameter, and interception by Brownian motion. Later "orthokinetic" collisions caused by fluid motion, induced by the action of mixing, become more significant. The importance of induced shear, in practical flocculation devices, for the production of stable clusters is discussed in Chapter 5, along with the different molecular mechanisms involved in coagulation and flocculation [Akers, 1986].

Briefly, in coagulation, charge neutralisation and compression of the effective range of repulsive forces are related to factors such as pH, salt valency and ionic composition. In flocculation, it is believed that physical attachment of one part of the long-chained polyelectrolyte, followed by further attachments to other particles, causes a bridging action from particulate to particulate. To be effective, the polymer has to be extended in length during the flocculation step. After bridging, the polymer coils together, thereby producing a large particle from several smaller units.

The reduction in filtration resistance, referred to above, may also be achieved by adding a certain amount of free-filtering particles to the feed suspension. Filter aids are naturally occurring materials such as diatomaceous earths (kieselguhr), perlites (volcanic ash), cellulose, and carbon which are characterised by a low resistance to the flow of fluids and, when deposited as filter cakes, are relatively incompressible. The filter cakes produced from such substances are open, porous structures of high permeability. Various grades of filter aids are available, with the finest grades required for the retention of sub-micrometre particles [Akers, 1986].

1.7 Clarification Filtration

Perhaps the most common application of filter aids is in the clarification of valuable liquid products, in circumstances where the latter are contaminated with small quantities of suspended impurities. Thus in beverage products, shelf-life and product appearance will require stringent control of product clarity. Here the filter aid must be suitable, chemically and physically, for use in food products.

Filter aids are applied in two ways: (a) as a pre-coat, where a thin (approximately 2 mm) layer of aid is deposited on the filter medium, before the commencement of filtration, or (b) as a "body feed" where a certain quantity of aid is added to the feed slurry. The latter procedure is adopted in cases where the suspended solids have poor filterabilities and would form an impervious layer on the surface of the filter medium, or even the filter aid pre-coat. The combined deposit of filter aid and suspended impurity possesses a much reduced filtration resistance. Usually an optimum dosage of body feed is sought, in overall cost terms, as discussed in Chapter 5.

Clarification of liquids often calls for the use of depth filtration equipment. The process techniques used and the filtration mechanisms involved in depth filters are outlined in Chapter 6, which includes information on sand filters and cartridges.

The effectiveness of the simple sand filter can be improved by superimposing a layer of anthracite on the surface of the sand layer. This is shown in Figure 1.9 where a dual layer of anthracite/sand is contained in a pressure vessel. The anthracite particles, being coarser than the sand, serve to prevent the formation of surface deposits on the sand surface. In low-pressure systems, avoidance of these deposits leads to longer operating cycles. In some cases, dense solids, e.g., alumina will be used as a third layer, situated beneath the sand. A general guide to the performance of these filters is provided in Table 1.3.

Cartridge and diatomaceous earth pre-coat filters are often used as second-stage polishing filters downstream of sand/gravel units. The information in Table 1.3 should be taken as a general guide only, since specific separations may involve quite different results; the data below pertain to the cleaning of seawater. In some cases, the above clarifiers are supported by upstream screens or strainers which are designed to remove larger detritus from the feeds. Some information on screening operations is included in Chapter 6.

The relatively high efficiency of deep-bed filters in removing fine particles present in low concentration has generated interest in industrial processes where water reclamation or polishing free of suspended inorganics is a principal concern. For example, the removal of suspended salts from brine prior to electrolysis is an essential feature in chlorine production; the reduction in suspended solids from factory effluents reduces sewage costs and in some cases, provides recycle water to the process. In the latter case, a reduction in supplied water to the factory produces further economies. The high capacity available in the modern sand filter, particularly when polyelectrolytes are used to enhance capture, offers the possibility of use in cases where conventional equipment could not produce a separation, or would be too expensive.

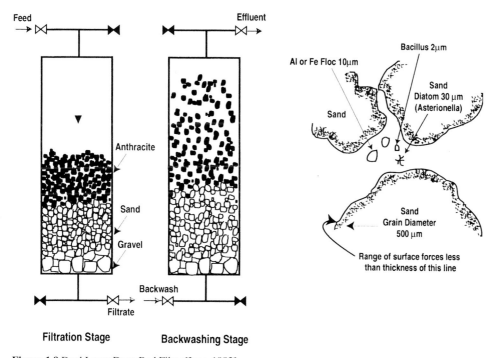

Figure 1.9 Dual Layer Deep Bed Filter [Ives, 1982]

Table 1.3 Clarifying filter performance data on seawater

Filter	Design flow rate, m/h	Maximum suspended solids in feed, ppm	Average suspended solids, after filter, ppm	Retained particles >μm
Sand	9	100	3	10
Anthracite	9	100	75	15
Cartridge	14	3	0.5	2
Pre-coat	1.5	30	0.2	7

Water is passed through a granular bed of sand and anthracite, (Fig. 1.9) in which the diameter of the pores is 100–10 000 times larger than the diameter of some of the suspended particles. Apart from mechanical interception of the particles, determined by their size relative to the pore size of the bed, interception due to electrostatic and London/van der Waals attractive forces also takes place. The latter mechanism dominates for finer material. Flow may be directed downwards or upwards, with typical liquid velocities of 40 and 15 m/h, respectively.

During the percolation, the filter bed is contaminated with the substances filtered off. At fixed times, the filter has to be rinsed clean in counterflow with water (back-wash); the filter bed is fluidised by the latter, quickly releasing accumulated particles deposited during the filtration stage.

The size of particles which are mechanically intercepted by granular bed filters can be estimated. The size depends on the smallest size of the filter bed material and the pore size of the filter cake formed. During the initial stages of filtration, when no filter cake is formed, it may be expected that particles with median diameters larger than 1/3 of the median pore diameter of the finest filter bed will bridge and form an external cake on the surface of the filter. Particles smaller than 1/3 but larger than 1/7 of the filter bed pore size will form an internal cake, while the smaller particles will pass through the filter, or be retained by surface forces.

Clarification of water in deep layers of granular material has been studied extensively and methods are available for the design and optimisation of these high-capacity units. The enormous capacity of the deep-bed filter is related to the large surface available between the solids (sand, anthracite, gravel) and the flowing fluid. The removed particulates (sand, algae, clay, etc.) cause an increased resistance to flow; in systems where a limited fluid pressure is available, this clogging manifests itself in the gradual drop in flow rate. In the cases where fluid is pumped under a substantial pressure, deposition causes a gradual buildup of pressure over the system; this in turn can cause solids to be scoured forward into the interior of the sand. Optimisation design procedures have been based on the model that the breakthrough of contaminants into the filtrate should coincide with the attainment of an upper, limiting available pressure differential. This involves simultaneous calculation of the changing concentration and pressure profiles in the liquid; these topics are developed in Chapter 6. Designs vary, and must be related to the particular filtration problem in hand.

Cartridge filters are used widely throughout process industries in the clarification of liquids. The media used include: yarns, papers, felts, binder-free and resin-bonded fibres, synthetic fibres, woven wire, sintered metal powders and fibres, ceramics, etc. The inclusion of membrane materials, Chapter 10, in cartridge constructions has extended the range of application of these ubiquitous elements so that particles from approximately 500 μm down to 0.1 μm are separated.

These filter elements are usually constructed in cylindrical form, as shown in Figure 1.10. Filter housings, which may contain single or multiple cartridges, have to be designed with particular attention to sealing of the media, and to the avoidance of "dead spots" in the filter. Accumulated solids in the latter may be spontaneously discharged by sudden flow or pressure changes.

Great care is taken in cartridge preparation and testing, in order to guarantee the performance of the medium in its capability of removing particles of a declared size from the filtered liquid. Cartridges are rated in terms of their capacity to remove particles; these ratings may be "absolute" or "nominal", as discussed in Chapter 6. The passage of particles through the cartridge into the filtrate is known as "bleeding"; cartridges are guaranteed to prevent the bleeding of solids greater than a certain size. In fibrous media, an equivalent interest is given to the prevention of fibre shedding. In poorly prepared media, fibres may be torn off the cartridge under the influence of fluid drag during filtration. It follows that a poorly sealed assembly will present similar problems with bleeds.

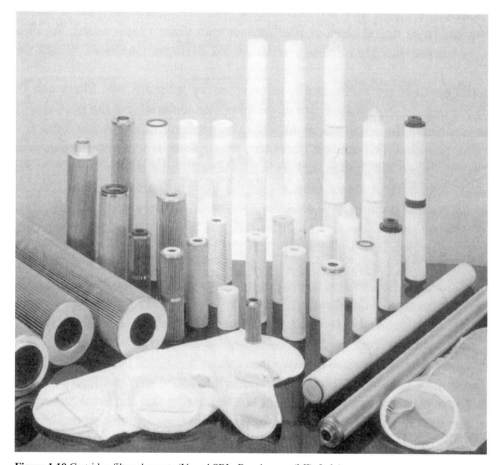

Figure 1.10 Cartridge filter elements (Vessel SRL, Buccinasco, (MI), Italy)

Some of the thinner media, e.g. paper, woven cloths, etc. may be designed to separate solids on the surface of the medium. Such filters are often pleated, in order to increase the available filter area per cartridge, and, therefore, to increase the "solids holding capacity" of the filter. Depth cartridges can be constructed with radial variations in fibre density; so that filtration ensues throughout the depth of the filter. Particle removal efficiency will depend upon, inter alia, the relative sizes of particles and fibres, as discussed in Chapter 6. Quite often, mixed fibres are used, of varying diameter, in order to improve particle capture. Activated carbon which adds an adsorptive capability to the cartridge, finds widespread use for colour removal and odour control.

The flow of liquid through the cartridge creates a pressure differential, ΔP across the filter; this differential increases with deposition of particles, which tend to block the pores in the filter, thereby increasing filter resistance. Solids deposition continues up to an acceptable ΔP, after which the cartridge is disposed of and replaced. The solids deposited at this point, e.g. 20 g per 40 cm long cartridge, is an important design feature of such filters. High solids loadings will require the use of multiple parallel units. The frequency of cartridge replacement is controlled by limiting the application of these units to low concentrations of solids (<0.01 wt%) in the feed, with a corresponding small volume (100 l) of filtrate. Filter testing, therefore, centres upon: (1) filter integrity; (2) pressure differential-time relationships (solids loading/capacity) and (3) particle removal rating.

Filter integrity specifications refer to the availability of effecting sealing and the absence of fibre shedding. The ability of the media to withstand large pressure differentials and high temperature is also of importance in some industries, e.g. polymer processing. Here, molten polymer of very high viscosity and temperature must be processed at extremely high ΔP. It follows, in these cases, that rugged constructions are required, e.g. the sintered metal fibre or powder media referred to above. Cartridges of this type are quite expensive, and cannot be considered as disposable. Such units are recycled frequently, after rigorous cleaning.

New, or recycled filters, are subjected to physical nondestructive testing, e.g. permeability and bubble-point tests. In the latter, the gauge pressure required to cause the passage of air bubbles through the filter is measured whilst the cartridge is submerged in a "wetting fluid", e.g. isopropyl alcohol. This measurement may be used to calculate the diameter of the filter pore d_p from the measured pressure difference ΔP from the equation:

$$d_p = \bar{k}\gamma / \Delta P \tag{1.5}$$

where γ is the surface tension of the "wetting" liquid and $\bar{k}$ is a constant which depends on the shape of the pore; see Equations 6.11–6.14, Chapter 6. This measurement and that of the fluid flow through the filter at a prescribed ΔP (used to calculate the "permeability" of the cartridge) can be compared with acceptable standards of the filter in question.

As described in Chapter 6, particle capture efficiencies are determined by "destructive" flow tests involving the removal of standard particles from flowing fluids. These tests may accompany others to measure bacterial retention, release of materials (extractables), and chemical compatibility.

These subjects are of great importance to the food, pharmaceutical and biopharmaceutical industries. Filter media continue to be developed to meet the needs of these industries. These needs are soon followed by related regulatory procedures, attached to quality assurance, integrity testing, etc., with a concomitant expansion in dedicated literature [Meltzer, 1987; Meltzer and Jornitz, 1998]. The latter publications present a vast amount of detailed information on the role of media required by certain demanding process circumstances. Some of this information is discussed in Chapter 4 on Filter Media, Chapter 5 on Pretreatment Techniques and Chapter 6 on Clarifying Filtration.

1.8 Sedimentation and Flotation

Chapter 7 returns to the separation of solids by sedimentation methods, dealing with the practical details of settling tests required in dilute and concentrated suspensions. Sedimentation processes can be classified into four types:

Type I   Clarification of dilute solids
Type II   Clarification of dilute flocculated solids
Type III Mass settling of high concentrations of flocculant solids
Type IV   Compression of settled solids

Type 1 settling systems contain dilute suspensions of discrete particles which are stable, dispersed and characterised by the absence of flocs. These systems may be designed using the process equations presented in Chapter 3. Type II suspensions require experimental measurements of the clarification rate; this follows from the fact that the removal rate depends on the depth of the settler. Larger flocs possess higher settling velocities than smaller flocs; collision of flocs tends to increase floc size, with a corresponding increase in settling velocity. This leads to the use of long-tube tests to simulate the action of clarifiers, as described in Chapter 7. Short-tube tests can be used in circumstances where the separation is governed by the settling rate of the solids only. In these tests, small batches of the suspension are contained in vertical glass cylinders. The height of the interface between the settling solids (type III) and the clear supernatant liquid is measured as a function of time, as shown in Figure 1.11.

Most of the difficulties encountered in the development of a sound design basis for sedimentation units handling type III and IV systems, centre on the problem of applying batch test information to continuous separation. Chapter 7 presents a review of these methods and their application to the specification of continuous sedimentation equipment. Obviously, the use of continuous, small-scale clarifier–thickeners provides a sounder technical base for process scale-up to industrial size [Khatib and Howell, 1979].

A detailed description is also presented on the reverse operation of flotation. Some solid materials may be present in the feed suspension which are less dense than the surrounding fluid. Thus waste water is often contaminated with quantities of oil and grease. These substances will have a natural tendency to float on the surface of the feed water. Advantage may be taken of this effect by providing scrapers to remove the floating phase. The flotation process may be accelerated by the introduction of air bubbles into the feed; attachment of air bubbles to the contaminants will cause the latter to rise more quickly. In mineral processes, solids which would normally sink in water can be caused to float by bubble attachment, after chemical modification of the solid surfaces.

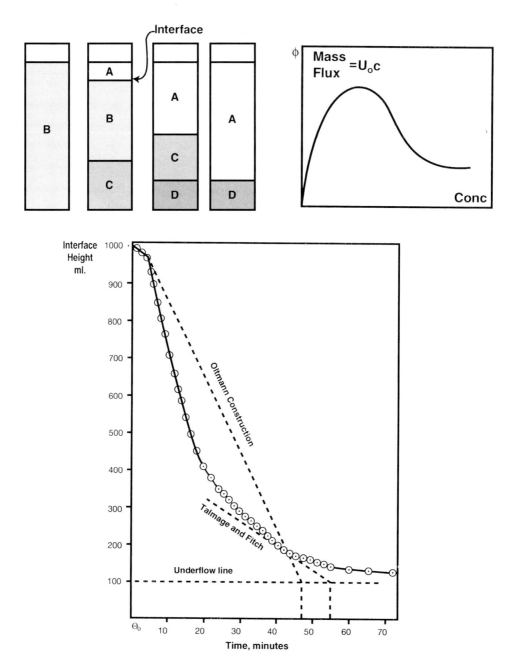

Figure1.11 Measurement of type III settling velocity

The efficiency of the attachment process depends upon the size of the gas bubbles. Process equations are available for predicting the course of such separations; these design equations require the support of laboratory tests to provide relationships between flotation velocities, the volume of air per unit mass of solids, etc. This chapter also deals with the design of inclined "lamella" separators which tend to intensify the sedimentation effect, thus requiring smaller vessel volumes when compared with conventional equipment.

It is of interest to record the modern trend to the use of process systems containing two process functions in one vessel. Thus flotation vessels are now available which are combined with settlers, as shown in Chapter 7, or deep-bed filters, as depicted in Figure 1.12. The use of centrifugal separation, in augmenting the sedimentation of small particles and in providing another means of separation by filtration, is outlined in Chapter 8. This section divides naturally into two: (a) centrifugal sedimentation and (b) centrifugal filtration.

In the specification of sedimentation machinery information is required on the volumetric throughput of the feed suspension Q (m³/s), the settling characteristics of the particles u_t (m/s), as described by the Stokes relationship, and the separating "power" of the sedimentor. The latter is provided by a so called, sigma factor Σ (m²) [Ambler, 1952], which, in turn, contains variables such as rotational speed ω (1/s), machine dimensions, etc. Typical solid-bowl units, with a scroll discharge device rotating coaxially with the bowl are shown in Figures 1.13a and b. These units are sometimes called decanters.

An elementary approach to deriving the process relationship between Q, u_t and Σ starts with the replacement of the gravitational constant g in Stokes equation by an expression for the centrifugal acceleration $\boldsymbol{a}$ resulting from rotation at a speed ω and radius r. Thus $\boldsymbol{a} = \omega^2 r$ replaces g to give:

$$u_t = x^2 \omega^2 r (\rho_s - \rho) / 18\mu \tag{1.6}$$

This basic equation is developed in Chapter 8 to give:

$$Q = u_t \Sigma \tag{1.7}$$

which forms the basis of the sigma theory of solid-bowl centrifugation. The power of such machines relates to the high values of Σ which are available, either from the high rotational speed of such devices, or from the inclusion of settling plates or lamellae, in the high-powered disc centrifuges.

Solid-bowl decanters find widespread use in process industry, particularly in food processing and waste-water treatment. Two-phase and three-phase separators are available. In the latter, e.g. in the separation of solids-oil-water, Leung [1998] lists applications in: oil refining, petroleum production, palm oil separation, rendering, low fat milk production, fish meal and oils and wool grease preparations.

Two-phase separators are even more widely used in pharmaceuticals, biotechnology, waste treatment and many other systems involving solid-fluid separations and dewatering.

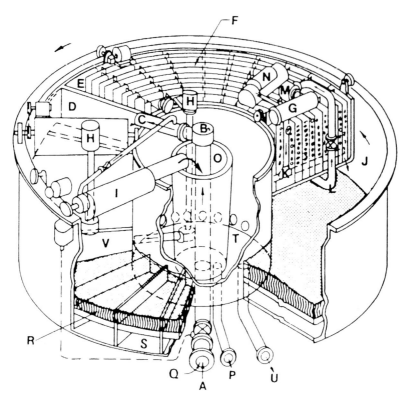

Figure 1.12 Combined flotation cell/ deep- bed filter (Courtesy: Krofta Eng. Corporation, USA)

A) Raw water inlet;   I) Spiral scoop;   P) Sludge outlet;
B) Hydraulic joint;   J) Flotation tank;   Q) Chemical addition;
C) Inlet distributor;   K) Dissolved air addition;   R) Sand filter beds;
D) Rapid mixing;   L) Bottom carriage;   S) Individual clear wells;
E) Moving section;   M) Pressure pump;   T) Centre clear well;
F) Static hydraulic flocculator; N) Air Compressor;   U) Clear effluent outlet;
G) Air dissolving tube;   O) Centre sludge collector;   V) Travelling hood
H) Back-wash pumps;

Several varieties of centrifugal filter are described in Chapter 8. Both batch-operated and continuous filters are available in the range of machines listed in Table 1.4, the latter also refers to the use of solid-bowl sedimentors for high-resistance filter cakes, e.g. sludges.

(a)

(b)

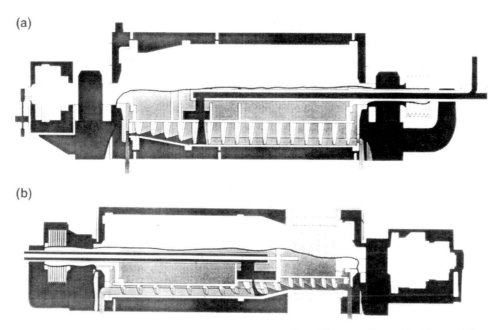

Figure 1.13 Solid-bowl (a) and screen bowl (b) decanter centrifuges (Courtesy: Thomas Broadbent and Sons Ltd., Huddersfield, UK)

Table 1.4 Centrifugal filter and sedimentors

Machine	Operational mode B: batch, C: continuous	Specific particle filterability α, m kg^{-1}
Solid-bowl	C	$>5 \times 10^{11}$
Basket	B	5×10^{9}–5×10^{11}
Peeler	B/C	5×10^{8}–5×10^{9}
Pusher	C	5×10^{8}

In the table the "specific" filterability of the cake α, is related to the overall cake resistance R_c (Eq. 1.2) as discussed in Chapter 2. Some of these machines are also suitable for filter cake dewatering and washing processes.

1.9 Washing and Deliquoring

Post-treatment of filter cakes, or settled sludges, is of great importance in process industries and often occupies much process time. The removal of residual liquid from packed assemblies of particles, or the reduction in solute contamination in the residual liquid phase can be a slow, inefficient processes. The main difficulties arise when the material to be removed is trapped in inaccessible parts of the deposit.

Thus in dewatering by gas blowing, or suction downstream of the cake, air flow through the bed will be of a random nature, thereby creating the possibility of bypassing "residual" liquor in the cake. Chapter 9 details such processes and presents process equations for the calculation of the amount of liquid remaining in the filter cake during air displacement processes.

The residual liquor is reported in terms of the "saturation" of the filter cake. Saturation S is defined as:

$$S = \frac{\text{Volume of liquid in filter cake}}{\text{Volume of pores in filter cake}} \qquad (1.8)$$

Obviously, a completely saturated cake will have a value $S = 1.0$. Again, a bone dry cake takes the value $S = 0$. In gas-blowing operations, S decreases with time, as shown in Figure 1.14, approaching a "residual" level, S_∞, the latter reflects the proportion of liquid retained in pockets and crevices in the cake, or within the micropores of the particles constituting the bed. Such liquid cannot be removed by gas flow; further decrease below S_∞ will proceed by evaporative drying, which involves slow, diffusional mass transfer into the gas phase. In large-scale, continuous filtration systems, use of a heating medium (superheated steam) is sometimes proposed as a means of improving the dewatering (drying) kinetics. Thus in horizontal belt filters (Chapter 11), the moving filter cake is surrounded by a blanket of steam, prior to cake discharge.

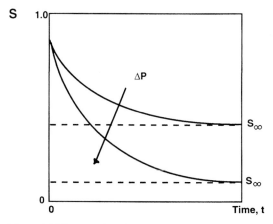

Figure 1.14 Desaturation-time curve

The latter chapter also reports machines which are designed to squeeze the filter cake, after or instead of gas dewatering. In these circumstances, the actual volume or porosity of the filter cake is reduced, thereby expelling residual liquor. Thus in batch operated, variable-chamber filters, the process of squeezing *and* blowing can result in material reductions in S_∞ levels. Combined vacuum and compression devices are available for continuous vacuum filters, whilst belt presses provide continuous squeezing of the cake.

The process of filter cake washing is depicted in Figure 1.15, where the concentration of solute in the liquor emanating downstream of the filter has been recorded as a function of time. The volume of the initial liquor will be determined by the porosity of the filter cake. Delivery of the same volume of wash fluid equates to "one void volume" of wash. In an ideal system, this amount of wash would remove all the dissolved contaminant; a step function would be recorded on the washing curve shown in Figure 1.15. In practice, breakthrough of the flowing wash liquor leads to void volume requirements of 2 or 3, even in favourable circumstances. Where solute is trapped, even larger void volumes of wash may be necessary; this leads to long wash times and lower plant productivity.

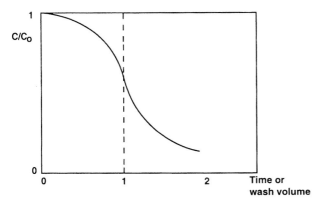

Figure 1.15 Concentration–time washing curve

Fundamental considerations point to the importance of dispersion in its effect on wash time and wash volume requirements. The influence of system parameters such as cake thickness, particle size and distribution, solute diffusivity, etc. combine to produce either the optimal step function in the washing curve, or an inefficient "tail". This information is of great importance, in view of the dominant role played by washing operations, where the latter are necessary.

As would be expected, the conditions at the particle surface–fluid interface have important effects in dewatering and washing operations. Thus the deliquoring rate is known to be a function of the pH level in the system. The presence of surfactants can also alter these processes quite seriously, as discussed in Chapter 11.

1.10 Membrane Filtration

Membrane technology has expanded at an incredible rate over the past thirty years, and now embraces a multi-billion dollar industry. Applications can be broadly divided into processes for (a) suspended particles and (b) dissolved solids. Thus the filtration of particles in the size range 0.1–10 μm, using relatively open membranes is described as microfiltration (MF).

Membranes for the removal of dissolved species are necessarily tighter, in pore size terms, than the microporous variety. Table 1.5 contains information on membranes used for ultrafiltration (UF) (0.001–0.02 μm) and reverse osmosis (RO) (1–10 Å)

Table 1.5 Membrane filtration media

Process	Species separated	Size range	
		μm	Å
Microfiltration	Suspended particles	0.02–10	$200–10^5$
Ultrafiltration	Colloids macromolecules in solution	0.001–0.02	10–200 (or 300–3 000 MW "spherical proteins"
Reverse osmosis	Dissolved salts	0.0001–0.001	1–10

In RO processes, the pumping system has to overcome the osmotic pressure of the salt in water. This leads to the necessity for large pressure drops (25–70 bar) across RO membranes in order to achieve acceptable filtrate rates. In contrast, MF and UF processes operate at relatively low pressures (0.07–7 bar).

Chapter 10 describes the various module configurations into which membranes are fitted: flat sheet, tubular and capillary type. The latter, in hollow fibre form are illustrated in Figure 1.16.

Conventional filtration involves slurry flow "dead-end" into the filter, i.e. the flow of fluid is perpendicular to the surface of the medium. The subsequent accumulation of colloidal or submicrometre-sized particles are particularly difficult to filter and retention of such particles often culminates in plugging of the filter medium.

Dead-end filtration in UF and RO processes, results in an accumulation of dissolved solutes on the surface of the membrane, thus producing an effect called "concentration polarisation". The overall effect is to produce a gradual decline in filtrate flow with time.

The increased concentration of dissolved species or colloids at the membrane surface can be ameliorated by causing the feed to move *across* the filter surface, rather than normally towards it. The accompanying fluid shear at the membrane surface results in a sweeping away of accumulated deposits. Unfortunately, even at a high cross-flow velocity (3–5 m/s), some deposit persists on the surface.

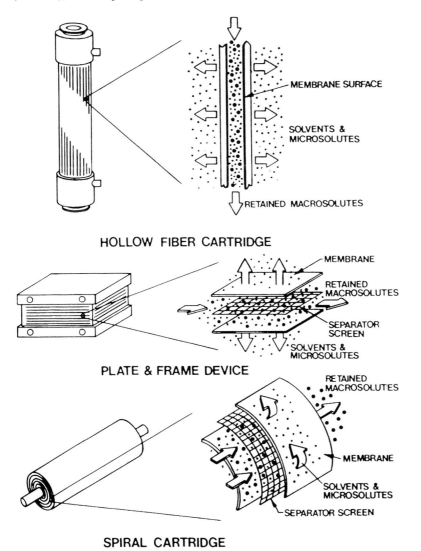

HOLLOW FIBER CARTRIDGE

PLATE & FRAME DEVICE

SPIRAL CARTRIDGE

Figure 1.16 Design of hollow fibre, plates and frame and spiral cartridge ultrafiltration modules (Ostermann, 1982)

In MF separations, overall fluxes in excess of 100 litres m^{-2} h^{-1} are generally sought after, in the interest of process economy, but are rarely attained without proper attention to membrane selection and/or cleaning. The latter process has often to be effected, in situ, perhaps by periodic back-flushing. This creates a "saw-toothed" flux curve (Figure 1.17) which in practice can gradually decay to low fluxes. This effect is due to changes in the membrane permeability despite back-flushing. Membrane separations are particularly useful for biochemical separations demanding sterile operations, in spent liquor removal, diafiltration, etc. In the latter process, removal of spent liquor is accompanied by the introduction of wash liquid into the bioreactor. This causes a gradual removal of unwanted contaminants.

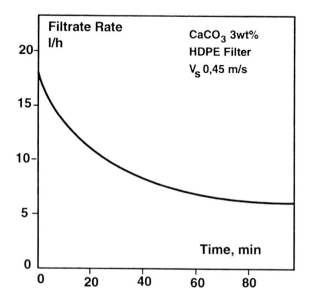

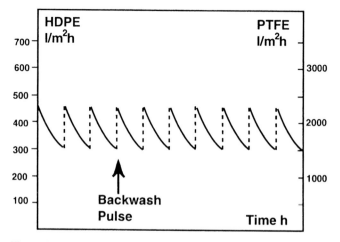

Figure 1.17 Filtrate flux curves in cross-flow microporous filtration

These applications and the successful use of membranes in waste water treatment, potable water production, etc., are cases where new technologies have replaced older, traditional purification methods.

1.11 Filtration Process Equipment and Calculations

This section deals with the description and specification of large-scale filters, and includes aspects of plant selection and-scale up. Several attempts have been made to develop a methodology in SLS filtration equipment selection. For example, the decision tree in Figure 1.2 is based on the speed at which a filter "cake" is formed from the suspended solids [Tiller, 1974]. The latter publication also relates the rate of filtration to equipment types, as shown in Table 1.6.

Table 1.6 Plant selection and rate of filtration

System characteristic	Cake thickness build-up rate	Equipment
Rapid separation	cm/s	Gravity pan Screens Top-feed vacuum filters Filtering centrifuge Vibratory conical bowl
Medium separation	cm/min	Vacuum drum, disc belt, Solid-bowl centrifuge
Slow separation	cm/h	Pressure filters Disc and tubular centrifuge
Clarification	Negligible cake	Cartridges Granular beds Filter aids, precoat filters

Again, it must be realised that all such generalisations are merely tentative suggestions on plant selection. Many types of filter (pressure, vacuum, centrifugal) are capable of similar performance and could be interchanged. Plant failure or process difficulties in a particular situation is often attributable to the absence of operational monitoring of process variables, such as solids concentration, percentage fines in feed, temperature, flow rate, etc. Changes in such variables can lead to plant inefficiencies such as media blinding, filter cake cracking, enhanced cake resistance and so on. Each of these effects would seriously reduce the performance of the separator. However, the rate of cake increase in simple laboratory vacuum trials is useful information in an approach to a suitable design.

Comprehensive considerations of the selection processes have been published elsewhere [Purchas and Wakeman, 1986; Wakeman and Tarleton, 1999]. Here the cake formation time (filterability), level of production required, filtrate clarity specification, sludge proportions (if any), operational modes (constant pressure, constant flow rates, etc.) are all taken, inter alia, into the consideration of filter selection. In the earlier work, the concept of a "standard cake formation time" (SCFT) was developed and defined as the time required for the formation of one centimetre thickness of filter cake under moderate pressure differential. The SCFT was proposed as a measure of filterability which could be used in plant selection processes. This concept received some support from filtration practitioners [Gaudfrin and Sabatier, 1978] in classifying the filtration process field into six areas of: (a) simple cake filtration, (b) filtration followed by air drying, (c) combined filtration and thickening, (d) filtration with simple washing, (e), filtration with multi-stage counter-current washing, and (f) systems where the SCFT was not measurable, i.e. in dilute systems requiring clarifying filters.

These considerations point to the importance of filter cake thickness in the optimisation of a particular process. A filter cake may be too thick, leading to an uneconomic lengthening of the filter cycle. On the other hand, thin filter cakes may be difficult to remove from the equipment, again increasing the filter cycle time.

Generally, in a filtration cycle, productivity may be described as the quotient:

$$\text{Productivity} = \frac{\text{Volume of filtrate per cycle}}{\text{Cycle time}}$$

In turn, the cycle time includes periods for filtration, dewatering and washing (if required), cake discharge, filter cleaning and possibly reassembly of the filter. Elementary theory points to thin-cake advantages in economic terms, as long as the specified processes of washing, discharge, etc. are attainable. Machine developments continue, as described in Chapter 11, with a principal objective of improving process profitability.

Chapter 11 deals with the application of these concepts to process calculations in large-scale vacuum and pressure filters used in continuous and batch processes. In all the examples presented, the importance of the consideration of the full filter cycle is stressed. Along with the development of modern filters capable of operating in thin-cake conditions, units are also included for the dewatering of cakes by gas displacement and squeezing.

The overall contents of this book are aimed at the development of a better understanding of the complex processes involved in solid liquid separation. Many of the mathematical models developed for separators (hydrocyclones, floatation cells, compression filters) are empirical in nature, being based on experimental studies with actual separators. Such models [Oja, 1996] have been included in the application of expert, or knowledge-based systems [Jamsa-Jounela, 1998] to some of the processes described in this book. These methods will continue to play an important role in the control of complex filtration systems.

1.12 References

Akers, R.J., (1987) Chap.5, Filtration Principles and Practices. 2nd Edn, M.J. Matteson, (Ed.), Marcel Dekker, New York.
Ambler, C.M., 1952, Chemical Engineering Progress, 48, p. 150.
Anlauf, H. and Müller, H.R., 1990, 5th World Filtration Congress, Nice, 2, p. 211.
Gaudfrin, G. and Sabatier, E., 1978, Int.Symposium KVI and Belgian Filtration Society, Liquid–Solid Filtration, pp. 29-47.
Ives, K.J., 1981, Chap.11, Solid–Liquid Separation, 2nd Edn, L. Svarovsky (Ed.), Butterworth.
Ives, K.J., 1982, Water Filtration Symposium, Koninklijke Vlaanse Ingenieursvereiniging, Desgutnlei 214, B-2018 Antwerpen Belgium, 1.
Jamsa-Jounela, S.K., 1998, C.S.T. Workshop in Separation Technology, Lappeenranta University of Technology, Lappeenranta, Finland.
Khatib, Z. and Howell, J., 1979, Trans. I.Chem.E., 57, p. 170.
La Mer, V.K. and Healey, T.W., 1963, J.Phys.Chem., 67, p. 2417.
Leung, W.F., 1998, Industrial Centrifuge Technology, McGraw Hill, New York.
Meltzer, T.H., 1987, Filtration in the Pharmaceutical Industry, Marcel Dekker, New York.
Meltzer, T.H. and Jornitz, M.W., 1998, Filtration in the Biopharmaceutical Industry, Marcel Dekker, New York.
Oja, M., 1996, Pressure Filtration of Mineral Slurries, Research Papers No. 53, Lappeenranta University of Technology, Lappeenranta, Finland.
Ostermann, A.E., 1982, World Filtration Congress III, Downingtown, USA, 2, p 496, Sponsored by the Filtration Society.
Pierson, W., 1981, Chap.20, Solid–Liquid Separation, 2nd Edn, L. Svarovsky (Ed.), Butterworth.
Purchas, D., 1987, Filtech Conference, U.K. Filtration Society, Utrecht, Holland, 2, p. 273.
Purchas, D., and Wakeman, R.J., 1986, Solid–Liquid Separation-Equipment Scale Up, Uplands Press, Croydon.
Svarovsky, L., 1981, Chap.1, Solid–Liquid Separation, 2nd Edn, L. Svarovsky (Ed.) Butterworth.
Shirato, M. and Tiller, F.M., (1987) Chap.6, Filtration Principles and Practices. 2nd Edn, M J Matteson, (Ed), Dekker, New York.
Tiller, F.M., 1974, Chemical Engineering, 81, p 116 April 29 and p 98, May 13.
Wakeman, R.J., 1994, Filtration Soc. 30th Anniversary Symposium, Birmingham, pp. 257– 271.
Wakeman, R.J. and Tarleton, E.S., 1999, Filtration: equipment selection modelling and process simulation, Elsevier, Oxford, UK.

1.13 Nomenclature

a	Centrifugal acceleration	m s^{-2}
C	Solid concentration by volume fraction	–
d_p	Pipe or pore diameter	m
g	Gravitational constant	m s^{-2}
ΔP	Pressure Differential	N m^{-2}
Q	Volume flow rate	m^3 s^{-1}
R_c	Filter cake resistance	m^{-1}
R_m	Medium Resistance	m^{-1}
r	Radial position	m
S	Saturation: volume of liquid in cake/volume of cake pores	–
S_∞	Residual Equilibrium (irreducible) saturation, t = ∞	–
t	Time	s
u_t	Terminal settling velocity	m s^{-1}
U_o	Sedimentation interface velocity	m s^{-1}
v_f	Filtrate velocity after cake formation	m s^{-1}
v_o	Filtrate velocity before cake formation	m s^{-1}
x	Particle diameter	m

Greek Symbols

γ	Surface tension	N m^{-1}
μ	Liquid viscosity	Pa s
ρ	Fluid density	kg m^{-3}
ρ_s	Solid density	kg m^{-3}
ω	Angular velocity	s^{-1}

2 Filtration Fundamentals

2.1 Introduction

A fundamental and numerical understanding of solid–liquid separation operation is important, not only when there is a need to specify or design a new piece of equipment to fit into a flowsheet, but also to evaluate the performance of existing facilities and in order to control the process. A few simple tests, not necessarily involving laboratory equipment, and some calculations should enable the engineer who has this understanding to be able to optimise, control or design with some degree of confidence.

This chapter starts with a description of fluid flow through porous media. This fundamental basis is then developed into working filtration equations. Filtration models require fundamental equations that are an accurate representation of the physical processes involved in fluid flow. The first filtration models discussed are those of cake filtration, as these are better understood. There is a substantial amount of operating data which supports the cake filtration approach.

Mechanisms and design models for filtration other than that forming a filter cake are also described, and so is non-Newtonian filtration. It is well known that a filtration cycle displays characteristics of these other filtration modes prior to the occurrence of cake filtration. This period is usually very short even at low concentrations of slurry. The other filtration modes are particularly relevant to clarifying filtration, at low to very low concentrations (less than 1% solids by volume). This type of filtration is important industrially, and is typical of the deep-bed filtration discussed in Chapter 6. Filtration of non-Newtonian liquids is somewhat more specialised, but can be found in the filtration of oils and polymer melts.

The final section describes simple laboratory apparatus used to characterise the behaviour of slurries to be filtered and includes comments on procedures to be followed and results obtained. Also included in the final section is a brief description of more specialised laboratory equipment which could be used outside the research laboratory, but has yet to find application away from that environment.

2.2 Fluid Flow Through Porous Media

The fundamental relation between the pressure drop and the flow rate of liquid passing through a packed bed of solids, such as that shown in Figure 2.1, was first reported by Darcy in 1856. The liquid passes through the open space between the particles, i.e. the pores or voids within the bed. As it flows over the surface of the solid packing frictional losses lead to a pressure drop.

The amount of solids inside the bed is clearly important; the greater this is the larger will be the surface over which liquid flows and, therefore, the higher the pressure drop will be as a result of friction. The volume fraction of the bed available for fluid flow is called the porosity or voidage, and this is defined below.

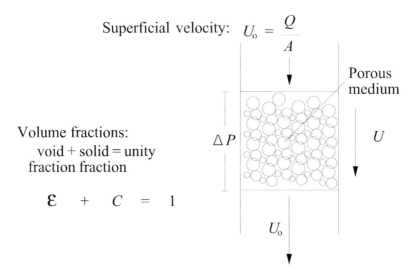

Figure 2.1 Schematic diagram of porous media

$$\text{Porosity} = \varepsilon = \frac{\text{volume of voids}}{\text{total bed volume}} \qquad (2.1)$$

In many solid–liquid separations porosity is often replaced by solid concentration which is usually the volume fraction of solids present within the bed C; porosity is the void volume fraction so these two fractions sum to unity. Hence solid volume fraction concentration is:

$$C = 1 - \varepsilon \qquad (2.2)$$

Darcy discovered that the pressure loss was directly proportional to the flow rate of the fluid [Darcy, 1856]. This is shown in Figure 2.2.

The constant of proportionality in the figure is dependent on the permeability k (m^2) of the porous network. Darcy's law can be regarded as the fluid flow through porous media analogue of the Ohm's law for flow of electric current.

The analogy is as follows: the driving force behind flow is potential difference or pressure drop per unit length, the flow is current or liquid velocity and the constant of proportionality is electrical resistance or the ratio of viscosity to permeability. Increasing the viscosity or decreasing the permeability increases the resistance to fluid flow.

In any situation involving fluid flow it is possible to consider the drag on the solid surface as being due to two phenomena; "skin" friction (viscous drag) and "form" drag.

The former is a result of the stationary layer of liquid that occurs at the surface of the solid object, the drag or pressure loss is, therefore, due to friction between the solid and liquid. Form drag is due to turbulent eddies which occur at higher flow rates and fluctuate in intensity and direction. Form drag leads to additional pressure losses over that of skin friction and a breakdown of the linearity between flow rate and pressure drop. It is usual to distinguish between the flow regimes, or the relative importance of the drag types, using a modified Reynolds number. Most filtration operations occur at low flow rates through the porous media, thus form drag will not be discussed further and the assumption of "streamline" flow conditions will be made. Darcy's law is valid only under these conditions.

Ohm's law:

$$V = R\,I$$

where V is the potential difference, I is the current and R is resistance

Darcy's law:

$$\frac{\Delta P}{L} = \frac{\mu q}{kA} \tag{2.3}$$

where ΔP is the pressure drop, L is the bed depth, μ is the liquid viscosity, q is volume flow rate and A is the cross-sectional area of the bed.

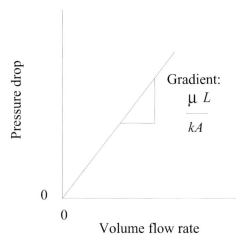

Figure 2.2 Proportional relation between pressure drop and flow rate clean liquid flowing through fixed porous media

2.3 Permeability

The permeability characterises the ease with which liquid will flow within a porous medium, including filter cakes. It is, therefore, an important parameter which has been investigated by many research workers. Factors which affect permeability include the size of particles making up the porous medium and the porosity. The best known equation for permeability is due to Kozeny [1927]:

$$k = \frac{\varepsilon^3}{K(1-\varepsilon)^2 S_v^{\,2}}$$
(2.4)

where S_v is the specific surface area per unit volume of the particles and K is the Kozeny constant which normally takes the value of 5 in fixed or slowly moving beds and 3.36 in settling or rapidly moving beds. The specific surface is discussed in greater detail in Appendix A. Substituting Equation (2.4) into Darcy's law gives the Kozeny–Carman equation:

$$\frac{\Delta P}{L} = \mu \left(\frac{5(1-\varepsilon)^2 S_v^{\,2}}{\varepsilon^3} \right) \frac{q}{A}$$
(2.5)

Other research workers have investigated the relation between concentration, particle size and permeability [Happel & Brenner, 1965].

In the permeability models the particles are assumed to be rigid, in a fixed geometry and in point contact with each other. Furthermore, liquid drag and pressure were the only forces considered. Thus any other body force present could cause deviation from the above simplified approach. Such forces are generally significant for particles less than 10 µm; care should be exercised when applying these relations to material less than this size. This is a fact which Kozeny recognised. Despite his warning many researchers have applied the Kozeny–Carman equation to smaller particles and even macromolecules undergoing ultrafiltration. Most have been disappointed with the accuracy of the resulting numerical model.

In filtration it is usually possible to deduce an empirical permeability from simple laboratory tests or from existing operating data. In practice filter cake permeability is a function of porosity, particle shape and packing, particle size and distribution, rate at which the cake was formed, concentration of the slurry being filtered, etc. Thus the theoretical relations for permeability should be used only as a guide to estimate permeability in the absence of operating data; measured permeabilities may easily be one or two orders of magnitude lower than that given by the above equations. This is a consequence of the many factors that the above equations do not recognise. It is also generally true that the finer the particles, and the wider the size distribution, the greater will be the deviation from the theoretical permeability relation.

2.4 Cake Filtration

The deposition of solids on a filter medium or septum is shown schematically in Figure 2.3. It is generally accepted that filtration resulting in a filter cake takes place by a bridging mechanism over the surface pores within a filter medium, cloth, septum or support. This helps to prevent the medium from clogging with fine particles. The filter medium plays a crucial role in initiating the filtration, and the medium can have long-lasting effects on the filter cake structure and properties throughout the filtration cycle. Chapter 4 is dedicated to the correct selection and description of filter media.

The mathematical description of the process starts with the neglect of septum resistance and the use of Darcy's law (Equation 2.3) to relate filtrate flow rate to pressure drop (and expanding $q = dV/dt$ where dV is filtrate volume in time dt):

$$\frac{\Delta P}{L} = \frac{\mu}{kA}\frac{dV}{dt} \tag{2.6}$$

During filtration the cake depth increases due to deposition of solids at the filter cake surface. The change in cake depth is accompanied by changes in fluid flow rate and pressure differential, as filtration time increases. Thus Equation (2.3) contains four variables, even for materials of constant cake permeability, constant liquid viscosity and constant filter area. A material which displays constant cake permeability usually does so as a consequence of a constant filter cake concentration. This is consistent with the permeability equations, which showed that this is a function of the solids type (size) and porosity (or concentration). A material displaying constant cake concentration is, therefore, *incompressible* and this type of filtration is known as incompressible cake filtration. Most of the following concepts are also relevant to compressible cake filtration, which is generally regarded as an elaboration of this basic theory. This is discussed in greater detail in Section 2.6.

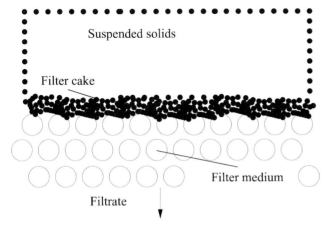

Figure 2.3 Schematic illustration of filter medium and cake

For incompressible filtration the cake concentration remains constant, thus for each unit volume of suspension filtered the filter cake volume increases by a uniform and constant amount. Likewise, the filtrate volume is uniform and constant with respect to each volume of suspension filtered. However, when filtering at constant pressure, the rate of cake deposition with respect to time will not be uniform because each new element of filter cake increases the total resistance to the passage of filtrate from the new cake layer. Thus the rate of filtration declines, as shown in Figure 2.4. The uniform relation between cake volume and filtrate volume is illustrated in Figure 2.5. The constant of proportionality β can be used to give an equation for cake depth, at any instant in time:

$$L = \frac{\beta V}{A} \tag{2.7}$$

this can then be substituted into Equation (2.6) to give:

$$\frac{dV}{dt} = \frac{A^2 \Delta P k}{\beta V \mu} \tag{2.8}$$

Thus Equation (2.8) contains one less variable than Equation (2.6) but one more constant (β).

Clearly, the volume ratio of cake to filtrate still needs to be calculated before Equation (2.8) can be evaluated. This is achieved by means of a mass balance on the solid and liquid entering the filter system.

Mass of solids in the filter cake:	$C A L \rho_s$	(kg)
Mass of liquid in the filter cake:	$(1 - C)AL \rho$	(kg)
Mass fraction of solids in the slurry feed:	s	
Total mass of slurry (solid and liquid):	M	(kg)

Mass balance on the liquid:

$$(1-s)M = V\rho + (1- C)\rho LA$$

Mass balance on the solids:

$$sM = C\rho_s LA$$

which when combined with Equation (2.7) and rearranged gives:

$$\beta = \frac{s\rho}{(1- s)C\rho_s - s(1- C)\rho} \tag{2.9}$$

Use of Equation (2.9) assumes knowledge of two constants (excluding densities): slurry concentration by mass fraction and filter cake concentration by volume fraction. The former is readily obtained by means of sampling, weighing, drying and reweighing. The latter is usually obtained by a similar route, but with the necessity of converting the data to volume fraction from mass fraction. This is illustrated in Section 2.4.2.

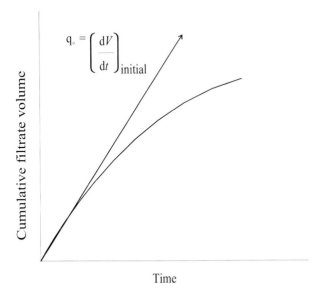

Figure 2.4 Declining filtrate rate during constant pressure filtration

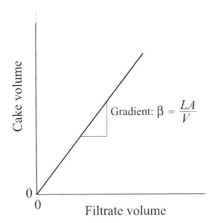

Figure 2.5 Proportional relation between cake and filtrate volumes

2.4.1 Mass Cake Deposited per Unit Area and Specific Resistance

Equations (2.8) and (2.9) may be used in mathematical descriptions of cake filtration processes but the expressions produced are unwieldy. It is more convenient to combine the cake permeability, cake volume fraction concentration and solid density, which will all be constants for an incompressible filtration, in a single constant called "specific resistance" α which has units m kg^{-1}:

$$\alpha = \frac{1}{kC\rho_s} \tag{2.10}$$

Note that as permeability is analogous to electrical conductivity it is logical to use the term resistance as being one of inverse fluid conductivity, or permeability. Furthermore, the cake thickness, solid density and cake volume fraction can be combined in a variable w, which is the mass of dry solids deposited per unit area:

$$w = LC\rho_s \qquad \text{and} \qquad \frac{L}{k} = \alpha w \tag{2.11}$$

The transformation of Equation (2.8) into a general filtration equation is achieved by first multiplying both the numerator and denominator by the product of solid concentration and density to give, after rearrangement:

$$\frac{dV}{dt} = \left(\frac{A}{\beta V C\rho_s} \right) \left(\frac{kC\rho_s}{1} \right) \left(\frac{A\Delta P}{\mu} \right)$$

and since $L = \beta V/A$ this gives:

$$\frac{dV}{dt} = \left(\frac{1}{LC\rho_s} \right) \left(\frac{kC\rho_s}{1} \right) \frac{A\Delta P}{\mu}$$

Introducing w and α from the above leads to:

$$\frac{dV}{dt} = \frac{A\Delta P}{\mu w \alpha} \tag{2.12}$$

which is the form of the differential equation that will be used in the following mathematical descriptions of filtration. It should, however, be borne in mind that the following procedure could be exercised on Equation (2.8), retaining permeability and using Equation (2.9) for the volume ratio of filter cake to filtrate volume collected.

The physical significance of specific resistance is illustrated in Figure 2.6. If only the cake resistance R_c is considered in Darcy's law, where $R_c = L/k$:

$$\Delta P = \mu R_c \frac{dV}{dt} \frac{1}{A}$$

then overall resistance to filtration increases with time due to the increase in depth of the filter cake. The rate of increase in filtration resistance is, however, linear with respect to the mass of dry solids deposited per unit filter area in accordance with Equation (2.11). Thus specific resistance is the ratio, or gradient, of these two terms as illustrated in Figure 2.6.

Equation (2.12) can not be solved without some method of calculating the dry mass of solids deposited per unit filter area, and this is considered in the next section.

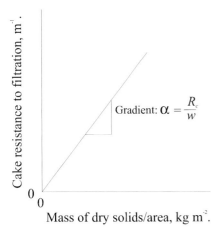

Figure 2.6 Specific resistance

2.4.2 Solid Concentration

An appreciation of the different expressions for solid concentration that may be quoted, and means to convert from one form to another, are essential in reconciling experimental data with the mathematical description of filtration. It is probable that experimental data will be obtained in the form of mass fraction; C_w or s, for the filter cake or slurry concentrations respectively, as this readily follows from drying and weighing the samples. Concentration by volume fraction comes from the definition of density and some cancelling:

$$C = \frac{\text{Volume solid in cake}}{\text{Volume solid in cake} + \text{Volume liquid in cake}} = \left(1 + \frac{(1-C_w)\rho_s}{C_w\rho}\right)^{-1} \qquad (2.13)$$

Concentration in terms of mass of solids per unit volume (C_{wv}) follows by a similar route:

$$C_{wv} = \frac{\text{Mass solids in slurry}}{\text{Volume solids in slurry} + \text{Volume liquid in slurry}} = \left(\frac{1}{\rho_s} + \frac{(1-s)}{s\rho}\right)^{-1}$$

and concentration in terms of mass of solids per unit volume of liquid C_{wvl}:

$$C_{wvl} = \frac{\text{Mass solids in slurry}}{\text{Volume liquid in slurry}} = s\rho/(1-s)$$

It is often assumed that w can be obtained from C_{wvl} (the slurry mass per unit volume liquid concentration) as follows:

$$wA = \text{Mass dry solids} = C_{wvl}V$$

which rearranges to give:

$$w = C_{wvl}\frac{V}{A} \tag{2.14}$$

but this neglects the liquid retained in the filter cake. Such an approach leads to underestimation of the value of w. The error in this assumption is small so long as the slurry concentration is low.

For a more rigorous treatment of the dry mass of cake per unit filter area it is usual to start with the cake moisture ratio, defined as:

$$m = \frac{\text{Mass of wet cake}}{\text{Mass of dry cake}} = \frac{\text{Mass of cake solids} + \text{Mass of cake liquid}}{\text{Mass of cake solids}}$$

Thus:

$$m = \frac{CAL\rho_s + (1-C)AL\rho}{CAL\rho_s}$$

and dividing by the cake solid content gives:

$$m = 1 + \frac{(1-C)}{C}\frac{\rho}{\rho_s}$$

Noting that w, mass of dry solids per unit area, is $\beta VC\rho_s/A$, using Equation (2.9) and rearranging gives:

$$w = \left(\frac{C\rho_s s\rho}{(1-s)C\rho_s - s(1-C)\rho}\right)\frac{V}{A}$$

which cancels to:

$$w = \left\{ \rho \middle/ \left[\frac{1}{s} - 1 - \frac{(1-C)\rho}{C\rho_s} \right] \right\} \frac{V}{A}$$

Substituting in the expression for cake moisture and rearranging leads to Equation (2.15):

$$w = \left(\frac{s\rho}{1 - sm} \right) \frac{V}{A} \tag{2.15}$$

Using the definition of C_{wvl} Equation (2.15) reduces to Equation (2.14) when the cake moisture ratio is unity. Both equations are of a similar form:

$$w = c \frac{V}{A} \tag{2.16}$$

where c is either:

$$c = C_{wvl} \tag{2.17}$$

or:

$$c = \left(\frac{s\rho}{1 - sm} \right) \tag{2.18}$$

depending on the available data and degree of complexity of the solution required. In both cases the concentration term on the right-hand side of the equation (C_{wvl} and s) refers to that of the slurry, and c will be a constant for incompressible cake filtration. Physically, the term c represents the mass of dry cake deposited per unit volume of filtrate. Equation (2.18) will, however, vary in compressible cake filtration as m is no longer a constant. The following mathematical treatment will use the general equation, i.e. Equation (2.16), for further manipulation of the variable w. The choice of which equation to use for c is left to the reader's own circumstances.

The above analysis demonstrates that a knowledge of the filter cake concentration is, therefore, not essential, as it was to evaluate Equation (2.9), provided that the liquid retained in the cake is small. This is a useful result for two reasons: obtaining a representative sample of the filter cake is much more difficult than that of obtaining a sample of a well-mixed slurry, and in the analysis of test data which employs a step increase in pressure (see Section 2.6) where obtaining a cake sample may well be impossible. For this reason Equation (2.17) is the most commonly met form in the literature.

2.5 Forms of Cake Filtration Equation

Substituting Equation (2.16) into Equation (2.12) gives:

$$\frac{\mathrm{d}V}{\mathrm{d}t} = \frac{A^2 \Delta P}{\mu c V \alpha} \tag{2.19}$$

Equation (2.19) contains three variables: time, filtrate volume and pressure; and four constants: filtration area, viscosity, dry cake mass per filtrate volume and specific resistance. The last two are constant only if the filter cake is incompressible. The equation can be solved analytically only if one of the three variables is held constant. This reflects the physical mode of operation of industrial filters; vacuum filtration tends to be at constant pressure and pressure filtration is often under constant rate, at least until some predetermined pressure has been achieved. Thus the following mathematical models are very relevant to these filtrations.

In deriving Equation (2.19) any pressure loss due to the flow of filtrate through the filter medium has been neglected. This assumption can be removed by assuming that the pressure drop in the medium ΔP_m can be added to the pressure drop over the filter cake ΔP_c, to give the total or overall pressure drop:

$$\Delta P = \Delta P_c + \Delta P_m \tag{2.20}$$

Darcy's law can then be applied to both terms:

$$\Delta P = \frac{\mu c \alpha}{A^2} V \frac{\mathrm{d}V}{\mathrm{d}t} + \frac{\mu}{A} \frac{L_m}{k_m} \frac{\mathrm{d}V}{\mathrm{d}t}$$

where L_m and k_m are the medium depth and permeability, respectively. If the medium resistance and depth remain constant during filtration these two constants can be replaced by a single constant known as the medium resistance R_m, with units of m^{-1}:

$$R_m = \frac{L_m}{k_m}$$

Hence,

$$\Delta P = \frac{\mu c \alpha}{A^2} V \frac{\mathrm{d}V}{\mathrm{d}t} + \frac{\mu}{A} R_m \frac{\mathrm{d}V}{\mathrm{d}t} \tag{2.21}$$

2.5.1 Constant Pressure Filtration

Under these conditions Equation (2.21) can be rearranged and integrated using the limits: zero filtrate volume at zero time, V volume filtrate after time t, thus:

$$\int_0^t dt = \frac{\mu c \alpha}{A^2 \Delta P} \int_0^V V dV + \frac{\mu R_m}{A} \int_0^V dV \qquad (2.22)$$

After integration and rearrangement the following equation, known as the linearised parabolic rate law results:

$$\frac{t}{V} = \frac{\mu c \alpha}{2 A^2 \Delta P} V + \frac{\mu R_m}{A \Delta P} \qquad (2.23)$$

Equation (2.23) is a straight line, where t/V is the dependent and V is the independent variable. Thus a graph of the experimental data points of t/V against V permits calculation of the gradient and intercept of Equation (2.23) [Ruth, 1935]:

The gradient and intercept are as follows:

$$\text{Gradient} = \frac{\mu c \alpha}{2 A^2 \Delta P} \qquad (2.24)$$

and:

$$\text{Intercept} = \frac{\mu R_m}{A \Delta P} \qquad (2.25)$$

Thus if the liquid viscosity, filter area, filtration pressure and mass of dry cake per unit volume of filtrate, either from Equation (2.17) or (2.18), are known, the graphical values can be used to calculate the cake specific resistance and filter medium resistance.

Worked Example: the data shown in Figure 2.7 were obtained from the constant pressure period of a pilot scale plate and frame filter press. Calculate the cake resistance given: filter area 2.72 m^2, viscosity 10^{-3} Pa s, mass of dry cake per unit volume filtrate 125 kg m^{-3} and filtrate pressure 3 bar. The specific resistance by Equation (2.24) is 5.4×10^{11} m kg^{-1}. The apparent medium resistance is 2.9×10^{12} m^{-1} by Equation (2.25). However, in this instance the medium resistance is a composite term including the resistance to filtrate flow due to the cake formed during the preceding constant rate filtration period on the filter press, in addition to the true medium resistance.

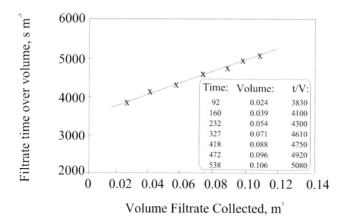

Figure 2.7 Linearised parabolic rate law plot

It is important to realise that the above equations are applicable to both small-scale laboratory test data and to full-scale industrial filters operating under conditions of constant pressure filtration. Thus specific resistance and medium resistance are regarded as two design variables which can be used to optimise filter throughput, or to scale up laboratory data, as will be discussed further in Chapter 11. Further important details regarding the filter medium resistance are discussed in Section 2.9, and these should be read before any further conclusions are made regarding this variable.

There are various forms of the parabolic rate law in common use. One alternative is to plot $(t-t_1)/(V-V_1)$ against V_1 where t_1 and V_1 are some arbitrary datum values of time and filtrate volume. This is useful if a long time has been taken in reaching the final and constant value of pressure across the cake and cloth. It is also useful in the analysis of data in which a step change of the filtration pressure was performed, such as detailed in Section 2.6. The value of the medium resistance calculated from Equation (2.25) under these circumstances, has dubious meaning, however, as it represents both the medium resistance and the resistance due to the cake deposited prior to the datum. Some investigators work with the differential form of Equation (2.23), plotting the reciprocal of the instantaneous filtrate rate q against volume of filtrate, which again produces a straight line in accordance with equations (2.26) and (2.27):

$$\frac{dt}{dV} = \frac{\mu c \alpha}{A^2 \Delta P} V + \frac{\mu R_m}{A \Delta P} \tag{2.26}$$

i.e.:

$$\frac{1}{q} = \frac{\mu c \alpha}{A^2 \Delta P} V + \frac{1}{q_0} \tag{2.27}$$

where q_0 is the filtrate rate at the start of the filtration when no cake is present. It is not, however, the same as the medium resistance in the total absence of slurry to be filtered, as will be discussed in Section 2.9. Use of Equations (2.26) and (2.27) presents some practical difficulties as the instantaneous filtrate flow rate usually has to be determined by graphical differentiation (i.e taking tangents) of the filtrate volume–time curve. Such a procedure is notoriously inaccurate.

Finally, one method which is readily applied to computer spreadsheet use is to consider the incremental version of Equation (2.23). The filtration data is arranged into equal volumes of filtrate, and the corresponding time for each increment is calculated. The filtration starts at t_0 and V_0, the next increment is t_1 and V_1, and subsequent times are measured after equal volumes of filtrate (ΔV) have been collected. The incremental equation for the first increment can be represented as:

$$t_1 = a(\Delta V)^2 + b(\Delta V)$$

subsequent increments occur at $(2\Delta V)$, $(3\Delta V)$, etc.
If the difference between consecutive increments is taken then the resulting, and general, first difference equation is:

$$\Delta t_n = (2n - 1)a(\Delta V)^2 + b(\Delta V)$$

The second-difference equation is obtained from the difference between two consecutive first difference equations, this will be:

$$2a(\Delta V)^2$$

Thus the second difference in the times required to achieve the filtrate volumes can be equated with the above, and rearranged to provide values of a and ultimately specific resistance. This technique is best illustrated by the application of a difference table, which is given below.

The equation for b, and hence R_m, is more complex, but again does not require any graphical construction.

$$b = \frac{(2n - 1)\Delta t_{n-1} - (2n - 3)\Delta t_n}{2(\Delta V)}$$

The advantage of the above procedure is that values of both specific resistance and medium resistance can be calculated for each data point (or at least $n-2$ data points where n is the total number of points). This helps to highlight any erroneous data points, and provides some indication of the spread of experimental values for specific and medium resistances. Clearly, the mean value of specific and medium resistance can be used in further computations, possibly combined with the lowest and highest values taken from the difference table as an indication of the measure of spread on these values. The routine appears very complex on first viewing but lends itself to computer spreadsheet application with, therefore, the minimum amount of repetitive calculation once set up.

Table 2.1 Difference table for constant pressure filtration

Incremental equation	First difference	Second difference
$t_1 = a(\Delta V)^2 + b(\Delta V)$		
$t_2 = a(2\Delta V)^2 + b(2\Delta V)$	$\Delta t_2 = 3a(\Delta V)^2 + b\Delta V$	
$t_3 = a(3\Delta V)^2 + b(3\Delta V)$	$\Delta t_3 = 5a(\Delta V)^2 + b\Delta V$	$\Delta(\Delta t) = 2a(\Delta V)^2$
: : :	: : :	: : :
$t_{n-1} = (n-1)^2 a(\Delta V)^2 + (n-1)b(\Delta V)$	$\Delta t_{n-1} = (2n-3)a(\Delta V)^2 + b(\Delta V$	$\Delta(\Delta t) = 2a(\Delta V)^2$
$t_n = n^2 a(\Delta V)^2 + nb(\Delta V)$	$\Delta t_{n-1} = (2n-1)a(\Delta V)^2 + b(\Delta V)$	$\Delta(\Delta t) = 2a(\Delta V)^2$

2.5.2 Constant Rate Filtration

This type of filtration commonly occurs when an efficient positive displacement pump is used to feed a pressure filter. The pump delivers a uniform volume of slurry into the filter; hence the filtration rate remains constant when filtering an incompressible material. In order to achieve this constant rate the pressure delivered by the pump must rise, to overcome the increasing resistance to filtration caused by cake deposition. It is usual for such a pumping system to include a pressure relief valve; if the filtration cycle is long enough, the constant rate period will be followed by a constant pressure period after the pressure relief valve opens.

Constant rate filtration is easily observed on a plot of filtrate volume against time, as illustrated in Figure 2.8.

Under these circumstances:

$$\frac{dV}{dt} = \frac{V}{t} = Constant$$

and Equation (2.21) can be rearranged to give:

$$\Delta P = \left(\frac{\mu c \alpha V}{A^2 t} \right) V + \left(\frac{\mu R_m}{A} \frac{V}{t} \right)$$ (2.28)

which is a straight line on a plot of filtration pressure against filtrate volume, as illustrated in Figure 2.9. Suitable rearrangement of the gradient and intercept equations coupled with the values taken from the graph will provide values for specific and medium resistances.

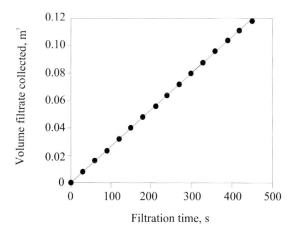

Figure 2.8 Constant rate filtration

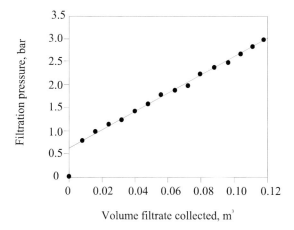

Figure 2.9 Pressure variation during constant rate filtration

Worked Example: the data shown in Figures 2.8 and 2.9 were obtained on a pilot scale plate and frame filter press, calculate the specific and medium resistances given: filter area 2.72 m^2, viscosity 10^{-3} Pa s, mass of dry cake per unit volume filtrate 125 kg m^{-3}.

The constant filtrate rate is obtained from Figure 2.8 as 2.6×10^{-4} m^3 s^{-1}, the specific and medium resistances are 4.5×10^{11} m kg^{-1} and 6.2×10^{11} m^{-1}, respectively, from Equation (2.28).

If a period of constant pressure filtration then follows, the parabolic rate law (Equation 2.23) can be applied, but care should be taken in interpreting the meaning of the intercept from this figure. It is no longer solely a function of medium resistance, but is now a function of medium resistance and the resistance due to filter cake formed under conditions of constant rate. Useful data can still be obtained from the intercept but an average cake resistance value must be calculated first, which can then be subtracted from the resistance obtained from the new intercept.

2.5.3 Variable Pressure and Rate Filtration

A pressure filter fed by a pump other than a positive displacement type provides this kind of filtration. The most common example is that of a centrifugal pump feeding a plate and frame pressure filter. Under these circumstances the delivery pressure of the pump increases as the flow rate decreases. The flow decreases as a consequence of increasing total resistance to filtration due to the increase in cake depth. The mathematical description of this process depends on knowledge of the pump characteristic, such as that shown in Figure 2.10.

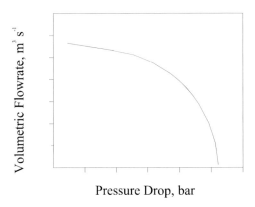

Figure 2.10 Typical centrifugal pump characteristic

The following argument assumes that tests have provided values of specific and medium resistances and the data is to be used to check, predict, control or design a pressure filter where information on the cycle time to filter a given volume of filtrate, or slurry, is required. The instantaneous flow rate q is assumed to be (neglecting the volume of the cake):

$$q = \frac{dV}{dt} \tag{2.29}$$

which can be substituted into Equation (2.21) and rearranged to give:

$$\frac{\Delta P}{q} = \frac{\mu c \alpha}{A^2} V + \frac{\mu R_m}{A} \tag{2.30}$$

The pump characteristic curve gives q as a function of pressure drop, and the value of the instantaneous flow rate given at an arbitrary pressure taken from the characteristic can be used in a rearranged and integrated form of Equation (2.29):

$$t = \int \frac{dV}{q} \tag{2.31}$$

The volume of filtrate for use in the above equation comes from Equation (2.30). The procedure is best illustrated by the following table.

A figure such as Figure 2.11 is drawn from the values shown in columns 3 and 4 of Table 2.2, and graphical integration is used to complete the last column of the table.

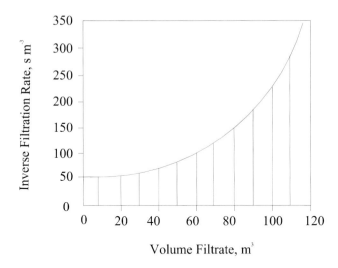

Figure 2.11 Inverse flow rate with filtrate volume

Table 2.2 Illustrative table headings for time required to effect a variable rate and variable pressure filtration

Pressure drop (ΔP)	Flow rate (q)	Volume filtrate collected (V)	Inverse flow rate (q^{-1})	Time (t)
Pump characteristic		Equation 2.30	From column 2	Fig 2.11

If specific and medium resistances are required from operating data then Equation (2.30) shows us that a plot of $\Delta P/q$ against V should be linear with the resistances contributing to the gradient and intercept.

2.6 Effect of Pressure on Cake Filtration

The most economic way of removing liquid from an incompressible filter cake is to filter at the lowest pressure required to overcome the fluid drag within the cake. Increasing the pressure will not lead to a drier cake (higher solid content) but might speed up the dewatering process. All filter cakes display some form of compressibility in practice; increasing the filtration pressure results in an increase in cake solid concentration. An increase in concentration leads to a decrease in cake permeability, see Section 2.3, thus the benefit of increased solid concentration has to be assessed against the disadvantage of operating at a lower cake permeability, or higher specific resistance, when deciding on a pressure to use during compressible cake filtration.

Many materials will give a filter cake of roughly constant concentration, when filtering under conditions of constant pressure, as would be expected of an incompressible filtration. Increasing the filtration pressure results in another cake of roughly uniform concentration but one higher than the first. Thus compressibility is displayed between filtrations but not, necessarily, within a filtration. This observation has led to the almost universal adoption of the concept of *average* cake concentration and permeability or specific resistance, when modelling compressible cakes on an industrial scale. The average values replace permeability or specific resistance in the equations given in Section 2.5.

A useful check on the compressible characteristics of a material under investigation follows from manipulation of Equation (2.23): taking filtration pressure into the dependent variable gives:

$$\Delta P \frac{t}{V} = \left(\frac{\mu \alpha c}{2 A^2} \right) V + \frac{\mu R_m{}'}{A} \tag{2.32}$$

If the pressure is increased by increments during a filtration and then held constant until the next incremental increase, one of the following figures results, depending on the compressibility of the material [Purchas, 1981; Murase et al, 1989].

The bracketed term in Equation (2.32) contains only constants in the case of incompressible filtration, and the pseudo medium resistance term ($R_m{}'$) includes a contribution to the flow resistance due to the layers of cake deposited prior to increasing the filtration pressure. Some time should be allowed before judging the parallel nature of the lines shown on Figures 2.12, to allow for equilibrium to be established.

If a figure resembling Figure 2.12b results from testwork, but the linearity is reasonable, then the material is clearly compressible, but with a constant average specific resistance and cake concentration values during filtration. When filtering material that forms compressible cakes a zero, or negative, intercept may be found on the parabolic rate law plot. This is illustrated on Figure 2.13.

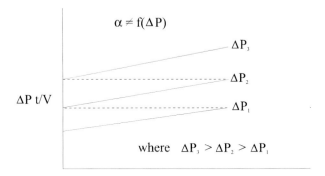

Figure 2.12a Modified parabolic rate law plot to show effect of incompressible cakes

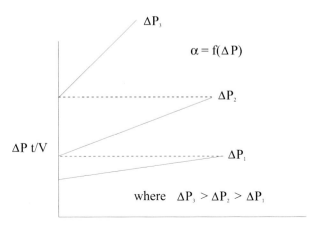

Figure 2.12b Modified parabolic rate law plot to show effect of compressible cakes

Cumulative filtrate volume is reset to zero after each step change in pressure. The filtration should commence with the lowest pressure and proceed in steps up to the highest. Note two pressure steps plus the original pressure are illustrated above.

Equations (2.23) or (2.32) suggest that a negative intercept indicates an additional driving force for the filtration other than filtration pressure! This is not the case. The relation, however, if followed closely at the start would show considerable concave

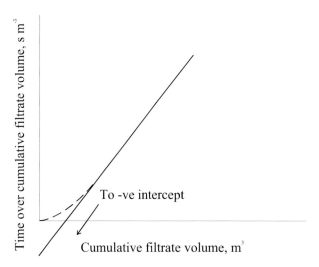

Figure 2.13 Parabolic rate law plot showing negative intercept compressible filter cake

upwards curvature at the origin. Under these circumstances the medium resistance must be calculated by applying Darcy's law to the initial stages of filtration, as shown in Figure 2.4, and using Equation (2.33):

$$R_m = \frac{A\Delta P}{\mu q_0} \tag{2.33}$$

The pressure drop over the filter medium ΔP_m can then be evaluated at any later stage during filtration from Darcy's law, using Equation (2.34):

$$\Delta P_m = \frac{\mu R_m}{A} \frac{dV}{dt} \tag{2.34}$$

Thus it is possible to calculate the pressure drop over the cake ΔP_c by a rearranged form of Equation (2.20):

$$\Delta P_c = \Delta P - \Delta P_m \tag{2.35}$$

There are many theoretical equations or developments which are only applicable to the pressure drop over the filter cake, or assume that the drop over the medium is negligible. Thus the above approach to calculate the pressure drop over the filter cake at any instant in time is an important one in the development of these models.

The relation between specific resistance and pressure drop over the cake is often expressed by the following simple expression [Grace, 1953]:

$$\alpha = \alpha_0 \Delta P_c^n \tag{2.36}$$

or:

$$\alpha = \alpha_0 \left(1 + \frac{\Delta P_c}{P_0} \right)^n$$

Similarly the variation of concentration can be correlated thus:

$$C = C_0 (\Delta P_c)^u \tag{2.37}$$

or:

$$C = C_0 \left(1 + \frac{\Delta P_c}{P_0} \right)^u$$

where u, C_0, P_0, α_0 and n are empirical constants. Similar equations have been proposed to correlate porosity with pressure. The empirical constants can be obtained from a graph such as Figure 2.14.

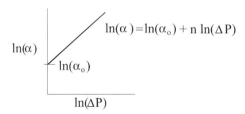

Figure 2.14 Logarithmic variation of specific resistance with pressure

The ratio of the pressure drop over the full filter cake to the average specific resistance is regarded as being equal to the integral of the differential amounts of these quantities:

$$\frac{\Delta P_c}{\alpha_{av}} = \int_0^{\Delta P_c} \frac{d\Delta P}{\alpha}$$

Substituting Equation (2.36) produces:

$$\frac{\Delta P_c}{\alpha_{av}} = \frac{1}{\alpha_0} \int_0^{\Delta P_c} \frac{d\Delta P}{\Delta P^n}$$

which integrates and rearranges to give:

$$\alpha_{av} = \alpha_0 (1 - n) \Delta P_c^n \tag{2.38}$$

This can be substituted into Equations (2.19) or (2.21) to give the general filtration equation for compressible cakes. Considering only the pressure drop due to the filter cake the general equation is:

$$\frac{dV}{dt} = \frac{A^2}{\mu c \alpha_0 (1 - n) \Delta P_c^{(n-1)} V} \tag{2.39}$$

An expression for the average concentration in a filter cake can be deduced by a similar approach to that used in the derivation of Equation (2.38). This leads to:

$$C_{av} = C_0 (1 - u) \Delta P_c^u \tag{2.40}$$

where C_0 and u are empirically derived constants. Equation (2.40) is particularly useful in the estimation of cake height by Equation (2.47) or by means of a mass balance with the solids filtered in the slurry. This is of special relevance to pressure filters in which a limited clearance is available for the cake. See Section 2.6.4 for an example of this.

2.6.1 Constant Pressure Filtration

If Equation (2.39) is integrated for time and volume filtrate, under conditions of constant pressure, an expression identical to that obtained for constant pressure incompressible filtration is produced in terms of average specific resistance, after back-substituting Equation (2.38). This result, together with the concept of constant pressure filtration, leads to the superficial conclusion that the mathematical description of compressible cakes is identical to that of incompressible cakes. If data provide a figure similar to that shown in Figure 2.12(b) then Equation (2.32) should be rewritten as:

$$\Delta P \frac{t}{V} = \left(\frac{\mu c \alpha_{av}}{2 A^2} \right) V + \frac{\mu R_m}{A} \tag{2.41}$$

where the gradient increases as a consequence of increasing α_{av} with cake formation pressure. If c can be calculated by Equation (2.17) this parameter remains constant, but if Equation (2.18) has to be used then account should also be taken of a varying value of m due to cake compression. This can be calculated if a sample of filter cake can be taken. The resulting data can be used to obtain values for the constants α_0 and n by plotting:

$$\ln \alpha_{av} = \ln(\alpha_o (1 - n)) + n \ln(\Delta P_c) \tag{2.42}$$

in preference to that shown in Figure 2.14.

It should be apparent that the major difference between compressible and incompressible cakes is that α_{av} (or k_{av}) becomes a function of pressure. Further significant differences between these types of filtration are discussed in Section 2.6.3.

2.6.2 Constant Rate Filtration

Again using the concept of a uniform specific resistance throughout the filter cake, rearranging Equation (2.39) and using the definition of constant rate ($dV/dt = V/t$):

$$\Delta P_c^{n-1} = \frac{1}{V}\frac{t}{V}\frac{A^2}{\mu c \alpha_0 (1-n)}$$

Substituting $q = V/t$ into the above and rearranging provides:

$$\Delta P_c^{1-n} = \left(\frac{\mu c \alpha_0 (1-n) q^2}{A^2} \right) t \tag{2.43}$$

Thus a logarithmic plot of pressure drop over the filter cake against time yields a straight line if average values of specific resistance and cake concentration exist.

2.6.3 Analysis of Flow Inside a Cake

A true description of the processes involved in a compressible filter cake can only be obtained by a mathematical investigation reflecting the microscopic level of detail. The solid concentration is a function of the solids stress or pressure gradient throughout the cake. Thus the specific resistance or permeability varies with position. The liquid flow rate also varies through the cake, increasing from the cake-forming surface to the filter septum. Note that the liquid velocity does not increase uniformly from zero at the cake-forming surface; in fact the majority of the filtrate comes from that surface. It is also generally true that specific resistance is a function of solid content of the slurry being filtered [Rushton & Hameed, 1969; Shirato & Okamura, 1959]. This is generally attributed to the speed with which a cake is deposited and the ability of the finer particles to migrate into positions of relatively higher flow resistance. Thus concentrated slurries provide greater hindrance to this migration and, therefore, provide lower specific resistance cakes. A more comprehensive illustration of a compressible filter cake is shown in Figure 2.15.

The following equation has been provided by many research workers and in a variety of forms [Holdich, 1990]:

$$\frac{d\Delta P}{dh} = gC(\rho_s - \rho) - \mu \alpha \rho_s C\left(\frac{q}{A} - v_s\right)$$

(2.44)

where v_s is the solids velocity in the cake towards the medium and q is the filtrate flow rate through the medium. The first term on the right-hand side of Equation (2.44) represents solids pressure due to the weight of the cake, the second term is due to the fluid drag acting on the cake. It is derived from the force–momentum balance (neglecting inertia terms) in conjunction with the liquid and solid mass balances.

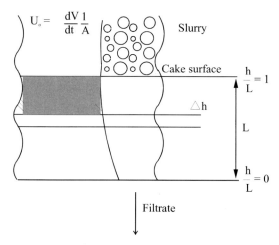

Figure 2.15 Internal structure of a compressible filter cake

The solids weight is usually small compared to the pressure due to liquid drag, the solids velocity is also normally neglected and the superficial filtrate flow rate is assumed to be constant throughout the cake, i.e all the filtrate comes from the new cake layer (q/A is uniform), thus [Tiller and Cooper, 1962; Shirato and Okamura, 1959; Wakeman, 1978]:

$$\frac{d\Delta P}{dh} = \mu \alpha \rho_s C \frac{q}{A}$$

(2.45)

Note that Equation (2.45) is similar to a differential form of Darcy's law with Equation (2.10) used to replace permeability. Equation (2.45) rearranges to give:

$$\int \frac{d\Delta P}{C\alpha} = \mu \rho_s \frac{q}{A} \int dh$$

which integrates over the full cake depth to give:

$$\int_0^{\Delta P_c} \frac{d\Delta P}{C\alpha} = \mu\rho_s \frac{q}{A} L$$

or over a limited range from an arbitrary position h within the filter cake to give:

$$\int_{\Delta P_h}^{\Delta P_c} \frac{d\Delta P}{C\alpha} = \mu\rho_s \frac{q}{A}(L-h)$$

An expression for the fractional pressure profile with respect to dimensionless height inside the filter cake is formed by taking a ratio of the last two equations.

$$1 - \frac{h}{L} = \left(\int_{\Delta P_h}^{\Delta P_c} \frac{d\Delta P}{C\alpha}\right) \Big/ \left(\int_0^{\Delta P_c} \frac{d\Delta P}{C\alpha}\right)$$

Equations (2.36) and (2.37) can be substituted into the above equation and the resulting equation integrated and rearranged to give:

$$\Delta P_h = \Delta P_c \left(\frac{h}{L}\right)^{1/(1-u-n)}$$

and if Equation (2.37) is used to replace pressure at some point inside the cake with concentration at that point:

$$C_h = C_0 \Delta P_c^u \left(\frac{h}{L}\right)^{u/(1-u-n)} \tag{2.46}$$

or in terms of porosity:

$$1 - \varepsilon_h = C_0 \Delta P_c^u \left(\frac{h}{L}\right)^{u/(1-u-n)}$$

Thus if the functional relation between the solids compressive pressure, specific resistance and solid concentration (Equations 2.36 and 2.37) is known it is possible to predict the solid concentration profile within the filter cake relative to the dimensionless, or fractional, distance h/L of that cake.

The filter cake concentration profile relative to actual height from the medium can be obtained from the above if the superficial filtrate flow rate through the medium q has been measured via a modified form of Darcy's law: Equation (2.47):

$$L = \frac{A}{\mu\rho_s q}\left(\frac{\Delta P_c}{\alpha_{av} C_{av}}\right) \tag{2.47}$$

A more accurate estimate of cake height comes from integrating the values of pressure, specific resistance and concentration throughout the cake, instead of assuming average values. This is represented by Equation (2.48):

$$L = \frac{A}{\mu\rho_s q} \int \frac{d\Delta P}{\alpha C} \tag{2.48}$$

If Equations (2.36) and (2.37) are used to replace specific resistance and concentration in the integration then:

$$\int_0^{\Delta P_c} \frac{d\Delta P}{\alpha C} = \frac{\Delta P_c^{(1-u-n)}}{\alpha_o C_o (1-u-n)} \tag{2.49}$$

and Equation (2.48) becomes:

$$L = \frac{A}{\mu\rho_s q} \frac{\Delta P_c^{(1-u-n)}}{\alpha_o C_o (1-u-n)} \tag{2.50}$$

Figure 2.16 shows the solid concentration profile in a filter cake predicted by means of Equations (2.46) to (2.50).

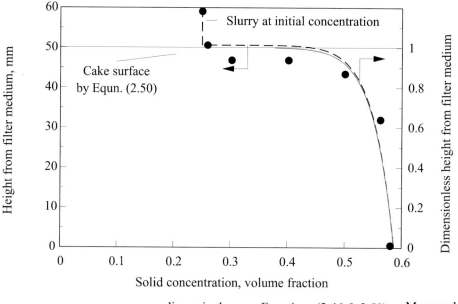

Calculated cake profiles: dimensionless   Equations (2.46 & 2.50)   Measured

Figure 2.16 Concentration profile formed during filtration: measured and calculated values by Equations (2.46)–(2.50)

Filtration conditions: 49.8 kPa pressure drop over cake, α_0 and n 9.3×10^8 and 0.334, C_0 and u 0.381 and 0.04, solid density 2877 kg m^{-3}, liquid viscosity 0.001 Pa s and filtrate flow velocity 2.7×10^{-5} m s^{-1}. Material filtered was a crushed dolomite of 24 μm median diameter.

A mass balance can be used to check the amount of solids present in the cake, and the cake height adjusted, if necessary [Holdich, 1993]. The extra computational effort is only worthwhile when the cake profile or height must be known to a high degree of accuracy.

The terms $L\,q$ are known as the rate×thickness product [Tiller and Green, 1973], and have been used to describe the filtration characteristics of several materials. It is a logical expression of the product of a filter, both filter cake (cake depth L) and filtrate (flow rate q) being that product. It is generally found that there is a logarithmic relation between rate×thickness product and filtration pressure. One notable exception is latex which displays a very sudden drop in filter cake porosity next to the filter medium. This results in a plateau on the rate×thickness–pressure plot. Thus increasing the filtration pressure does not increase either cake height or filtrate rate proportionally. This is an effect similar to that of a gel layer observed during ultrafiltration. The practical consequence of this observation is that high–pressure filtration is not appropriate to this material and mechanical expression has been recommended .

The introduction to this section included some comments on the variability of specific resistance. The physical model used in Equation (2.45) was that the liquid flow rate through the cake is constant. This can only be true when filtering an incompressible material. When dealing with compressible cakes the "average specific resistance" defined earlier is interpreted as being the resistance when filtering a slurry of zero solid content. This is, in fact, the resistance obtained from the compression permeability (CP) cell tests described in Section 2.9; the static condition of the solids in the test is assumed to be representative of the dynamic condition of the solids during filtration. Real slurries will contain suspended material and filter cakes will compress, the liquid flow rate will, therefore, increase towards the filter medium. It has been argued that the filtrate rate will be higher than that assumed from the CP tests. An increase in flow rate, for a given pressure drop, is equivalent to a reduction in resistance. Thus a fractional correction term has been applied to the average specific resistance in compressible cake filtration. This is known as the Tiller–Shirato J factor [Tiller and Shirato, 1964]. Removing solids weight from Equation (2.44) results in:

$$\frac{\mathrm{d}\Delta P}{\mathrm{d}h} = \mu\alpha\rho_s C\left(\frac{q}{A} - v_s\right)$$

which rearranges to:

$$\mu\rho_s \int\left(\frac{q}{A} - v_s\right)C\mathrm{d}h = \int\frac{\mathrm{d}\Delta P}{\alpha}$$

and:

$$\rho_s C\mathrm{d}h = \mathrm{d}w$$

where $\mathrm{d}w$ is the mass of dry solids per unit area deposited in distance $\mathrm{d}h$, thus:

$$\int_0^w \left(1 - \frac{v_s A}{q}\right) dw = \frac{1}{\mu} \frac{A}{q} \int_0^{\Delta P_c} \frac{d\Delta P}{\alpha} = \frac{1}{\mu} \frac{A}{q} \frac{\Delta P_c}{\alpha_{av}}$$

Dividing by the total mass per unit area of the filter cake (w) gives:

$$\int_0^1 \left(1 - \frac{v_s A}{q}\right) d\left(\frac{w_h}{w}\right) = \frac{1}{\mu} \frac{A}{q} \frac{\Delta P_c}{w \alpha_{av}}$$

where w_h is the mass of dry solids deposited per unit area up to position h within the filter cake. The left-hand side of the equation is known as the J factor. Putting Equation (2.16) and the J factor into the equation and rearranging gives (including medium resistance):

$$\frac{dV}{dt} = \frac{A^2 \Delta P}{\mu(c\alpha_{av} JV + AR_m)} \tag{2.51}$$

where $0 \ll J < 1$, and is nearer unity for dilute slurries.

An increasing flow of liquid throughout the filter cake will have a consequent effect on the liquid pressure gradient within the cake. Measurement of this has been used experimentally to infer the concentration within the cake, and to obtain information on the local specific resistance. A full mathematical description of the internal pressure and concentration profiles within a filter cake, and a comprehensive review of the work of Tiller and Shirato has been provided by Wakeman [1978].

In a later paper Wakeman solved Equation (2.45) using incremental time and position within a forming filter cake [Wakeman, 1981]. A theoretical permeability model and a measured value for medium resistance were used. By splitting the cake into time and position increments it was possible to predict the pressure, concentration and specific resistance within the forming cake; this is one of the most comprehensive treatments of cake formation. This procedure does, however, require considerable computing power and expertise. The results obtained provided values for:

i)   Pressure loss across the medium
ii)   Filtrate flow rate through the medium
iii) Tiller–Shirato J factor
iv)   Mean cake porosity
v)   Mean specific resistance of the cake
vi)   Solids concentration in the slurry next to the cake
vii) Pressure loss with respect to position in the cake
viii) Filtrate flow rate within the cake
ix)   Mass of solids in each element
x)   Depth of each layer (considering constant mass layer of solids)
xi)   Local specific resistance
xii) Local porosity

In most practical industrial filtrations involving compressible cakes, filtration equations (2.41)–(2.43) are applied using an average value for cake resistance and concentration which are usually measured. For a more detailed understanding of the cake structure solids concentration profiles have to be measured or assumed, thereby Equations (2.45)–(2.50) can be used to determine the overall filtration resistance, throughput rate, cake depth, etc. If it is possible to measure these functions and profiles then confident equipment scale-up can be performed from a comprehensive knowledge of the filter cake properties.

2.6.4 Variable Rate and Pressure Filtration for Compressible Cakes

The pump characteristic is again required to evaluate variable rate and pressure filtration. The result normally required is a calculation of the time taken to filter a known volume, or mass, of slurry. The cake depth is also important as most filters are of the batch type and this depth must be less than the clearance within the filter.

Equation (2.50) can be rearranged with constants grouped to give:

$$L = \frac{k_1}{q} \Delta P_c^{k_2} \tag{2.52}$$

where the constants in Equations (2.50) and hence (2.52) should come from pressure leaf tests, or may be estimated from those provided in Table 2.3. It should be apparent that the mass of solids in the filter cake must be equal to that added to the slurry, thus:

$$A\rho_s \int_0^L C\mathrm{d}h = sM \tag{2.53}$$

The procedure is best summarised as follows:

i) Obtain the flow rate at a value of pressure from the pump characteristic
ii) Calculate L from Equation (2.52)
iii)Calculate C as a function of h/L from Equation (2.46)
iv)Convert (iii) to C as a function of true height using Equation (2.50)
v) Integrate using Equation (2.53) to give mass of solids in cake
vi)Increment onto another flow rate until the desired mass of solids has been filtered

The time required to filter can again be calculated by graphical integration of a plot such as Figure 2.11. The cumulative volume of filtrate at each increment is obtained after calculating the mass of solids filtered and, therefore, the mass of slurry filtered up to that increment M' thus:

$$V = (1 - s)M'/\rho - LA(1 - C_{av})$$

Stages (iii) – (v) can be simplified considerably by assuming an average cake concentration provided by Equation (2.40). Both procedures can be entered readily into a standard computer spreadsheet program.

2.6.5 Simulation of Cake Filtration by Incremental Analysis

When filtering materials giving rise to incompressible filter cakes the simulation of cake formation is straightforward and can be achieved using the equations and methods described in Section 2.5. However, in the case of compressible cakes the cake properties will alter as the pressure drop forming the filter cake changes during the process. This is even true for apparently constant pressure filtration: when no cake exists at the start of the cycle the pressure drop is entirely over the filter medium, at the end most of the pressure drop will be over the filter cake. Between these two limits the pressure forming the cake will gradually increase. Hence, the overall pressure drop remains constant but the pressure influencing the cake properties varies substantially. During constant rate filtration the applied pressure and cake forming pressure drops are designed to increase in order to overcome the increasing resistance to filtration.

A method of approach to the simulation of cake formation for compressible cakes that has become well established in recent years is to break the cake formation cycle into many increments. The cake properties described by Equations (2.38) and (2.40) can be assumed to be constant during the time increment, but are recalculated at the next time increment. There are many instances of this incremental approach to filtration simulation [Wakeman, 1981; Theliander and FathiNajafi, 1996; Stamatakis and Chi Tien, 1991; Holdich, 1994] and the following worked examples demonstrate how it is applied. The examples form part of the spreadsheet files included in Appendix C: files CONRATE and FILTER for constant rate and constant pressure filtration respectively.

In the case of constant rate filtration increments in operating pressure are considered: at each increment the cake properties are considered constant, but allowed to vary between each increment. Any error associated with this approach can be minimised by reducing the step size (in pressure) between the increments. The solution uses arbitrary values of overall pressure drop, which increase during the constant rate filtration, and calculates the time interval corresponding to the increase in pressure. The starting point is a similar form of Darcy's law to Equation (2.45):

$$\frac{\Delta P_n}{L_n} = \mu \alpha_{av} \rho_s C_{av} \frac{\delta V}{\delta t} \frac{1}{A}$$

where ΔP_n and L_n are the pressure drop over the newly formed layer of filter cake during an increment and the depth of that layer. A material balance on the depositing solids gives:

$$L_n A \rho_s C_{av} = QC_s \delta t \rho_s$$

where Q is the feed rate of suspension and C_s is the feed concentration as a volume fraction. This equation can be rearranged for L_n and substituted into Darcy's law to give:

$$\Delta P_n = \mu \alpha_{av} \rho_s QC_s \delta V / A^2$$

The incremental filtrate volume is equal to the product of the filtrate volume flow rate (q) and the incremental time difference:

$$\delta V = q \delta t$$

which can be substituted into the above equation to provide an expression for the increment in time required to form the new cake layer:

$$\delta t = \Delta P_n (A^2 / \mu \alpha_{av} \rho_s qQC_s) \tag{2.54}$$

Note that q is the filtrate flow rate and Q is the feed flow rate, thus $q < Q$. The filtrate flow rate can be deduced from the feed flow using a knowledge of the filter cake and slurry volume fraction concentrations and a mass balance. It is Equation (2.54) that is solved for time increments using the selected pressure increments. The full solution scheme is illustrated in the flow diagram given in Figure 2.17.

The flow chart illustrated in Figure 2.17 is relevant to a spreadsheet implementation of the simulation, where circular references to cells are permitted. Thus the incremental time is required for the calculation of the new total cake height which, in turn, is used to calculate the incremental time. The iteration is easily performed on a spreadsheet and requires no intervention by the user, whereas a computer program would require a loop in order to iterate a converged solution for the correct incremental time. The most significant assumptions in this method of solution are:
1. R_m remains constant throughout the filtration,
2. cake pressure drop comes from the total pressure drop minus that over the medium at the previous time increment (this limits the step size), and
3. the properties of the newly formed cake layer (concentration and specific resistance) are similar to the average filter cake properties.

Figure 2.18 illustrates the volume of filtrate collected with respect to time for a simulation following the above scheme. The operating conditions were a constant feed rate of 70 litres per minute, total filter area of 9.4 m^2, slurry concentration of 0.05 by mass fraction, liquid viscosity 1mPa s, liquid density 1000 kg m^{-3}, solid density 2800 kg m^{-3}, n=0.5, u=0.08, α_o=4.5x10^8, C_o=0.15 and cloth resistance of 1x10^{11} m^{-1}.

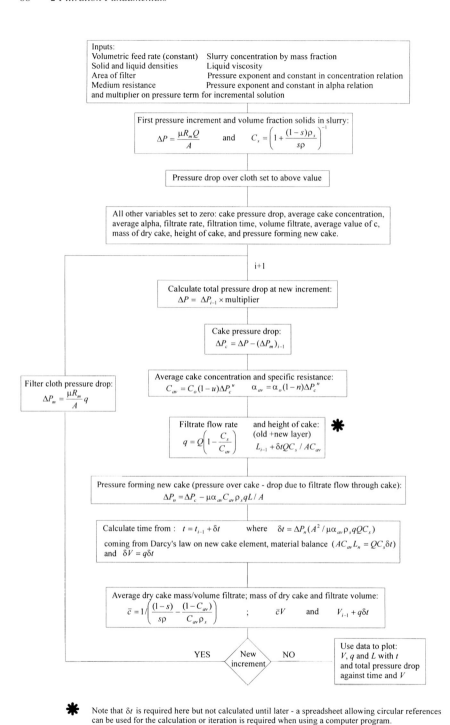

Figure 2.17 Flow diagram for the calculations required during compressible cake constant rate filtration

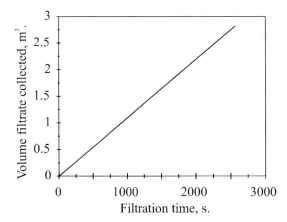

Figure 2.18 Constant rate filtration of a compressible cake

It is apparent that the filtration rate deduced from Figure 2.18 is substantially constant, in a similar way to the rate that can be deduced from constant rate filtration of incompressible materials; such as that illustrated in Figure 2.8. However, the applied pressure required to maintain the filtration does not follow a similar trend to the incompressible case. This is illustrated in Figure 2.19 and the result should be contrasted with Figure 2.9.

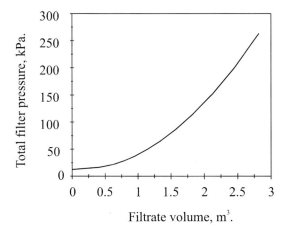

Figure 2.19 Pressure variation during constant rate filtration of a compressible cake

The spreadsheet example described in Appendix C (CONRATE), and downloadable from the World Wide Web site, uses the operating parameters employed in the above figures.

The constant pressure compressible cake simulations included in Appendix C (CONPRESS and PROFILE), use increments in time and calculate the cake properties at each increment. Again Equations (2.38) and (2.40) are used for average cake properties, which are assumed to be constant at each increment in time. Equation (2.41) can be rearranged to give a quadratic with respect to filtrate volume:

$$\frac{\mu \alpha_{av} \overline{c}}{2A^2 \Delta P} V^2 + \frac{\mu R_m}{A \Delta P} V - t = 0 \qquad (2.55)$$

The average mass of dry cake per unit volume filtrate is obtained from the average cake concentration and the mass fraction of solids in the feed slurry by

$$\overline{c} = \frac{1}{(1-s)/(s\rho) - (1-C_{av})/(C_{av}\rho_s)} \qquad (2.56)$$

The iterative solution on the computer spreadsheet is started by using the full filtration pressure in Equations (2.38) and (2.40) to calculate the specific resistance and cake concentration and then the average dry solids mass per unit volume of filtrate. These values are then used in Equation (2.55), solving for filtrate volume as the positive root of the quadratic equation.

The instantaneous filtrate rate (q) follows from Equation (2.55) in its differential form

$$q = \frac{dV}{dt} = \left[\frac{\mu \alpha_{av} \overline{c}}{A^2 \Delta P} V + \frac{\mu R_m}{A \Delta P} \right]^{-1} \qquad (2.57)$$

which is then used to calculate the pressure drop over the filter cake by deducting the pressure drop over the medium from the total applied pressure

$$\Delta P_{cake} = \Delta P - \frac{\mu R_m}{A} q$$

The new value of pressure drop over the filter cake is then used in Equations (2.38) and (2.40) to calculate specific resistance and cake concentration. The procedure is repeated until convergence of all pressures and flow rates, then another time increment is investigated. Thus Equation (2.55) is assumed to be valid for any instant in time, but the values of α_{av} and $\overline{c}$ vary at each time increment. During the iterative processes the cake forming pressure converges to a value less than the full filtration pressure, i.e. the full pressure is used only to start the process off and the overall filtration behaviour described may then be non-parabolic in nature.

At any instant in time cake depth can be calculated from Equation (2.50) and the local volume concentration inside the filter cake from Equation (2.46). Hence, the local concentration at any height and time within the filter cake can be calculated by this incremental approach. Figure 2.20 is an example of the simulated cake concentration profile at three filtration times under the operating conditions stated. The example shown is particularly pertinent to the simulation of a rotary vacuum filtration.

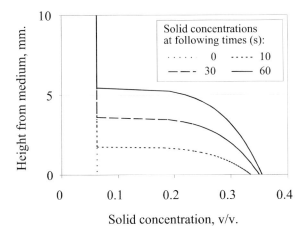

Figure 2.20 Simulation of local concentrations during compressible cake constant pressure filtration

operating conditions: total filter area 1 m², total filtration pressure 65 kPa, liquid viscosity 1 mPa s, liquid density 1000 kg m⁻³, slurry concentration 0.15 by mass fraction, n=0.5, u=0.08, α_o=4.5x10⁸, C_o=0.15, cloth resistance 8x10¹⁰ m⁻¹, solid density 2650 kg m⁻³.

The height of the filter cake after just 10 seconds of filtration can be seen to be only 2 mm, which is unlikely to give satisfactory cake discharge. Furthermore, the rheology of cakes, pastes and suspensions is such that it is usual for cakes to be very loose up to a threshold value of concentration. Above that threshold the cake would be firmer and discharge more readily. In the absence of some testwork it would be impossible to say if the material illustrated in Figure 2.20 is above or below that threshold after 10 seconds filtration, but if that value is known, or can be estimated, then the simulation could be used to investigate suitable operating conditions required to achieve that condition. Two further filtration times are considered in the figure: cake form times of 30 and 60 seconds, corresponding to drum speeds of 1/2 and 1/4 rpm respectively at 25% submergence. When filtering a material such as that illustrated it can be seen that slow drum speeds are required in order to obtain filter cake heights suitable for discharge by a scraper. The slow drum speed will also result in a more concentrated cake more likely to have the right rheological properties for discharge. This simulation was performed using the spreadsheet file PROFILE, the same operating data is also used in the simulation FILTER to investigate operating conditions further.

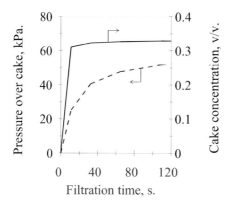

Figure 2.21 Simulation of compressible cake constant pressure filtration

Figure 2.21 shows how the pressure drop over the cake increases from zero; i.e. all the pressure drop is over the filter medium to start with, rising to a value of just over 50 kPa after 2 minutes. Thus the total filtration pressure may remain constant at 65 kPa but the distribution of that pressure drop will vary significantly during the early stages of filtration. The filter cake is compressible, hence the average cake concentration will also vary over the same time period. However, as the exponent value in Equation (2.40) is relatively low, only 0.08 in the example used, the cake concentration reaches an approximately constant value well before that of the pressure drop over the cake. However, only a limited cake depth is achieved at short cycle times, as illustrated in Figure 2.20. It is worth reflecting that under these conditions a maximum of 77% of the applied vacuum for the filtration is actually going to form the filter cake, the rest draws the filtrate through the cloth.

In order to use the simulations provided by the profile and filter files the operator will have to know the values for the constitutive relations: Equations (2.38) and (2.40) and the cloth resistance. Thus some experimental work or plant data are required. In the absence of test data an estimate of equipment performance could be obtained using published values of these constants, selecting the ones most similar to the application under investigation. Table 2.3 illustrates the range of values for some previously investigated materials [Wakeman and Tarleton, 1999].

Table 2.3 Published values for constants used in Equations (2.38) and (2.40)

Material	Resistance at 1 bar (m kg^{-1})	α_o (m kg^{-1} Pa^{-n})	n	C_o (Pa^{-u})	u
100 μm glass beads*	0.064 x10^9	0.064 x10^9	0	0.60	0
10 μm glass beads*	6.4 x10^9	6.4 x10^9	0	0.60	0
Alumina	7.5x10^9	2.37x10^8	0.3	0.07	0.066
Calcium carbonate in salt solution	20.9 x10^9	46.9 x10^8	0.13	0.22	0.03
Calcium carbonate in water	89.3 x10^9	89.3 x10^8	0.2	0.09	0.1
…end of vacuum filtration range					
Coarse grade kaolin	142 x10^9	4.5 x10^8	0.5	0.15	0.08
Talc in salt solution	250 x10^9	7.1 x10^8	0.5	0.008	0.29
Talc in water	250 x10^9	7.1 x10^8	0.5	0.088	0.144
Titanium dioxide	506 x10^9	127 x10^8	0.32	0.06	0.13
1 μm glass beads*	640 x10^9	640 x10^9	0	0.60	0
Cement	686 x10^9	222 x10^8	0.3	0.33	0.05
Kaolin (Hong Kong pink)	1720 x10^9	64.8 x10^8	0.5	0.17	0.08
zinc sulphide pH 3	1800	1000	0.25	0.19	0.06
…end of pressure filtration range	x10^9	x10^8			
Magnesium hydroxide	3020 x10^9	135 x10^8	0.47		
Aluminium hydroxide	16600 x10^9	3320 x10^8	0.34		
Stable zinc sulphide	58900 x10^9	14.8 x10^8	0.92	0.002	0.39
Colloidal clay	165000 x10^9	1440 x10^8	0.61		

 * assuming a random close packing and using Equations (2.4) and (2.10)

The constants shown in Table 2.3 have no physical meaning, but are empirically derived over a limited pressure range for use in Equations (2.38) and (2.40). However, the constant 'n' has become known as the compressibility coefficient, and highly compressible materials have high values of n and tend to be more difficult to filter. A high value for the constant 'u' also indicates a high degree of compressibility. A high value of C_o combined with a low value of u indicates a very low degree of compressibility, but this does not imply little filtration resistance; see the position of cement in Table 2.3. Thus filtration resistance is due to the nature of the material to be filtered (e.g. more than one mineral and wide size distribution), the ionic state of the continuous phase as well as the solids in suspension. This is discussed further in Chapter 5. Hence, the filtration performance of any material may be altered by the nature of the solution due to surface chemistry effects [Tarleton and Willmer, 1998]. Thus the above table should be read as providing nothing more than a rough estimate for the materials listed and more accurate design should entail an investigation of these constants in the appropriate medium for the material.

2.7 Other Modes of Filtration

There are many filtrations which do not result in the formation of a filter cake and cannot, therefore, be described by the above models. These usually occur at dilute suspended solids concentration and are, therefore, regarded as clarifying filtrations. It is usual for the liquid to be the important component of the suspension, as it is often impossible to recover the solids in uncontaminated form from the filter. Chapter Six describes clarifying filtration and modelling, but the fundamentals are discussed here as they complement the above discussion on cake filtration.

In 1936 Hermans and Bredée published the results of their investigation into the various filtration modes. This was later comprehensively reviewed, summarised and tested by Grace [1956]. The different modes of filtration reflect not only the concentration of suspension but also the mechanism pertaining. These include:

a) Direct sieving action at the medium pores or cake surface
b) Gravity settling
c) Brownian diffusion onto filter
d) Interception
e) Impingement
f) Electrokinetic forces.

In the filtration of particles greater than 0.5 μm from liquids the dominating mechanisms are (a) and (d). The direct interception of particles from liquid streamlines in contact with the pore walls (*not* inertial interception) is more important when filtering finely powdered material; direct sieving is more relevant to suspensions containing coarse particles. It is generally true that direct interception becomes more important as the viscosity is increased and some of the above mechanisms are much more relevant in the filtration of particles from gas streams. Another important distinction between gas and liquid filtration is the porosity of the filter medium. In the former the medium provides a series of targets for particles to attach to, and has a porosity of 90–99%. In liquid filtration the medium can be regarded as a barrier with an interconnected pore structure, which the particles must pass through if they are not collected, and the corresponding porosity is much lower at 30–70%. The lower porosity also increases the mechanical strength of the medium which must be much higher in high-viscosity fluids to resist the forces caused by the higher pressure drops experienced over the filter.

The cake filtration mechanism has already been explained in Section 2.4. The following discussion refers mainly to other types of filtration which can occur very early on in the otherwise cake filtration cycle, or over a prolonged period during clarifying filtration.

Some of the filtration mechanisms are illustrated in Figures 2.22–2.24.

The mathematical description of the different filtration modes is based on a common differential equation with two constants n' and K':

$$\frac{d^2 t}{dV^2} = K' \left(\frac{dt}{dV} \right)^{n'}$$

(2.58)

for constant pressure filtration, and:

$$\frac{\mathrm{d}(\Delta P)}{\mathrm{d}V} = K'(\Delta P)^{n'} \tag{2.59}$$

for constant rate filtration. The exponent is given by the type of filtration occurring and the proportionality constant depends on the filtration conditions (viscosity, specific resistance, etc). The mathematical description of the various modes is given in Table 2.4.

In the case of cake filtration n' = zero and integration of Equation (2.58) provides:

$$\frac{t}{V} = \frac{K_c V}{2} + \frac{1}{q_o} \tag{2.60}$$

which is similar to Equations (2.23) and (2.27). In an investigation of filtration where mechanisms other than cake filtration are suspected the experimental data can be displayed on diagrams which reflect the equations given in Table 2.4.

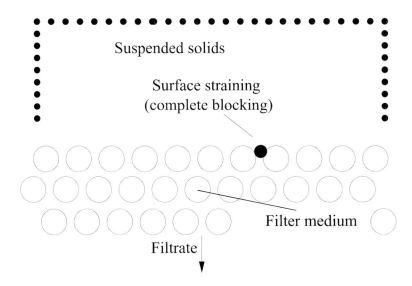

Figure 2.22 Schematic diagram of surface straining: complete blocking mechanism

For example if the data provides a straight line on a plot of inverse instantaneous flow rate q^{-1} against time t then an intermediate blocking model is apparent, or a linear relation on a plot of time over cumulative filtrate volume t/V against time t suggests standard blocking.

Standard blocking has been reported to be an appropriate filtration model for cartridge filtration of liquids and a form of the standard blocking equations shown in Table 2.4 has been used to correlate the filtration of viscose and cellulose acetate [Grace, 1956]. The model assumed in the derivation of the standard blocking equations is that particles build

Table 2.4 Filtration laws of Hermans and Bredée as given by Grace

1. For constant-pressure filtration

Function	Complete blocking	Standard blocking	Intermediate blocking	Cake filtration
$\dfrac{d^2 t}{dV^2} = \left(\dfrac{dt}{dV}\right)^n$	$n = 2$	$n = \dfrac{3}{2}$	$n = 1$	$n = 0$
$V = f(t)$	$V = q_o(1 - e^{-K_b t})$	$\dfrac{t}{V} = \dfrac{K_s}{2} t + \dfrac{1}{q_o}$	$K_i V = ln(1 + K_i t q_o)$	$\dfrac{t}{V} = \dfrac{K_c}{2} V + \dfrac{1}{q_o}$
$q = f(t)$	$q = q_o e^{-K_b t}$	$q = \dfrac{q_o}{\left(\dfrac{K_s V}{2} + 1\right)^2}$	$K_i t = \dfrac{1}{q} - \dfrac{1}{q_o}$	$q = \dfrac{q_o}{(1 + K_c q_0^2 t)^{1/2}}$
$q = f(V)$	$K_b V = q_o - q$	$q = q_o\left(1 - \dfrac{K_s V}{2}\right)^2$	$q = q_o e^{-K_i V}$	$V K_c = \dfrac{1}{q} - \dfrac{1}{q_o}$

Table 2.4 Continued

2. For constant-rate filtration

Function	Complete blocking	Standard blocking	Intermediate blocking	Cake filtration
$\dfrac{\mathrm{d}(\Delta P)}{\mathrm{d}V} = k(\Delta P)^n$	$n = 2$	$n = \dfrac{3}{2}$	$n = 1$	$n = 0$
$\Delta P = f(V)$	$\dfrac{\Delta P_o}{\Delta P} = 1 - \dfrac{K_b V}{q_o}$	$\left(\dfrac{\Delta P_o}{\Delta P}\right)^{0.5} = 1 - \left(\dfrac{K_s}{2}\right)V$	$\ln\left(\dfrac{\Delta P}{\Delta P_o}\right) = K_i V$	$\dfrac{\Delta P}{\Delta P_o} = K_c q_o V + 1$
$\Delta P = f(t)$	$\dfrac{\Delta P_o}{\Delta P} = 1 - K_b t$	$\dfrac{1}{q_o}\left(\dfrac{\Delta P_o}{\Delta P}\right)^{0.5} = \dfrac{1}{q_o} - \left(\dfrac{K_s}{2}\right)t$	$\dfrac{1}{q_o}\ln\left(\dfrac{\Delta P}{\Delta P_o}\right) = K_i t$	$\dfrac{\Delta P}{\Delta P_o} = K_c q_o^2 t + 1$

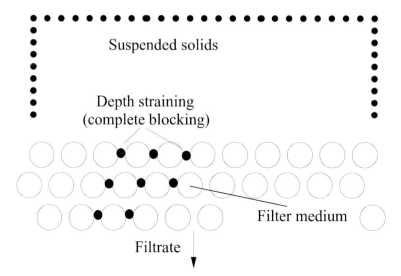

Figure 2.23 Schematic diagram of depth straining: complete blocking mechanism

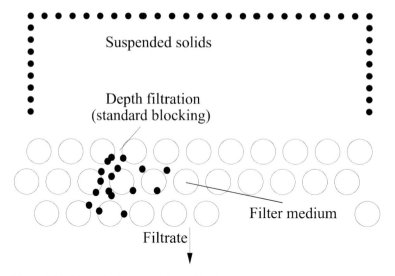

Figure 2.24 Schematic diagram of depth filtration: standard blocking mechanism

up at the surface of the pore walls, leading to a diminishing pore size and eventual plugging. The constants in the standard blocking model, which are obtained from the graphical construction, are related to the properties of the filter medium and solids in a similar way to that already shown for cake filtration. The constants contain contributions due to the number of pore openings, pore channel depth, pore radius, viscosity, filter area, filtration pressure, suspended solid concentration and packing porosity of the deposited solids. The model is derived by an adaptation of Poiseuille's law for fluid flow through channels. Complete blocking is a consequence of pore plugging, with the number of channels available for fluid flow diminishing in proportion to the volume of filtrate passed.

In most filtrations it is highly probable that more than one filtration mode occurs. When filtering suspensions of suspended solids content above some threshold (usually 1% by volume) cake filtration is rapidly established. Below this value other mechanisms may be present for some time. An example of this is illustrated in Figure 2.25.

The data illustrated in Figure 2.25 display standard law filtration characteristics over the first 400 seconds, and cake filtration characteristics from 800 to 1600 s. The intercept of this t/V–V plot is negative and illustrates the care with which the procedure to determine medium resistance from the intercept during cake filtration should be exercised, as discussed in the previous section.

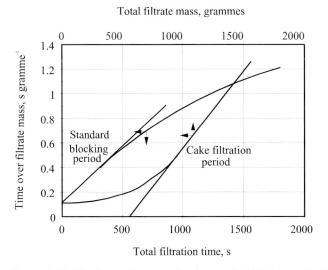

Figure 2.25 Filtration performance showing standard blocking and cake filtration [Grace, 1956]

Filtration conditions: filter media Orlon taffeta, not calendered, filtering a 23 ppm suspension of modal diameter 6 μm, by weight, at 1 bar differential pressure. Note that mass is directly proportional to volume.

2.8 Filtration with Non-Newtonian Fluids

In the previous sections the suspending liquid has been assumed to display Newtonian rheological characteristics, see Appendix B for further details. This section considers the effects of filtering solids from liquids displaying non-Newtonian flow characteristics.

Constant-pressure cake filtration with non-Newtonian suspending fluids has received considerable attention [Kozicki, 1990]. The average specific resistance has been found to vary considerably as a function of the flow behaviour index N even during the filtration of apparently incompressible materials [Shirato et al, 1977]. Later papers extended the analysis to compressible filter cakes and constant rate and variable pressure and rate filtrations [Shirato et al, 1980a, b].

The development of the basic working equation under the conditions of constant pressure is as follows.

Darcy's law can be written in the following form for the flow of Newtonian fluids if Equation (2.10) is used to replace permeability with specific resistance:

$$\frac{dV}{dt}\frac{1}{A} = \frac{1}{\mu\alpha\rho_s C}\frac{d\Delta P}{dh}$$

The equivalent equation for the flow of power law non-Newtonian fluids is:

$$\left(\frac{dV}{dt}\frac{1}{A}\right)^N = \frac{1}{k'\rho_s\alpha C}\frac{d\Delta P}{dh} \tag{2.61}$$

where N and k' are the flow behaviour index and consistency coefficient.

The equivalent non-Newtonian form of Equation (2.12) is:

$$\left(\frac{dV}{dt}\frac{1}{A}\right)^N = \frac{A\Delta P_c(1-sm)}{k'Vs\rho\alpha} \tag{2.62}$$

after using Equations (2.16) and (2.18) for mass of dry cake deposited per unit area. Some research workers use the cumulative volume of filtrate per unit filter area V' instead of the actual filtrate volume:

$$V' = \frac{V}{A}$$

Thus Equation (2.62) becomes:

$$\left(\frac{dV}{dt}\frac{1}{A}\right)^N = \frac{\Delta P_c(1-sm)}{k'V's\rho\alpha}$$

and if an average value of specific resistance can be assumed the equation rearranges to give:

$$\left(\frac{\mathrm{d}V}{\mathrm{d}t} \frac{1}{A} \right)^{N} = \frac{k'V's\rho\alpha_{av}}{\Delta P_{c}(1- sm)} \qquad (2.63)$$

In both Newtonian and non-Newtonian filtration a method to determine the filter cake depth at an instant in time is useful as a means to calculate the ratio of wet to dry cake mass m, and a method for this has been described [Murase et al, 1989]. The apparatus is shown in Figure 2.31a, Section 2.9, and it consists of a cylindrical filter containing a disc of much narrower diameter than the filter chamber. When the top of the filter cake reaches the height of the disc the filtrate rate rapidly diminishes due to the reduction in available filter area, and this effect can be seen on Figure 2.26 for a non-Newtonian suspending liquid. The ratio of wet to dry cake mass is [Murase et al, 1987]:

$$m = \frac{\rho_{s}sV + \rho_{s}L - \rho sV}{\rho_{s}s(V + L)} \qquad (2.64)$$

assuming that the ratio remains constant as the cake forms.

This analysis is relevant to the cake filtration mode. Filtration of non-Newtonian liquids is common in clarification such as oil or polymer filtration by cartridge and candle filters. In these instances an alternative mode of filtration is likely to occur.

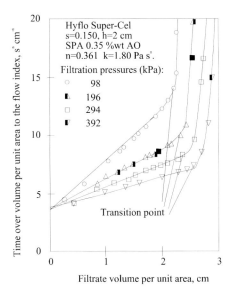

Figure 2.26 Inverse superficial filtrate rate to the power of the fluid behaviour index plotted against volume of filtrate per unit area [Murase et al, 1989a]

Under conditions of constant-pressure filtration the equations describing all the modes of filtration in Section 2.7 are modified for non-Newtonian fluids as follows [Hermia, 1982]

Table 2.5 Constant-pressure filtration laws modified for non-Newtonian fluids
[Hermia, 1982]

Filtration mode	Form of general filtration equation: Equation (2.58) and clogging constant
Complete blocking	$$\frac{d^2t}{dV^2} = K_b\left(\frac{dt}{dV}\right)^2$$ $$K_b = \sigma\frac{N}{3N+1}\left(\frac{\Delta P}{2k'L'}\right)^{1/N} r_o^{(N+1)/N}$$
Standard blocking	$$\frac{d^2t}{dV^2} = \frac{3N+1}{4N}K_s q_o^{2N/(3N+1)}\left(\frac{dt}{dV}\right)^{(5N+1)/(3N+1)}$$
Intermediate blocking	$$\frac{d^2t}{dV^2} = K_i\left(\frac{dt}{dV}\right)$$ i.e. no differences
Cake filtration	$$\frac{d^2t}{dV^2} = \frac{K_c}{N}\left(\frac{dt}{dV}\right)^{1-N}$$ $$K_c V = \left(\frac{1}{q}\right)^N - \left(\frac{1}{q_o}\right)^N$$

where σ is the blocked area per unit filtrate volume and L' is the filter medium thickness.

2.9 Laboratory Tests

Equipment to test the filtration characteristics of slurries in the laboratory is usually small, compact and fairly easy to construct. It is, therefore, well worth the constructional effort to enable the characterisation of slurries which are being, or will be, filtered on a larger industrial scale. Most of these tests are conducted at constant pressure and the data analysis has been introduced already in Section 2.5.1.

2.9.1 Vacuum Filter Leaf

One of the simplest and most useful test apparatus is the laboratory vacuum filter leaf, illustrated in Figure 2.27.

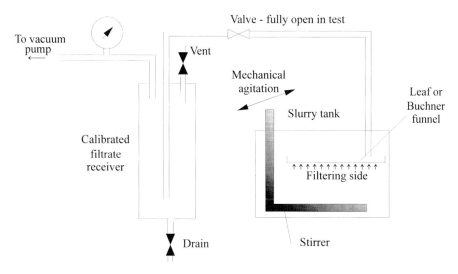

Figure 2.27 Laboratory vacuum filter leaf

The filter leaf includes a filter cloth retained over an open metal mesh support with adequate sealing of the cloth around the edges. Filtrate can therefore permeate through the cloth, metal support and into the evacuated tube connecting the leaf to the filtrate receiver. A vacuum gauge is used to indicate the level of vacuum achieved and, therefore, the pressure differential forming the filter cake. The line connecting the leaf to the vacuum/filtrate receiver usually contains a valve so that the receiver can be evacuated prior to initiating the filtration. This valve must be fully open during the test as it is assumed that the resistance to liquid flowing into the receiver is due only to the filter cake and medium, Equation (2.21); a partially restricted valve or too small bore tubing would also contribute to the resistance to liquid flow. A preliminary test with clean liquid sucked into the receiver can be used to assess the importance of these resistances. During the filtration test the cumulative volume of filtrate with respect to time is recorded. Data analysis is achieved by plotting a graph similar to that shown in Figure 2.7. The gradient and intercept are then used to provide values for specific resistance and filter medium resistance, as discussed in Section 2.5.1. Any consistent set of units can be used, but if S.I. units are adopted the specific resistance will be in m kg^{-1} and the medium resistance will be m^{-1}.

It is recommended that the position of the filter leaf in the slurry should reflect the geometry of the intended large-scale filter. Thus in the upward filtering position shown in Figure 2.27, the results obtained are applicable to upward filtering machines, such as the rotary vacuum filter. To simulate the performance of a downward filtering machine, such as a horizontal belt filter, a similarly facing test filter is recommended. Such considerations are important because of sedimentation of the larger particles away from, or towards, the filter medium under the action of gravity. Both the medium and cake resistances can be influenced by particle segregation by sedimentation, leading to errors in filter design.

The vacuum filter leaf is limited to a maximum cake-forming pressure of less than 1 bar, thus investigations of the effect of pressure on filtration are limited. It is very useful for characterising fairly incompressible materials such as inorganic precipitates or minerals with particle sizes of 10 μm and above, which could be filtered on rotary vacuum or leaf filters. For pressures in excess of 1 bar the vacuum leaf can be replaced with a similar filter operating under pressure, supplied usually by compressed air or a gas bottle. The leaf is mounted in a pressure vessel with some continual bleed of gas permitted through the vessel to agitate the slurry being filtered. The gas rate should be the minimum required to achieve this without scouring off the forming filter cake. Pressures of up to 10 bar are often employed with this equipment. The data analysis is similar to that already described for the vacuum filter leaf, but with an investigation of the effect of pressure on cake filtration as described in Section 2.6.

2.9.2 Compression Permeability Cell

An alternative technique to investigate the variation of specific resistance and cake solid content with applied pressure, and a method which is typically used up to pressures considerably in excess of 10 bar is a compression permeability (CP) cell. Such a device is illustrated schematically in Figure 2.28.

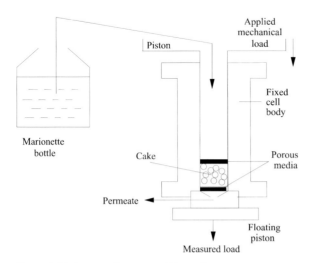

Figure 2.28 Compression permeability cell

A typical graph of porosity and specific resistance variation with applied pressure in the cell is shown in Figure 2.29 [Murase et al, 1987].

The constants in Equations (2.38) and (2.40) readily follow from the figures, thus the slurry is fully characterised and its filtration performance can be deduced from Equations (2.41)–(2.50).

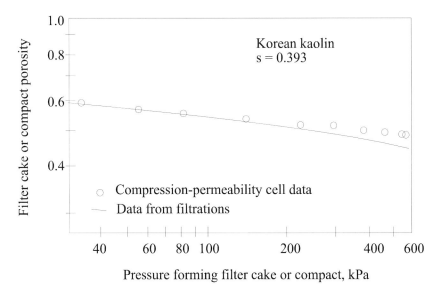

Figure 2.29a Porosity with applied pressure by means of a compression permeability cell [Murase et al, 1987]

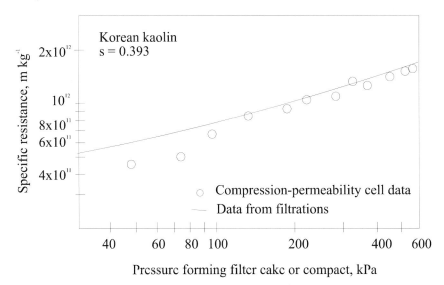

Figure 2.29b Specific resistance with applied pressure by means of a pressure permeability cell [Murase et al, 1987]

Considerable potential for a comprehensive description of filtration behaviour based on some simple CP tests is possible. Extensive sets of such data were first published by Grace [1953], and many researchers have investigated the CP cell and its use in characterising slurries. Some shortcomings of the technique have been recognised and a cell length to diameter ratio less than 0.6 is often recommended. A discussion of the various attempts to overcome the shortcomings of the technique has been published [Tiller and Lu, 1972]. The most significant shortcomings being side wall friction of the cake leading to nonhomogeneous cakes and pressure profiles, and the static condition of the cake which may not represent the dynamic condition of the filter cake.

Expressions which account for side wall friction in CP cells have been published [Shirato et al, 1971; Tiller et al, 1972], they are of the general form:

$$\Delta P_s = \left(\Delta P_c + \frac{C_f}{k_o f} \right) \exp\left(-\frac{4k_o f}{D} z \right) - \frac{C_f}{k_o f} \tag{2.65}$$

where k_o is defined in soil mechanics as the coefficient of earth pressure at rest, f is the coefficient of friction, C_f is the cohesive force between the side wall of the cell and the compressed cake, D is the inside diameter of the cell and z is the distance measured from the top of the compressed cake. It is assumed that $k_o f$ and C_f are constants for a given slurry and side wall material. These constants can be deduced from a series of tests by a numerical method [Shirato et al, 1968].

2.9.3 Capillary Suction Time (CST)

The CST test was designed to quickly assess the improvement in filtration characteristics due to flocculent addition to sewage sludges [Baskerville and Gale, 1968]. The method relies on the capillary pressure created by special filter paper to suck filtrate from a slurry contained in a small cylindrical reservoir placed on the surface of the paper. It is believed to be equivalent to a constant-pressure filtration. A picture of a capillary suction time device is given in Figure 2.30, together with a schematic diagram of the test cell.

The time taken for the filtrate to travel from one ring encompassing the reservoir to a second is measured, using electrical conductance to start and stop a timer; when filtrate passes these rings the conductance of the saturated paper increases, if an electrically conducting liquid is used. Sludges with fast CST are easier to dewater. In tests related to sedimenting centrifuges a CST of less than 12 s is sought. A dimensionless CST value can be deduced by ratioing the CST of the slurry with that for water under the same experimental conditions, the nearer to unity the easier the filtration.

The amount of material required in the CST cell is very small, less than 50 ml and cleaning between tests is simple. Hence the technique is well suited to a series of tests investigating the effect of flocculent dosage, conditioning chemicals, etc. The tests can be used in conjunction with some vacuum leaf or pressure bomb filtrations in order to correlate CST with specific resistance. Such a correlation tends to be specific to the material and conditions investigated, and a universal correlation should not be assumed. In general, correlation between specific resistance and capillary suction time appears successful only for sludge-like suspensions, i.e not for granular ones.

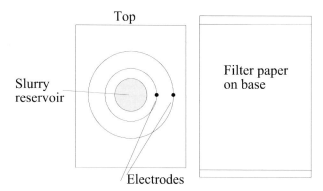

Figure 2.30a Capillary suction time test-schematic of test cell

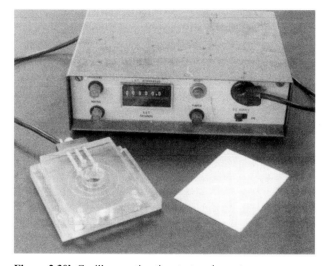

Figure 2.30b Capillary suction time test equipment

2.9.4 Other Laboratory Tests and Procedures

Permeation tests on preformed filter cakes are a useful means of investigating the functional relation between specific resistance and cake concentration. If the cake-forming pressure is known then the constants of Equations (2.38) and (2.40) readily follow. This is a procedure similar to that used in the CP cell, but without the need for that piece of specialised equipment. Care needs to be exercised to ensure that the filter cake is not further compressed during the permeation test, and if the cake permits the migration of fines the specific resistance will increase with volume of liquid permeated. A separate permeation test conducted on the clean filter medium, with a view to establishing the medium resistance for use in Equation (2.21) should *not* be attempted. In all filtrations the true medium resistance is due to both the filter medium and the fine solids which enter and bridge over it. Thus medium resistance must be determined in situ.

A means of identifying the filter cake height during a filtration, coupled with a mass balance on the solids and liquid filtered, would enable the average cake concentration to be determined from Equation (2.64). If the test is repeated at various known cake-forming pressures the constants of Equation (2.40) follow. One such device to establish this using a disc with a hole inserted into the filter chamber, thus reducing the available filtration area at a known height, has been described [Murase et al, 1987]. The time, or filtrate volume, at which the filter cake reaches the disc can be determined from the plot of the reciprocal filtrate rate against cumulative volume filtrate (both per unit filter area) as shown in Figure 2.31, together with the filter cell used.

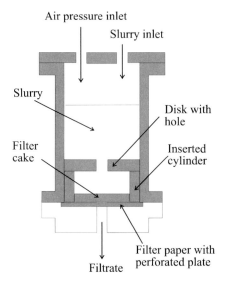

Figure 2.31a Test filter with disc to reduce filter area [Murase et al, 1987]

This filtration equipment has also been used in the investigation of the filtration characteristics of a non-Newtonian slurry, see Figure 2.26.

Cake height has also been measured using six pressure probes positioned inside a filter cell. The hydraulic pressure was measured and remained constant until the filter cake surface reached the probe, when the hydraulic pressure started to fall. Thus the height of the filter cake surface was given from the known distance of the probe from the medium [Murase et al, 1989b]. A schematic diagram of the experimental equipment is given in Figure 2.32.

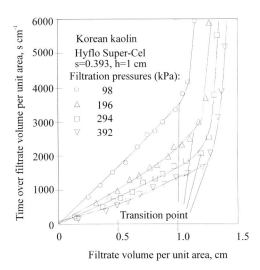

Figure 2.31b Relation between reciprocal filtrate rate and filtrate volume per unit medium area using cell illustrated in Figure 2.31a

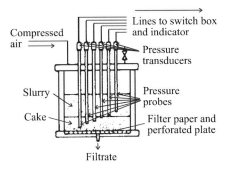

Figure 2.32 Test filter with pressure probes to infer cake height [Murase et al, 1989 b]

A straight line plot of cake height against volume filtrate at a given cake-forming pressure implied that the cake was uniform throughout the filtration, confirming the validity of Equation (2.7). The cake moisture content was calculated from a knowledge of the cake height and filtrate volume. The technique was then applied to a "step up" pressure filtration, similar to that illustrated in Figure 2.12; the pressure was increased when the filter cake reached a probe. The pressure was increased again on reaching another probe, etc. Thus the experiment yielded all the information necessary to investigate Equations (2.38) and (2.40) without the drawbacks and complications of the CP cell.

The approach of using an experimental technique to provide or measure the cake height with time, and to increase the cake-forming pressure by increments is likely to become the major method of measuring filtration properties in the future. The procedure is quick and more reliable than a CP cell and yields much information on the filtration characteristics of the slurry and cake.

Many research workers have investigated the solid concentration profile of the forming filter cake using electrical resistance, or conductivity, pins positioned diametrically opposite in a nonconducting filter cell [Reitema, 1953; Baird and Perry, 1967; Shirato et al, 1971; Wakeman, 1981; Holdich 1990]. One such measured concentration profile has already been reported in Figure 2.16. Electrical resistance or conductivity is a function of temperature and liquid electrolyte concentration (the solids are assumed to be nonconducting). Furthermore, electrolyte concentration is often a function of temperature. Thus it is assumed that temperature must be maintained constant, or at least monitored so that a conductivity correction factor can be applied. This is not necessary, however, as an efficient filtration will provide a stream of continuous phase (filtrate) at the same electrolyte concentration and temperature as the suspending liquid in the saturated filter cake and slurry. Thus if the solid concentration is related to the ratio of the conductivity of the mixture phase (k_m cake or slurry) with the conductivity of the continuous phase (k_c), both measured simultaneously, then fluctuations due to electrolyte concentration or temperature are not relevant [Holdich and Sinclair, 1992]. One well-known relation between the conductivity ratio and concentration is Equation (2.66):

$$\frac{k_m}{k_c} = (1 - C)^{u'} \tag{2.66}$$

where u' is an empirical constant.

Electrical conductivity studies can also be used to determine the variation in cake height with time, and the cake-forming pressure could be stepped up during a filtration in a similar way to that described earlier. This experimental technique also provides information on the forming filter cake structure and has considerable potential as another method to replace the CP cell in investigating the constants needed to apply Equations (2.38) and (2.40). It is, however, slightly more complicated to apply than the technique illustrated in Figure 2.31.

2.10 Developments in Filtration Modelling and Understanding

The basis of this chapter has been to report the current accepted theory behind filtration, and to show how the models can be used to construct simple simulations of filter performance. The models are amenable to application on computer spreadsheet packages, and examples of these are given in Appendix C and are downloadable *via* the Internet. The models also form the basis for Chapter 11: 'Filtration Process Equipment and Calculations', which looks in greater detail at process simulation and provides a fuller description of the relevant equipment. However, there are numerous different techniques for filtration modelling, such as a multiphase approach [Willis, 1983] and many numerical solutions to the governing equations such as Equation (2.44) [Landman *et al*, 1995]. The accepted approach of Tiller and Shirato reduces Equation (2.44) to Equation (2.45) in order to effect a simple solution but, in doing so, solid velocity within the filter cake is neglected. Such a situation can only be reliable for incompressible filter cakes. Hence this approach should be restricted to cakes of limited compressibility. There are additional important inadequacies in the numerical solutions such as the invariable occurrence of sedimentation onto, or away from, the forming cake surface [Tiller *et al*, 1995], and the significant change in filter medium resistance at the start of the filtration cycle. The simple simulations described in Section 2.6.5 rely on a known and constant value for the medium resistance; they could be amended to include this as a variable but the function defining the medium resistance would have to be known. This subject has recently received significant quantitative analysis [Koenders and Wakeman, 1996; Koenders and Wakeman, 1997].

The greatest shortcoming of the theories underlying Equations (2.49) and (2.50), and elsewhere, is the occurrence of highly compressible compacts in which $u+n>1$, and exhibit significant solids velocity within the cake. Under these circumstances equations based on the exponent $(1-u-n)$ become negative and inapplicable. In practice, precipitated and crushed materials, clays and some flocculated solids can usually be modelled by the approach described, but solids of a biological origin at high concentration need a more specialised approach [Cleveland *et al*, 1996; Tiller and Kwon, 1998]. There is also experimental evidence [Tarleton and Willmer, 1998] to suggest that filtration on even well defined geometries, and under laboratory conditions, can not be acceptably modelled using one dimensional models; i.e. with concentration and other filter cake properties varying only with distance from the filter medium in addition to time. Nevertheless, the modelling approaches described in the first six sections of this chapter have found considerable acceptance and utility in the understanding of existing filter performance, and in assisting in the specification of new equipment based on existing or laboratory generated data.

2.11 References

Baird, R.L. and Perry, M.C., 1967, Filtration and Separation, 4, p 471.
Baskerville, R.C. and Gale, R.S., 1968, J. Inst. Water Pollut. Cont., 2, p 3.
Cleveland, T.G., Tiller, F.M. and Lee, J.B., 1996, Water Science and Technology, 34, pp 299–306.
Darcy, H.P.G., 1856, Les Fontaines Publiques de la Ville de Dijon, Victor Dalmont, Paris.
Grace, H.P, 1953, Chem. Eng. Progr., 49(6), pp 303–367, and 49(7), pp 367–367.
Grace, H.P., 1956, AIChE Journal, 2(3), pp   307–336.
Happel, J. and Brenner, H., 1965, Low Reynolds numbers hydrodynamics, Prentice-Hall Englewood Cliffs.
Hermia, J., 1982, Trans. IChemE., 60; pp 183–187.
Holdich, R.G., 1990, Chem. Eng. Comm., 91; pp 255–268.

Holdich, R.G. and Sinclair, I., 1992,   Powder Technology, 72; pp 75–85.

Holdich, R.G., 1993, Int. J. Mineral Processing, 39; pp 157–171.

Holdich, R.G., 1994, Filtration and Separation, 31; pp 825–829.

Koenders, M.A. and Wakeman, R.J., 1996, Chem. Eng. Sci., **51**, pp 3897–3908.

Koenders, M.A. and Wakeman, R.J., 1997, Trans. IChemE., **75**, Part A, pp 309–320.

Kozeny, J., 1927, Sitz-Ber. Wiener Akad, Abt. IIa, 136, p 271.

Kozicki, W., 1990, Can. J. Chem. Eng., 68; pp 69–80.

Landman, K.A., White, L.R. and Eberl, M., 1995, AIChEJ, 41, pp 1687-1700.

Murase, T. Iritani, E., Cho, J.H., Nakanomori, S. and Shirato, M., 1987, Chem Eng. Japan J. 20, pp 246–251.

Murase, T. Iritani, E., Cho, J.H. and Shirato, M., 1989a, Chem Eng. Japan J. 22; pp 65–71.

Murase, T., Iritani, E., Cho, J.H. and Shirato, M., 1989b, Journ. Chem. Eng. Japan, 22(4), pp 373–378.

Purchas, D.B., 1981, Solid/Liquid Separation Technology, Uplands Press, Croydon.

Reitema, 1953, Chem. Eng. Sci., 2, p 88.

Rushton, A. and Hameed, M.S., 1969, Filtrat. and Separat., 6; pp 136–139.

Ruth, B.F., 1935, Ind. Eng. Chem., 27; pp 708–723.

Shirato, M. and Okamura, K., 1959, Chem Eng. (Japan) 23; p 226.

Shirato, M., Aragaki, T., Mori, R. and Sawamoto, K., 1968, Chem Eng. Japan J. 1; pp 86–90.

Shirato, M., Aragaki, T., Ichimura, K, and Ootsuji, N., 1971, Chem Eng. Japan J. 4; pp 172–177.

Shirato, M., Aragaki, T., Iritani, E., Wakimoto, M., Fujiyoshi, S. and Nanda, S., 1977, Chem Eng. Japan J.   10, pp 54–60.

Shirato, M., Aragaki, T. and Iritani, E., 1980a, Chem Eng. Japan J. 13; pp 61–66.

Shirato, M., Aragaki, T., Iritani, E. and Funahashi, T., 1980b, Chem Eng. Japan J. 13; pp 473– 478.

Stamatakis, K. and Chi Tien, 1991, Chem.   Eng. Sci., 46, pp 1917-1933.

Tarleton, E.S. and Willmer, S.A., 1997, Trans. IChemE., **75**, Part A, pp 497-507.

Tiller, F.M. and Cooper, H., 1962, AIChE J., 8, pp 445–449.

Tiller, F.M. and Green, T.C., 1973, AIChE J., 19, pp 1266–1270.

Tiller, F.M. and Kwon, J.H., 1998, AIChE J., 44, pp 2159–2167.

Tiller, F.M. and Lu, W.-M., 1972, AIChE J., 18, pp 569–572.

Tiller, F.M. and Shirato, M., 1964, AIChE J., 10, pp 61–67.

Tiller, F.M., Haynes, S. and Lu, W-M., 1972, AIChE J., 18, pp 13–20.

Tiller, F.M., Hsyung, N.B. and Cong, D.Z., 1995, AIChE J., 41, pp 1153–1164.

Theliander, H. and FathiNajafi, M., 1996, Filtration and Separation, **33**, pp 417-421.

Wakeman, R.J., 1978, Trans. IChemE. 56, pp 258–265.

Wakeman, R.J., 1981, Trans. IChemE. 59, pp 260–270.

Wakeman, R.J. and Tarleton, E.S., 1999, Filtration: equipment selection modelling and process simulation, Elsevier, Oxford, UK.

Willis, M.S.,1983, A multiphase theory of filtration. In Progress in Filtration and Separation 3 (Ed. R.J. Wakeman), Elsevier, Amsterdam.

2.12 Nomenclature

A	Area	m^2
C	Solid concentration by volume fraction	–
C_h	Local solid concentration by volume fraction	–
C_o	Empirical coefficient – Equation (2.40)	Pa^{-n}
c	Dry mass of solids per unit filtrate volume	$kg\, m^{-3}$
g	Gravitational constant	$m\, s^{-2}$
h	Position or height in filter cake	m
J	Tiller/Shirato "J" factor - see Equation (2.51)	
K	Kozeny constant	–
k	Filter cake permeability	m^2
k'	Rheological power law consistency coefficient	
L	Filter cake or bed thickness	m
m	Ratio of wet to dry cake mass - "moisture ratio"	–
M	Total mass of slurry	kg
N	Rheological power law flow index	
n	Empirical exponent - "compressibility coefficient"	
n'	Exponent on filtration laws - see Section 2.7	
ΔP	Pressure Differential	$N\, m^{-2}$
ΔP_c	Pressure Differential over filter cake	$N\, m^{-2}$
ΔP_m	Pressure Differential over filter medium	$N\, m^{-2}$
Q	Filter feed rate	$m^3\, s^{-1}$
q	Filtrate flow rate	$m^3\, s^{-1}$
q_o	Initial filtrate flow rate	$m^3\, s^{-1}$
R_c	Filter cake resistance	m^{-1}
R_m	Medium Resistance	m^{-1}
s	Mass fraction of solids in feed mixture	–
S_v	Specific surface area per unit volume	$m^2\, m^{-3}$
t	Time	s
u	Empirical exponent	
V	Volume of filtrate	m^3
V'	Volume of filtrate per unit area	$m^3\, m^{-2}$
v_s	Velocity of solids inside filter cake	$m\, s^{-1}$
w	Mass of solids deposited per unit area	$kg\, m^{-2}$
x	Particle diameter	m

Greek Symbols

α	Local specific cake resistance	m kg^{-1}
α_{av}	Average specific cake resistance	m kg^{-1}
α_{o}	Empirical coefficient – Equation (2.38)	$\text{m kg}^{-1}\,\text{Pa}^{-n}$
β	Filter cake to filtrate volume ratio	–
ε_{h}	Local porosity	–
μ	Liquid viscosity	Pa s
ρ	Fluid density	kg m^{-3}
ρ_{s}	Solid density	kg m^{-3}

3 Sedimentation Fundamentals

Sedimentation of Dilute and Concentrated Suspensions

Sedimentation is the separation of particles from fluids due to the effect of a body force, which may be either gravity or centrifugal, on the buoyant mass of the particle. This Chapter deals with gravity sedimentation and centrifugal sedimentation is discussed in Chapter 8. The mass of the particle through its size and density and the volumetric concentration of the particles in the suspension both play a major part in descriptions of the behaviour of a suspension under these conditions. Dilute sedimentation is the case where the particles are able to settle as individuals. Hindered settling or thickening are terms used to describe behaviour at higher concentrations where sedimentation rates are largely related to concentration rather than to particle size.

The behaviour of a settling suspension is determined mainly by two factors. First the concentration of the particulate solids and second by the state of aggregation of the particles. If the solids are present in a dilute concentration and are not aggregated in any way, i.e. in a "particulate" condition then the particles will settle as individuals and the motion can be described by Newton's or Stokes' laws. However if the particles are so concentrated that they are almost touching, the level of interference is such that the particle size has less bearing on the settling rate which will relate more to concentration than to any other property. The general relationship between these factors is shown in Figure 3.1.

It is important to realise that the separation mechanisms are different in the various areas of the illustration. In the consolidation/compression area, for example, the ways in which the suspension becomes concentrated are more like filtration mechanisms than settling ones and the development of channels is very significant.

3.1 Dilute Sedimentation

The general equation of motion of a small solid particle in a liquid can be obtained by applying Newton's second law of motion to the physical situation. A listing of the forces that may be experienced by such a particle has been given by Soo [1967].

$$A - D + F + P - L - B = 0 \tag{3.1}$$

where

A = inertia force on the particle
D = drag force on the particle

F = field force on the particle
P = pressure force due to the pressure gradient in the liquid
L = force to accelerate the apparent mass of the particle relative to the liquid
B = force to take into account deviations in the flow pattern due to unsteady state

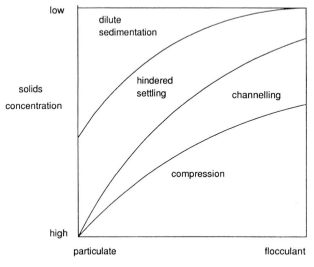

Figure 3.1 Settling systems [Fitch, 1962]

This equation, when set out fully, is too complex to be solved readily and is usually simplified before solution is attempted. For example if the pressure gradient in the liquid is not large P may be ignored and if the flow pattern is steady then B is zero. Force L is normally ignored when the density of the liquid is low compared with that of the solid. If these simplifications are made, Equation (3.1) reduces to:

$$A - D + F = 0 \qquad (3.2)$$

If the particle is spherical of diameter x, the acceleration force A is given by:

$$A = \frac{\pi x^3}{6} \rho_s \frac{du}{dt} \qquad (3.3)$$

where ρ_s is the solid density, u is the relative velocity of the particle to the liquid and t is time. The effect of force L is often included in this term by using $(\rho_s + k\rho)$ instead of ρ_s. The symbol k is a factor which takes into account the presence of a liquid envelope surrounding the particle and ρ is the liquid density. The value of k is generally to be taken to be 0.5.

Drag forces are described generally by Newton's law, i.e.:

$$D = C_D(Re_p)A_p\frac{\rho}{2}u^2 \tag{3.4}$$

where A_p is the projected area of a particle in the direction of flow and C_D is a drag coefficient, which for any shape of particle, is a function of the Reynolds number *for the particle:*

$$Re_p = \frac{xu\rho}{\mu} \tag{3.5}$$

which characterises the nature of the liquid flow around the particle.

The drag coefficient may also be regarded as the ratio of τ, the force per unit projected area of the particle, measured perpendicular to the direction of motion, to the kinetic energy of the liquid so:

$$C_D = \frac{2\,\tau}{\rho\,u^2} \tag{3.6}$$

The relation between C_D and Re_p , the standard drag curve, is shown on a log–log plot in Figure 3.2.

The value of the drag coefficient decreases with increasing Reynolds number. Various workers (see Clift et al [1978]) have produced equations to describe the relation between C_D and Re_p, usually over a limited range of Re_p. For example, for Re_p values up to 10^5, Khan and Richardson [1989] suggest that:

$$C_D = (2.249\,Re_p^{-0.31} + 0.358\,Re_p^{0.06})^{3.45} \tag{3.7}$$

Note that for Re_p values between 10^3 and 2×10^5 the value of C_D is constant at 0.44. It falls sharply at an Re_p value of about 2×10^5 to a constant value of 0.1. This drop in value is due to a change in the flow regime in the boundary layer of the fluid around the sphere from laminar to turbulent with the fluid beginning to separate away from the solid surface at the rear of the particle.

In gravity sedimentation the dominant field force is due to the gravitational acceleration acting on the buoyant mass of the particle so:

$$F = \frac{\pi x^3}{6}(\rho_s - \rho)g \tag{3.8}$$

Thus the force balance Equation (3.2) becomes:

$$\frac{\pi x^3}{6}\rho_s\frac{du}{dt} - C_D\left(\frac{\pi x^2}{4}\right)\left(\frac{\rho}{2}\right)u^2 + \frac{\pi x^3}{6}(\rho_s - \rho) = 0 \tag{3.9}$$

and, if the acceleration term is deemed to be so small in value as to be negligible, the value of the velocity u which now becomes the terminal settling velocity u_t, is given by the equation:

$$\frac{\pi x^3}{6}(\rho_s - \rho)g = C_D \left(\frac{\pi x^2}{4} \right) \left(\frac{\rho}{2} \right) u_t^2$$

(3.10)

giving:

$$u_t = \left(\frac{4(\rho_s - \rho)gx}{3\,\rho C_D} \right)^{1/2}$$

(3.11)

which can be evaluated if a value for C_D is known.

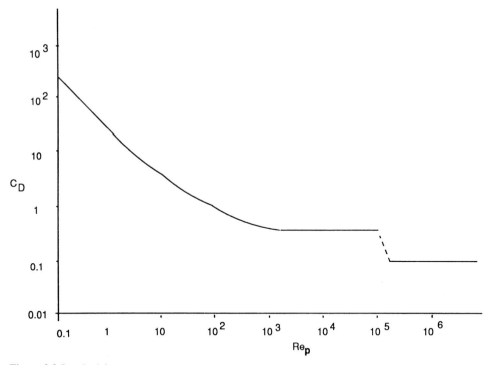

Figure 3.2 Standard drag curve

In dilute sedimentation the particle Reynolds numbers are usually low (< 1.0) and in this case the C_D v Re relationship can be described simply. Stokes solved the Navier–Stokes equation, which describes the behaviour of an infinitesimal element of an incompressible fluid experiencing only gravity as a body force, by assuming the inertia term to be negligible, and produced:

$$D = 3\pi\mu x u \quad or \quad C_D = \frac{24}{Re_p} \tag{3.12}$$

These equations apply with a maximum error of about 4% up to Re_p values of 0.2. The maximum particle size to which they may be applied in a given system can be obtained from the relation:

$$x_{max} = \left(\frac{3.6\,\mu^2}{(\rho_s - \rho)\rho g} \right)^{(1/3)} \tag{3.13}$$

When the Stokes drag expression is used the simple force balance Equation (3.2) becomes:

$$\frac{\pi x^3}{6} \rho_s \frac{du}{dt} - 3\pi\mu x u + \frac{\pi x^3}{6} (\rho_s - \rho)g = 0 \tag{3.14}$$

After the initial acceleration period which is usually short the terminal settling velocity u_t is obtained from Equation (3.14) by taking the acceleration term as zero and putting $u = u_t$, giving:

$$u_t = \frac{x^2}{18\mu}(\rho_s - \rho)g \tag{3.15}$$

This is Stokes' law, defined for spherical particles.

There are two restrictions on the applicability of Stokes' law. One concerns the effect of concentration, which in practical terms should be so low that the behaviour of the settling particle is unaffected by the presence of any neighbours and the other is that the particle Reynolds number should be less than 0.2. If these conditions are met the settling is often called "free Settling", and may be reliably predicted by the use of Equation (3.15). Many devices for particle size analysis use Stokes' settling equation, see Table 1 in Appendix A, one example being the Andreasen pipette which is pictured in Figure 3.3.

Figure 3.3   Photograph of free settling taking place in an Andreasen pipette

Evidence of free settling, which may be observed in Figure 3.3, includes: no distinct interface between settling solids in suspension and the overlying liquid (i.e. a cloudy suspension that slowly clears with time) and an observable layer of sediment at the base of the vessel.

   If the concentration effects are significant, the settling behaviour cannot be described in the way set out here and other approaches have to be used, see Section 3.2. The situation where the particle Reynolds number is greater than 0.2 requires a solution to a force balance equation where the drag force is expressed by Newton's law (above) and is complicated by the interrelation between the drag coefficient and the settling velocity through the Reynolds number.

   Oseen obtained an improved solution for the case where the inertia force is not negligible and, for a sphere, this equation is:

$$C_D = \frac{24}{Re_p}(1 + \frac{3}{16} Re_p) \tag{3.16}$$

which is accurate up to a Reynolds number of 1.0.

   Solutions of equations involving the drag coefficient can be made using an iterative procedure or by means of functions separating the variables, u and x. For example:

$$C_D Re_p^{\,2} = \frac{4\,(\rho_s - \rho)\rho x^3 g}{3\mu^2} = P^3 x^3 \qquad (3.17)$$

and:

$$\frac{Re_p}{C_D} = \frac{3\rho^2 u^3}{4\,(\rho_s - \rho)g\mu} = \frac{u^3}{Q^3} \qquad (3.18)$$

The term $[(\rho_s-\rho)\rho x^3 g]/\mu^2$ is often called the Galileo number *Ga* or Archimedes number *Ar* and these symbols may be found in correlations.

It is clear from Equations (3.17) and (3.18) that *P* and *Q* are dependent only on the system properties. Graphs of:

$$\left(\frac{Re_p}{C_D}\right)^{1/3} \quad vs \quad \left(C_D Re_p^{\,2}\right)^{1/3}$$

(see Figure 3.4 or Tables 3.1 and 3.2) enable easy solution of such problems.

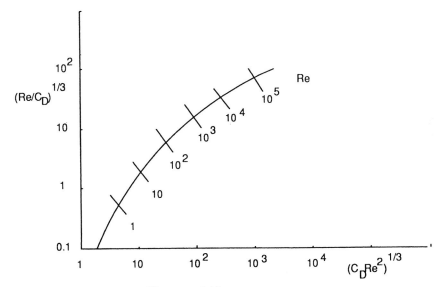

Figure 3.4 Plot of $(Re_p / C_D)^{1/3}$ vs $(C_D Re_p^2)^{1/3}$

Table 3.1 Values of $\log_{10} u_t/Q$ in terms of $\log_{10} Px$ for spherical particles
[Heywood, 1962]

$$P = \left(\frac{4(\rho_s - \rho)\rho g}{3\mu^2} \right)^{1/3} \qquad\qquad Q = \left(\frac{4(\rho_s - \rho)\mu g}{3\rho^2} \right)^{1/3}$$

x = particle diameter u_t = terminal velocity

$\log_{10} Px$	0.00	0.01	0.02	0.03	0.04	0.05	0.06	0.07	0.08	0.09
−0.2	−1.780									
−0.1	−1.580	−1.600	−1.620	−1.640	−1.660	−1.680	−1.700	−1.720	−1.740	−1.760
−0.0		−1.402	−1.422	−1.442	−1.461	−1.481	−1.501	−1.521	−1.541	−1.560
0.0	−1.382	−1.362	−1.343	−1.323	−1.303	−1.283	−1.264	−1.244	−1.225	−1.205
0.1	−1.185	−1.166	−1.146	−1.126	−1.106	−1.087	−1.068	−1.048	−1.029	−1.010
0.2	−0.990	−0.971	−0.952	−0.932	−0.912	−0.893	−0.874	−0.855	−0.836	−0.817
0.3	−0.799	−0.780	−0.762	−0.743	−0.725	−0.707	−0.688	−0.670	−0.652	−0.634
0.4	−0.616	−0.598	−0.580	−0.562	−0.544	−0.527	−0.510	−0.492	−0.475	−0.457
0.5	−0.440	−0.423	−0.406	−0.389	−0.373	−0.357	−0.341	−0.325	−0.308	−0.292
0.6	−0.276	−0.260	−0.245	−0.229	−0.213	−0.198	−0.183	−0.168	−0.153	−0.138
0.7	−0.123	−0.109	−0.095	−0.080	−0.066	−0.052	−0.038	−0.024	−0.011	0.003
0.8	0.017	0.030	0.043	0.057	0.070	0.083	0.096	0.109	0.122	0.135
0.9	0.148	0.161	0.173	0.186	0.199	0.211	0.224	0.236	0.248	0.261
1.0	0.273	0.285	0.297	0.309	0.321	0.333	0.345	0.356	0.368	0.380
1.1	0.391	0.402	0.414	0.425	0.436	0.447	0.458	0.469	0.480	0.491
1.2	0.502	0.513	0.523	0.534	0.545	0.555	0.565	0.576	0.586	0.596
1.3	0.607	0.617	0.627	0.637	0.647	0.657	0.667	0.677	0.686	0.696
1.4	0.706	0.715	0.725	0.734	0.744	0.753	0.762	0.772	0.781	0.790
1.5	0.800	0.809	0.818	0.827	0.836	0.844	0.853	0.862	0.870	0.879
1.6	0.887	0.895	0.904	0.912	0.920	0.928	0.936	0.944	0.951	0.959
1.7	0.967	0.974	0.981	0.989	0.996	1.004	1.011	1.018	1.026	1.031
1.8	1.040	1.048	1.055	1.062	1.069	1.076	1.083	1.090	1.097	1.104
1.9	1.111	1.118	1.125	1.132	1.139	1.146	1.153	1.160	1.167	1.174
2.0	1.180	1.187	1.194	1.200	1.207	1.214	1.220	1.227	1.233	1.240
2.1	1.246	1.253	1.259	1.265	1.272	1.278	1.284	1.290	1.296	1.302
2.2	1.307	1.313	1.319	1.324	1.329	1.335	1.340	1.345	1.350	1.355

Table 3.1 Continued

2.3	1.360	1.364	1.369	1.374	1.378	1.383	1.388	1.392	1.397	1.401
2.4	1.406	1.411	1.415	1.420	1.424	1.428	1.433	1.437	1.441	1.445
2.5	1.450	1.454	1.458	1.462	1.466	1.470	1.474	1.478	1.482	1.486
2.6	1.490	1.494	1.498	1.502	1.506	1.510	1.514	1.518	1.521	1.525
2.7	1.529	1.533	1.537	1.541	1.545	1.549	1.553	1.557	1.561	1.565
2.8	1.569	1.573	1.578	1.582	1.586	1.590	1.594	1.598	1.603	1.607
2.9	1.611	1.616	1.620	1.624	1.629	1.633	1.637	1.642	1.646	1.651
3.0	1.655	1.660	1.665	1.669	1.674	1.679	1.684	1.689	1.694	1.698
3.1	1.703	1.708	1.713	1.718	1.724	1.729	1.734	1.740	1.746	1.751
3.2	1.757	1.763	1.770	1.776	1.782	1.788	1.795	1.801	1.808	1.814
3.3	1.821	1.828	1.834	1.841	1.848	1.854	1.861	1.868	1.875	1.881

Table 3.2 Values of $\log_{10} Px$ in terms of $\log_{10} u_t/Q$ for spherical particles
[Heywood, 1962]

$\log_{10}(u/Q)$	0.00	0.01	0.02	0.03	0.04	0.05	0.06	0.07	0.08	0.09
−1.7	−0.160									
−1.6	−0.110	−0.115	−0.120	−0.125	−0.130	−0.135	−0.140	−0.145	−0.150	−0.155
−1.5	−0.060	−0.065	−0.070	−0.075	−0.080	−0.085	−0.900	−0.095	−0.100	−0.105
−1.4	−0.009	−0.014	−0.019	−0.024	−0.029	−0.034	−0.040	−0.045	−0.050	−0.055
−1.3	0.041	0.036	0.031	0.026	0.021	0.016	0.011	0.006	0.001	−0.004
−1.2	0.093	0.087	0.082	0.077	0.072	0.067	0.062	0.057	0.052	0.046
−1.1	0.143	0.138	0.133	0.128	0.123	0.118	0.113	0.108	0.103	0.098
−1.0	0.195	0.190	0.185	0.179	0.174	0.169	0.164	0.159	0.154	0.148
−0.9	0.246	0.241	0.236	0.231	0.226	0.221	0.216	0.211	0.206	0.200
−0.8	0.299	0.293	0.288	0.283	0.278	0.272	0.267	0.262	0.257	0.252
−0.7	0.354	0.348	0.343	0.337	0.332	0.326	0.321	0.316	0.310	0.305
−0.6	0.409	0.404	0.398	0.392	0.387	0.382	0.376	0.370	0.364	0.359
−0.5	0.465	0.460	0.454	0.448	0.442	0.437	0.432	0.426	0.420	0.414
−0.4	0.524	0.518	0.512	0.506	0.500	0.494	0.488	0.483	0.477	0.471
−0.3	0.585	0.579	0.573	0.567	0.561	0.555	0.548	0.542	0.536	0.530

Table 3.2 Continued

−0.2	0.649	0.642	0.636	0.629	0.623	0.616	0.610	0.604	0.597	0.591
−0.1	0.716	0.709	0.702	0.695	0.688	0.682	0.675	0.668	0.662	0.656
−0.0		0.781	0.773	0.766	0.759	0.752	0.745	0.738	0.730	0.723
0.0	0.788	0.795	0.802	0.810	0.818	0.825	0.832	0.840	0.848	0.856
0.1	0.863	0.871	0.879	0.886	0.894	0.902	0.910	0.917	0.925	0.933
0.2	0.941	0.949	0.957	0.965	0.973	0.981	0.989	0.997	1.006	1.014
0.3	1.022	1.031	1.039	1.048	1.056	1.064	1.073	1.082	1.090	1.099
0.4	1.108	1.117	1.126	1.135	1.144	1.153	1.162	1.171	1.180	1.189
0.5	1.198	1.208	1.217	1.227	1.236	1.245	1.255	1.265	1.274	1.284
0.6	1.294	1.303	1.313	1.323	1.333	1.343	1.353	1.363	1.373	1.384
0.7	1.394	1.404	1.415	1.425	1.436	1.446	1.457	1.468	1.479	1.490
0.8	1.500	1.511	1.522	1.533	1.545	1.557	1.568	1.580	1.592	1.604
0.9	1.616	1.628	1.640	1.652	1.665	1.678	1.691	1.704	1.718	1.731
1.0	1.745	1.759	1.773	1.786	1.800	1.813	1.827	1.841	1.855	1.870
1.1	1.884	1.899	1.913	1.927	1.941	1.956	1.970	1.985	2.000	2.015
1.2	2.030	2.045	2.060	2.075	2.090	2.106	2.122	2.138	2.154	2.170
1.3	2.187	2.204	2.222	2.241	2.260	2.280	2.300	2.321	2.343	2.365
1.4	2.387	2.409	2.431	2.454	2.477	2.500	2.524	2.549	2.574	2.600
1.5	2.626	2.651	2.677	2.703	2.728	2.753	2.778	2.803	2.827	2.851
1.6	2.874	2.897	2.920	2.943	2.966	2.988	3.010	3.032	3.053	3.073
1.7	3.093	3.113	3.133	3.152	3.170	3.188	3.204	3.220	3.236	3.252
1.8	3.268	3.283	3.298	3.313	3.328	3.343	3.358	3.373	3.388	3.402

In the viscous flow regime an irregularly shaped particle will settle in such a way that the line joining the centres of reaction and mass is parallel to the direction of gravity, giving a preferred orientation and thus a definitive settling velocity. The use of Stokes' law (Equation 3.15) to obtain an equivalent spherical diameter x_{sE} for an irregular particle through a measurement of its terminal settling velocity is the basis of a common particle size analysis method. Other particle size equivalent diameters can be related to x_{sE} through the sphericity shape factor ψ, which is defined as the ratio of the surface area of a sphere of equal volume to the actual surface area of the particle:

$$x_{sE} = x_v \psi^{1/4} \tag{3.19}$$

$$x_{sE} = x_{sA} \psi^{1/2} \tag{3.20}$$

where x_v = diameter of a sphere with the same volume as the particle, which may be obtained for example by Coulter Counter size analysis and x_{sA} is the diameter of a sphere of equal surface area, which may be obtained by permeability or permeametry tests. Heywood [1962] measured the effect of shape on the terminal velocities of a large number of irregular mineral particles over a wide range of Reynolds numbers and these results, together with a tabulated method of obtaining accurate terminal settling velocities, can be found in the literature.

Example 3.1

Calculate the terminal settling velocity of a 70 μm diameter sphere, density 2.6×10^3 kg/m^3 in water at 18^0C (density 1.0×10^3 kg/m^3 and viscosity 1×10^{-3} N s/m^2).

Solution

From Equation (3.13) the maximum particle size for which Stokes' law can be strictly applied can be calculated:

$$x_{max} = \left(\frac{3.6 \times 10^{-6}}{(2.6 - 1.0) \times 10^3 \times 1.0 \times 10^3 \times 9.81} \right)^{1/3} = 61.2 \ \mu m$$

Thus for a 70 μm diameter Stokes' law is not perfectly accurate but a reasonable estimate of the velocity can be obtained:

$$u_t = \frac{(70 \times 10^{-6})^2 \times (2.6 - 1.0) \times 10^3 \times 9.81}{18 \times 10^{-3}} = 4.27 \times 10^{-3} \ m/s$$

The particle Reynolds number can be calculated for this velocity:

$$Re_p = \frac{70 \times 10^{-6} \times 4.27 \times 10^{-3} \times 1.0 \times 10^3}{1.0 \times 10^{-3}} = 0.2989$$

C_D, calculated from Oseen's Equation (3.16), gives:

$$C_D = \frac{24}{0.2989} \left(1 + \frac{3}{16} \times 0.2989 \right) = 84.79$$

and, from Newton's law (Equation 3.10):

$$u_t = \left(\frac{4(\rho_s - \rho)gx}{3\rho C_D} \right)^{1/2} = \frac{4 \times 1.6 \times 10^3 \times 9.81 \times 70 \times 10^{-6}}{3 \times 1.0 \times 10^3 \times 84.79}$$

so

$$u_t = 4.16 \times 10^{-3} \ m/s$$

1st Iteration

The particle Reynolds number for this latest estimate is calculated:

$$Re_p = \frac{70 \times 10^{-6} \times 4.16 \times 10^{-3} \times 1.0 \times 10^3}{1.0 \times 10^{-3}} = 0.2912$$

Similarly the new C_D is 86.92 and $u_t = 4.10 \times 10^{-3}$ m/s.

2nd Iteration

$$Re_p = 0.287, \quad C_D = 88.12, \quad u_t = 4.077 \times 10^{-3} \text{ m/s}$$

3rd Iteration

$$Re_p = 0.2854, \quad C_D = 88.593, \quad u_t = 4.066 \times 10^{-3} \text{ m/s}$$

4th Iteration

$$Re_p = 0.2846, \quad C_D = 88.829, \quad u_t = 4.061 \times 10^{-3} \text{ m/s}$$

The solution converges rapidly. Use of Stokes' law is about 5% in error and the initial drag coefficient calculation is about 2.5% out. The improvement between the 3rd and 4th iterations is 0.12%.

Example 3.2

Calculate the terminal settling velocities of spherical particles of sizes 50, 100 and 1000 μm respectively. Solid density is 2.8×10^3 kg/m³, liquid density is 1.0×10^3 kg/m³ and liquid viscosity is 1.0×10^{-3}N s/m².

Solution

Calculate the parameters P and Q from Equations (3.17) and (3.18):

$$P = \left(\frac{4 \times 9.81 \times 1.8 \times 10^3 \times 10^3}{3 \times 10^{-3} \times 10^{-3}} \right)^{1/3} = 2.8661 \times 10^4$$

$$Q = \left(\frac{4 \times 9.81 \times 1.8 \times 10^3 \times 10^{-3}}{3 \times 10^3 \times 10^3} \right)^{1/3} = 2.866 \times 10^{-2}$$

Set up a table with columns for the various stages of the calculation. Use Tables 3.1 and 3.2 to obtain $\log_{10} (u/Q)$ from $\log_{10} Px$.

x, μm	x, m	Px	$\log_{10} Px$	$\log_{10} (u_t /Q)$	u_t /Q	u_t, m/s
50	5×10^{-5}	1.433	0.15625	-1.076	0.0839459	2.406×10^{-3}
100	10^{-4}	2.866	0.45729	-0.515	0.306193	8.776×10^{-3}
1000	10^{-3}	28.661	1.45729	0.759	5.741165	0.1645

Calculate from Equation (3.13) the maximum particle size at which Stokes' law will apply:

$$x_{max} = \left(\frac{3.6 \times (10^{-3})^2}{(2.8 - 1.0) \times 10^3 \times 10^3 \times 9.81} \right)^{1/3} = 58.9 \; \mu m$$

Hence Stokes' law could be used for the 50 μm sphere:

$$u_t = \frac{(5 \times 10^{-5})^2 (1.8 \times 10^3) \times 9.81}{18 \times 10^{-3}} = 2.453 \times 10^{-3} \, m/s$$

which is within 2% of the value provided by the Heywood Tables.

3.2 Hindered Settling

When the particle concentration is sufficiently high the particles no longer settle as individuals. In general, settling behaviour changes with increasing concentration, rapidly going through a transition region in which clusters of particles develop and settle as clouds, and leading to hindered settling where the particles sediment en masse. In this situation the particles or flocs are not touching, but are prevented from behaving as individuals by their close proximity to each other. An example of a suspension undergoing hindered settling is shown in Figure 3.5. The measuring cylinders contain the same suspension at different stages of settlement. Initially, the suspension is uniformly mixed; an interface develops between the solids settling en masse and the clear liquid displaced by the sedimentation. Finally, the interface comes to rest when the sedimentation is complete.

Figure 3.5 Photograph of suspension undergoing hindered settling at different times

3.2.1 Voidage Functions

For non-flocculated systems Richardson and Zaki [1954] compared sedimentation and fluidisation and showed that the settling rate is linked to the terminal settling velocity of the particles by the voidage or porosity, raised to a power that is a function of the particle Reynolds number. These relationships are described by the equation:

$$U = u_t \varepsilon^n \tag{3.21}$$

where U is the settling velocity of the particle suspension (which is equivalent to the superficial velocity during fluidisation), u_t is the terminal settling velocity of the particle in an infinite fluid and ε is the voidage or porosity of the system. The exponent n varies with the particle Reynolds number and also with D the diameter of the vessel in which the sedimentation is taking place. The effect of the latter is usually only significant in laboratory experiments. The relations are given in the table.

$Re_p = \dfrac{xu_t\rho}{\mu}$	n for small tubes	n for large tubes
< 0.2	$4.65 + 19.5\ x/D$	4.65
$0.2 < Re_p < 1$	$(4.35 + 17.5\ x/D)\ Re_p^{-0.03}$	$4.35\ Re_p^{-0.03}$
$1 < Re_p < 200$	$(4.45 + 18\ x/D)\ Re_p^{-0.1}$	$4.45\ Re_p^{-0.1}$
$200 < Re_p < 500$	$4.45\ Re_p^{-0.1}$	$4.45\ Re_p^{-0.1}$
$Re_p > 500$	2.39	2.39

3.2.2 Batch Settling: Kynch Theory

An important feature of hindered settling is the development of a distinct interface between the clear liquid and the settling solids. This boundary is usually more marked for flocculated or coagulated systems.

It is important to consider behaviour at the base of the vessel. Here the first higher concentration layers appear as illustrated in the Figure 3.6. In the first elemental increment of time δt, the concentration $C + \delta C$ is formed at the base of the vessel as new settled solids arrive. In the next time increment further solids accumulate to give a concentration of $C + 2\delta C$ at the base and $C + \delta C$ in the layer above. Similarly in the third time increment the base concentration becomes $C + 3\delta C$, the next uppermost layer is now $C + 2\delta C$ and the one above that is $C + \delta C$. Since the highest concentration present at any point in time must be at the base of the vessel, layers of constant concentration appear to move upwards.

Whilst all this occurs at the bottom of the vessel, the interface between the clear liquid and the settling solids is moving downwards. The change in the height of this interface with time is known as the batch settling curve, and characterises such a system.

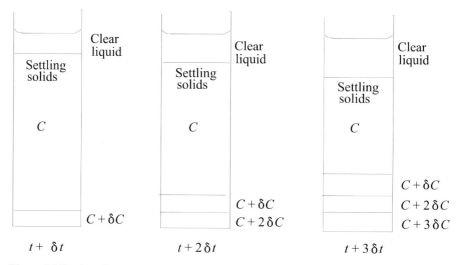

Figure 3.6 Batch settling

The features of a typical batch settling curve are illustrated in Figure 3.7.

The initial period, up to point (a), represents an induction period in which the suspension recovers from initial disturbances or, if it is a flocculating or coagulating suspension, in which the loosely aggregated particles called flocs are formed. From points (a) to (b) a constant rate of fall of the interface is observed. At point (b) there is a transition to a first falling rate section which ends at point (c), the "compression point", beyond which is a second falling rate section. Initially all the particles settle at apparently the same velocity and higher concentrations than the initial one appear, first at the base of the settling vessel. Eventually a solid concentration in excess of the original will have risen from the base and is now present at the settling interface – point (b). The settling rate is inversely related to the solid concentration, thus the observed settling rate begins to slow down. At the compression point it is believed that the particles are now touching and settling as such is taking place no longer. Thus this stage marks the end of the hindered settling regime.

The theory of batch settling is mainly due to Kynch [1951] and begins with the postulation that the settling rate U is solely a function of the solids concentration C. The particle flux G is defined as:

$$G = UC \tag{3.22}$$

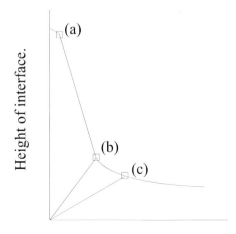

Figure 3.7 Batch settling curve

Consider an elemental layer between heights h and $h + dh$ as shown in Figure 3.8. In a time interval δt the accumulation of particles in the layer is given by the difference in flux:

$$input \quad - \quad output \quad = \quad accumulation$$

$$UCA\rho_s - \left[UC + \frac{\delta(UC)}{\delta h} dh \right] A\rho_s = \frac{\delta C}{\delta t} A\rho_s dh \qquad (3.23)$$

$$(kg\ s^{-1}) \qquad\qquad (kg\ s^{-1}) \qquad\qquad (kg\ s^{-1})$$

Thus,

$$-\frac{\delta(UC)}{\delta h} dh = \frac{\delta C}{\delta t} dh$$

giving:

$$\frac{\delta C}{\delta t} = -\frac{\delta(UC)}{\delta h} \qquad (3.24)$$

and by rearrangement:

$$\frac{\delta h}{\delta t} = -\frac{\delta(UC)}{\delta C} \qquad (3.25)$$

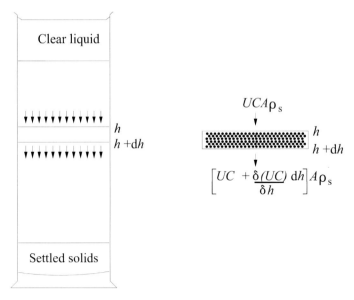

Figure 3.8 Flux variation during batch sedimentation

Equation (3.24) is the relation between the rate of change of concentration, at a fixed height, and the change of solids flux with distance in the vessel. Equation (3.25) is the rate at which a fixed concentration propagates through the sedimentation column (dh/dt), as a function of the change in solid flux relative to solid concentration. If the settling velocity (U) is a unique function of concentration, then Equation (3.25) shows that the value of propagation velocity will also be a fixed value – giving rise to a line of constant concentration propagating from the origin to the settling interface. These lines are called 'concentration characteristics' and are illustrated on Figure 3.7. Both Equations (3.24) and (3.25) can be converted into differentials, and then integrated, in order to arrive at expressions for the solid settling velocity of concentrations between that of the initial and final sediment. The resulting equations are:

$$U = -\frac{1}{C}\int_{0}^{h}\frac{\partial C}{\partial t}\,dh \qquad\qquad (3.26)$$

and

$$U = -\frac{1}{C}\int_{C_{max}}^{C}\frac{\partial h}{\partial t}\,dC \qquad\qquad (3.27)$$

In order to evaluate Equations (3.26) and (3.27) the local solid concentration profile below the settling interface is required. There are many experimental techniques that may be employed to provide this, such as ultrasonic, electrical current and X-ray attenuation [Williams *et al*, 1990, Holdich and Butt, 1997, Gaudin and Fuerstenau, 1962]. These equations are useful when the settling material demonstrates compression and Equation (3.27) is easier to numerically integrate as the function dh/dt changes smoothly with concentration. However, during sedimentation of incompressible materials, a much simpler method may be used to determine the settling velocity of concentrations between the initial and the final as demonstrated in the next section. A knowledge of the settling velocity and concentration enables the batch settling flux curve to be determined.

The concentration characteristics propagate upwards from the base of the batch settling vessel with a velocity shown as the slope of the appropriate line, illustrated in Figure 3.7. For incompressible materials the settling and propagation velocities for each solid concentration are independent of vessel geometry. This conclusion is particularly useful when scaling batch settling tanks from laboratory data. This is discussed further in the next section.

3.2.3 Batch Flux

Conventionally batch flux is defined as the product of the settling velocity and the solid concentration by volume fraction. Hence the SI units of batch flux will be m s^{-1}. In most instances the solid density and the vessel area are constants, and a flux balance is required, thus the area and density cancel. Thus true solid flux is defined as:

$$G' = UCA\rho_s \qquad\qquad (3.28)$$

but for convenience area and density are ignored and Equation (3.22) is applied. Now, if the settling rate is a unique function of concentration, such as Equation (3.21), then it is possible to perform a set of batch sedimentations at different starting concentrations and to measure the settling rates, as illustrated in Figure 3.9, where $C_1 < C_2 < C_3$. The product of the settling velocity and concentration gives the batch sedimentation flux.

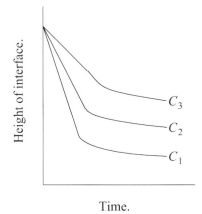

Time.

Figure 3.9 Settling curves during batch sedimentations

Talmage and Fitch [1955] extended this approach further to show that full data for a batch flux curve may be obtained from a single batch settling curve for the lowest of the solids concentrations in the range of interest. Their technique is best illustrated by the diagram provided in Figure 3.10 and a simple mass balance. Initially, there is a homogeneous suspension of concentration C_1 of height H_1 that is allowed to settle. After time t_2 there is a significant amount of supernatant liquid, some of which is decanted off to give a new height of H_2. If the sediment and remaining supernatant are then mixed again to give a homogeneous suspension it will have a higher concentration than the original. However, the mass of solids present will be the same – the liquid mass has reduced. The new concentration (C_2) can be deduced from the mass balance on the solids:

$$C_1 H_1 A \rho_s = C_2 H_2 A \rho_s \tag{3.29}$$

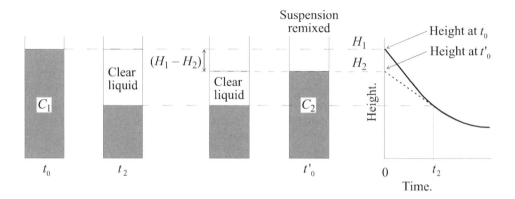

Figure 3.10 Graphical projection to obtain flux data from a single batch sedimentation

The new starting concentration is greater than the original and the settling velocity is, therefore, slower. After the period in time labelled t_2 the new settling curve will meet the old one and follow it from thereafter. Note that at time t_2 the concentration at the interface is slightly greater than C_2, and this concentration has propagated up from the base of the column in accordance with the discussion in Section 3.2.2. These propagation velocities are a function of solid concentration, hence the rate at which they meet the settling interface will be the same when sedimenting from concentrations of C_1 and C_2. Thus the settling curves are the same from time t_2 onwards. Following this theory it is possible to generate a set of data for settling velocity at different concentrations by decanting off supernatant liquid from the previous experiment. However, the significance of this technique is that the experiments do not need to be performed – the data can be obtained by simple geometry from the settling curve illustrated in Figure 3.10. Several tangents can be drawn to the settling curve and the height of the intercepts read off. Equation (3.29) is then used to predict what the concentrations would be, and the settling velocities come from the gradients of the tangents drawn to the settling curve. Thus a full set of settling velocities, concentrations and fluxes can be obtained from a single batch sedimentation. This procedure is demonstrated in Example 7.3 in Chapter 7.

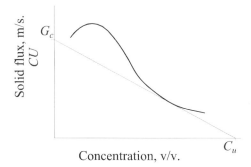

Figure 3.11 Batch flux curve

The batch flux curve is illustrated in Figure 3.11. The flux is zero when the solid concentration is zero and when the settling velocity is zero, i.e. the concentration is at a maximum. Between these two limits the flux has finite values and must, therefore, display a maximum value as illustrated in Figure 3.11. However, it is often found that the maximum occurs at very low concentrations that may not be experimentally measured. The batch flux curve is very useful: propagation velocities can be deduced from it in accordance with Equation (3.25) and it may be used in the design of continuous thickeners as discussed in Chapter 7. In the latter case the design procedure seeks to find the limiting solids flux that the settling device permits. An additional flux comes from the method of operation, so this limit is not simply the lowest value of batch flux. However, a tangent to the batch flux curve going through the underflow concentration (C_u) provides this limiting flux (G_D), as illustrated by worked Example 7.3.

3.2.4 Use of Batch Flux Curve for local concentration

If an experimentally derived batch flux curve exists it can be differentiated in accordance with Equation (3.25) to provide values for the propagation velocity of concentrations during sedimentation. The deduced settling and propagation velocities can then be used to predict the shape of the batch sedimentation curve under any operating conditions; e.g. investigate the effect of increasing the height of the vessel, or change in feed concentration.

An estimate of the hindered settling performance of a material may be made from its particle size distribution, just as it is possible to estimate the free settling velocity using the Heywood tables. In the settling of concentrated suspensions the following force-momentum balance applies [Holdich and Butt, 1997]

$$\frac{\partial P_s}{\partial x} = Cg(\rho_s - \rho) - \frac{\mu}{k}U \tag{3.30}$$

where P_s is the solids compressive pressure and k is the permeability of the assembly of particles. Hence the left hand side of Equation (3.30) may be considered to be the reaction force due to solid support from network contact between the particles, the first term on the right side is the weight force and the remaining term is the liquid drag force on the particles. In the case of sedimentation of materials not exhibiting network contact the stress gradient is zero and Equation (3.30) can be rearranged to give:

$$U = Cg(\rho_s - \rho)k / \mu \tag{3.31}$$

An expression for permeability can be substituted into Equation (3.31), such as that due to Kozeny, Equation (2.4), using solid concentration by volume fraction in preference to porosity, and multiplying by concentration provides an expression for solid flux:

$$UC = \frac{g(\rho_s - \rho)(1 - C)^3 x_{sv}^2 \psi^2}{36K\mu} \tag{3.32}$$

where x_{sv} is the Sauter mean diameter for the size distribution and ψ is the sphericity; see Appendix A for a description of these. The Kozeny constant K usually has the value of 5 for packed beds and 3.36 for moving ones. Differentiating Equation (3.32) in accordance with Equation (3.25) provides an expression for propagation velocity of a characteristic:

$$\frac{dh}{dt} = -\frac{g(\rho_s - \rho)3(1 - C)^2 x_{sv}^2 \psi^2}{36K\mu} \tag{3.33}$$

Thus, if the particle size distribution is known, it is possible to estimate the batch settling velocities from Equation (3.31), the batch flux from Equation (3.32) and the propagation velocities from Equation (3.33). This enables a full simulation of batch settling under any condition to be performed. An example is shown in Figure 3.12, using spheres with a Sauter mean diameter of 25 μm and density of 2650 kg m^{-3}, settling in water from a height of 0.34 m and employing a Kozeny constant of 3.36.

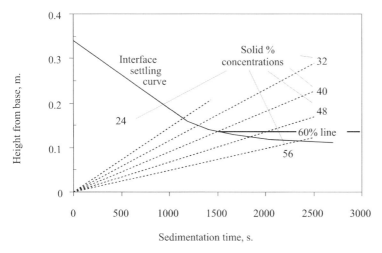

Figure 3.12 Simulated batch sedimentation of a 24% v/v suspension using Equations (3.31) and (3.33)

The interface settling velocity is determined from Equation (3.31), using the starting concentration of 0.24 by volume. When the first characteristic, at a concentration slightly higher than the initial, reaches the settling interface (deduced from Equation (3.33) the interface settling rate slows. The settling rate is again determined using Equation (3.31) but by using successive values of increasing concentration as the characteristics meet the settling interface. The example shown in Figure 3.12 uses five characteristics in this way, the lines representing the settling rates are drawn on the characteristics, and all the lines are joined together to give a smooth settling curve.

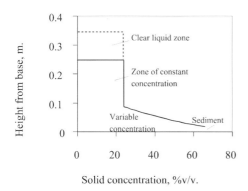

Figure 3.13 Simulated local concentration profile in batch sedimentation

The local concentration profile at any instant in time is deduced by reading off the height corresponding to each of the concentrations shown in Figure 3.12. A graph of height and concentration can then be produced, as illustrated in Figure 3.13 for a sedimentation time of 600 seconds. The different zones evident in Figure 3.13 are in accordance with the description of batch sedimentation by Kynch, and discussed in Section 3.2.2.

The most significant problem in the simulation of batch sedimentation of incompressible materials is determining the final settled sediment concentration. In both Figures 3.12 and 3.13 this is left unanswered. Thus characteristics of increasing concentration are drawn from the origin. However, it is unrealistic for a sediment concentration to be in excess of 60% by solids, and most commonly encountered inorganic sediments are more likely to be 20 to 30% by volume. Organic sediments are likely to be even lower in solid content. Thus a more realistic simulation is obtained if the final sediment height is deduced by a material balance such as Equation (3.29) using the known initial suspension concentration and height, and an estimated maximum sediment concentration. This height can then be marked on the sedimentation simulation, as indicated by the 60% line included in Figure 3.12.

In the above analysis the propagation velocities are constant throughout the entire sedimentation, in accordance with flux theory and Equation (3.33). However, towards the end of a batch sedimentation the characteristics have been observed to curve to become parallel to the time axis. This has been attributed to compressive forces becoming significant, thus P_s cannot be ignored in Equation (3.30), and represents the start of the sediment zone. It has also been argued that the concentration characteristics originate from the top of the sediment zone and not from the base of the vessel, and Kynch's analysis of batch sedimentation has been revised accordingly [Tiller, 1981]. This is more significant during sedimentation of materials showing appreciable compression effects, and is discussed in the next section.

3.3 Sedimentation with significant compression effects

If significant compression effects are present then the solids stress gradient cannot be neglected in Equation (3.30). Hence the solids settling velocity is no longer a unique function of solid concentration and the batch flux approach to settling is not valid. There have been numerous numerical solutions to the settling equations represented by Equations (3.30) and (3.24) or (3.25). In some instances the permeability is replaced by a resistance function [Landman and White, 1994] and many different constitutive equations linking solid concentration and pressure have been suggested. However, the simplest one to apply is that already covered in Chapter 2, Equation (2.40):

$$C = C_o P_s^u$$

noting that pressure drop across the cake ΔP is not relevant to solid concentration by sedimentation, instead it is the compressive solids stress P_s due to the weight of overlying solids in network contact that is. Hence the term P_s has been substituted for ΔP in the equation. However, the constants C_o and u are the same as those given in Chapter 2, and the coefficients provided in Table 2.3 are relevant to the batch sedimentation of the listed solids as well as filtration. One of the earliest numerical solutions to batch settling involving compression was by Shirato *et al* [1970], who converted the problem from Cartesian into material co-ordinates before solving the resulting partial differential equation. The material co-ordinate system is illustrated schematically in Figure 3.14.

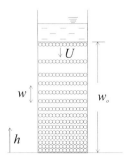

Figure 3.14 Material co-ordinate system in batch sedimentation

Defining the mass of solids per unit area to be 'w' then

$$w = C\rho_s \delta h \tag{3.34}$$

The total mass of solids up to a height *h* is

$$w_0 = \int_0^h C\rho_s \, \mathrm{d}h$$

Equation (3.34) can be used to transfer the system to material co-ordinates. In this system the total mass of solids in a batch sedimentation vessel is split into a number of elements, each containing the same mass per unit area of material (Δw_i). Only at the start of the sedimentation are the elements of equal volume, i.e. area and height.
Darcy's law, written in a material co-ordinate system and combined with a liquid force balance gives

$$U = \frac{kC\rho_s}{\mu} \frac{\partial P_L}{\partial w} \tag{3.35}$$

where P_L is the liquid pressure in excess of the hydrostatic pressure due to the presence of suspended solid material. The continuity equation in material co-ordinates is

$$\frac{1}{\rho_s} \frac{\partial C}{\partial t} \frac{1}{C^2} = -\frac{\partial U}{\partial w} \tag{3.36}$$

Combining Equations (3.30) and (3.34) to (3.36), employing the Chain Rule and rearranging provides the following equation for the variation in the excess liquid pressure, as a function of time, due to the settlement of a compressible compact

$$\frac{\partial P_L}{\partial t} = -\frac{\rho_S C^2}{dC / dP_S} \frac{\partial}{\partial w} \left[\frac{kC\rho_S}{\mu} \frac{\partial P_L}{\partial w} \right] \tag{3.37}$$

which is a parabolic equation in excess liquid pressure with non-linear coefficients. A suitable choice of transforming coefficients can be made to make Equation (3.37) non-dimensional, and readily solvable by means of a finite difference solution [Holdich and Butt, 1997]. Note that the use of a constitutive equation between pressure and concentration, such as Equation (2.40), leads to an ordinary differential equation becoming one of the coefficients in the above equation: solids pressure is uniquely dependent on liquid pressure and concentration is uniquely dependent on solids pressure – hence the full differential. The relation between liquid and solids pressure gradients is:

$$\frac{\partial P_s}{\partial h} = -\frac{\partial P_L}{\partial h}$$

Thus a knowledge of the liquid pressure at any height and time can be used to provide the solids pressure and the solid concentration. The total height of the suspension undergoing compression is calculated from:

$$h = \sum_{i=1}^{m} \frac{\Delta w_i}{\rho_s (C_i + C_{i-1}) / 2} \tag{3.38}$$

where Δw_i is the mass per unit area that the sedimentation column is divided into and m is the number of divisions. The height of the settling interface with respect to time comes from Equation (3.38). However, before Equation (3.37) can be evaluated the dependence of permeability on solid concentration must be known. The Kozeny equation can be used, as it was in Section 3.2.4, but in compressible systems there is considerable evidence to suggest that the Kozeny constant should be regarded as a coefficient dependent on concentration rather than a constant [Davis and Dollimore, 1980]. In many cases the Kozeny coefficient has a simple linear dependency on the solid concentration.

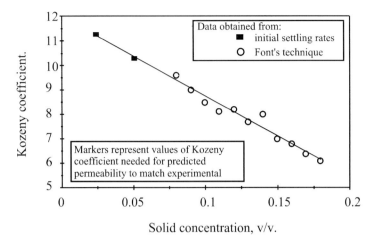

Figure 3.15 Variation of Kozeny coefficient during batch sedimentation of talc

Experimentally determined values of permeability can be obtained using Equation (3.30) applied to the observed initial settling rates of batch sedimentations at low solid concentration. Under these conditions the solids stress gradient is negligible and Equation (3.30) can be rearranged to give observed permeability. At higher concentrations the permeability of systems exhibiting compression can be analysed by a method proposed by Font [1994], which takes into account the characteristics rising from the top of the sediment zone as discussed in Section 3.2.4. Figure 3.15 shows the empirical values required for the Kozeny coefficient for talc using these two methods for permeability determination. Figure 3.16 illustrates the results from a numerical solution using Equation (3.37), and the above methods, and compares the result with experimental observations.

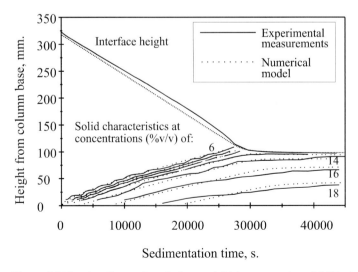

Figure 3.16 Batch sedimentation of talc at an initial concentration of 5.2% v/v [Holdich and Butt, 1997]

The interface height was recorded from observations and the local solid concentrations below the settling interface were measured using a sedimentation vessel equipped with electrodes to measure the local electrical resistance. A calibration between electrical resistance and solid concentration was used to convert the resistance readings to concentration as a function of height and time. There is good agreement between the numerical model and the experimental observations.

The mathematical basis for the solution of the equations for compressible sedimentation, given above, has also been applied to the continuous thickening of compressible materials and is discussed further in Chapter 7. The spreadsheet used to construct the solution used in Figure 3.16 is downloadable from the Internet, together with those mentioned in Chapter 2 and Appendix C, so that the reader may further investigate the simulation of batch sedimentation of compressible compacts.

3.3.1 Stirring and channels during sedimentation

The preceding sections described methods for analysing and predicting the behaviour of solids settling in the regions of dilute sedimentation, hindered settling and compression. The remaining region included in Figure 3.1 is that where channelling is prevalent. The behaviour of material in this region is more difficult to quantify. Channels during batch sedimentation are often observed [Vesilind and Jones, 1993]: they are directly visible when the sedimentation occurs in transparent vessels and the settling interface may display little mounds of solids ejected from the top of the channels that resemble volcanoes. The interface settling curve may exhibit a linear rate of descent at the start, such as that illustrated in Figure 3.7, but often accelerates just before the falling rate period is observed. This phenomenon is due to the channels providing an easy path for the liquid entering the supernatant compared to flowing past the surrounding settling solids. The channels form initially in the lower regions of the batch sedimentation vessel then propagate towards the settling interface. When the top of the channel comes close to the settling interface, and the liquid flowing into the supernatant experiences lower flow resistance as it flows through a channel, the solid settling rate close to the interface can increase because it is subjected to a lower liquid drag force. Visible evidence of channelling are shown and illustrated in Figures 3.17a and 3.17b.

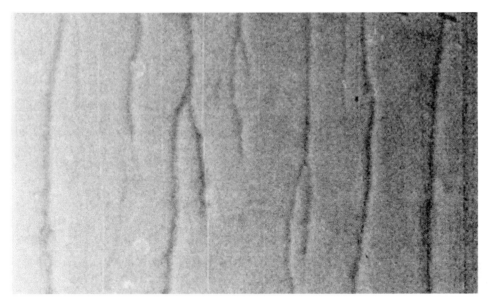

Figure 3.17a A picture of channelling taken during batch sedimentation in a transparent vessel

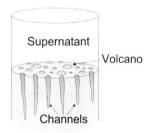

Figure 3.17b Schematic illustration of channelling and volcanoes during batch sedimentation

It has been argued that as channels provide a means to augment the rate of production of supernatant liquid they should enhance the rate of sedimentation and be beneficial in commercial settling equipment. Settling under inclined surfaces, see the next section, is one successful application of this principle. However, materials that display channelling characteristics also tend to exhibit significant solids network formation: e.g. highly flocculated materials, organic solids and very fine precipitates. Channels can only form naturally with these materials in the absence of stirring. Sediment height is an important variable with these materials, as the weight of overlying solids can be used to compress the solids network to a higher sediment concentration. Thus a new thickener design based on a tall sedimentation vessel without stirring has been proposed [Chandler, 1983].

Stirring breaks up channel formation, but also helps to break up the formation of loose aggregates, flocs and the solids network. Thus stirring can be used to achieve higher sediment concentrations than would form naturally; sediment concentrations of calcite suspensions have doubled when employing stirring [Holdich and Butt, 1995]. Hence, channelling is desirable near to the settling interface but stirring is desirable within the sediment [Dixon, 1979].

Stirring is recommended during batch sedimentation tests of flocculated materials designed to obtain design data for thickeners, as it leads to more reproducible data. A minimum vessel diameter of 150 mm is also recommended for these tests. However, stirring and vessel diameter are less important when settling materials that exhibit true zone settling behaviour. Stirring is also employed when the settling material may otherwise become buoyed by the presence of gas bubbles, for example when settling biological material that has undergone aerobic or anaerobic digestion.

3.4 Settling Under Inclined Surfaces

Gravity settlers which use inclined surfaces in the form of flat plates (lamella) or tubes have received increasing interest because of the compactness of the process unit that they comprise, since the use of a large number of closely packed settling surfaces gives a large horizontally projected settling area in a relatively small volume thus achieving a high measure of process intensification.

The presence of an inclined surface in a sedimenting system has the effect of increasing the settling rate of the particulates. (Indeed in order to ensure accuracy it is important that batch settling tests should be done in vessels that are perfectly vertical.) In 1920 some research was reported in which it was noticed that the vertical sedimentation velocity of defibrinated blood corpuscles in tubes was increased as the tube was tilted and that the settling was quicker in smaller bore tubes and in those where the initial suspension height was greatest. This was interpreted at the time as a Brownian motion effect. Later, simpler mathematical models which use a geometrical rearrangement of the separated fluid in the tube were used to describe the settling behaviour. A further advantage of the presence of additional surfaces is in the potential available for the control of liquid flow pattern and the reduction of convection effects within the fluid.

The additional settling area is easily calculated. For example in Figure 3.18 the plane ABCD inclined at angle α to the horizontal will have a projected area onto a horizontal plate AEFD. This projected area will equal $LW \cos \alpha$ where L and W are the length and width of the inclined plane.

3.4.1 Nakamura–Kuroda Equation

The settling behaviour was described by Nakamura and Kuroda [1937], who assumed that only the downward facing surfaces accelerated sedimentation and that particles in the settling suspension tend to keep the same distance apart until they alight on a solid surface or upon other particles. The model describes the settling process in a tilted square section tube on its edge.

Consider Figure 3.19a: at the start of settling all particles on a surface denoted by the line CAB settle with an initial velocity v for an elemental time interval dt and reach a hypothetical surface DFH. The velocity v has the same value at all points and thus the vertical distances represented by AF, BH and CD are equal. Neglect the volumes represented by triangles CDE and BGH. The volume of clarified liquid freed from particles in time dt is represented by the area ABGFEC. However the particles will not take up the surface shown as EFG because of the density and height difference between the suspension at plane FG and the liquid at point E. An instantaneous rearrangement will take place giving a new clear liquid–suspension surface at plane A' B' as in Figure 3.19b. The two volumes of clear liquid shown must be equal and thus area AA' BB' must equal area ABGFEC. If the initial height of the interface AB is h and this falls to a final height h–dh after the elemental time dt, then by equating the areas:

$$-\,\mathrm{d}h = v\mathrm{d}t\left(1 + \frac{h}{b}\cos\alpha\right) \tag{3.39}$$

$$-\frac{\mathrm{d}h}{\mathrm{d}t} = v\left(1 + \frac{h}{b}\cos\alpha\right) \tag{3.40}$$

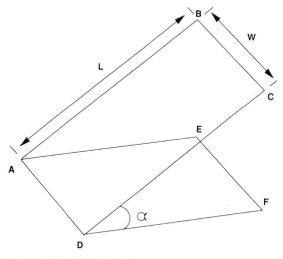

Figure 3.18 Enhanced settling area

Thus the model may be extended to other geometries to give the general equation:

$$-\frac{1}{v}\frac{\mathrm{d}h}{\mathrm{d}t} = 1 + k'\frac{h_o}{b}\cos\alpha \tag{3.41}$$

where v is the settling velocity in the vertical tube, h is the height of the solid–liquid interface, h_o the initial value of h, b is the plate separation distance normal to the inclined surfaces, α is the angle of inclination of the surface to the horizontal and k' is a constant which has the value of 1.0 for inclined plane surfaces, $4/\pi$ for a circular tube

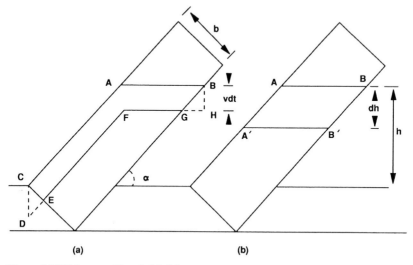

(a)   **(b)**

Figure 3.19 Nakamura–Kuroda Model

and $\sqrt{2}$ for a square tube resting on a corner. It seems that these equations represent an upper limit to the rate of sedimentation since rarely have settling velocities been reported which are in excess of those predicted by this model. More recently, it has been suggested that three mechanisms may well be causing settling convection. First, suspension density gradients and particle concentration gradients are not aligned with hydrostatic pressure gradients. Second, the return flow of displaced liquid passes through regions of low particle concentration and, finally, particles collect and tend to flow down along the upward facing inclined surfaces. These convection currents may be disturbed by flow instabilities which occur under certain conditions and thus reduce the beneficial effect of the presence of the inclined surfaces on the settling rate.

3.4.2 Grashof Number and Sedimentation Reynolds Number

The sedimentation of particles under inclined surfaces may be characterised by two dimensionless numbers; G_R, a sedimentation Grashof number which represents the ratio of gravitational forces to viscous forces in a convective flow and R a sedimentation Reynolds number representing the significance of inertial forces to viscous forces in a convective flow [Acrivos and Erbholzeimer, 1979, 1981].

These parameters are defined by the equations:

$$G_R = \frac{h^3 g \rho (\rho_s - \rho) C_f}{\mu^2} \tag{3.42}$$

and

$$R = \frac{h v \rho}{\mu} \tag{3.43}$$

where h is a characteristic length dimension such as the height of the suspension, C_f is the initial volume fraction of the particles and v is the settling velocity of those particles, at concentration C_f measured in a vertical vessel.

The ratio G_R: R is Λ, which represents the significance of gravitational forces to inertial forces and is typically very large. Mathematically Λ is given by:

$$\Lambda = \frac{h^2 g (\rho_s - \rho) C_f}{\mu v} \tag{3.44}$$

If Λ is in the range 10^4–10^7 and R in the range 0–10, then the Nakamura–Kuroda equation can be expected to predict adequately the settling rate for both batch and continuous lamella separators provided steady state conditions can be maintained and separator dimensions chosen so that flow instabilities are avoided.

3.5 References

Acrivos, A. and Herbholzeimer, E., 1979, J. Fluid Mech., 92, (13), pp 435–457.
  idem, ibid, 1981, 108 pp 485–499.
Chandler, J.L., 1983, Filtration and Separation, 20, pp 104-106.
Clift, R., Grace, J.R. and Weber, M.E., 1978, Appendix B in Bubbles, Drops and Particles, Academic Press, New   York.
Davis, L. and Dollimore, L., 1980, J. Phys. D, 13, p 2013.
Dixon, D.C., 1979, in Wakeman, R.J. ed Progress in Filtration and Separation, Vol 1, Elsevier, Amsterdam.
Fitch, E.B., 1962, Trans. Amer. Inst. Min. Engrs., 223, p 129 - 137.
Font, R., 1994, Ind. Eng. Chem. Res., 33, p 2859 - 2867.
Gaudin, A.M. and Fuerstenau, M.C., 1962, Trans. Amer. Inst. Min. Engrs., 223, p 122 - 129.
Heywood, H., 1962, Symposium on the Interaction between Fluids and Particles pp. 1–8, I.Chem.E., London.
Holdich, R.G. and Butt, G., 1995, Trans. Inst. Chem. Eng. 73, p 833 - 841.

Holdich, R.G. and Butt, G., 1997, Sep. Sci and Tech.. 32, p 2149 - 2171.

Khan, A.R. and Richardson, J.F., 1989, Chem. Eng. Comm. 78, p 111 - 130.

Kynch, G.J.,1951, Trans. Faraday Soc., 48, p 166 - 177.

Nakamura, H. and Kuroda, K., 1937, Keijo J. Med., (8), pp 256–296.

Landman, K.A. and White, L.R., 1994, Adv. Colloid Interface Sci., 51, pp 175 – 246.

Richardson, J.F. and Zaki, W.N.,1954, Trans. Inst. Chem. Eng. 32, p 35 - 53.

Shirato, M., Kato, K., Kobayashi, K. and Sakazaki, H., 1970, J. Chem. Eng. Japan, 3, pp 98 – 104.

Soo, S.L., 1967, Fluid Dynamics of Multiphase Systems, Blaisdell, Waltham, Massachusetts.

Talmage, W.P. and Fitch, E.B., 1955, Ind. Eng. Chem. 47, p 38 - 41.

Tiller, F.M., 1981, AIChE J., 27, pp 823 - 829.

Vesilind, P.A. and Jones, G.N., 1993, Water Science and Technology, 28, pp 59 – 65.

Williams, R.A., Xie, C.G., Bragg, R. and Amarasinghe, W.P.K., 1990, Colloids and Surfaces, 43, pp 1 – 32.

3.6 Nomenclature

A_p	Projected area of particle	m^2
C	Solid concentration by volume fraction	–
C_D	Particle drag coefficient	–
D	Vessel diameter	m
g	Gravitational constant	$m\,s^{-2}$
G	Solids settling flux	$m\,s^{-1}$
H	Height	m
h	Height co-ordinate	m
k	Permeability	m^2
n	Exponent in Richardson & Zaki Equation (3.21)	
P	Coefficient defined in Table 3.1	m^{-1}
P_s	Solids compressive pressure	Pa
Q	Coefficient defined in Table 3.1	$m\,s^{-1}$
Re_p	Particle Reynolds number	–
t	Time	s
u	Single particle velocity relative to fluid	$m\,s^{-1}$
u_t	Terminal settling velocity	$m\,s^{-1}$
U	Sedimentation velocity of suspension	$m\,s^{-1}$
v	Characteristic settling rate	$m\,s^{-1}$
v_o	Overflow rate	$m\,s^{-1}$
v_s	Apparent settling rate	$m\,s^{-1}$
v_2	Upward propagation velocity of concentration characteristic	$m\,s^{-1}$
w	Mass of solids per unit area	$kg\,m^{-2}$
x	Particle diameter	m

Greek Symbols

α	Angle of lamella plate to horizontal	degree
ε	Local porosity	–
μ	Liquid viscosity	Pa s
ρ	Fluid density	kg m^{-3}
ρ_s	Solid density	kg m^{-3}
ψ	Particle sphericity	–
τ	Shear stress on particle surface	N m^{-2}

4 Filter Media

4.1 Introduction

The principal role of a filter medium is to cause a clean separation of particulate solids from a flowing fluid with a minimum consumption of energy. Media may be broadly classified as (a) those designed to recover a valuable solid product and (b) those used in the clarification of a fluid, e.g. deep-packed beds of sand in water clarification. In (a) attempts are made to create surface deposition of the solids in a recoverable form. Of course, certain media cannot be described as belonging to the (a) or (b) classification and create the required separation by a combination of surface deposition and particle capture in the internal interstices of the medium. *The successful performance of a filter is largely dependent on the selection of a suitable filter medium.* Despite the vast amount of technical information on the performance and selection of media, in the present state of development, media choice usually follows experimental trials with the solid–fluid mixture. Such experiments have to be conducted with great care, since, as will be discussed below, small changes in process conditions can have an important effect on the outcome. The task of media selection is made difficult by the multitude of media available. It is rarely clear, by inspection of the solids, which type of filter medium will prove most suitable for the separation. Exceptions here are extremely coarse solids (> 100 µm) which, in certain conditions, could be handled with perforated plate or edge filters, and extremely fine material (< 0.05 µm), which require membrane filtration. Even in these situations, changes in the particle concentration and shape can present practical difficulties. Examples of some of the woven fabrics available are shown in Figure 4.1.

A qualitative review of the principal forms of filter media has been provided in the literature [Dickey, 1961]. This early information was later developed into comprehensive treatments of filter media properties and performance characteristics [Purchas, 1981 & 1998]. The wide range of media received early attention [Kovacs, 1960] in descriptions of wire cloth, metal edge filters, fused metallic fibres, ceramics and impregnated cellulose. This wide spectrum was later extended [Shoemaker, 1975] in practical suggestions on media selection and classification. Structural and materials aspects of filter module construction [Loff, 1981], including media aspects, further extended this array. The range of filter media available continues to expand. Of particular interest is the recent massive expansion in the use of ceramics in (a) conventional solid-liquid filtration and dewatering processes [Allenius, 1999] and (b) membrane operations [Schloemer, 1999]. The latter development points to important economic gains in the use of small-pore, highly resistant media and membranes created from ceramics. Similar recent interest has been exhibited in reports on the widely-used textile filter fabrics [Hardman, 1994] for solid-liquid separation processes. Both woven and nonwoven filter cloths constructed from polymeric materials (polypropylene, polyethylene, polyester, polyamides, polyvinyl chloride, etc.) add to the traditional materials such as wool, cotton and paper.

FIG 1 (a)

FIG 1 (b)

FIG 1 (c)

FIG 1 (d)

Figure 4.1 Typical woven filter media [Ward, 1972].
(a) Plain-weave monofilament; (b) Reverse-plain-dutch weave monofilament; (c) Twill-weave monofilament;
(d) Plain-weave, multifilament, continuous fibre; (e) Plain-weave, multifilament, staple fibre.

Much of the information reported in this chapter also relates to woven media. Some mention is made on the equally important nonwoven fabrics; the latter, are further discussed in Chapter 6, which also deals with other media such as deep sand filters and filter aids. The rapidly expanding use of membrane media is reported in Chapter 10.

The inherent problems associated with media selection are illustrated quite clearly when one considers the woven fabric field. Variations in weave pattern, materials of construction and type of yarn produce important changes in the applicability of a particular medium to a filtration problem. The importance of correct media choice can also be illustrated graphically. In Figure 4.2, the yield of a constant-pressure filtration process for incompressible cakes has been calculated at various levels of filter medium resistance, as defined in:

$$V^2 + \frac{2AR_mV}{\overline{\alpha}c} - \frac{2A^2\Delta Pt}{\overline{\alpha}\mu c} = 0 \qquad (4.1)$$

in which $\overline{\alpha}$ is the average filter cake resistance at the applied pressure and R_m is the resistance of the medium to fluid flow. This equation is derived in the section dealing with filtration fundamentals.

In those cases where R_m may be assumed negligible in comparison with the cake resistance, simple yield equations of the form:

$$V_R = \left(\frac{2\Delta P}{\overline{\alpha}\mu ct} \right)^{0.5} \tag{4.2}$$

may be derived; V_R is the volumetric filtration rate per unit time per unit area.

As may be seen, in Figure 4.2, which depicts diagrammatically the effect of changes in weave and resistance on the system, the medium resistance has an important effect in determining the yield from such a process. It will be noted that the above equations are derived by assuming that solids are surface-deposited, with no penetration into the internal pores of the medium, and with no change in R_m during the process. These simplifying assumptions are rarely justified since, in practice, passage of fine particles into and through the medium cause increases in the resistance of the medium. These processes can create conditions of media blinding, in which the permeability of the medium is reduced to zero. Internal deposition mechanisms are the principal cause of poor application of the theoretical parabolic rate law to practical situations. Some of the benefits obtained by a correct cloth or medium specification include:

(a) A clean filtrate, with no loss of solids by bleeding or passage through the medium
(b) An easily discharged filter cake
(c) An economic filtration time
(d) No media deterioration by sudden or gradual blinding, stretching, wrinkling, etc.
(e) An adequate cloth lifetime, with the first, clean performance reproduced by
      back-washing of the medium

This list of requirements may be extended in practice when the construction and fitting of the medium must be considered. Thus, gasket performance, chemical and/or biological stability, strength, abrasion resistance, etc. may have to be considered in machine applications.

In view of the complexity of the situation, it will be realised that a system of classification of media action, linked to particulate properties, would be of great value in the process engineering of filtration systems.

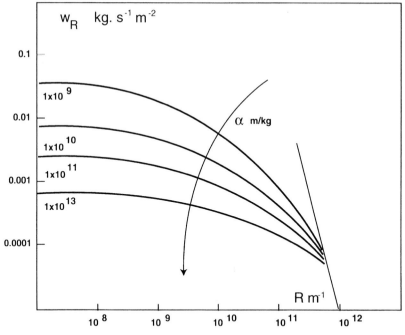

Figure 4.2 Effect of medium resistance on filter productivity

A classification has been suggested [Purchas, 1967] which is based on the rigidity of the media; the main features of this classification are presented in Table 4.1 below.

Table 4.1. Media classification

Type	Example	Minimum trapped particles (μm)
Edge filters	Wire-wound tubes Scalloped washers	5–25
Metallic sheets	Perforated plates	100
	Woven wire	5
Woven fabrics	Woven cloths Natural and synthetic fibres	10
Cartridges	Spools of yarns or fibre	2

It will be observed that such a classification gives only the broadest guide to the types of media available. As will be discussed in later sections, the information in the table on the minimum size of particle retained by a particular medium must be viewed with reservation. Retention of particulates depends on many process variables and lower minima than reported in the table can be realised in practical separations. Thus in woven fabrics, changes in particulate concentration can produce particle capture much below 10 μm.

It is vital to have information on the particle size range to be processed in view of the interaction between the solids and the pores during the first few moments of cake growth.

Most manufacturers report the permeability of their fabric measured by the flow of air as a measure of particle retention. Thus the flow (ft^3 min^{-1} or m^3 s^{-1}) per unit area (ft^2 or m^2) under a constant pressure differential (0.5 inch w.g. or 1693 Pa) is taken as a measure of the permeability; the latter is directly related to the porosity (free space per unit volume of fabric) of the medium. Thus a high permeability is taken as an indication of high porosity and, in turn, low particle retentivity.

Similar information has been collected for water permeabilities [Rushton & Griffiths, 1971].

In terms of the resistance of the medium R_m, i.e. the inverse of permeability, it has been shown that the resistance to the flow of water is sometimes higher (after allowance for the viscosity change) than that for gas flow. Again, rises in R_m in use must be expected, up to a reasonable limit. Several standards are used (Frazier, DIN, etc.) in reporting air permeabilities. The latter, even though related to the openness of the cloth, and its particle-stopping potential, give no indication of the retention capacity of media. This is demonstrated in Figure 4.3, where two cloths which remove all particles greater than 40μm, have quite different removal capacities for smaller particles. The blinding characteristics of these cloths will be related to the curves in Figure 4.3. The physical characteristics of the filter media listed above are catalogued by by Loff [1981], who also discusses the question of "filter media rating" in relation to particle retention.

Typical values of the "permeability" of fabrics to air:

Fabric Type	m^3/m^3 s
Nylon multifilament	0.030–1.52
Polypropylene monofilament	0.015–1.52
Polypropylene multifilament	0.005–0.51
Nonwoven cloth	0.002–1.27

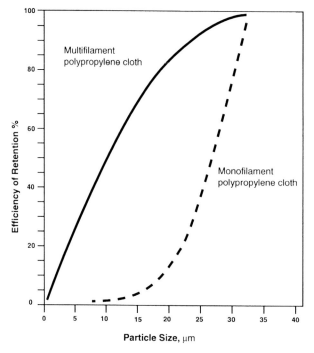

Figure 4.3 Retention characteristics of woven media [Purchas, 1967]

4.2 Woven Cloths

The structure of woven filter cloths depends on the type of yarn used in weaving. Yarns are available in several forms: monofilament, multifilament, staple fibre and mixtures of the same.

4.2.1 Monofilaments and Multifilaments

The yarns in these cloths are composed of solid polymeric material (polypropylene, polyester, polyamide, etc.). Cloths of different patterns are produced on the loom by varying the manner in which the warp and weft yarns are woven together. The warp yarn is stretched in the machine or longitudinal direction and the weft yarn lies at right angles to the warp. Yarns are usually cylindrical, although other shapes are available.

In the illustration shown in Figure 4.4, the plain-weave monofilament cloth has been produced by warp and weft yarns of the same diameter, woven together in a simple one-under, one-over pattern. These cloths are available in a wide range of pore sizes from

5 000 to about 30 µm, the lower limit being determined by the size of fibre available for the weaving process. These cloths are characterised by pores of an open type which create little flow resistance and many applications are found in areas where high flows are required, e.g. in oil, paint, and water filtration and screening. Such cloths are readily cleaned by back-flushing.

The surface of the fabric can be modified by finishing; this involves heat treatment and calendering in order to flatten the surface, reduce pore size, but preserve the construction by way of reducing any tendency to shrink or stretch in service.

Unfortunately, these fabrics are light in weight and would be easily damaged if used directly in pressure filters. Thus the advantages of high throughput and ease of cleaning must be weighed against the fragile nature of the medium. Modern trends are to produce composite weaves from fine and coarse monofilaments in the production of a surface layer with good release properties and nonblinding characteristics. The underside layer of coarser fibres provide support, assist drainage and promote attachment or caulking of the cloth onto the filter plate.

In effect these cloths are designed to simulate the combination of a top filtering cloth supported by a backing cloth. This subject is discussed below.

In order to produce apertures finer than mentioned above, changes in the weave are available to alter the size (and shape) of the cloth pores. Thus in the production of various twill cloths, weaves such as one over-two under, three over-one under, two over-two under, can be arranged to produce different cloths, as shown in Figure 4.4. Cloth of this type, e.g. sateen weaves, possess very smooth surfaces which are optimal for cake release.

The effect of changes in weave pattern on flow through the fabric will be discussed in the section dealing with mathematical modelling of flow through monofilaments.

An interesting variant is where the diameter of the warp yarn is different from the weft, in the production of Dutch weaves of strong construction. A Dutch weave produces two openings or pores in the fabric, one on the surface between parallel yarns and a second, triangular pore in the plane of the cloth. These cloths produce excellent performances in the separation of solid, crystalline materials, but are less successful in filtering slimy colloids which can lodge in the triangular pores.

New fabrics are continually being developed by manufacturers aiming to improve particle retention, cake release and filtrate drainage.

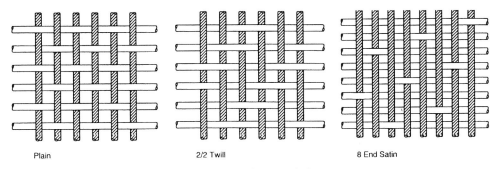

Plain   2/2 Twill   8 End Satin

Figure 4.4 Weave patterns in woven mono- and multifilament cloths.

A relatively new type of closed twill weave is available for colloidal separations. Whilst such fabrics have lower clean permeabilities than plain or Dutch weaves, they are more easily cleaned and withstand heavy use.

These media are described as two-sided, with an upper slick surface for filtration being produced by a sateen weave. This surface is further modified by calendering. Another important feature of this design is that ultrafines are not retained by the cloth, being released downstream by the underside funnel-shaped pores. A published industrial report [Technical Textiles International, 1992] refers to success with this type of cloth (monofilament polyester, calendered to screen size 30 μm) in high-pressure variable chamber pressing of dyestuffs, hydroxides, stearates, etc.

Multifilament yarns are produced by twisting together a number of fine filaments, as shown in Figure 4.1d. The filaments may be continuous or short (staples) in length. Continuous filaments can be twisted tightly, to simulate a monofilament yarn of zero permeability. A monofilament character is thus induced whilst also creating a flexible, strong medium, suitable for high pressure separations. Detailed discussions of the benefits of high twist cloths will be dealt with below, along with the effects of flow through low twist multifilaments. Staple fibres can be short or long and made up into yarns of a directed or disoriented character. Such yarns have high particulate retentivity, but poor cake release properties. Combined hybrid monofilament and multifilament cloths are available with characteristics to suit filtration conditions in belt presses and large automatic filters. These machines require heavy, dense monofilaments.

4.3 Cloth Selection

4.3.1 Effect of Yarn Type and Weave Pattern

The filter cloth is the deciding factor in the success or failure of all press operations. In view of the wide range of process variables involved in the filtration process, it is virtually impossible to select a filter medium that will satisfy all process requirements and the usual limited time scale available for cloth selection is used to find an acceptable medium, i.e. one that will satisfy most, if not all of the requirements. In this respect, one particular requirement (e.g. filtrate clarity) may have to be relaxed, if other specifications (e.g. filtrate rate, absence of blinding) are to be maximised. Thus the more open weave fabrics will be superior in nonblinding characteristics, but may have poor particle retention. The latter will improve in the order monofilament < multifilament < staple fibre. Tabulated information is presented in Tables 4.2, 4.3 and 4.4 below on the effect of yarn properties, weave patterns, etc. on the processes of cake release, productivity, resistance to blinding, etc..

Further information of this type, on the effects of fibre materials and weave pattern, is provided in the literature [Purchas, 1988].

Cloth selection from tabulated information is impossible without reference to the slurry being processed. The tables give general guidelines to the behaviour of media; interaction of the latter with particles can only be determined by practical trials. The best test method, of course, is the installation of potential media in an operation unit. This type of study will produce relevant information on wear characteristics, cloth life expectancy and other factors (which are difficult to predict with certainty from other test methods).

The next-best option is the use of a pilot-scale model of the filter. Here the fluid flow patterns at the surface of the medium will, at least, be similar to the large-scale unit. The pilot filter cannot produce information on wear properties, e.g. produced by the effect of movement of large, heavy plates. Cloth behaviour must at least be studied experimentally using laboratory Buchner filters. The latter low-pressure test units will provide information on the resistance of the used medium, tendency to blind, etc. However, the filtration process conducted downwards on the surface of the medium, under a pressure differential of 0.5 bar, cannot be expected to simulate exactly the processes occurring inside large recessed-plate or plate and frame filters where particle movement is a complex mixture in vertical and horizontal directions.

Such effects are discussed in later sections of this Chapter. Filter cloth respecification often occurs in developing processes. In these circumstances, on-line information on such matters as periodicity of cloth replacement or causes of taking the cloth out of service is of great value. Samples of used cloth are useful for inspection purposes and in providing information on any change in mechanical properties, e.g. serious decline in stretching resistance. The latter may be the result of chemical contact, high temperatures and/or clogging; the latter gives the cloth a cardboard-like appearance and weakens it mechanically.

Table 4.2 Effect on performance yarn type listed in descending order [Purchas, 1981]

	Maximum retention	Maximum production	Maximum Cake moisture reduction	Maximum cake discharge	Maximum life	Maximum resistance to blinding
Fibre Form	Spun staple	Monofilament	Monofilament	Monofilament	Spun staple	Monofilament
	Multifilament	Multifilament	Multifilament	Multifilament	Multifilament	Multifilament
	Monofilament	Spun staple	Spun staple	Spun staple	Monofilament	Spun staple

Table 4.3 Weave patterns. Effect on performance listed in descending order [Purchas, 1981]

	Maximum retention	Maximum production	Maximum Cake moisture reduction	Maximum cake discharge	Maximum life	Maximum resistance to blinding
Weave Pattern	PRD	Satin	PRD	Satin	PRD	Satin
	Plain	Twill	Satin	PRD	Twill	Twill
	Twill	Plain	Twill	Twill	Plain	PRD
	Satin	PRD	Plain	Plain	Satin	Plain

PRD: Plain Reverse Dutch

Table 4.4 Weave to meet filtration requirements [Harrison;1975]

Property	Weave			
	Plain	Chain	Twill	Sateen
Rigidity	1	2	3	4
Bulk	4	3	1	1
Initial flow rate	4	3	2	1
Retention efficiency	1	2	3	4
Cake release	2	3	4	1
Resistance to blinding	4	3	2	1

Key: 1 Best, 2 Good, 3 Satisfactory, 4 Poor

Whilst the financial investment attached to filter media is relatively low, compared with other items of plant, the correct specification of the medium remains the key to successful and profitable filtration processes.

4.3.2 Criteria of Choice

Three criteria by which a medium may be judged are:

1) The permeability (or, inversely the resistance) of the clean, unused medium
2) The particle-stopping power of the medium
3) The permeability (or resistance) of the used, or deposited medium

The filtration process involves two principal resistances: (a) the resistance of the filter cake α and (b) the resistance of the medium R_m. The relative importance of α and R_m in determining the overall resistance of the system is highlighted in Figure 4.2.

At high levels of α, e.g. greater than 1×10^{12} m/kg (characteristic of sludge-like material) changes in R_m have little influence on overall productivity—at least in the range $1 \times 10^8 < R_m < 1 \times 10^{11}$. Thus a partially blinded medium may still function quite satisfactorily in a system controlled by α. Obviously, improved productivities in these circumstances will follow a reduction in α, e.g., by conditioning of the slurry. On the other hand, rapid productivity losses will be noticed for relatively small R_m increases, in those systems of low α ($< 1 \times 10^9$).

The concept of a negligible medium resistance has attractions, in a mathematical sense, in leading to simple process equations for filtrate productivity, e.g. Equation 4.2 above. It has been estimated [Tiller, 1985] that for this assumption to be justified (with less than a 5% error in productivity calculations) the quantity $(\alpha \, w)/R_m$ should be numerically greater than 40. In this concept, w is the mass of solids deposited per unit area. This leads to the following observations. At equivalent α and R_m levels, a considerable surface deposit (e.g. a precoat) would be required to eliminate the effect of the medium. On the other hand, for high α (sludges), extremely small cake deposits are all that are required to reduce the medium effect to negligible levels. Thus in the separation of sludge-like materials, it is acceptable to ignore R_m in process calculations. In these circumstances the choice of a tightly woven cloth of relatively high initial R_m value is often recommended if the easily cleaned cloth resists blinding mechanisms and can produce good cake release.

4.4 Operational Aspects of Woven Media in Filters

4.4.1 Loading of Yarns with Solids

The space between filaments in multifilament media will become filled with solids if the particle size involved is commensurate with the pore sizes in the fabric, and the shape of the particle causes it to become firmly lodged in the pore. When such solids are carried into the yarn they become trapped and are difficult to remove by back-flushing. This contrasts with the inevitable non-plug filling of inter yarn pores which are created by the weave pattern used. Unless these pores are firmly plugged with relatively larger material, back-flushing can remove the deposits. The pore plugging effect of commensurately sized particles is depicted in Figure 4.5 These sites become areas of retention of smaller, slimy materials which normally would bleed through the medium.

Successive filtering cycles add solids to the intra-yarn pores until they become filled; subsequent filtered solids tend to adhere to the entrapped particles and this leads to increasing difficulties with cake discharge. It is claimed [Smith, 1951] that there is little adhesive action between solids and the fabric of polymeric cloths. Yarn loading can be detected in tests where, although gradual reductions in filtration rates are recorded from test to test, the used cloth rates are still acceptable from a productivity standpoint. On the other hand, simultaneous deterioration in cake discharge with increase in yarn loading can become a more serious effect than productivity.

4.4.2 Bacterial Growths

Media blinding may be caused by the onset of organic growths such as fungi, bacteria and algae. This is borne out by the well-known "filter effect" which can be observed in long-term permeations of cloths in laboratory conditions using superficially clean tap water [Heertjes, 1957]. A rapid decline in the permeability of clean media can be created by the continued permeation of unsterilised water; long-term tests require the use of some fungicidal agent in the test system.

Again, it is reported that these growths do not occur on the surface of the polymer, but on small or sub-micrometre particulates captured in the interstices of the filter. Such growths can occur on solids at all points in the system, e.g. on the slurry during its movement in pipelines. Stationary or low-flow conditions promote the production of gelatinous growths, produced by the organisms. An early indication of such attack is the impossibility of the formation of even cake deposits, with cake growth occurring randomly on the surface of the medium. This blinding process may continue up to the point where filter cake formation ceases altogether. In operations where deliquoring and cake washing follow the filtration stage, increasing process time will be required to effect these operations. Washing difficulties start to appear long before cake formation becomes a problem.

The cloth may be tested using a Buchner filter and flask; the presence of organic growths will be demonstrated by forming a thin filter cake and decanting supernatant slurry. Random deposits of solids on the cloth surface are an indication of blinded areas. Use of a fungicide may then restore the medium to its original permeability. An example of used cloth, before and after cleaning is shown in Figure 4.5.

4.4.3 Precipitation from Solution

The effect of this phenomenon is similar in action to organic growths; washing of solubles from the filtered cake will become increasingly difficult and finally cake deposition will reduce to zero. Microscopic examination of the fabric or an ash test on the fabric will reveal the presence of internal solids. The latter may be further identified by chemical analysis.

4.4.4 Inadequate Drainage

Poor support of the filter medium either in the absence of a porous backing cloth or in the selection of incorrect drainage channels on the surface of the filter plate will lead to processing difficulties, e.g. in the formation of sloppy, sticky filter cakes. In the particular case of pressure filters, undersizing of the drainage ports from the filter will produce the same effect. A principal effect of the latter is to restrict the escape of ultrafine solids which bleed through the filter in the first moments of pressing.

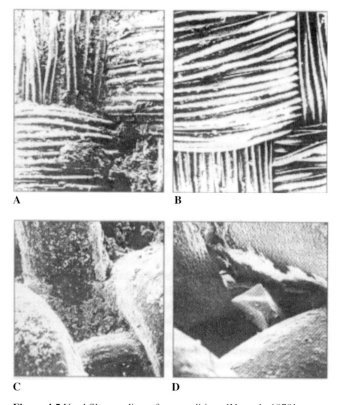

Figure 4.5 Used filter media surface conditions [Nemeth, 1979].
A) Blinded multifilament cloth: B) After cleaning ; C) Inter–yarn plugging by fine solids; D) Inter–yarn pore plugging by larger solids

Since it is virtually impossible to deal with all the cloth problems attaching to a particular separation, the compromise referred to above is adopted, between filtrate clarity and blinding of the medium by the smaller particles in the feed. Early fore–runnings can be recycled, since the first bleeds will then be blended with the feed and will filter on the initial cake layers.

Restriction of the first fines by poor drainage will increase the probability of retention by the medium, with concomitant blinding. On the other hand, the modern dimpled filter plate used in conjunction with properly selected filter media has reduced the necessity of using backing cloths. An example of the latter is described below in the section dealing with industrial information. An alternative to backing cloth is the use of reinforced filter fabric in the flange and stay-boss areas of the filter. See Section 4.4.11 below.

4.4.5 Critical Concentration

Problems are usually encountered when pressure systems designed to operate at reasonable levels of particulate concentration are fed with dilute suspensions at the start of filtration. Low concentrations of solids prevent the bridging effect which ensues when concentrated swarms of solids are directed towards the pores in the filter medium. This effect is discussed quantitatively elsewhere in this Chapter. Failure to bridge the cloth pores will lead to deposition of particles inside the fabric.

4.4.6 Critical Pressure

It has been suggested that a critical pressure exists for each system above which a collapse of particulate bridges will ensue, with the same effects as discussed above. The existence of such a critical pressure must be tested in the laboratory by conducting a series of separations at increasing pressure levels, coupled with analysis of the test data in the manner described in the theoretical section of the Chapter.

For constant levels of pressure, experimental data plotted in the form of $t/V, V$ will exhibit either an increasing intercept on the ordinate of the graph in circumstances where particle penetration is occurring or a nonlinear, initial portion of the graph. Of course, here it is necessary to use new cloth material in each pressure test in order to isolate possible causes of blinding other than bridge collapse. The state of the cloth pores after filtration can also be examined microscopically to check for particle penetration.

4.4.7 Classification of Particles

In separations involving a suspension of particles of a wide size range, the possibility always exists that the particles may become classified into fine and coarse fractions before deposition on the filter cloth. This phenomenon is caused by the relative importance of gravity and the fluid drag forces which direct the suspension towards the medium. In some cases the condition is a result of the geometry of the filter, e.g. in plate and frames, or recessed plates, where fluid velocities fall below a level necessary for homogeneity in the suspension. In centre feed recessed plates, fine solids may be scoured to the edges of the plate with an overall result of maldistribution of particulates, in size terms, on the surface of the medium. Other evidence of such maldistributions was collected by injecting carbon black particles into a plate and frame unit during the processing of white powders. These studies are reported below [Rushton & Metcalfe, 1973].

Other filters operate in such a way as to maximise the deleterious effects of gravity, e.g. in upward-directed rotary vacuum filters. Downward filtering tends to maximise the arrival of coarse solids with a general trend towards thicker cakes and absence of blinding.

In some circumstances, arrival of the fines fraction on the surface of the medium can lead to penetration of the yarns, etc. as outlined above. The bridge collapse effect will be aggravated by low concentrations of solids. On the other hand preferential arrival of the coarse fraction will maximise the probability of pore bridging.

Selective deposition is indicated by poor separation of the solids from the filter cloth. The presence of a heavy residual cake on the surface, after cake discharge, may be the result of preferential deposition of fines. One control mechanism here is to increase the effective solids concentration before filtration, e.g. by settling or hydrocyclones. As discussed in later sections dealing with particulate bridging increases in concentration generally lead to the creation of stable bridges.

4.4.8 Effect of Gas Bubbles

Suspensions containing air or gas bubbles can lead to particle classification, as a result of the tendency for colloidal materials to concentrate at the surface of the gas bubble. Highly aerated suspensions, e.g. from centrifugal separators or in process circumstances where a sudden drop in fluid pressure creates cavitation in the flowing suspension, can lead to blinding. Bubbles can persist on the surface of the medium (vacuum filters) or where collapse of the bubble takes place a preferential deposition of surface colloids will occur on or in the fabric of the cloth (Figure 4.6) [Nemeth,1979].

Figure 4.6 Filter effect phenomenon: Gas bubble formation [Nemeth, 1979]

Deaeration by vacuum prior to filtration is a solution here; filtration rates have been reported to double in those applications where degassing of the feed is practised.

4.4.9 Evaporative Effects

In open filters, e.g. rotary, disc and horizontal belt units, cake cracking may occur during the dewatering period, after cake deposition. Prolonged high-velocity passage of air through the areas of cracking can lead to evaporation of the liquid phase in the cloth with deposition of dissolved solutes.

Modern developments aimed at avoiding such effects involve the use of high-porosity, small-pore ceramic filters which use the inherently high capillary pressure of such materials to avoid gas entry into the medium. This materially reduces the cost of compression power in filters dewatered by compressed air.

4.4.10 Effect of Fabric Construction

It has long been realised that, in fluid flow through multifilament fabrics, a considerable proportion of the filtrate can pass through the yarn strands in the cloth.

The relative amount of flow through and around the yarns in such cloths will depend on the degree of twist imparted to the yarn and the size of the apertures between yarns. The aperture size will, in turn, depend on the weave pattern; plain, twill, sateen, etc.

This division of flow has technical importance in those circumstances where the fine particles present are small enough to follow the flow into the yarn. Once inside the yarn, removal of these particles is effected with great difficulty. A tightly twisted, closely woven cloth will approach the constructional and filtration characteristics of a monofilament cloth but also will generally exhibit better mechanical properties. The effect of cloth construction is discussed in detail below in the section dealing with mathematical models of flow in filter fabrics.

Swelling of fabrics can change the nature of flow in that closure of the cloth pores can force an increasing amount of flow through the yarns. This has been reported in studies using cotton, wool, linen and natural fibres [Smith, 1951]. In the latter it was recorded that up to 98% of the flow takes place through the yarns of such fabrics after swelling.

The early blinding difficulties experienced with plastic yarns, caused by penetration of solids into the fabric led to the adoption of monofilaments, despite the tendency to bleed. This trend was reported in mining operations where cloth replacement was dramatically improved with monofilament cloths. In these cases, filtrate clarity was not a major requirement. At the same time, monofilament characteristics are not without problems. In circumstances where coarse solid exactly fits the pores in the monofilament or high-twist yarn, high initial filtrate velocities can produce a permanent plugging of the pore, particularly by particles of a nonspherical, elongated shape. If these solids resist removal by back-cleaning, they obviously reduce the effective permeability of the fabric and become a site for continued plugging by even finer solids which normally would have passed through the clean cloth, (Figure 4.5 C, D).

Again, the relatively poor mechanical properties of monofilaments such as stretching, wrinkling or tearing, promoted an interest in high-twist multifilament cloths woven in such a way as to ameliorate the associated problems of cake release.

The effects of twist on the percentage of fluid flow through the yarns and the reduction in cloth life by permanent blinding is recorded below [Smith, 1951]: These data apply to 60/70 denier yarns, where the denier is the mass in grams of 9000 m of yarn.

Twists or turns per m	%Flow through yarns	Yarn solids loading rate
60–120	95–98	Fast
590	70	Reduced
1380	2	Minimal

The number of yarn turns per metre necessary to produce an impervious yarn diminishes with increase in yarn denier. A practical example of these effects is shown by the filtration of clay slip containing 40% solids (clay, bonding agents, algae):

	Cleaning cycle, h	Cloth life weeks
Light cotton	24	2
Nylon 60 denier 3 turns/in	24	6
Nylon 60 denier 35 turns/in	168	22

High-twist yarns are advantageous if compressed air is used for cake discharge. In the absence of plugging by large solids, which just fit the cloth pores, multifilament high-twist cloths are usually returned to the original porosity (permeability) by back-flushing. This simulates monofilament behaviour. Further information of the division of flow through and around cloth yarns is presented later.

4.4.11 Effect of Cloth Underdrainage

Where a cloth is resting on a flat impervious surface, the filtrate and associated particles will have a certain distance to travel before entering a drainage port or channel.

Thus for a fabric of thickness X cm resting on a slotted surface with drainage ribs $8X$ cm wide, particles may become trapped in the tortuous pores existing between the underside of the cloth and the plain solid support. Generally undrained zones greater in width than four times the cloth thickness should be avoided; this principle is inherent in the relatively more efficient performance of dimpled or pipped drainage plates compared with vertical channels in pressure filter plates.

The presence of a coarse backing cloth, of woven metal or plastic, under a thin cloth will improve the escape of fluid and particles from the underside of the medium.

4.5 Aspects of Filter Cloth Selection and Performance

The critical importance of filter cloth selection in filter press systems has been stressed in the literature [Regan, 1977]. As discussed above, methods of filter cloth selection have been based on "selection factors" [Purchas, 1967, 1981]. These factors include: textile fibre type, yarn type, fabric geometry and weave pattern. Other factors include: cloth shrinkage and stretching, filter cake release, cloth surface characteristics and cleaning. These added factors are considered below.

4.5.1 Cloth Shrinkage

Shrinkage of the cloth can produce severe problems particularly with plate and frames and the larger recessed plates. In the latter, it is reported [Regan, 1977] that top-feed plates give less trouble from this phenomenon than centre-feed. Repeated cloth washing/drying cycles aggravate the shrinkage problems, particularly for polyamides. It is sometimes recommended not to dry out cloths, after washing, storing wet if possible.

Preshrinkage of fabrics is, therefore, widely practised in order to retain dimensional stability in service. Preshrinkage can be effected in a number of ways:

a) Use of hot (boiling) water with medium in a relaxed state.
b) Heat setting in an oven, with fabric under tension in warp and weft direction. This will maintain the original levels of cloth porosity, permeability, etc.
c) Oven treatment of the relaxed, or mildly-tensioned cloth through an oven.

The shrinkage process has to be carefully controlled, in view of the large structural changes of up to 15% shrinkage which can ensue in relaxed conditions.

Spun staple yarns are reported to shrink less than comparable fabric woven from filament yarns [Bosley, 1977]. This follows the slipping action of short fibres, one over the other, without major change in the overall yarn dimension.

4.5 2 Cloth Stretching

Absorption of liquids causes swelling of fibres and yarn. The increase in fibre diameter and length causes dimensional change in the cloth, with serious consequences for closely fitted plates. Obviously the fewer ports, etc., in the basic design of the plate the better. This is particularly true for those yarns with poor absorption characteristics (nylon). The latter may absorb up to 4% by weight; this may be compared with 0.4% for terylene.

Filter development, in terms of complete automation, has placed further demands on media characteristics. Recessed plates are preferred to plate and frames, in order to facilitate mechanisation. Discharge of cakes is usually of high priority and is generally easier in recessed plate filters; cake discharge from a plate and frame unit does not necessarily follow immediately after the opening of the filter.

The trend to larger plate sizes, e.g. up to 2×2 m accompanied by different drainage patterns and various drainport arrangements has also placed further demands on the mechanical strengths of the medium. Thus increased cloth distortion, resulting from the change from plate and frames to recessed plates has called for fabrics more resistant to stretching, under load.

Automation, which may involve discharge of a relatively thin cake by peeling apart the cloth and solids, usually also involves high-pressure squeezing during the dewatering cycle. The complete cycle of filling, pressing, air blowing, discharge and cloth washing may be short (5–6 minutes) in the modern auto-variable chamber unit [Barlow, 1983]. Thus the cloth may be subjected in modern units to as much stress in days, compared with months in manually operated systems.

All these considerations call for cloths of great strength whilst retaining high-level filtration characteristics. For example, a traditional strong polyester fabric suitable for use on large plates (1×1 m) in slow or manual systems would have, say, breaking load figures of 900 N/cm (warp) and 350 N/cm (weft). These figures may be compared with cloths of the same size in mechanised units: 1800 N/cm (warp) and 1600 N/cm (weft), respectively.

The variable chamber filter has recently [Young, 1991] been demonstrated as the economic option in sludge dewatering, when compared with centrifuges and belt presses. Cake solids above the 24% DS autothermic incineration limit can be achieved, albeit at higher flocculation doses than normally used in press operation.

4.5.3 Filter Cake Release

Adequate cake release is a fundamental pre-requisite in efficient pressing operations, in maintaining a low down-time, t_d in the overall 'batch' time. The overall productivity is given by the quotient (V_f/t_c) where V_f is the filtrate produced per cycle and the cycle time $t_c = t_f + t_w + t_d$; t_f and t_w are the filling/filtration time and wash time, respectively.

An understanding of the failure of the release mechanism follows consideration of the balance between the forces causing adhesion of the cake to the medium and the discharge forces. Rumpf [1977] and Shubert [1977] conclude that whilst a great many studies have been published on particle adhesion, both theoretical and practical information has limited generalised application. The adhesion of particles dispersed in liquids is mainly the result of electrostatic and van der Waals interactions; chemical bonding also plays in important role.

Chemical bonding is characterised by direct interaction of atoms and molecules . The surface forces involved are, therefore, short range compared with other forces of attraction. Chemical bonds are so substance specific that it is virtually impossible to generalise an outcome which is so sensitive to the state, type and structure of the surface layers on the particles.

High temperatures (approx 60/75% of the absolute melting point of the adherents) may result in a sinterbridge, at a point of contact. In these adhesions, similar solids are required and the process is unlikely to occur, say, between an organic solid filtered on a polymeric medium. On the other hand, if particles of the type shown in Figure 4.5 have become firmly lodged in the medium, these particles may be responsible for sinter adhesion of fines. Sugar crystals, separating above 60°C serve as an example of such effects. Recrystallization of solutes can produce a material connection between the depositing solute and filtered solids. The latter may be the same material as the solute. This effect can be produced by evaporative or cooling processes which occur during the drying of filter cake in dewatering.

As a wet cake dries, liquid bridges form between solid surfaces; dissolved materials tend to concentrate in the residual liquors and, at a certain stage, will precipitate out in the liquid bridge (Figure 4.7). Crystallization within such liquid bridges can produce considerable adhesion between solids; removal may be possible only by redissolution of the bridges. It follows that in precipitative systems, drying processes must be controlled above the moisture level concomitant with crystal nuclei formation. This suggests limiting the cake drying to relatively high moisture levels, or prefacing the drainage process by efficient washing out of the solutes.

The adhesion force created by liquid bridges has received attention; the ratio of adhesion force F to surface tension γ is related to particle size, particle separation and volume of liquor in Figure 4.8.

When a liquid bridge is extended by increasing the distance between the solids, the bridge becomes unstable and breaks at a particular separation. On the other hand, these studies show how the liquid bridge force changes with drying.

These considerations point to the importance of a residual filter cake liquor in its influence on the cake discharge mechanism. It follows that the cake discharge process is materially affected by the cake drainage or dewatering efficiency. Industrial reports [Hongxiang, 1991] refer to a cake moisture reduction to less than 25% as being necessary for effective cake discharge. Low moistures in variable chamber pressing of dyestuffs, metallic hydroxides and stearates followed the use of monofilament polyester cloth calendered to reduce the screen size to 30 µm [Reid, 1974], as mentioned above. Carleton and Heywood [1983] have discussed the fundamental aspects of filter cake properties and cake discharge mechanism. The effectiveness of discharge will depend on:

1) The strength of the bond between the cake and cloth; this is influenced by cake: stickiness and yarn/weave characteristics (Tables 4.3). The bonding force depends on the mode of deposition of the first layer.

2) The internal strength of the cake. If the cohesion of the latter is less than the adhesion to the cloth, the cake will fail internally and leave solids on the cloth. The moisture content will vary across the depth of the cake.

3) The applied discharge force (e.g. gravity discharge from a vertical surface).

Optimal plate discharge requires a cake of high internal cohesion and low cloth adhesion; cakes which have been compressed and re-expand slightly on pressure release meet these requirements. In some circumstances, air drying can produce moisture variations which aggravate the discharge mechanism. In this respect discharge from pressure filters (large-scale) is particularly difficult to simulate on the small scale.

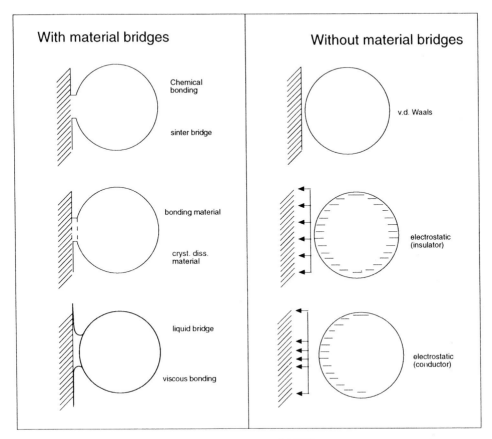

Figure 4.7 Bridge formation between solid surfaces [Rumpf, 1977]

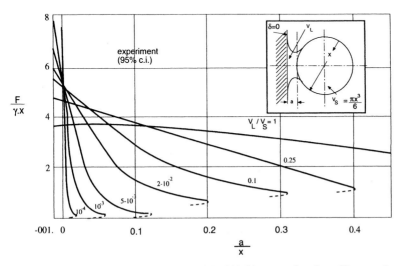

Figure 4.8 Dimensionless adhesion force of liquid bridges as a function of liquor volume [Rumpf, 1977]

4.5.4 Cloth Structural Effects

Each fabric has its own particular characteristic which can be used to advantage in the filtration process. Thus the smooth-surfaced mono- and multifilament (high-twist) cloths are particularly suited to the liquid environment, in the separation of sticky, clogging substances such as sludges or colloids. Retention of such small material depends on a rapidly formed surface layer of particulates in the reported use of monofilament polyamides for effluent treatment plants [Grove & Daveloose, 1982]. Apparently in this application, change from spun fibre to woven cloth increased the cloth lifetime from 3 months to 2.5 years! The adhesive nature of such solids may be modified by treatment with polyelectrolytes. Quite opposite results can also be experienced, however, where the three-dimensional capture characteristics of staple fibre material is the best option.

A type of closed twill fabric has been reported [Muller, 1983] aimed at a reduction in blinding when filtering colloidal/slimy particles. Although of low initial permeability compared with Dutch twills and plain monofilaments, the closed twills are more readily cleaned. These cloths are referred to as two-sided, with a slick upper surface formed by warp fibres passing over several weft threads before passing under. This weave design permits small particles to pass freely through the cloth, bleeding through the funnel-shaped openings on the underside of the fabric.

Stratified structures, or asymmetric media, have been used successfully in gas and membrane filtration systems. Woven cloths combining mono-, multifilament and spun yarns, made up in special weaves have been produced [Carleton & Heywood, 1983]. Thus cloths are available which combine retention efficiency, low pressure drop and depth filtration characteristics of spun fibres with the smooth surface features of calendered monofilaments. Inducing a staple fibre character to a press cloth reduces edge leakage problems, in view of the compressible nature of such fibres.

Of course, the asymmetric or stratified medium can be effected by the use of a backing cloth (e.g. felt) on top of which a tight filtering cloth is superimposed. This construction has been successfully applied in variable chamber presses (tube-press) of 8–10 000 cycles between cloth changes [Johns, 1991], with pressure differentials up to 100 bar. The higher pressure reduces the importance of cloth resistance so a high particle retention medium may be applied. This system is particularly suited to clay-like solids (50–80% particles < 2 µm) but is not recommended for fibrous or highly compressible sludges, starches, bentonite, etc. With these materials filtration ceases upon the formation of a skin-like first layer of high resistance.

The use of barrel-necked media for use in recessed plate filters points to a 30–40% reduction in re-clothing time [Hongxiang, 1991]. The latter report refers to the advantages of correct cloth specification which leads to sufficient cake discharge simply by movement of the filter plates. Lower plate mass reduces cloth damage; backing cloths again can reduce the effect of the closure of two 40 CI plates (400 lb each). In this respect, overhung press plates are less prone to cloth damage; the latter design tends to ensure parallel plate movement with the elimination of plate twisting. The latter is a major cause of cloth damage.

Successful cloth selection often follows a long, practical trial period, e.g., as reported in filter pressing of coal tailings [English & Radford, 1977]. During development, changes in media for the backing cloth through hessian, sisal, PVC and polypropylene were coupled with changes in the top cloth through cotton, nylon, courlene and rilsan. Eventually a 1900 polypropylene covered with a 650 courlene was accepted, despite the tendency of this combination to produce slightly turbid fore-runnings. This defect was offset by the excellent cake discharge characteristics of the courlene surface.

4.5.5 Cloth Cleaning Processes

Particular attention should be given to cloth washing operations; serious reductions in the effective life of a cloth will follow poor washing practice. High-pressure hosing or cleaning by a built-in cloth washer (Figure 4.9), is satisfactory for some systems. The washer in the latter diagram is automatically controlled and indexes between plates. Water at 50 bar pressure may be used in these systems. The latter are particularly advantageous in mining operations.

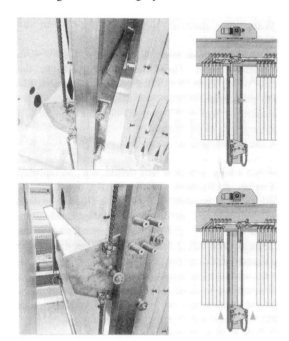

Figure 4.9 Automatic cloth washing plant [Bosley, 1977]

Particle size distribution effects have, however, to be taken into consideration since materials containing high percentages of fine solids may, in addition to being difficult to filter, cause problems if high-pressure direct hosing is used for cloth washing. Here the fluid pressure may force the surface-deposited fines into the fabric, thereby aggravating rather than ameliorating the blinding phenomenon. In these circumstances, a cleaning-in-place (CIP) method, using back-washing with suitable cleaning fluids will prove more effective. The cleaning fluids used will depend on the particular system and may range from water alone, through acid/alkali treatments, detergents, etc. Many proprietary cleaning fluids are described in the patent literature.

An important aspect of back-wash cloth cleaning is to ensure that the medium in the large-scale plant is irrigated over the complete surface. This is essentially a hydraulic design problem, similar to that attaching to maldistributions of flow in cake deposition and washing. Another factor is to consider the pH of the washwater.

A typical, practical washing facility using membrane filters on a fill-squeeze-discharge cycle time of 4 hours may involve a monthly cloth wash with high pressure water. The cycle can be varied (1-3 washes per month) for different applications.

4.6 Nonwoven Filter Media

The above section has concentrated on the use of woven filter media in pressure filters. It will be realised that a vast interest lies in the application of nonwoven, random fibre filter media to gas *and* liquid filtration systems [Fletcher, 1982]. Plate and frame presses covered with a heavy felt (nylon and polypropylene mixed fibres) are used in viscose dope clarification. These processes, involving the slow removal of gel-like substances from the liquid phase, are better served by nonwovens since surface blinding (promoted by the use of woven cloths) may be avoided. This applies also to the removal of gels from molten polymers by high pressure filtration through nonwoven, metallic fibre filters. Nonwovens are less sensitive to process changes, e.g. particle precipitation and/or concentration changes, typified by protein removal in corn syrup filtration.

Applications into the traditional woven cloth area are increasing in view of the improved performance (better gasket action around edges/bosses of plate) following surface treatments for pore size and cake release control. Skinned felts, for presses, are sometimes supplied with a surface finish on both sides of the cloth.

Polymeric cloths of the type mentioned above are produced by dry processes, e.g. where extruded, molten polymer filaments are dispersed by hot air streams and deposited on moving belts. Paper is an example of a nonwoven produced from a wet process, in the separation of cellulosic fibres from their dispersion in water, by impingement on a high speed, moving monofilament belt.

The trend is towards improved nonwoven performance, by way of fibre size reduction. Most woven-cloth manufacturers also supply a range of nonwovens, for specific press applications.

Despite the antiquity of filtration processes involving nonwovens, e.g. felts, laps, paper etc. media developments continue to appear involving modern fibres prepared from polymers, glass, fluorocarbons, etc. These are used alone, or in mixtures with traditional fibres such as wool, cotton, cellulose, etc., to produce a vast array of media aimed at solid-liquid separation. A detailed listing of the relevant physical properties of these fibres is available in the literature [Purchas, 1981].

A nonwoven fabric, or felt, is essentially a random, entangled assembly of fibres, of relatively high porosity. The physical strength of the fabric depends upon the mechanical interlocking of the fibres. The looser form, the mechanically weak lap, is sometimes strengthened by inclusion of a scrim of another strong, open weave fabric, e.g. muslin. This material finds application in air filtration. Densification of the felt can be achieved, to a degree, by further mixing, using barbed needles. The later are pushed into and withdrawn from the felt a number of times, thereby bringing about a close packing of the fibres. This process produces the so called needlefelts which again are widely used in large scale gas filtration. The limiting density of 0.2 gm/c.c. attainable by needling can be further increased, by heat- setting calendering of the felt, to produce a medium more useful for wet filtration. These techniques [Fletcher, 1980] include: surface singeing, lamination (usually with an PTFE film), singeing followed by calendering and finally, calendering with heat and pressure. The last treatment is most favoured for media to be used in liquid systems.

The high porosity of nonwovens is an advantage in producing a high retention capacity for depth filters. These filters usually contain bonded fibres. Thus the wet strength and overall resistance to fibre shedding, etc., can be improved by sealing the fibres, one to the other, to produce a rigid network. Bonding can be effected by the inclusion of adhesives or by heat setting. The latter is inherent in spun-bonded fabrics. These involve the extrusion of molten polymer into cylindrical filaments which are dispersed by hot gas flow into a tortuous, random array. The fibre mixture may include a small proportion of low melting point material.

A wide variation in fibre diameter exists. Examples are:

i)   10 µm Polyester (1.3 dtex where the latter unit is the weight in grams of 10 000 m of fibre).
ii) 40 µm Polypropylene (13.3 dtex).
iii) 30 µm Cellulose (may be fibrillated to produce fine fibre attachments or fibrils).
iv) 0.03–8 µm Glass (100% glass media used in laboratory liquid separation and in gas filtration).

Both the permeability and filtration characteristics of nonwovens are dependent on the felt porosity and fibre diameter. A medium which as been heavily calendered on both sides will possess the lowest porosity. Surface treatments and/or use of laminations of different porosities, are aimed at improving cake filtration performance and cake release. Generally speaking, the filtration efficiency at a particular particle size is inversely proportional to the fibre diameter, other factors being the same.

Quantitative relationships for the pressure differential created by liquid flow through nonwovens are presented below. The question of filter efficiency is also discussed, with reference to laboratory tests for particle retention.

4.7 Mathematical Models of Flow Through Filter Media

4.7.1 Permeability of Clean Media

Flow through the clean medium will be determined by the geometric characteristics imposed on the medium by the weaver, in forming various patterns (plain, twill, sateen, etc.) from basic yarns. In woven cloth, the latter are either solid monofilaments, or are multifilamanets (which can be further subdivided into continuous or staple-fibre constructions, depending on the type of filament used). In some cases, the surface of the medium may be modified to improve its ability to release the filter cake, etc. Nonwovens are paper-like, random arrays of fibres which can be obtained in many forms: uniform fibres, mixed and composite pads, etc. These media, like wovens, can be supplied surface modified.

Since overall productivities are related to the (V, t) or $(\Delta P, t)$ characteristics of the system it is of direct interest to have an understanding of the effect of the construction of the medium on filtrate flow. From this base, it may then be possible to identify the causes of malfunction (e.g. blinding) when the cloth is used.

The flow through the clean fabric can be described in terms of the pressure differential ΔP, flow rate (dV/dt) and filter area A by:

$$\left(\frac{1}{A}\right)\frac{dV}{dt} = v = \frac{\Delta P}{\mu R_m} \qquad (4.3)$$

where v is the filtrate velocity.

Alternatively, many published reports refer to the *permeability* of the medium, $B\ \mathrm{m}^{-2}$, defined by:

$$v = B\frac{\Delta P}{\mu L} \qquad (4.4)$$

in which L is the thickness, or depth of the medium. The latter equation has its origins in the early work of Darcy and his studies of flow through thick layers of sand (Chapter 2).

The permeability of the clean medium has importance in determining power requirements, e.g. fan size in gas filter stations, and in deciding the initial flow rate of fluid through the medium. The initial flow rate has an effect on the cake structure of particles deposited near the medium, and, in upward filtering systems influences the size of particles deposited.

Where successful cake filtration is obtained, the medium resistance will generally be a low percentage ($< 10\%$) of the mean filter cake resistance. Another way of expressing this principle is that the medium should represent an extra "equivalent cake thickness" in the range $0.02 < L < 0.15$.

Perhaps the most important criterion of performance relates to the permeability of the used medium. Failure to release solids after an initial deposition also has serious economic consequences, except in those cases where disposable filter elements are intentionally used. In these cases, fluids containing extremely small amounts of solids are processed, the particles are trapped internally and no attempt is made to clean the element. An important characteristic of such elements is their solids-holding capacity.

4.7.2 Particle-Stopping Power

The particle-stopping power of the medium is, of course, of prime importance in deciding the course of a successful filtration. Much work has been reported in the literature which is aimed at the description of the medium in terms of an "equivalent pore size", which can be related to the particle size in the deposit. Media efficiency tests are made using dilute suspensions of particles; the concentration of particles in the fluid before and after passing through the medium is measured and attention is given to proper sampling techniques and particle dispersion. The latter is of great importance since particle concentration in the fluid has a great influence in determining the probability of bridging a pore and producing a sieve-like filtration mechanism. As was pointed out above, since most of the process difficulties encountered in practice ensue when the sieve-like mechanism breaks down, specification of media pore-size is of fundamental importance. The effect of particle size and concentration on the probability of bridging a pore of a certain size is discussed below.

The size and shape of the pore in the medium will determine the feasibility of a complete separation by sieving, particularly with media of the edge, perforated plate, simple wire or monofilament type filters. Where random fibres, sintered or porous elements, or staple fibre cloth are used, the pore size of the medium has less significance or use in predicting media behaviour. In simple woven cloth; the projected "square" opening is directly calculable from mesh counts and the diameter of wire, and such data are used to predict the smallest spherical particle which can be retained on the mesh. Such microscopic count methods are attractive, because of their simplicity and have been compared with "pore diameters" measured by more complicated techniques such as (a) bubble-point tests and (b) permeability tests. In (a) a sample of filter medium is submerged in a wetting liquid and the air pressure necessary to force air through the fabric is measured. For a circular-shaped pore, the radius of the latter may be calculated from:

$$r_{bp} = \frac{\gamma \cos\theta}{2\Delta P} \tag{4.5}$$

where r_{bp} is the bubble-point radius, γ is the surface tension of the fluid, ΔP the applied pressure and θ the contact angle.

The appearance of the first bubbles, at a measured pressure differential can thus be related to an equivalent pore size. Where a range of pore sizes exists, continued increases in ΔP, with associated increases in gas flow, can be measured in porometers. These automated analysers print out histograms of the pore size and frequency; such tests are continued up to the point of dry gas flow through the medium, see Figure 6.24.

Bubble point tests are used extensively in determining the largest and mean pore, in new and recycled filters. Simultaneously, the integrity of the filter can also be established [Johnston, 1986] in detecting wear and tear of the filter fabric.

The pore radius r may also be inferred from permeability tests and the use of Equation (4.6):

$$ r = \left(\frac{K_o B}{\varepsilon} \right)^{\frac{1}{2}} \tag{4.6}$$

where K_o is the Kozeny Constant and ε is the porosity of the medium.

Experimental determinations of r, r_{bp} and r_c (pore radius by microscopic count) have established [Rushton & Green, 1968] the following simple relations for woven wire of monofilament cloths:

$$ r_c = 1.26 \, r \tag{4.7}$$

$$ r_{bp} = 1.58 \, r \tag{4.8}$$

No such relationships are available for multifilament cloths since the permeability is not accounted for by inspection or bubble-test. In the latter the larger inter-yarn pores are measured. Calculations on the multifilament yarns [Rushton & Griffiths, 1987] show that, in such cases, the poresize within the fibre r_f is generally smaller than the pore size between the yarns and $r_f < r < r_y$.

The above tests are "non–destructive" since the filter medium is unchanged after completion of the test. Challenge tests, involving the filtration of particles of known size are "destructive" since after testing, it is virtually impossible to recover the original, unused characteristics of the medium. These challenge tests are considered in the next section.

4.7.3 Nonwoven, Random Fibre Media

The problem of correlating the permeability of filter media with the basic dimensions of the materials composing the septum has received attention [Davies, 1952]. Basically, the problem is to relate B, as defined in Equation 4.4 to variables such as fibre diameter, weave construction, etc. Whilst the permeability of random system is not the principal interest here, it is important to record some work reported in this area, in view of the growing application of nonwovens in the area of pressure filtration.

Perhaps the best-known equation for describing the required relationship in packed beds of particles or fibres is the Kozeny–Carman equation in terms of the porosity of ε and the specific surface S_v of the medium:

$$\frac{\mu v L}{\Delta P} = B = \frac{1}{K_o S_v^2} \frac{\varepsilon^3}{(1-\varepsilon)^2} \qquad (4.9)$$

The so-called Kozeny Constant, K_o has been shown to be a variable [Johnson,1986, p.156] which is dependent on the porosity of the deposit; rapid increases in K_o when $\varepsilon>0.7$ have been reported. Fibrous structures of wool, cotton, rayon, glass and steel wool have received attention and semi–empirical equations of the form:

$$B = \frac{d_f^2}{70\,\varepsilon\,(1-\varepsilon)^{1.5}(1+52\,(1-\varepsilon)^{1.5})} \qquad 1 \quad (4.10)$$

have been reported in terms of the fibre diameter d_f and the porosity ε [Davies, 1952]. Information on the permeability and porosity can be used to estimate the fibre diameter. In turn, d_f can be used in theroretical estimates of filter pad efficiency in particle retention [Stinson, 1990].

For felted materials and air flow, the equation:

$$\Delta P = k^* \mu\, v \qquad (4.11)$$

has been recommended, where $k^* = 4.29 \times 10^6 \times W_c$, and W_c is the cloth weight in grams per square centimetre (wool, rayon and cotton) [Cunningham, 1954].

Fibrous filters are often used in clarification processes, e.g. in hydraulic systems, in the electronics industry to ensure the absolute clarity of rinse water or in the beverage industry prior to bottling.

Stringent specifications are made on the retentivity of the filter media for particles above a certain size; other requirements are an absence of fibre shedding and a high solids holding capacity at a specified upper pressure drop level. The range of materials used in the construction of filter cartridges include paper, felts, nonwoven polymerics, sintered metal powders, metal fibres, etc. Generally the pressure drop-solids deposit characteristics of these media are widely different; those designed to avoid surface deposition and to allow entry of the particles into the depth of the filter exhibit the best capacity.

Filter media efficiency tests are made using dilute suspensions of "standard" particles. A wide variety of test powders is available, in fine and coarse grades. Industrial users will tend to use a test mixture typical of the products of interest, e.g. yeast in the brewing industry. In the gaseous field, both solid (sub-micron sodium chloride crystals) and liquid particles (di-octyl phthalate dispersion) are used.

In view of the random nature of the pores in nonwoven materials, or filters containing packed spheres, care has to be exercised in the interpretation of the filter rating provided by the supplier. Ratings may be absolute or nominal. The former term is relatively easy to understand as the particle size above which the medium is declared to be 100 percent efficient in particle retention. In nominally rated media, specifications are of the type: "removal of 90-95% of particulates larger than the nominal pore rating". These statements are intended as a guide to the user, rather than a guarantee, as in absolute ratings. Spherical latex particles can be generated in various narrow size ranges and are popular in test work [Bentley and Lloyd, 1992]. Glass spheres can be used in "absolute" filter trials aimed at determining the largest pore size in the medium. Traditionally, irregular particles of known size distribution, have been used in liquid tests. These powders are known as air cleaner test dusts and are available as fine (ACFTD) and coarse (ACCTD) powders. In the pharmaceutical industry, organisms such as Pseudomonas Diminuta are used in absolute trials of filter media and membranes [Meltzer, 1987] down to the 0.2 µm level.

Tests are performed in equipment shown in Figure 4.10 which depicts the arrangement of a multipass test. Suspensions of particles are circulated through the filter with measurements being made of the particle concentration before and after the filter. The ratio of these concentrations is recorded at a particular size x:

$$\beta_x \text{ Ratio} = \frac{\text{Concentration of particles size } x \text{ upstream}}{\text{Concentration of particles size } x \text{ downstream}}$$

Obviously a high β ratio indicates a high recovery of particles at size x. The cumulative efficiency of capture, E, of particles larger than a certain size can be calculated from:

$$E\% = [(\beta-1)/\beta]\times100$$

This relationship leads to the numerical values relating $E\%$ and β:

β	2	10	100	1000	5000
$E\%$	50	90	99	99.9	99.98

These figures prompt the suggestion [Williams and Edyvean, 1995] that the β ratio concept is of use only in absolute filter applications. At lower efficiencies, the simple particle removal efficiency $E_x = (1/\beta)$ is preferred.

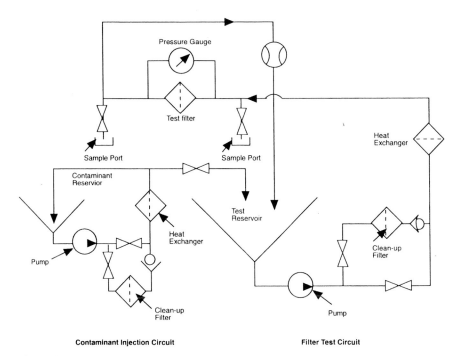

Figure 4.10 Multi-pass test flowsheet

Details of the multipass test and the simpler single pass test are reported in the literature (a) on cartridges [Williams and Edyvean, 1995] and (b) for porous metallic materials [de Bruyne, 1979].

Another method of recording the retention efficiency of particles in a "log reduction value" (LRV), which is given by LRV $= \log_{10} (N_1/N_o)$. Here N_1 and N_o are the concentration of particles at the inlet and outlet of the filter [Bentley and Lloyd, 1992]. The latter authors also discuss the difficulties relating non-destructive tests such as the bubble point value to destructive efficiency tests using particle suspensions. These tests are obviously a subject of controversy and development. The use of single point descriptions of performance has been rejected in favour of a complete profile in specific test conditions. This may be compared with recommendations [Verdegan, 1992] which involve specifying the particle micron levels at which $\beta = 2$, 20 and 75 respectively.

These β values correspond to $E\%$ values of 50, 95, 98.7 % respectively. The dirt holding capacity of depth filters at pressure drops eight times the clean ΔP is usually quoted along with the efficiency information.

The capture of suspended particles during the passage through a fibrous filter depends on processes of interception diffusion or Brownian motion, and surface forms of adhesion/attraction. In liquid environments, mathematical models for the prediction of filter efficiency usually ignore diffusional processes [Stinson, 1990]. However, the strong influence of adsorption in clean water systems has been observed in the removal of small positively charged particles of latex and bacteria [Raistrick, 1980].

4.7.4 Woven Media

4.7.4.1 Multifilament Cloth Permeability

In multifilament cloths, fluid flow may occur through or around the permeable yarns. The degree of this flow division inter-yarn or intra-yarn has been shown to explain certain dyeing characteristics of such cloths [McGregor, 1965]. If we define B_o as the permeability of the porous yarns, and B_1 as the permeability of the cloth if the yarn were solid, i.e. monofilament, it may be shown that:

$$\Omega = \frac{B}{B_1} = \left(1 + 1.80\left(\frac{B_o}{B_1}\right) + 2.68\left(\frac{B_o}{B_1}\right)^{1/2} \text{ for } \frac{B_o}{d_y^2} < 0.0017 \right) \tag{4.12}$$

where B is the overall permeability of the cloth and d_y is the yarn diameter. The Ω index has been shown to vary in the range $1 < \Omega < 20$ within the order of accuracy of the experimental measurements necessary for the determination of B and B_o. A monofilament cloth will have a coefficient of unity since:

$$\Omega = \frac{\text{Permeability of cloth}}{\text{Permeability of cloth composed of monofilament yarn}} \tag{4.13}$$

A high Ω value points to a large percentage of flow passing through the yarn.

These equations were used in an extensive study of the effect of multifilament characteristics in filtration, including such aspects as tendency to bleed, filtration mode (Chapter 2), [Rushton & Hassan, 1980]. The basis of this work was the premise that where particles are small enough to enter cloth pores <u>and</u> in the circumstances where yarn flow is possible ($\Omega > 1$), the possibility follows that such particles will follow the filtrate flow <u>into</u> the yarns. These particles become embedded within the yarns; this tends to diminish yarn in porosity with W approaching unity. Removal of the particles by back-washing becomes difficult, since back-wash will tend to follow the path of least resistance, i.e. around the plugged yarns.

The problem of correlation of the permeability of multifilament materials is aggravated in media made from natural materials such as wool, cotton, etc. In these cases, the smooth character of the so-called continuous filament (CF) is replaced by the hairy-random staple fibre (SF). Mathematical description of such systems is difficult, although much progress is being made by means of computer simulations [Davies et al, 1995].

The division of flow in multifilaments, with associated increase in cloth density (with tight weaving) is shown below in Figure 4.11. Typical multifilament media properties [Rushton and Griffiths, 1987] are recorded in Table 4.5 below

Table 4.5 Typical Multifilament media properties [Rushton & Griffiths, 1987]

Cloth description	$B \times 10^9$ cm	Ω	d_p, μm	d_{py}, μm	d_{pf}, μm
Polyester Plain 1/1; 100% CF	101.8	1.24	20.0	22	
Polyester Twill 2/2; 100% SF	13.7	1.34	7.2	9.8	0.8
Nylon mixture Twill 2/2; 88% SF	80.0	2.0	17.0	21.0	4.2
Polyester Twill 2/2; 100% SF	64.2	2.66	14.6	19.6	4.4
Polyproylene Plain 1/1; 100% CF	16.25	16.0	7.8		
Nylon Plain 1/1; 100% CF	6.5	16.6	5.0		
Nylon Plain 1/1; 100% CF	0.92	52.2	2.2		
Polyproylene Plain 1/1; 100% CF	1.73	4 550	3.0		
Polyester	10.2	10 400	6.0		
Nylon mixture Twill 2/2; 88% SF	9.1	20	6.0		

Media were supplied by P&S Filtration Ltd., Haslingden, U.K.; CF = Continuous Filament; SF = Staple Fibre

4.7.4.2 Monofilament Cloth Permeability

In the monofilament area much more success in correlating permeability with cloth structure has followed the suggestions of Pedersen [1969], who adopted orifice-type formulae to correlate pressure-drop-flow information for various weave patterns.

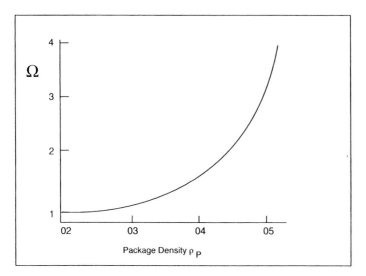

Figure 4.11 Effect of Package Density on the Division of Flow Through a Cross–Wound, Cotton Yarn Package

A discharge coefficient was defined as:

$$C_D = \left(\frac{v^2}{2\Delta P} \frac{(1 - a^2)}{a^2} \right)^{0.5} \tag{4.14}$$

where a, the effective fraction open area of the pore is:

$$a = A_o \, (ec) \, (pc) \tag{4.15}$$

in which

(ec) = warp yarns per centimetre
(pc) = weft yarns per centimetre
A_o = effective area of orifice

The discharge coefficient was anticipated to be a function of the Reynolds number within the fabric.

Figure 4.12 depicts the four warp yarn configurations which are possible in a single-layer monofilament fabric. Pederson tested the analysis by comparison with air permeability data on plain and 2/2 twill fabrics; a successful correlation was obtained, producing better results than those based on the projected pore size.

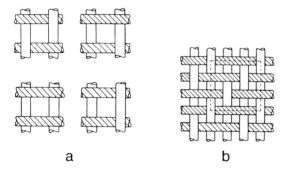

Figure 4.12 Basic configurations in monofilament weaves
(a) Basic pores: plain; twill; plain (type I); plain (type II); (b) Twill (2/1): repeating cloth cell

In later work [Rushton & Griffiths, 1971] the more complicated 2/1 twills and 5/1 sateens were analysed and flow data extended to water flow. The results are depicted in Figure 4.13A. Use of the simple projected diameter approach (pore viewed from above) gives the poor results presented in Figure 4.13B. The flow data in the range $1 < Re < 10$ has been shown to be represented by:

$$C_D = 0.17 \, (Re)^{0.41} \tag{4.16}$$

for water flow, in plain and twill fabrics, with a maximum error of $\pm$ 18%. In the above equation Re = pore flow Reynolds number = $(v_p \, d_p \, \rho)/\mu$.

In order to correlate the 2/1 twill and 5/1 sateen data it is necessary to define a "flow cell", which is repeated in the pattern of the cloth. In the case depicted (2/1 twill) the cell is made up to 6 twill pores and 3 plain pores; a weighting procedures is proposed so that:

$$a = \frac{1}{3} a_T + \frac{1}{3} a_p \tag{4.17}$$

where a_T, a_p are the effective fraction open areas of the twill and plain pores. The orifice perimeter calculation is treated similarly. Computed monofilament pore characteristics are reported in Table 4.6. These results can be applied to the analysis of any flow cell

Anlauf and Muller [1990] report the effect of monofilament pore size and shape on the cake formation process, and also on the passage of air through the system during dewatering. In the latter, the advantages of small filter medium pores are demonstrated.

Table 4.6 Monofilament cloth properties

Cloth weave	Nominal aperture μm	ec, cm^{-1}	pc, cm^{-1}	d_1, cm×10^2	d_2, cm×10^2	B, cm^2×10^7	W, cm×10	a, ×10	A, cm^2×10^5	d_p, μm
Plain	188	29.5	29.5	1.53	1.53	62.4	1.54	3.51	40.3	105
Plain	155	40.2	40.2	1.05	1.05	42.0	1.11	3.77	23.3	84
Plain	100	57.5	57.5	0.75	0.75	14.9	0.78	3.79	11.5	59
Plain	75	75.6	75.6	0.61	0.61	12.4	0.60	3.74	6.1	40
Plain	60	97.2	97.2	0.44	0.44	7.9	0.46	3.76	3.9	33
Plain	41	126	126	0.37	0.37	3.6	0.37	3.36	2.1	22
Plain	25	185	185	0.30	0.30	1.2	0.27	2.77	81	12
5/1 S		35.1	15.8	2.72	2.66	132.0	2.53	2.81	50.7	80
2/2 T		7.9	17.8	2.67	2.52	91.9	2.43	3.16	63.6	104
2/1 T		40.0	17.3	1.96	1.93	115.2	1.78	3.37	48.7	109
2/1 T		44.1	18.5	2.11	1.70	59.6	2.45	3.24	39.1	65

S = Sateen; T = Twill; d_p: Calculated pore size from $d_p = 4A/W$; ec,pc = end count and pick count respectively.
d_1,d_2 = Weft and Warp diameters, respectively
The above cloths were supplied by P & S Filtration Ltd, Haslingden U.K.

4.7.5 Filter Cloth Pore Bridging

Whilst the experimental information contained in this section refers to woven media, it is considered that the various relationships reported will also find use in the quantification of the bridging of surface pores in non woven fabrics. The formation of a stable particle deposit on the surface of a fabric (cake filtration) will follow if the first layer of solids arriving can effectively "bridge" the cloth pores. Early work by Hixson et al [1926], using fine capilliaries, pointed to a relationship between the pore size, d_p, and the particle size, d, for bridging with a highly concentrated suspension of quartz:

$$d_p = K_1 (d)^{0.25}; s > 20\% \text{ w/w} \tag{4.18}$$

A high value of K_1 indicated a good bridging characteristic as demonstrated in later work [Rushton & Hassan, 1980] on a range of solids. Here the concentration effect was extended to extreme dilution with the result:

$$d_p/d = K_2 s^i \tag{4.19}$$

Table 4.7 below contains values of K_1, K_2 and i for filter aids, etc. A graphical representation of this form of equation is given in Figure 4.14. Bridging conditions are obtained below the curve; lowering the concentration results in bleeding, at constant pore size. Materials with poor bridging characteristics can be improved by the body addition of substances with high K values, e.g. filter aids.

Table 4.7 Bridging characteristics

	K_1	K_2	i
Kieselguhr	500	21	0.26
Calcium carbonate	438	10	1.04
Magnesium carbonate	249	13	0.35
Sand	175		

The graph in Figure 4.14 illustrates that the Hixson relationship refers to the upper limiting pore size, above which bridging is impossible, for the particular average particle $\bar{d}$ irrespective of particle concentration. Again, it must be noted that Hixson used single

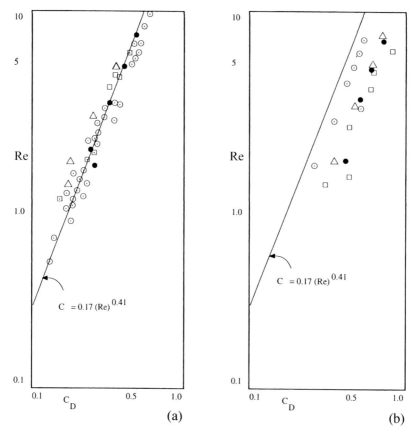

Figure 4.13 A) Water flow through several plain- and twill-weave monofilament fabrics.
    ○, Plain weave (1/1); ●, 5/1 Sateen weave; □, 2/1 Twill weave, Δ, 2/2 Twill weave
    B) Deviations produced by the use of projected open area weave;
    ○, Plain weave (1/1); ●, 5/1 Sateen weave; □, 2/1 Twill weave, Δ, 2/2 Twill weave

capillaries in his work; single-pore bridging concentrations may be as high as 35% solids, whereas cloths usually bridge in the range 0.01–2%.

Failure to form a bridge on the surface of a cloth will lead to pore penetration and possible bleeding of particles through the septum. Factors which complicate this problem include flow rate and flow division, effective pore sizes, shape, particle size distribution and particle type. On the latter point, certain particles are composed of agglomerates of fine precipitates which in areas of high fluid shear can become detached from the particle, scouring through the filter cake and medium, giving a cloudy filtrate. Equally, particles which deform readily can produce unusual results when viewed in terms of a pore–particle size ratio. Earlier work suggested that flow division, as measured by the Ω ratio, had an influence on the resistance of media in the used condition and on the process of cleaning the cloth by laundering.

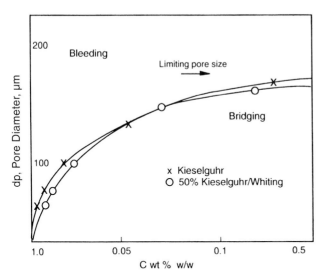

Figure 4.14 Bridging concentration as a function of pore diameter

Along with particle concentration, the filtration velocity has an influence on the bridging process [Rushton & Hassan, 1980]. Simultaneous arrival of sufficient particles to form a bridge will be influenced by the fluid drag on the particles, as they approach the filter medium. Experimental measurements have been reported in terms of the particle flux ϕ calculated from the product of the particle concentration and approach velocity:

$$(d_p/\bar{d}) = K_p \, \phi^\eta \tag{4.20}$$

Values for K_p and η are reported in Table 4.8 below for depositions on plain pore monofilament cloths. These are calculated using nominal and minimum values of the cloth pore size required for bridging. In equations (4.19) and (4.20) the "constants" K_3 and K_p can be considered a measure of the "bridging effectiveness" of the particle/pore combination; in this respect K_3 or K_p represents the pore/particle size ratio for bridging at unit concentration/particle flux respectively.

The variations in K_p and η are also attributed to particle size distribution parameters; σ (standard deviation) and SF (skewness factor) are recorded in Table 4.8 below. Data reported in Figure 4.15 on $MgCO_3$, using a multifilament cloth shows that the smaller particles in sample A ($x = 29 \, \mu m$) not only bridged at a lower concentration than sample B ($x = 51 \, \mu m$) but also less materials was lost through the cloth by bleeds. Apparently the views of Hixson [1926] who stated that the probability of bridge formation is related to the number of large particles in a particular population find support in these limited experiments.

In addition to relative size, velocity, etc. the type of pore will also have an important bearing on the bridging process. In the plain pores reported above all the pores are the same shape and are bounded by four yarns which are at least partly in the same plane, Figure 4.12. This feature does not apply to for cloths woven in other pore orientations; in the twill and plain pore type 2, three yarns (two warp and one weft) are coplanar whereas the plain pore type has only two warp yarns similarly situated. These factors have been considered [Rushton & Hassan, 1980] in defining a "bridging pore" for complex weaves as a cell which has four boundary yarns in the plane first contacted by the impinging solids.

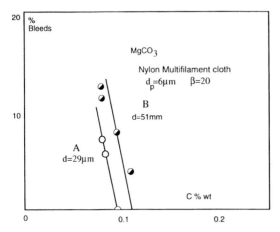

Figure 4.15 Bleeds % and Concentration

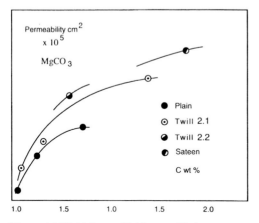

Figure 4.16 Bridging on Multiweave Cloth

As may be noted there appears no real advantage in using the minimum pore, in view of the widespread K_p values. The latter may also be viewed as a "bridging effectiveness factor" representing the pore to particle ratio with unit flux.

The available data, whilst insufficient to provide a complete calculus of the phenomenon, do point to the important variables. In particular the fluid velocity has a clear bearing in that increased velocities lead to lower bridging concentrations; this influence reflects the enhanced concentration of particles at the pore by fluid drag on the particles and the overall result depends upon the relative magnitudes of fluid drag and particle inertia. This type of behaviour is inherent in the separation of particulates on membranes of the Nuclepore variety.

Table 4.8 Effect of particle flux on bridging: monofilament cloths

Substance	Density, kg m^{-3}	Mean Particle size, μm	SD, μm	SF	K_p minimum pore	K_p nominal pore	η
Kieselguhr	2200	12	6	+0.6	86	159	0.30
Dicalite special speedflow	2300	15	13	+1.2	25	47	0.24
Dialite Speedex	2300	25	15	+1.0	18	30	0.26
Calcium carbonate	2630	2	0.3	+2.8	61	103	0.24

Experimental data on the bridging 2/1, 2/2 twills and 5.1 sateen cloth are compared with plain weave monofilaments in Figure 4.16. In the above table, σ and SF are the Standard deviation and skewness factor associated with the particle size distribution.

4.7.6 Bridge Failure and Particle Bleeding

When the process conditions are such that bridging of the surface pores of the medium is impossible, particles may be deposited within the interstices of the cloth or, particularly for monofilament, bleed through the pores. In multifilament or nonwoven media, internal deposition in the cloth may prove to be permanent, despite attempts to clean the filter by laundering, back-flushing, etc.

The appearance of solids in the filtrate may be persistent for coarse cloths and fine particles, but usually bleeds are observed only for a short period of time at the start of the separation. In some circumstances the presence of a small quantity of solids in the filtrate may not be serious and the advantages attaching to a low-resistance filter medium may outweigh the cost of bleeds; the latter can also be recovered by recycling the fore-runs. In other circumstances penetration of fine material into the cloth pores is associated with a sudden or gradual increase in medium resistance which cannot be alleviated. Eventually in these circumstances the blinded medium must be replaced. Cloth replacement constitutes an important economic problem in the filtration industry and it is of obvious interest to have information relating to the onset of cloth blinding.

From the information collected in the previous section, the important variables affecting bridging on well-defined monofilament pores were the pore–particle size ratio and the flux of particles towards the pore.

4.7.7 Flow Resistance of Used Media

Hatschek [1926] published a study of the throttling of the free area in a medium, following the deposition of solids. The effect is most pronounced in the deposition of the first layer of particles and this early work demonstrated that the structure of subsequent deposits will be influenced by the initial throttling process. The resistance of a cloth plus a thin layer of cake is much greater than would be expected from the sum of the resistance of the clean cloth and the resistance of the layer.

Later studies [Rushton & Hassan,1980] involved repeated filtrations of dilute suspensions of fine particles on cloths of various types. Between each filtration, the cloths were back-washed thoroughly, then checked for permeability to pure water. It was demonstrated that the Ω ratio appeared to influence the change in permeability with use. The results below refer to the change in permeability after fifteen filtration–cleaning cycles.

Ω	1	16	52	4550
$\dfrac{\text{Final permeability}}{\text{Initial permeability}}$	0.91	0.57	0.44	0.032

The data collected suggest that for low concentrations and where the particle is small enough to enter the interfibre pores, enhancing the flow through the yarns by tight weaving will promote cloth blinding. The higher relative flow rates encountered with the low-Ω cloth will produce better performance in this respect; this conclusion was also reached by McGregor [1965] with reference to the dyeing of twisted yarns. In multifilament or nonwoven cloths, the extent of bleeding occurs over wider concentration ranges, e.g. in felted material where particles are retained in the random pores created by tangled fibres. The latter process has been described by Davies [1984] in a series of computer studies and simulations of flow/particle deposition in media.

Thus in Figure 4.17, the computer model of a fibrous pad represents a good image of the real filter. These models are then bombarded with computer generated distributions of particles and the blinding rate predicted from the loss of available flow area [Davies et al,1995].

Quantitative and reproducible measurements are made difficult by the fact that the degree of pore plugging is related to the shape of the particle; the latter tends to orientate itself in the flowing fluid so that it represents minimum resistance to fluid drag force.

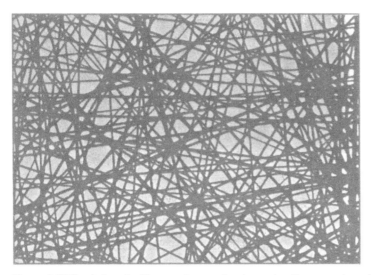

Figure 4.17 Simulation of a fibre membrane using the random line network model [Davis, 1984]

Grace [1958] described the filtrate flow–time relationship in "standard law conditions" in terms of the filter medium properties; for constant pressure and constant filtrate rate (see Section 2.7):

Constant Pressure

$$\frac{t}{V} = \frac{Ct}{\pi N (1- \varepsilon_p) A h r_0^2} - \frac{8h \mu}{\pi N r_0^4 A \Delta P} \qquad (4.21a)$$

Constant Flow Rate

$$\left(\frac{\Delta P}{\Delta P_o}\right)^{0.5} = 1 - \frac{CV}{\pi N (1- \varepsilon_p) A h r_0^2} \qquad (4.21b)$$

Here ε_p is the porosity of the packed particles, h the pore length and r_0 the clean pore radius. The group $(\pi N h r_0^2)$ was termed the "clogging" value and was related to the pore size distribution of the medium. It was freely admitted that such quantifications would find rare application to real separations, where particle depositions will occur mainly in mixed mode, particularly at the start of the separation. Thus Heertjes [1957] demonstrated the change in mode with change in particle concentration as shown below in Figure 4.18

Dmitrieva and Pakshver [1951] relate the tendency to blind to the sedimentation velocity of the suspension. A parameter $\Theta = 18 \, \Delta P/R_m \, x^2 \, (\rho_s-\rho) \, g$ was suggested for classification of the blinding tendency of particles of size x in a medium of resistance R_m. Pore plugging occurs for Θ values greater than 1000; cake filtration ensues for $\Theta < 100$.

Between these limits $100 < \Theta < 1000$ an intermediate type of deposition takes place. Kehat [1967] developed the relationship below for a partly plugged medium of total resistance R_T:

$$R_T = K_a R_o - R_b - R_c \quad \text{(complete blocking)} \tag{4.22a}$$

$$R_T = I R_o \qquad \text{(standard law)} \tag{4.22b}$$

In the above relationships: R_o is the effective resistance at the start of the cycle, R_b the reduction in R_o caused by bleeding or scouring, R_c the reduction in R_o due to bridging. The "constants" I and K_a depend on the particles and filtration conditions.

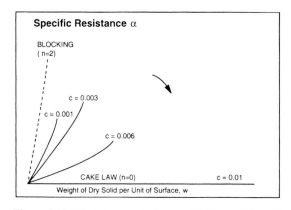

Figure 4.18 Effect of concentration on filtration and deposition mechanisms

Experimental studies of many systems [Rushton & Wakeman, 1977] have demonstrated that the resistance of a cloth to liquids is at least equal to the air resistance. In some cases an increase R_o is recorded, after allowance for viscosity, up to values of twice the gas flow resistance. This effect is attributed to poor wetting of the medium in some liquid systems. Use of the medium in filtering solids which tend to "surface filter" produces increases in R_m up to thirty times the clean water resistance. This value can increase materially $(100 \rightarrow \infty)$ if the particles are penetrating. The associated decrease in permeability may be gradual over many uses or dramatic, in a single use of the medium.

4.8 References

Allenius, H., 1999, Filtech'99, Dusseldorf, Filtration Society, Horsham, UK.
Anlauf, H. and Muller H.R., 1990, Vth World Congress Proceedings, Nice, 2, p 211.
Barlow,G., 1983, Shirley Institute Conference, S49, Manchester; p 91.
Bentley, J.M.B. and Lloyd, P.J.; 1992, Filtrat. Separat., 29, p 333.

Bosley, R., et al, 1977, Batch Filters, Solid-Liquid Separation Equipment Scale-up, 2nd Ed.

Carleton, A. and Heywood, N., 1983, Filtrat. Separat. 20, p 357.

Cunningham, C.E., et al, 1954, Ind. Eng. Chem., 46, p 1196.

Davies, C. N., 1952, Proc. IMechE., B, p 185.

Davies, G. A. D., 1984, Solid-Liquid Separation, Ellis-Horwood, Chichester, Chap 22.

Davies G. A. D., Bell, D. I. and Jackson, N. M.,1995, Sep. Sci. and Tech.30, p 1529.

de Bruyne,R., 1979 Second World Filtration Congress, UK Filtration Society, London.

Dickey, G.B., 1961, Filtration, Reinhold, New York.

Dmitrieva, T. F., 1951, Khim Prom, 11, p 20.

English, R.R. and Radford B.c., 1977, Filtrat. Separat. 14, p 492.

Fletcher, E.; 1980, Intl. Symposium: Filter Media, Flemish Filtration Society, Bruges, Belgium, 233 EFCE Event.

Fletcher,E., 1982, I,Ch.E., (UK), NW Branch Symposium No, 3,4, pp 1–7.

Grace, H. P., 1958, AIChE J., 2, p 316.

Grove, M. and Daveloose, F., 1982, ibid, p 424.

Hardman, E., 1994, Filtrat. Separat., 31, p 813.

Harrison, J., 1975, Shirley Institute Publication 519, Shirley Institute, Manchester.

Hatschek, E., 1928, J. Soc. Chem. Ind., 27, p 528.

Heertjes, P., 1957, Chem. Eng. Sci., 6, p 269.

Hixson, A., et al, 1926, Am. I. Min. Met. Eng., 73, p 225.

Hongxiang, G., 1991, Filtech Conference, Karlsruhe, Germany (UK Filtration Society), p 239.

Johns, F.E., 1991, ibid, p 183.

Johnston, P., 1986, Fluid Filtration Liquids Volume II, ASTM Special Technical Publication 975, Publication Code 04–975002–39, p 59.

Johnston, P., 1986, ibid, p 156.

Kehat, E., 1967, Ind. Eng. Chem. Proc. Dev., 6, p 48.

Kovacs, J.P., 1960, Chemical Engineering.

Loff, L., 1981, Chap 17, Solid-Liquid Separation, 2nd Ed, Svarovsky, L., Butterworth (Ed).

Meltzer, T.H., 1987, Filtration in the Pharmaceutical Industry.

McGregor, R., 1965, J. Soc. Dyers. Colour., 81, p 429.

Muller, H., et al, 1983, Filtrat. Separat, 20, p 134.

Nemeth, N., 1979, Chap 1, Filtration Principles and Practices, Pt 1 I, Dekker.

Pederson, G. C., 1969, AIChE 64th Ann. Meeting, New Orleans.

Purchas, D. B., 1967, Industrial Filtration of Liquids, Leonard Hill, London.

Purchas, D. B., 1981, Solid-Liquid Separation Technology, Chap 3, Uplands Press, Croydon.

Purchas, D.B., 1998, Handbook of Filter Media, Elsevier Advanced Technology, Kiddlington, Oxford, U.K.

Raistrick, J.H. 1980, International Symposium, Filter Media, Flemish Filtration Society, Bruges, Belgium, 233rd EFCE Event.

Regan, J. M., 1977, Filtrat. Separat. 14, p 485.

Reid, D. A., 1974, Filtrat. Separat. 11, p 293.

Rumpf, H., 1977, Particle Adhesion, in Agglomeration 77, K. Sastry, (Ed) AIME, New York.

Rushton, A. and Green, D., 1968, Filtrat. Separat., 5, p 213.

Rushton, A. and Griffiths P., 1971, Trans IChE, 49, p 49.

Rushton, A. and Griffiths P., 1987, Filtration Principles and Particles, 2nd Edition, Chapter 4, Marcel Decker, New York.

Rushton, A. and Metcalfe, M., 1973, Filtrat. Separat., 10, p 398.

Rushton, A., and Hassan, I., 1980, Filter Media, Intl. Symposium, Bruges, Belgium, KVIV. Royal Society of Flemish Engineers.

Rushton, A. and Wakeman, R.J. 1977, Powder Bulk Solids Tech, 1, p 58.

Schloemer, H.J., Filtech'99, Dusseldorf, Filtration Society, Horsham, U.K.

Shoemaker, W., 1975, Filtrat. Separat., 12, p 61.

Shubert, H., 1977, Chem. Ing. Tech., 48, p 716.

Smith, E. G., 1951, Chem. Eng. Progr., 47, p 545.

Stinson, J.A., 1990, Vth World Fitration Congress, Nice; France, French Filtration Society. Proceedings, 2, p 222.

Technical Textiles International, 1992, p 7, Elsevier.

Tiller, F., 1985, Chap. 2, Mathematical Models and Design Methods in Solid Liquid Separation, A. Rushton (Ed), NATO ASI Series E: Applied Science NO 88.

Verdegan, B.M., et al, 1992, Filtrat. Separat., 29, p 327.

Ward, A. S., 1972, Filtration Soc. Symposium, Filter Media, Manchester.

Williams, C. and Edyvean, R., 1995, Filtrat. Separat., 32, p 157.

Young, I., 1991, ibid., 28, p 145.

4.9 Nomenclature

A	Area	m^2
B	Medium permeability	m^2
c	Dry mass of solids per unit filtrate volume	$kg\, m^{-3}$
F	Adhesion force	N
g	Gravitational constant	ms^{-2}
L	Thickness of medium	m
K	Kozeny constant	$-$
N	Number of pores per unit area	m^{-2}
ΔP	Pressure Differential	$N\, m^{-2}$
R_m	Medium Resistance	m^{-1}
S_v	Specific surface area per unit volume	$m^2\, m^{-3}$
s	Solids concentration	$-$
t	Time	s
t_c	Cycle time	s
V	Volume of filtrate	m^3
V_f	Volume of filtrate per cycle	m^3

V_R	Filtrate flow rate per unit area	m s^{-1}
v	Filtrate velocity	m s^{-1}
w	Mass of solids deposited per unit area	kg m^{-2}
w_R	Mass of filtered solids per unit area per unit time: $W_R = cV_R$	kg m^{-2} s^{-1}

Greek Symbols

α	Specific cake resistance	m kg^{-1}
ρ, ρ_S	Fluid & solid density, respectively	kg m^{-3}
β	Filtration performance ratio	–
ε	Porosity	-
γ	Surface tension	N m^{-1}
μ	Liquid viscosity	Pa s
θ, δ	Contact angle	
Ω	Cloth permeability ratio defined in Equation (4.13)	

5 Pretreatment Techniques

5.1 Introduction

Some solid–liquid systems defy easy separation by the straightforward methods described elsewhere; the combination of particle sizes, densities, surface properties, solids concentration and liquid viscosity give an extremely stable suspension which will not respond to "normal" methods of treatment. To solve this difficulty the process engineer has to change the behaviour of the system so that it becomes easier to separate the two phases. Occasionally the liquid properties of viscosity, density and surface tension may need to be modified in order to achieve major improvements to the solid–liquid separation process.

Liquid viscosity control can be effected in two ways.

1) By adjustment of temperature. For highly viscous or non-Newtonian liquids small temperature changes can produce large differences in viscosity. Even with aqueous suspensions heating, perhaps by using some low grade waste heat, may enable filtration to take place at an elevated temperature which can improve filtration rates significantly.

2) By dilution with a less viscous liquid. Reeve [1947] describes the technique used in oil dewaxing processes and shows that an optimum exists in that the gain in increased flow rate must not be lost in an excessive total volume.

Liquid density, although only little changed by temperature, can be reduced by mixing with a miscible liquid of lower density. Solid density may be affected by "ageing" or by chemical change. Coagulation and flocculation can alter the effective density of the particle by causing the formation of loosely bound units which behave as aggregates.

Surface tension is affected by temperature and to a considerable extent by the presence of surface active agents. In some operations such as the dewatering of coal fines it is common to add a surface active agent which produces a significant reduction in the moisture content of the dewatered cake. However it is more usual to need to change the properties of the solids and there are a variety of possible ways of doing so. Very often simply by allowing the suspension to stand for some length of time, ageing can be effective. In some systems ageing allows crystal growth, or adsorption of liquid or indeed disintegration, which can all produce a change in particle size.

An examination of an upstream process may be worthwhile in that a small but valuable change in particle or liquid properties might be possible at little cost. Processes such as precipitation, polymerisation or crystallisation are amongst those that would merit reconsideration if the output suspension were not readily separated.

Several methods have been successful in specific applications. Freezing and thawing has been used with sewage sludge, water treatment sludge and some radioactive sludges [Doe, 1948; Doe et al 1965]. Slow freezing is recommended. The process causes the particles to form aggregates which often contain ice crystals.

Subjecting the suspension to ultrasonic vibrations has been used, but conflicting results have been reported, not only by users of the technique but by workers investigating the fundamental aspects [Asai & Sasaki, 1958; Thompson & Vilbrandt, 1954; Lyon, 1951; Husmann, 1952; Toshima, Umezawa & Katow, 1961]. It has been suggested that ultrasonic vibrations give the particles the energy to surpass the potential barrier to flocculation. At a particular level of energy input cavitation will take place and the particles in the cavities are subjected to large forces causing them to break down or disintegrate.

The effective particle size distribution can be changed by classifying the suspension and feeding the coarse fraction to the filter ahead of the fine fraction. By doing this the finer material, which is the more difficult to filter is captured by depth and surface filtration mechanisms on the coarse fraction, which acts as a precoat.

Two methods which have found a much wider application involve either adding chemicals to aggregate the particles (coagulation and flocculation) and adding other powders (filter aids) to alter the effective particle size and size distribution of the solids to increase the permeability and porosity of the cake. These techniques merit a detailed consideration but it must be remembered that they do involve the adulteration of the solid–liquid system with extraneous material and this may not be acceptable in the process or product.

5.2 Coagulation and Flocculation

In the behaviour of many suspensions it has been observed that the particles show such stability in the liquid that they exhibit little if any settling rate at all despite the existence of a density difference between the phases. This problem of colloid stability has bedevilled many industries, especially those of beer and winemaking, for centuries and the existence of many tricks of the trade such as the use of finings in brewing and acidifying the suspension in clay treatment suggested to chemists that significant physicochemical changes could be induced in the suspension by relatively minor additions of additives.

There are two practical ways of bringing about changes in the state of aggregation of a suspension of particles and, although there has been synonymous use of the words in the past, it is now generally accepted that they be distinguished by the terms "coagulation" and "flocculation". Coagulation is reserved for those phenomena brought about by reducing the zeta potential of a particle suspended in an electrolyte by changing the nature and concentrations of the ions present. Flocculation is reserved for those processes where certain types of long-chain polymer or polyelectrolytes cause the particles to aggregate by forming bridges between them. Although in any actual pretreatment process both of these effects may be induced by the same reagent, it is usually clear which of them is dominant. The terms coagulation and flocculation have been used indiscriminately in the literature of colloid and surface chemistry, probably because flocculation in the sense of our definition has received much less attention than coagulation which has been the subject of much study for over a century. The terms coagulation and flocculation are also used in a different sense in the water treatment industry, coagulation being used there to describe the addition of reagents to the water to induce the reaction, and flocculation to describe the subsequent slow agitation during which floc growth occurs.

Treatment of a suspension with the necessary chemical will produce a faster settling system with a more bulky and more permeable sediment, but the final liquid content of the sludge or filter cake may well be significantly higher than if a nontreated suspension had been settled. The benefits therefore tend to be found in the enhanced settling rate and increased permeability, and sometimes the sediment is of such bulk as to prove troublesome and expensive in disposal. Other solid–liquid separation operations such as dissolved air flotation and centrifugation have different requirements that must be met by the pretreatment system.

Coagulation. Coagulation is the adhesion of particles by forces of molecular and atomic origin and the presence or absence of coagulation depends upon the balance between the attractive van der Waals' forces and the repulsive electrical double layer forces. These forces can be described by the DLVO theory, devised by Derjaguin and Landau [1941] and Verwey and Overbeek [1948].

Small particles suspended in a liquid are subjected to a random bombardment by the molecules of the liquid which gives rise to the irregular movement of the particles called Brownian motion. Such behaviour may bring some particles into close enough proximity to allow the attractive surface forces to bind them together into small agglomerates. However if the surfaces of the particles are charged electrically in some way the ensuing repulsive forces between the particles may be sufficient to prevent the spontaneous agglomeration brought about by the Brownian motion. The phenomenon is shown in Figure 5.1 in terms of the potential energy against interparticle distance as found on the surface of the particle. In part Figure 5.1A curve (a) is the van der Waals' energy which is an attractive force having an increasingly negative value at small interparticle distances, curve (b) is the repulsive electrical force acting in the opposite sense and curve (c) is the resultant of the two, showing as a maximum the energy barrier which results in the system becoming stabilised.

The added chemicals are intended to alter the surface charge to allow the agglomeration to take place. The effect of the coagulant reagents, shown in Figure 5.1B, is to reduce the level of curve (b) thus bringing the resultant curve (c) below the zero value of potential energy and thus enabling the particles to coagulate if they should become close enough.

Coagulation arising solely from Brownian motion is termed perikinetic coagulation and the basic rate equation for this is due to von Smoluchowski [1916] and developed by Camp and Stein [1943], who assumed that the particles will coagulate if their potential energy of interaction is suitable at the time of collision and that the rate of collision will be governed by their Brownian diffusion rate. The reduction in particle number concentration for mono-sized spherical particles is given by the equation:

$$-\frac{dN}{dt} = 4\pi DxN^2 \tag{5.1}$$

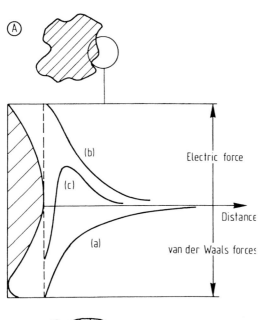

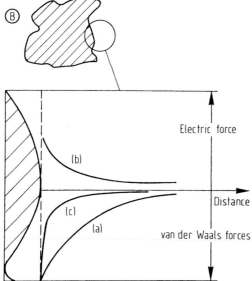

Figure 5.1 Potential energy curves: A) Stabilised particles; B) Coagulated particles

where D is the Brownian diffusion coefficient of the particles, N the particle concentration and x the diameter of the particles.

This equation can be integrated to give N, the number of particles present at time t, in terms of N_o, the number present initially as:

$$N = \frac{N_o}{1 + 4\pi DxN_o t} \tag{5.2}$$

When the number of particles has been halved, i.e. $N = N_o/2$ the corresponding time is:

$$t_{0.5} = \frac{1}{4\pi DxN_o} \tag{5.3}$$

The particle size term may be eliminated by substituting for the diffusion coefficient from the Stokes–Einstein equation:

$$D = \frac{K_B T}{3\pi x \mu} \tag{5.4}$$

where T is the absolute temperature, μ is the liquid viscosity and K_B is Boltzmann's constant (1.3805×10^{-23} J/K), giving:

$$t_{0.5} = \frac{3\mu}{4K_B TN_o} \tag{5.5}$$

A typical calculation for water at 15°C with particles at a number concentration of $10^5/m^3$ produces a value of 2.15×10^{12} s, which shows that this spontaneous process is very slow! However shear gradients set up by the return flow of the liquid that is displaced by settling particles or by convection in the liquid or by gentle agitation improve the rate enormously.

In the presence of sedimentation and shear gradients, which soon occur as the coagulation process develops, significant changes in coagulation rate are observed and this type of coagulation is known as orthokinetic. As particle sizes in a system grow due to perikinetic flocculation, particles begin to sediment and the coagulation rate will increase because of the resulting change to orthokinetic coagulation.

Coagulator Design. At low concentrations of particles orthokinetic coagulation can be induced by introducing shear via a stirrer system in a series of tanks. The shear rate in these mixers is an important parameter because large shear rates will have a deflocculating effect. The design of such a coagulator is based on the mean shear rate G' which can be readily developed for a Newtonian liquid or a power-law liquid.

Consider a small element of liquid (Figure 5.2) with dimensions δL, δy, δr which is subject to a shear stress τ. In a rotational system torsional work is done in shearing this cube and the power involved is the product of the torque and the angular velocity. For a small angular rotation $\delta\theta$ which produces a displacement δL at a radius r, $\delta\theta = \delta L/r$. The angular velocity is defined as $d\theta/dt$ and:

$$\frac{d\theta}{dt} = \frac{dL}{rdt} = \frac{v}{r}$$

which is the velocity gradient or shear rate. (v is the linear speed of the motion).

Hence the power P is given by:

$$\text{Torque} \times \text{ velocity gradient } = \tau\,\delta L\,\delta y\,\delta r\,\frac{dv}{dr}$$

But the volume V is $\delta L\,\delta y\,\delta r$ so: $\dfrac{P}{V} = \tau\dfrac{dv}{dr}$

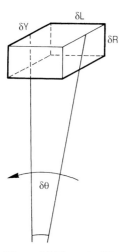

Figure 5.2 Element of liquid in rotational shear

A Newtonian fluid is defined as one in which the shear stress is linearly proportional to the shear rate, i.e.:

$$\tau = \mu\frac{dv}{dr}$$

so:

$$\frac{P}{V} = \mu\left(\frac{dv}{dr}\right)^2$$

But G' is defined as dv/dr hence:

$$G' = \left(\frac{P}{\mu V} \right)^{1/2} \tag{5.6}$$

where P is the power expended in agitating the fluid, V is the volume of the coagulator and μ the liquid viscosity. The product of G' and the mean residence time, $\bar{t}$ in the coagulator can be used to scale up coagulator designs and to compare different types.

The chosen value of G' is usually in the range 10–100 s^{-1} and the product, $G'\bar{t}$, between 10 000 and 100 000. It is better to use a lower shear rate with a higher residence time since flocs may be irreparably damaged by excessively high shear rates.

The choice of residence time and shear rate may affect the type of floc produced by the coagulator. Low values of shear rate and long times are likely to produce large, light, weak, "fluffy" flocs whilst the opposite combination will give smaller, denser, stronger agglomerates that are more likely to resist the shear forces present in some separation systems such as dissolved air flotation and in sedimenting centrifuges. Staged coagulators in which several mixers are used can be very effective. In the initial sections a high G', short $\bar{t}$ design is used to give a small dense floc which is coagulated again in a later part of the device, using a G' value of about half the initial one and a longer $\bar{t}$, to give a large denser floc which will have a high settling rate.

Various types of mixers can be used; in-line blenders, baffled continuous stirred tanks with propellor impellers, tanks with air spargers (pneumatic coagulators) and baffled tanks with flat blade slatted paddle mixers have all been used. The latter is commonly found in the water treatment industry and Peavy et al [1985] give some design rules for this version.

The power input P is calculated from the product of the paddle tip velocity relative to the liquid v_p and the drag force on the paddles D, which is given by:

$$D = C_D A_p \rho \frac{v_p^2}{2} \text{ and } P = C_D A_p \rho \frac{v_p^3}{2}, \tag{5.7}$$

where A_p is the area of the paddles, ρ is the liquid density and C_D is a drag coefficient, given as 1.8 for flat blades. The paddle area is the total area of the slats that is perpendicular to the cylinder of rotation and this area should not exceed 40% of the total area swept by the paddle. The value of v_p is about 75% of the actual linear speed of the paddle tip. The latter should be less than 1 m/s and a minimum distance of 0.3 m should be preserved between the paddle tips and all other obstructions in the vessel such as baffles and feed pipes. These precautions are to guard against the generation of high local shear rates.

Desired values for the average shear rate and the product $G'\bar{t}$ are chosen—these may be based on the results of laboratory or pilot plant tests—which lead to a value for the mean residence time and hence the tank volume. The latter is divided into the number of stages required and the respective values of G' assigned to each stage. The power P in each stage is calculated from Equation 5.6. Some practical values need to be assigned to the tank dimensions at this point and it is suggested that the flow area should be square in an attempt to minimise the spread of residence times. Next a paddle configuration is proposed to fit the tank dimensions, but with the paddle width w as an unknown. The latter is now calculated through the area A_p from the power and the assumed paddle velocity using Equation (5.7). The outcome is tank dimensions, paddle configurations, speeds and power consumptions.

Coagulation Reagents. The cause of colloid stability is the magnitude of the ionic repulsive forces that are found at the interface between the particle and the liquid. The behaviour may be explained by the double layer theory due to Stern [1924]. The surface of a particle suspended in an electrolyte solution has ions present and these attract ions of opposing charge from the liquid whilst repelling ions of like charge. Thus an electrical double layer of ions develops and concentration and potential gradients are established around the particle. The inner part of the double layer is composed of ions more strongly bound and the outer layer is weaker and more diffuse. The electrical potential exhibited by the particle is considered to be at the boundary between the two layers and is called the zeta potential.

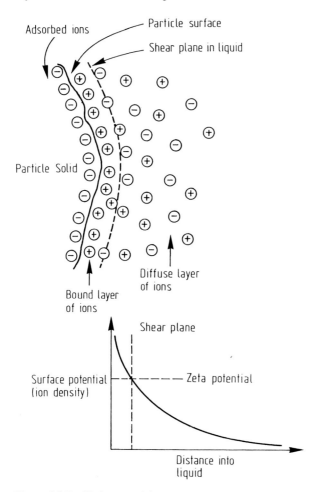

Figure 5.3 Double layer model

The ions of the coagulating chemical have to reduce the zeta potential and this is most effectively achieved by multivalent ions which compress the double layer owing to their greater charge concentration so that most of the charge between the double layer is brought within the plane of shear between the bound and diffuse layers.

The Schulze–Hardy rule describes the overall effect and it states that the minimum ionic concentration necessary to produce coagulation is approximately proportional to the sixth power of the reciprocal counter ion charge. The ions commonly used are Al^{3+}, Fe^{3+} and Ca^{2+} which are found in chemicals that are readily available. Aluminium is usually obtained as aluminium sulphate and supplied as either a solid or solution, or sodium aluminate, which is a soluble powder. Iron salts are cheaper and may be more easily obtained than aluminium salts, being the byproducts of many processes, e.g. pickle liquors and TiO_2 manufacture. Fe^{2+} is obtainable as ferrous sulphate (copperas) although it is often used as chlorinated copperas. Ferric ions can be obtained from ferric chloride, which may be hazardous to handle, or ferric sulphate. Most iron salts are corrosive and savings on the cost of materials may well be offset by higher initial plant and maintenance costs. Calcium in the form of limestone or slaked lime is mainly used in conjunction with other coagulants in water treatment processes, but has the disadvantage of a low solubility.

The major use of coagulating chemicals is the production of potable water, where the task is usually to remove very small concentrations of suspended solids and organic materials such as humic acids which impart odour, taste, and colour to the water and because of the low concentrations of particles present, the initial coagulation rates may be slow. However, if they are used at the correct pH, the common coagulating salts produce insoluble hydrolysis products which increase the particle concentration and improve the coagulation rate.

There are optimum pH conditions for the hydrolysis of these salts; for example, Al^{3+} forms soluble aluminates at high pH and will not hydrolyse at all below a critical pH and ion concentration and thus the economic use of any particular salt depends on the optimisation of pH level and reagent requirement. The optimum levels of pH and chemical concentration are found experimentally in the laboratory or pilot plant and must be controlled in the process environment.

Some additional side benefits of such treatment of water are in the adsorption of undesirable organic materials, anions such as phosphate, and heavy metal ions by the unstable hydrolysis intermediate products of the coagulant chemicals. These properties are also useful in physicochemical waste aqueous treatment processes, in which biological oxidation is replaced by a combination of coagulation, solids removal by conventional solid–liquid separation processes followed by adsorption to remove soluble components. Two less desirable side effects of the use of hydrolysing coagulation salts are that the hydrolysis products add substantially to the bulk of the dried sludge produced and that there is always the presence in the liquid of the counter-ion both of which may be unacceptable economically or environmentally.

The value of zeta potential may be measured [Akers and Ward, 1977] and the results used to select the best reagent and its optimum dosage for a particular application, but these aims may be readily achieved by the simpler tests described below. It is important not only to ascertain the optimum dose, but to arrange that this level is maintained during operation because overdosing produces a poorer coagulation effect and increases costs. In some applications the presence of hydrophilic colloids adsorbed onto the surface of the particles may prevent coagulation in situations where it might otherwise be expected and alternative approaches may have to be explored.

Flocculation. The mechanisms involved in flocculation are quite different from those in coagulation. The flocculant chemical is a long-chain polymeric molecule and some examples have been known for centuries, e.g. gelatin, isinglass, tannins, guar gum, but recently synthetic polyelectrolytes have displaced these natural materials owing to their more economical dosage rates and their ability to give more durable flocs. The technology of synthetic flocculant manufacture has become very sophisticated and reagents may now be "designed" to give the desired behaviour when applied to a particular problem.

The overall flocculation mechanism is thought to involve a molecular bridge or series of bridges between particles. The polymer chain is adsorbed from the solution on to one particle and when another particle comes within close enough range the extended polymer chain is adsorbed on to it. This elementary floc grows by bridging with other particles until an optimum floc size is formed. The flocculation reaction is irreversible, unlike the coagulation reaction, so an equilibrium size is not attained and high rates of shear need to be avoided since they will easily break the molecular bridges.

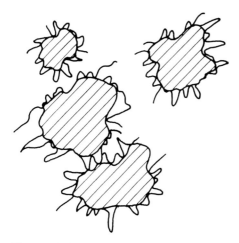

Figure 5.4 Flocculation model

Generally the coagulation effect produced by reducing zeta potential is not important in polyelectrolyte flocculation. This is demonstrated by the efficiency of the natural flocculants and the high efficiencies shown by some systems where the flocculant carries the incorrect electrical charge for the solid being flocculated. However charge effects may be more important for some very low molecular weight flocculants, where the polymer chain length is insufficient for bridging to occur to a significant extent, and they do play a part in the adsorption process for ionic flocculants.

The mechanism of flocculation may be considered as a sequence of reaction steps [Akers & Ward, 1977]. Firstly the flocculant is dispersed in the liquid phase, secondly the flocculant diffuses to the solid–liquid interface, then the chemical becomes adsorbed onto the solid surfaces, the coated particle collides with another particle, free polymer chains adsorb onto the second particle forming bridges and so on forming the multi-particle floc.

Practically all polymer is adsorbed at optimum dosage which is proportional to the solid surface area over a wide range of particle sizes and solution concentrations. It has been suggested that optimum flocculation occurs when half the area of solid is covered with polyelectrolyte. To simplify the argument this is the optimum coverage between the maximum number of extended polymer loops available to form bridges and the maximum uncovered area of solid available to permit the loops from other particles to adsorb and form bridges. Any detailed consideration of this point would involve making assumptions about the nature of the adsorption reaction, but it is purely academic as the dosage of polyelectrolyte required to bring this coverage about would be grossly uneconomic. This point is illustrated in Figure 5.5 where the economic optimum is to be found well to the left of the optimum determined by the maximum settling rate.

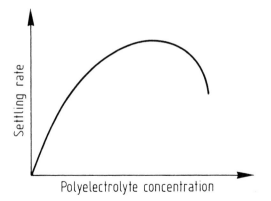

Figure 5.5 Optimum reagent dose

At high concentrations the degree of flocculation decreases, and this may reach such a point that the particles are completely protected by the presence of an adsorbed polymer layer. Thus overdosing can be a serious mistake in that it may create a well-stabilised suspension that is extremely difficult to separate.

It is necessary to agitate the suspension to effect the dispersal of the polymer and to increase the collision rate of the particles. Insufficient mixing will lead to excess flocculant adsorption on some of the solids, but excessive agitation will tend to rupture flocs and create haziness in the suspension due to the presence of colloidal material. This problem can be difficult to remove solely by reflocculation.

The effect of solution pH is complex. For some systems it has been shown that the optimum polymer dose decreases as the pH moves away from the isoelectric point, which is the pH corresponding to zero net charge. But in general it would be expected that the presence of high surface charge will inhibit the adsorption of a polyelectrolyte holding charge of the same sign and enhance that of a polymer bearing opposite charge. If the surface charge is very high it may inhibit the collisions required to produce the flocs but this effect would decrease with the increasing amount of polymer adsorbed thus lessening the need for a very close approach of the particles. Polymers with ionisable sites such as anionic and cationic flocculants will display an extended rigid polymer chain length if charges are present at these sites. Without charge effects the polymer molecules will tend to have randomly coiled chains.

As described earlier, the charge intensities on both a solid surface and on a polymer chain will also be affected by the concentrations of other ions present. Some ions, especially multivalent cations, will usually produce a decrease in the potential at the solid–liquid interface, which in turn will bring about a decrease in the mutual repulsion and hence an improvement in the efficiency of the flocculation reaction. The effect of salts on the overall reaction is also due to their contribution to the charges present and thus to the configuration of the polymer chain and to the interaction between the polymer chain and its solvent. Thus in some applications a two-stage coagulation–flocculation system may be necessary to achieve the floc size and strength required.

There is an optimum polymer molecular weight or chain length for efficient flocculation. If the molecular weight is too low the polymer chains will not be long enough for bridging to take place. If the molecular weight is too high the material becomes difficult to dissolve and disperse. For the polyacrylamide derivatives, the practical range of molecular weights is 10^5–10^6. Generally increasing molecular weight leads to increasing flocculation efficiency and floc strength, but at the expense of increased cost of polyelectrolyte owing to the greater quantities required.

Flocculant Reagents. Natural polymer flocculants have been known and used for centuries. Some examples are: isinglass used in beer fining, glues, gelatin and starches in the mineral industries, and alginates, from seaweed, used in water treatment. The market for these materials has been challenged by the rise of the flocculant chemical industry which has produced materials which are cheaper and more reliable .

Synthetic flocculant technology is now highly developed and uses several source monomers. Flocculants can be made which have a cationic, anionic or nonionic nature and a molecular weight range which is closely controlled to improve solubility and ease of use. Attempts have been made to improve natural flocculant properties by modifying their chemical structure. The earliest and most highly developed range of synthetic polyelectrolytes is based on polyacrylamide and its derivatives which may be made by the catalytic polymerization of either acrylamide or acrylonitrile to give a neutral polyelectrolyte or anionic and cationic derivatives.

The nonionic acrylamide flocculants are effective over a wider range of conditions and with greater variety of substrate materials than the anionic and cationic forms; however, the latter are more efficient where conditions of pH and the nature and amount of surface charge on the solid favours their use.

The residual monomer left in polyacrylamides is toxic, but may be reduced to an acceptable level by further processing. Special grades have been available especially for potable water treatment which are guaranteed to contain less than 0.025% of free acrylamide monomer and the use of these grades is likely to extend into environmentally sensitive applications.

Use and Application. Flocculant reagents are not readily dispersed in water and casual attempts to make up solutions are likely to produce a lumpy gelatinous mass that can only be dispersed with much time and effort. The chemicals are usually supplied as powders, but liquid and dispersion forms are available although they contain only about 50% active material compared with the solid form. The technique of dispersing the light fluffy powder is to wet it out slowly with water using a minimum amount of low-shear mixing and, having achieved a wetted paste, to add further liquid under low-shear mixing to produce the solution at the required concentration. Stock solutions made in this way are at up to 0.1% weight/volume and these solutions are diluted by about a tenth for use on the plant.

Addition of the flocculant solution to the process has to be made carefully. Although it is important to obtain a good distribution of the reagent it is vital that the long-chain molecules are not broken and the effectiveness of the chemical reduced. Low-shear mixing is required and this can be achieved in carefully designed mixing tanks as described above. Because of the low natural diffusion rates of these long-chain compounds addition of the reagent at several places is usually more effective than at a single point. Multipoint injection into a baffled open launder has been successful as has the injection of the chemical into the corners of a closed pipe ring dispersal system which uses the secondary flows at the changes of direction in the pipe to give the gentle mixing effect (see Figure 5.6).

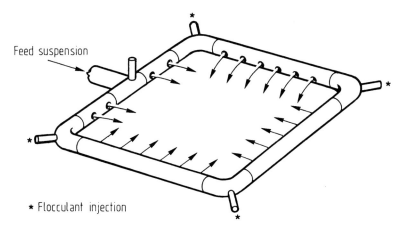

Figure 5.6 Multipoint injection system

In some applications it is usual to take the suspension from the mixing section to a further vessel where slow stirring can be maintained or the flocculation allowed to develop during flow along a pipe or launder; in others it is more effective to take the flocculating feed into a higher concentration of already-flocculated particles; whichever approach is used depends on the floc characteristics that are required for the subsequent solid–liquid separation process.

Effects of Surfactants. Many workers have observed effects in colloid systems, such as those involving clay minerals, that could not be explained fully by the classic DLVO theory and this has led to an extension of that theory to include polar surface forces. These forces are the result of electron acceptor-electron donor interactions, such as hydrogen bonding, which if sufficiently strong and asymmetrical ('monopolar'), produce an orientation of the water molecules adsorbed on the surface of the particles. The presence of oriented water molecules adsorbed onto these surfaces results in mutual repulsion and may prevent coagulation of the particles. Conversely, if these forces are reduced by some means or other, the particles may coagulate due to the attraction of the net van der Waals forces.

Such effects have been described variously as 'solvation' or 'hydration' effects, or due to 'hydrophobic forces' or 'hydration pressure'.

Van Oss, Giese and Constanzo [1990] investigated these effects using the clay mineral hectorite. When small (~1 micron) particles are suspended in sodium chloride solutions the DLVO theory predicts colloid stability at all salt concentrations but practical experience proves that the mineral coagulates at concentrations of 0.1M and greater. The DLVO calculations are illustrated in Figure 5.7(a) which indicates a clear energy barrier at small inter-particle distances. Including the polar forces in the prediction as shown in Figure 5.7(b) removes this energy barrier and gives negative energy values at NaCl concentrations of 0.1M and 1.0M, which is in accordance with the experimental observations.

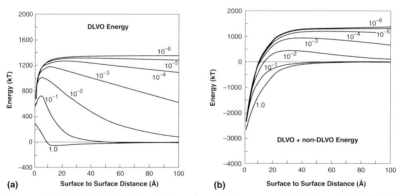

Figure 5.7 Hectorite in NaCl solutions; theoretical energy calculations (a) DLVO theory (b) DLVO + polar force.[Van Oss, Giese and Constanzo [1990]]

Obviously the presence of surface active agents and changes in solution pH can each be expected to alter this behaviour. Yoon and Ravishankar [1994, 1996] examined the hydrophobic forces between mica surfaces in dodecylammonium chloride (DAHCl) solutions at various solution pH values and in the presence of dodecanol.

The repulsive force F between two mica surfaces radius R was measured at a range of separation distances H using an atomic force balance. The effect of increasing concentrations of DAHCl in ultrapure water at pH 5.7 is illustrated in Figure 5.8(a) which shows that small concentrations of surfactant (10^{-5} M) can eliminate this force. In Figure 5.8(b), the effect of increasing pH from 5.7 through to 10.1 is a steady decline in value of the hydrophobic force. When a second surfactant, dodecanol, is present at low concentration (10^{-7} M) the reduction is more marked giving rise to an attractive force as shown in Figure 5.8(c).

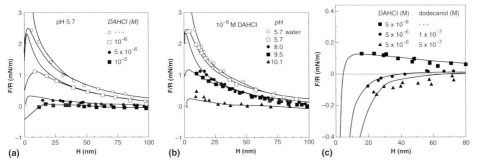

Figure 5.8 Mica in surfactant solutions; hydrophobic force measurements (a) DAHCl solutions (b) pH effects (c) DAHCl + dodecanol solutions. [Yoon and Ravishankar [1994, 1996]

The practical conclusion for the process engineer is that even trace concentrations of surfactants can have a significant effect on coagulation behaviour. An example of the application of surfactants and coagulants to dissolved air flotation is given in Chapter 7.

Laboratory Testing. In principle the best laboratory test for the suitability of a coagulant or flocculant reagent is one that most closely replicates the process conditions under which it will be used. However that statement assumes that the laboratory is equipped with the appropriate apparatus such a bomb filter, vacuum leaf filter, small scroll discharge centrifuge, deep-bed filter, air flotation rig, etc. which is not likely to be the case. It may have to be sufficient to make the appropriate measurements on laboratory equipment and to be prepared to make final adjustments to pH levels and dosage rates on the full-scale plant.

It is important to measure a parameter that is relevant to the process because although other measurements of behaviour will be related they may not have a close or linear relationship. Thus for a gravity sedimentation application it is obviously sensible to measure settling rate and for filtration it would be appropriate to examine permeability or specific cake resistance. In some processes such as sedimenting centrifugation and dissolved air flotation floc strength is of major importance and a large floc that settles quickly in a gravity test may not be robust enough to survive the shear fields in those processes.

The easiest and most convenient way of evaluating a range of coagulant or flocculant chemicals is to use a gravity settling test. The behaviour of a flocculated suspension is characteristically different from a non-flocculated one in that it quickly exhibits an interface or mud-line between the clear liquid and the settling suspension. Following the movement of that mud-line establishes the settling rate, as described more fully in Chapter 3. Although tests may be carried out in a variety of ways it is important that the technique used is as reproducible as possible. Graduated measuring cylinders of 1 litre or 500 ml capacity make useful sedimentation vessels and it is desirable that the cylinders be of equal diameter with parallel sides. They must be kept vertical because any inclination can enhance the settling rate. The cylinder is filled with suspension to the top mark and the appropriate quantity of flocculant added, usually as a 0.05% solution made up that day from a concentrated stock solution, and at the same time a timer is started and the contents of the cylinder are then thoroughly and slowly mixed. This may done either with an agitator such as a perforated metal disc attached to the end of a rod which is pushed up and down the cylinder, or by inverting the cylinder after closing with a stopper or the palm of the gloved hand. The mixing method must be thorough, rapid, and reproducible. After mixing, the cylinder is allowed to stand, and the height of the mud-line is noted at regular intervals until no further sedimentation occurs. The settling rate is calculated from the slope of the linear portion of the settling curve (Figure 5.7) and the volume of the settled sludge from the height of the final mud-line.

The choice of reagent is empirical and may depend as much on financial considerations as technical ones. As an example, the process economics may decide whether a relatively inefficient natural flocculant such as starch is used rather than a more efficient, but also more expensive polyacrylamide. If the feed stream to be treated is of very constant composition or if variables such as pH are not expected to change it may be more effective to use a specific flocculant. On the other hand, if conditions are more variable, a less specific but less efficient flocculant is probably better.

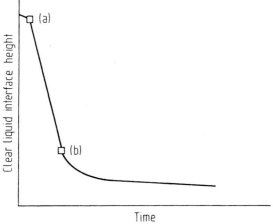

Figure 5.9 Settling rate curve

As a general rule amongst polyelectrolyte flocculants it is suggested that cationic grades are effective particularly with organic substrates or at low pH, such as in acid and aqueous leach metallurgical processes; anionic polymers are better for mineral or inorganic suspensions, especially under alkaline or neutral pH conditions and nonionics are most versatile with applications in a wide range of industries. The combination of an inorganic coagulant followed by an anionic polyelectrolyte may often give good results with more difficult problems.

A very useful test for filtration applications is the capillary suction time (CST) test [Baskerville & Gale, 1968], which was developed for the sewage industry to give plant operators a quick and effective method of establishing the optimum dose of reagent for filtration applications. It measures the permeability of a flocculated suspension by using the uniform capillary suction of a filter paper. A measured volume is taken and placed on the paper and the rate of movement of the liquid front between two radial electrical contacts is timed automatically to give the capillary suction time. For sewage sludges the product of the specific resistance of the filtered solids and the solids concentration of the suspension correlates with the CST to within about ± 30% The use of the technique and its application to rotary vacuum filter operation has been described by Gale [1972].

5.3 Filter Aids

Addition of another powdered insoluble solid, termed a filter aid, to the system can produce a significant improvement to a filtration operation. The filter aid may be used in two ways, either separately or in conjunction. The first of these methods is to precoat the filter medium with a layer of filter aid cake. Precoat aids must filter quickly without bleeding or penetrating through the cloth and must give a uniform thickness with a reproducible filtering surface. The suspension is then filtered onto the precoat by surface and depth filtration mechanisms. Thus surface properties are important in the choice of aid.

The second approach involves the admixture of the filter aid powder to the suspension. Such material is called "body aid". The object is to open up the pores of the filter cake and provide faster filtration. The size distribution may be made coarser. The object of "pretreating" a slurry prior to a separation process is to change the properties of the suspension so that a desired improvement results to the separation process. It is useful to consider the parameters that might be altered.

Voidage	depends on particle size and distribution
Specific surface	particle size and distribution
Specific resistance	particle size and distribution
Separating size and grade efficiency	particle size and distribution, solid and liquid density and liquid viscosity

Residual moisture   liquid surface tension, contact angle and
                                                  particle size and distribution

Slurry concentration   prethickening

Filter Aid Materials and their Properties. A successful filter aid must have a high permeability which can be obtained through a combination of particle size and shape distribution and, in some cases, because the particle itself is porous, which makes a significant contribution to the overall permeability. The surface properties of the filter aid may be a major factor in the successful applications of a filter aid in some processes. Other factors such as chemical resistance and compatability with the product specification are also important. Some of the major filter aid materials that are available commercially are described and discussed below.

Diatomite (kieselguhr or diatomaceous earth) is one of the most widely used filter aids especially in the commercial brewing of beer. It is mined by opencast methods from huge deposits of the skeletal remains of diatoms, which are a type of algae. There are several biologically distinguishable types of constituent particles, each with a different shape, some flat and cylindrical, others ellipsoidal and others fibrous or needle-like. The particles themselves have a high intrinsic permeability due to the skeletal structure and these features combine to give a powder of a very high overall permeability. Diatomite is refined by crushing, classifying, drying and calcining, which reduces the number of very fine particles and increases the permeability. Flux calcination in the presence of alkali gives an even narrower size distribution, but increases the soluble alkali metal content which may be undesirable. Diatomite consists mainly of silica so is soluble in both strong acids and strong alkalis. A wide range of grades is available and some typical specifications and properties are given in Tables 5.1 and 5.2.

In the brewing process diatomite is used as precoat and body aid on candle filters and horizontal leaf filters. Hermia & Brocheton [1993] give a comparison of these methods and cite typical precoat dosage as 1.5 kg/m^3 and body feed dosage as 1 kg/m^3 with a cake average specific resistance of 10^{11}m/kg. Other aspects of diatomite filtration are discussed comprehensively in a historical review by Cummins [1973].

Perlite is a glassy volcanic mineral, mainly aluminium silicate with some combined water, which expands on heating to give a highly porous particle. It is processed in a similar manner to diatomite, but in this case when heated to the softening point the occluded water in the perlite expands and the particles swell to produce hollow spherulites many times their original size. The crushing stage breaks down the spherulites and produces particles with a variety of irregular jagged shapes.The result is a powder with very high permeability and a low bulk density. A smaller range of grades are produced than is the case for diatomite. Perlite is used for rough filtrations where high flow rates are desirable with less significance given to filtrate clarity. Typical chemical analysis is given in Table 5.3. Note that expanded perlite has a higher alkali content than diatomite and is even more susceptible to acid and alkali attack.

Table 5.4 shows some particle size distributions for Turkish perlites measured by dry-sieving out the particles smaller than 90μm and using a Fritsch photosedimentograph on the fine fraction. Densities and bulk densities are also given.

Table 5.1 Properties of diatomites [Celite ®]

	Natural	Calcined	Flux-calcined
Relative permeability	1	3	20
Wet cake density, kg/m^3	260	280	280
Weight % finer than particle, size µm (laser sizer)			
196	100	100	100
128	100	99	99
64	93	87	91
32	78	75	67
16	54	52	36
8	27	30	16
4	10	12	7
2	4	4	3
1	2	2	2
Specific gravity	2.00	2.25	2.33
Median pore size, µm	1.5	2.5	7.0
pH	7.0	7.0	10.0

The very uneven distributions illustrate the effect of the complex shapes of perlite particles and indicate that particle size analysis is a poor method of characterisation for perlite. Most suppliers of perlite prefer to use permeability as a quality control standard for their products. Kalafatoglu et al [1994] report some measurements of specific resistance of incompressible cakes made by pressure filtration of borax solutions using these perlite grades as body aid at a ratio of 2 kg perlite per kilogram of insoluble solids. Some of these values are also given in Table 5.4. A reasonable direct correlation of specific resistance with bulk density was obtained.

Asbestos is a fibrous mineral with complex surface properties which make it particularly successful in many applications such as the removal of protein hazes from beverages. It was often used in mixtures with other filter aids made to form sheets, but contains soluble impurities of Fe, Mg and Ca. It is rarely used now because of health hazard problems.

Cotton cellulose is the "filtermass" used in brewing, but *wood cellulose* is also available. It is sometimes used as underlay for diatomite or perlite or as a blend with diatomite. Generally the fibres are about 20 µm diameter and 50–100 µm long, which will bridge wide openings in the filter medium. Cellulose is very pure, is combustible and soluble in solvents so that inert suspended solids may be recovered by either of these methods.

Table 5.2 Typical chemical analyses of diatomite filter aids [Celite [®]]

wt%	Natural	Calcined	Flux-calcined
SiO_2	85.8	91.1	89.6
Al_2O_3	3.8	4.0	4.0
Fe_2O_3	1.2	1.3	1.5
CaO	0.5	0.5	0.5
MgO	0.6	0.6	0.6
P_2O_5	0.2	0.2	0.2
TiO_2	0.2	0.2	0.2
$Na_2O + K_2O$	1.1	1.1	3.3
Ignition loss	3.6	0.5	3.3
Water	3.0	0.5	0.2

Unactivated carbon has the advantage of being chemically stable and can be used in both acid and alkaline conditions. It mainly used in alkaline systems where diatomite and perlite would be subject to chemical attack. A range of grades are available which compare roughly with the coarser grades of diatomite and perlite and generally it is more expensive than those materials.

Pulverised fuel ash, sawdust, spent lime and many other waste products can and have been used in some applications usually where a rough separation is required. Other specially prepared solid powder materials such as hydrated magnesium silicate, hydrated calcium silicate, bentonite, fuller's earth, or activated carbon are used in some applications principally because of their adsorptive properties but they may also act as a filter aid in the separation of the particulate solids from the liquid.

Filtration with Filter Aids. A wide range of pressure and vacuum cake filters can be used with both precoat and body aid filtration. Most of these operate in the batch mode and include the filter press, the vertical and horizontal leaf filters with sluice or centrifugal discharge and the vertical tubular filter. The rotary drum vacuum filter works in a semi-batch fashion in that the precoat is removed progressively as the fitration proceeds.

Table 5.3 Chemical analysis of a typical expanded perlite [Harborlite [®]]

	weight %
SiO	74.9
Al_2O_3	12.6
Fe_2O_3	0.8
CaO	0.6
MgO	0.1
P_2O_5	trace
TiO_2	0.10
Na_2O	4.6
K_2O	4.7
Ignition at 1000°C	1.3
Loss on drying at 105°C	0.3

Table 5.4 Particle size analysis of expanded perlite filter aids wt% [Kalafatoglu, 1994]

Grade	⏐90 μm	−90+40 μm	−40+30 μm	−30+20 μm	−20+10 μm	−10+5 μm	−5 μm	Specific gravity kg/m^3	Bulk Density, kg/m^3	Specific Resistance, m/kg $\times 10^{-9}$
Y.1				16	34	28	22	2074	254	8.39
Y.6			10	12	28	27	23	2193	258	6.51
S.0	23			9	22	24	22	2200	175	3.1
S.1	16				23	33	28	2178	202	2.85
S.2	25			14	27	22	12	2128	274	6.69

Batch Pressure Filtration. The general arrangement is shown in Figure 5.10 with two filter aid mixing tanks, one for precoat suspension and another for body aid suspension. Three pumps are shown; a precoat pump, a feed suspension pump and an injection pump for the body aid: but a single pump might double as the precoat/feed pump if the required characteristics are met perhaps by using a variable speed control.

With diatomite or perlite a coating of 1.5–3 mm in thickness is achieved by applying 5–10 kg of aid per 10 m^2 of filter surface area at a concentration of not less than about 0.5% wt/vol. It is important that the precoat concentration should be high enough to prevent excessive "bleeding" of the filter aid in the early stages of the precoat operation, but the concentration should be low enough for the suspension to have enough mobility to give a good distribution of precoat suspension in the filter. The flow rate has to be high enough to keep the solids in suspension which is a function of the density and viscosity of the liquid, but not sufficiently strong enough to erode the coating of filter aid as it is laid down. For aqueous suspensions about 7×10^{-4} to 14×10^{-4} m^3/s per m^2 of filter surface area is the usual range of flow rates.

The filtrate from precoating is recycled to the precoat tank and it will appear cloudy at first but clearing within 5 min or so. The precoating operation is continued until the prescribed amount of precoat has been laid down. At this point the body feed and suspension feed systems must be ready so that as the precoat flow is shut off the body feed and suspension feed flows are started simultaneously. The flow of liquid through the filter must be maintained through this changeover; otherwise there is a danger of the precoat slipping off the filter leaves. (In some filters it is possible to retain the precoat on the filter by means of air or vacuum whilst the filter is drained of precoat filtrate with the feed suspension and body aid filtration following under the appropriate pressure.)

Some experimentation is required to obtain the correct amount of body feed for a particular application. Figure 5.11 shows the dependence of overall throughput on the proportion of body feed. Too little may have a negative effect in that there is no improvement obtained in permeability of the cake to compensate for the increase in cake thickness due to the presence of the filter aid.

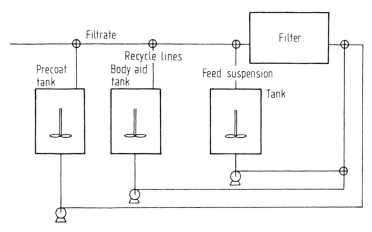

Figure 5.10 Filter aid filtration flowsheet

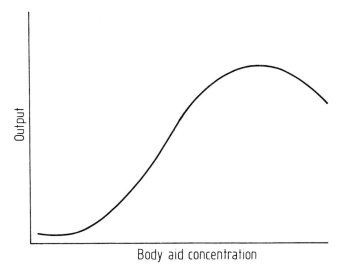

Figure 5.11 Throughput *vs* body feed concentration

As the proportion of body aid is increased the throughput improves rapidly and then goes over a peak in the graph which indicates an operational optimum. The behaviour of the filtration using different amounts of body aid can be followed from the rise in the pressure drop over the filter as the constant-rate filtration operation progresses. Figure 5.12 shows the pressure drop as a function of time.

The ideal is curve (a) which shows a steadily increasing rise in pressure with time. Curve (b) indicates a rapid build-up of pressure which reflects the high specific resistance of the cake with an inadequate proportion of filter aid and curve (c) shows a rapidly increasing pressure difference reflecting the rapid build-up of the cake to a maximum thickness before the end of the cycle.

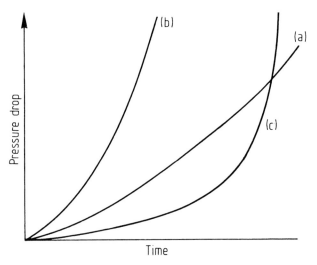

Figure 5.12 Pressure drop *vs* time

Rotary Drum Vacuum Filters. This filter can accommodate a large thickness of cake and can take a precoat of filter aid which is up to 150 mm in depth. The feed suspension is filtered onto the precoat and the filtered solids together with a thin layer of precoat is scraped off the cake surface by a horizontal knife situated just before the surface passes back into the feed suspension tank, thus exposing a fresh surface of precoat. The scraper knife advances automatically until it reaches a minimum distance from the cloth, usually 5–10 mm. At this point the cycle is finished and the filter is cleaned of used precoat.

This system is designed for fine slimy solids that would not be retained on a filter cloth. The capture mechanism is a combination of surface and depth filtration so some feed suspension particles will penetrate into the precoat layer. It is a matter of experimentation to find the optimum depth of precoat cut which will preserve a high filtration rate with a long overall cycle time.

5.4 References

Akers, R.J. and Ward, A.S., 1977 Chap. 2 in Filtration Principles and Practice, vol. 1, C. Orr, (ed.) Dekker, New York.

Asai, K. and Sasaki, N., 1958 3rd. Int. Coal. Prep. Congress Liege 1958 paper E.9 pp 518–525.

Baskerville, R.C. and Gale, R.S., 1968, J. Inst. Water Pollut. Contr., 2, p 3.

Camp, T.R. and Stein, P.C., 1943, J. Boston Soc. Civ. Eng., 30, p 219.

Cummins, A.B., 1973, Filtrat. and Separat. Jan/Feb p 53.
  idem, ibid, March/April, p 215.
  idem, ibid, May/June, p 317.

Derjaguin, B.V. and Landau, L., 1941, Acta Physicochim. USSR 14, p 633

Doe, P.W., 1948, J. Inst. Water Eng, 12, p 409.

Doe, P.W., Benn, D., Bays, L.R., 1965, J. Inst.Water Eng, 19, pp 251–291.

Gale, R.S., 1972, Filtrat. and Separat., 9, p 431.

Hermia, J. and Brocheton, S., 1993, Proc Filtech Conference, Karlsruhe, 21, Filtration Society, London.

Husmann, W., 1952, Gesundheitzung, 73, p 127.

Kalafatoglu, I.E., Ayok T. and Örs, N., 1994, Chem. Eng. Proc., 33, pp151–159.

Lyon, W.A., 1951, Sewage & Industr. Wastes, 23, pp 1084–1095.

van Oss, C.J., Giese, R.F. and Constanzo, P.M., 1990, Clays and Clay Minerals, 38, (2), pp151-159.

Peavy, H.S., Rowe, D.R. and Tchobanoglous, G.T., 1985, Environmental Engineering, McGraw-Hill, New York.

Reeve E.J., 1947, Ind. Eng. Chem. Industry, 39, (2) pp 203–206.

von Smoluchowski, M., 1916, Z. Phys., 17, pp 557–585.

Stern, O., 1924, Z. Electrochem, 30, p 508.

Thompson, D. and Vilbrandt, F.C., 1954, Ind. Eng. Chem. (Industry) 46, (6) pp 72–80.

Toyoshima, Y., Umezawa, M. and Katow, T., 1961, J. Kogyo Yosui 28, p 39–42.

Verwey, E.J.W. and Overbeek, J.Th.G., 1948, Theory of the Stability of Lyophobic Colloids, Elsevier, Amsterdam.

Yoon, R-H. and Ravishankar, S.A., 1994, J. Colloid Interface Sci., 166, pp 215-224
  idem, ibid, 1996, pp 39-402, 403-411.

5.5 Nomenclature

A_p	Projected area of paddles in coagulator	m^2
C_D	Blade drag coefficient	–
D	Particle diffusion coefficient	$m^2\,s^{-1}$
g	Gravitational constant	$m\,s^{-2}$
G'	Shear rate	s^{-1}
N	Number concentration of particles	m^{-3}
P	Power required to agitate fluid	$J\,s^{-1}$
r	Radial position	m
t	Time	s
V	Volume of coagulator	m^3
v	Tangential velocity	$m\,s^{-1}$
v_p	Paddle tip velocity relative to liquid	$m\,s^{-1}$
x	Particle diameter	m

Greek Symbols

μ	Liquid viscosity	Pa s
θ	Angle of rotation	
ρ	Fluid density	$kg\,m^{-3}$
τ	Shear stress	$N\,m^{-2}$

6 Clarifying Filtration

The usual objective of clarifying filtration is to separate solids at a very low concentration from a liquid stream. The liquid may be drinking (potable) water, wine, beer, oil, etc. and it is usually the liquid which is the valuable product. The techniques used in clarification processes include: deep-bed, precoat, candle and cartridge filtration all of which involve capture of particles inside the porous mass of the filter. Such techniques produce clearer filtrates than those obtained in clarification by sedimentation. The filtration techniques listed are often complementary; they are employed for similar duties, but usually operate over different conditions of feed flow rate, feed concentration and process economics. These operating conditions are summarised in Table 6.1.

Table 6.1 Comparison between clarifying filtration techniques

Technique	Typical face velocity	Volume filtrate before regeneration per unit filter area at $0.1g\ l^{-1}$ solids	Relative running cost per unit mass removed*	Best filtrate quality achievable*
	$m^3\ m^{-2}\ h^{-1}$	$m^3 m^{-2}$		
Deep bed	8	60	4	4
Precoat	50	1000	3	2
Candle	20	100	2	3
Cartridge	5	0.4	1	5
Screens	35	continuous	5	1

*A high value represents good performance or cost

The different techniques share similar capture mechanisms, and the reader should refer to Section 2.7 for an introduction to methods of correlating experimental data using semiempirically derived equations. The different capture mechanisms will be discussed in greater detail in the following sections and a more fundamental description of the techniques will be given where relevant.

6.1 Capture Mechanisms

All clarifying filtration techniques include at least one of the following mechanisms:

a) Straining
b) Sedimentation
c) Interception
d) Inertial impaction
e) Diffusion
f) Hydrodynamic interaction
g) Electrostatic interaction

The following mechanisms should also be considered in view of their importance in particle attachment/detachment mechanisms:

h) Electrical double layer repulsion
i) van der Waals forces

The various mechanisms are illustrated in Figure 6.1.

It is assumed that once a particle has touched the surface of the target grain it will remain attached to that surface, in the absence of unbalanced repulsive forces, and is effectively captured from the liquid flow. The particle may be made to detach later, during filter cleaning, but only after substantially altering the operating conditions. It may appear surprising that particles can be captured by means of straining within the porous network, rather than inertial interception; however, two important suspended particle parameters need to be considered: size and density. If the particle density is similar to that of the suspending fluid then inertial interception will not be important. In these circumstances, particle collection will be dominated by direct interception and straining. Small particles of low density will have little chance of capture by inertial or impaction means, but greater chance due to diffusion and hydrodynamic interactions. It should also be remembered that these mechanisms have been described separately, but it is likely that more than one mechanism acts in bringing the suspended particle into contact with the collecting grain.

In addition to the above mechanisms orthokinetic flocculation may be induced due to the shear produced inside the porous media. This may encourage entrapment as collection efficiency usually improves with increase in suspended particle size. The collection mechanisms are described in greater detail below.

Straining. This is the simplest of the collection mechanisms and occurs when the particle diameter is larger than the constriction through which the fluid flow streamlines pass. The grain size plays an important role in this mechanism as narrower passages are found with smaller grained collection media.

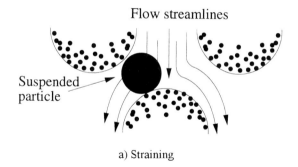

a) Straining

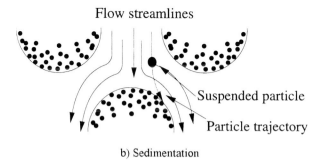

b) Sedimentation

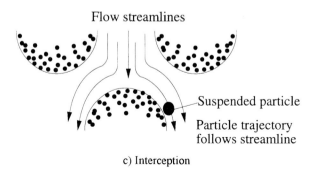

c) Interception

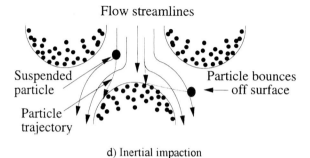

d) Inertial impaction

Figure 6.1a–d Schematic illustration of various collection mechanisms

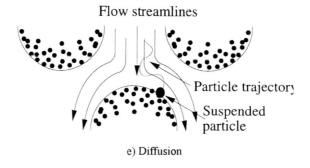

e) Diffusion

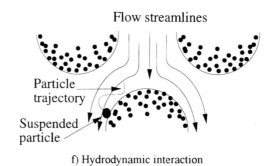

f) Hydrodynamic interaction

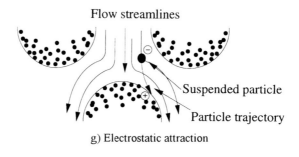

g) Electrostatic attraction

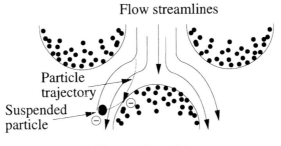

h) Electrostatic repulsion

Figure 6.1e–h Schematic illustration of various collection mechanisms

Sedimentation. When the fluid flow is directed downwards through a filter, gravitational sedimentation effects will cause particles to settle vertically through the flow streamlines, as the latter distort around the collector. The possibility of deposition can be characterised by a dimensionless number which is the ratio of the Stokes settling velocity and the approach velocity of the fluid:

$$\frac{x^2(\rho_s - \rho)g}{18\mu u}$$

A typical range for this parameter is 0–1.4, where zero refers to particles of equal density to the fluid.

Interception. If the suspended particle radius is greater than the distance between the flow streamline which contains the particle and the collecting media grain, then the suspended particle will contact the target, in the absence of any repulsive mechanisms.

Inertial Impaction and Bounce. Whenever fluid flow streamlines change direction it is possible that particles may be less able to alter direction because of their greater inertia. If a collection surface is nearby the particles may make contact with the target before becoming trapped in another flow streamline. The dimensionless term used to characterise the inertia of a system is called the Stokes number:

$$\frac{\rho_s x^2 u}{18\mu d_t}$$

where d_t is the characteristic linear dimension of the target (sand grain, filter aid, fibre, etc.). The Stokes number is derived from a force balance including particle weight, drag and inertia and should not be confused with the Stokes settling velocity which neglects inertia. Typical values for the Stokes number are 10^{-9} to 2×10^{-3}. The higher the Stokes number the greater the chance of inertial impaction. However, the particle may also bounce off a target at high Stokes numbers.

Diffusion. All particles suspended within a fluid are subject to bombardment by the surrounding fluid molecules. Small solid particles may acquire sufficient momentum from the fluid molecules to cause the solid particle to move. It will continue to do so until further bombardment causes the particle to change direction. This Brownian motion of the suspended particle may cause it to approach the surface of a target. It is generally accepted that particle diffusion is only applicable for particles of diameter less than 1 µm, and the process can be characterised by the dimensionless Peclet number:

$$Pe = \frac{xu}{D}$$

where D is the diffusion coefficient for the particle. The diffusion coefficient of molecular species can be calculated from the Stokes–Einstein equation:

$$D = \left(\frac{2RT}{6\pi\mu x N_A} \right)$$

where R is the Universal gas constant, N_A is Avogadro's constant and T is the Absolute temperature.

Typical values for the Peclet number are 10^5 to 10^8 for colloids. The lower the Peclet number, the greater the chance of diffusional collection.

Hydrodynamic Interaction. When considering fluid flow it is usual to apply a Reynolds number to distinguish the type of flow. In pipe flow the Reynolds number is:

$$Re = \frac{d\bar{v}\rho}{\mu}$$

where d is pipe diameter and $\bar{v}$ is average fluid velocity. A deep bed consists of many flow channels of differing sizes and it is common to find the Blake number used instead of the Reynolds number:

$$Bl = \frac{xU_o\rho}{\mu(1-\varepsilon)}$$

where x is the average particle diameter of the medium, U_o is the superficial liquid velocity and ε is the bed porosity. This approach is simplistic in its use of an average particle size representing the size-distributed bed. An improvement is to consider the "modified" Reynolds number (Re_1)

$$Re_1 = \frac{U_o\rho}{(1-\varepsilon)S_v\mu}$$

where S_v is the specific surface area per unit volume of the particle size distribution making up the bed. If the Sauter mean diameter, the particle size which has the same specific surface as the full particle size distribution, is used in the Blake equation then the value of the Blake number will be 6 times larger than the modified Reynolds number, for spherical particles. If any other "average" particle size is used in the Blake number equation then the proportionality between these numbers is lost.

If the modified Reynolds number is less than 2 the flow conditions within the bed are usually regarded as streamline. During cleaning there is considerable advantage in employing flow conditions with a modified Reynolds number greater than 2. However, in streamline flow, suspended particles may be seen to wander between flow streamlines even in the absence of the other forces described in this section. This is attributed to "inertial lift" and "tubular pinch" which can be described as an effect caused by a large particle covering several different flow streamlines. The different values of shear on the particle due to the difference in velocity of the streamlines causes the particle to rotate, which can lead to particle migration. The net effect on the particle is something similar to diffusion but with an enhanced diffusion coefficient over that described by the Stokes-Einstein equation.

Electrostatic Interaction. The Smoluchowski equation for electrophoresis (induced motion of a particle by an electrical field) is:

$$U_E = \frac{E\varepsilon_w\zeta_p}{\mu}$$

where U_E is the induced velocity, E is the electric field strength, ε_W is the electrical permittivity of water and ζ_p is the zeta potential of the particle.

Electrical Double Layer Repulsion. Double layer interaction theory has been extensively investigated independently by Deryagin and Landau, and Verwey and Overbeek, DVLO theory. There are many reviews on the application of DVLO theory to colloids [Shaw, 1989]. One expression for the repulsive energy between two small spheres is:

$$V_R = \frac{32\pi\varepsilon_w x_1 x_2 k^2 T\gamma_1\gamma_2}{(x_1 + x_2)e^2 z^2}\exp(-\kappa a)$$

where e is the charge on an electron, k is the Boltzmann constant, T is absolute temperature, z is counter ion charge number, κ is double layer thickness, a is the shortest distance between the particles, and:

$$\gamma = \frac{\exp(ze\psi_d/2kT) - 1}{\exp(ze\psi_d/2kT) + 1}$$

where Ψ_d is the relevant (particle 1 or 2) Stern potential.

Under certain circumstances the DVLO electrostatic repulsion energy can in fact become an attractive force, notably when the two Stern potentials are of opposite sign.

van der Waals and London Forces. van der Waals postulated the existence of an attractive force between neutral and chemically saturated molecules in order to explain nonideal gas behaviour. Such a force which is not dependent on ionic or electrical attraction is also assumed to exist for colloidal species. These universal attractive forces were first explained by London and are due to the fluctuating charge distribution of one particle inducing a polarised charge on the other. It is an extremely short-range force; inversely proportional to the sixth power of distance between particles but the forces are additive in a colloidal dispersion. Hamaker has provided the following theoretical description of the London dispersion interaction energy V_A in vacuo:

$$V_A = -\frac{H}{12}\left[\frac{Y}{X^2 + XY + X} + \frac{Y}{X^2 + XY + X + Y} + 2\ln\left(\frac{X^2 + XY + X}{X^2 + XY + X + Y}\right)\right]$$

where:

$$X = \frac{2d_s}{x_2 + x_2} \quad \text{and} \quad Y = \frac{x_1}{x_2}$$

and H is a Hamaker constant, d_s is the distance between the particles, x_1 and x_2 are the two particle diameters. The Hamaker constant varies with material, but is generally $10^{-20} - 10^{-19}$ J. The presence of water will alter the Hamaker constant and it is usual to apply an effective Hamaker constant based on the geometric mean of the individual constants. Thus to calculate the attraction between particle 1 and 2 in the presence of phase 3 the effective Hamaker constant is:

$$H_{132} = (H_{11}^{1/2} - H_{33}^{1/2})(H_{22}^{1/2} - H_{33}^{1/2})$$

where H_{33} is the Hamaker constant for the suspending phase, 5×10^{-20} J for water.

Collection Efficiency. The relative importance of the above nine listed forces and their interaction in particulate capture is difficult to assess. It follows that practical filtration modelling and design does not usually start from a consideration of these fundamental equations. However, a recognition of the various forces is still important if only as a qualitative tool in the interpretation of practical events. Thus an often reported graph relating deep-bed filtration efficiency and the size of suspended particles can be explained in terms of the relative importance of diffusion, inertia and straining. At low particle diameter removal efficiency is mainly due to diffusion. This effect becomes less relevant at higher diameters and a minimum is evident on Figure 6.2. However, as particle size increases inertial impaction becomes more relevant and efficiency increases again with size. Eventually straining, or sieving, becomes the dominant mechanism.

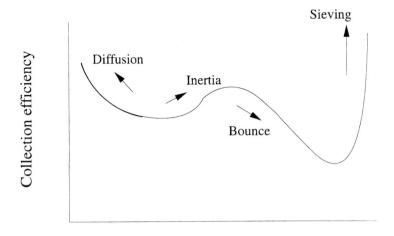

Figure 6.2 Capture efficiency as a function of particle size in a deep bed filter [Yao et al, 1971]

Most clarifying filtration modelling and design is based on kinetic and continuity considerations of the process rather than the more complex fundamental approach.

6.2 Deep-Bed Filtration

Deep-bed filters have been in common use for well over one hundred years. The principle involved is used in nature to filter dirty water through porous rock, in potable water wells and springs. Commercial deep-bed filters are of simple construction consisting of an enclosed cylindrical metal tank and a pressure vessel or open rectangular concrete tank containing a packed bed of solids through which the process fluid to be filtered is passed. The vessel design depends on the process fluid being filtered and the method used to effect the filtration: pumping or gravity. The suspended particles in the fluid are attracted to the packing in the bed by a mixture of mechanisms described in Section 6.1. Particle deposition leads to an increased pressure drop for the fluid flowing through the bed and may cause blockage, thus deep-bed filtration can be used only for very low concentrations of suspended solids; generally less than 0.5 g l^{-1}. Alternative clarifiers should be considered for suspensions containing solids at a greater concentration, or frequent bed cleaning should be anticipated. Their greatest application is in drinking water filtration and the final polishing of an effluent before discharge into, for example, a river where the suspended solids concentration should be below 0.025g l^{-1} (i.e. tertiary waste water treatment). They can be effective in the removal of sub-micrometre material which does not settle or filter readily, especially when used in conjunction with coagulants and flocculants. The common types of deep bed filter are summarised below.

Slow Sand Filter. Fine grains of sand are used which leads to efficient suspended particle removal, but slow filtration rates. Relatively shallow beds are employed and biological activity on the surface is common. The initial sand layers become clogged first and cleaning is by removal and replacement of these layers. The deposit formed on the surface is sometimes called "schmutzdecke".

Rapid filter. A coarser grain size, and sometimes a narrower size distribution than slow filters are used; often pressurised flow is employed. A variety of rapid filters exist, for example, downflow, upflow, mixed-media and continuous. These will be discussed in greater detail in the following sections.

Direct Filtration (Contact Flocculation). This is a mixture of filtration and flocculation in which the suspension entering the bed has been flocculated, but the flocs have not fully formed prior to bed entry. The technique is applied to suspensions of low concentration of colloidal matter. The shear induced inside the deep bed provides excellent conditions for orthokinetic flocculation. The main advantage of direct filtration is a reduction in the capital cost over the provision of a flocculator and separate vessel for solid–liquid separation. In some instances the treated effluent can also be of a higher standard than that obtained by a more conventional route.

Cut-away sections of two types of deep-bed filter are shown in Figure 6.3.

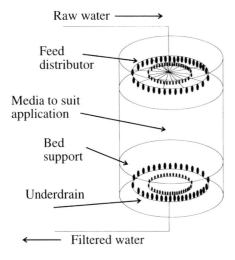

Cylindrical tank

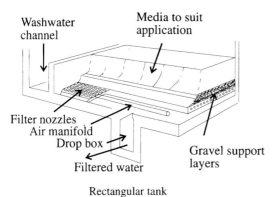

Rectangular tank

Figure 6.3 Diagrams of a deep-bed filter

Typical deep-bed filter dimensions are 0.5–3 metres in height with cylindrical diameters of 1 m. The packed bed is usually made up of sand, gravel, anthracite or a variety of other packings having a particle size of 0.4–5 mm. Smaller particles would have the advantage of providing a greater bed area and, therefore, probability of suspended solid capture, but the pressure drop over the bed, and the occurrence of clogging, would also increase. Gravity feed is usual for the rectangular tank design, but this limits the flow rate through the bed.

A common design is to use mixed media within the bed; either different types of media or different particle sizes making up the bed. In this case the design of each bed section can be made independently, as shown in Section 6.2.5, the feed concentration for one bed section coming from the effluent from the previous.

Biological sludges can also be grown on the bed packings in order to reduce the concentration of dissolved effluents at the same time as filtering out the suspended solids. This has found some application in the treatment of effluents arising from the woollen industry.

6.2.1 Performance

Typical flow rates through a filter with suspended solids > 1 μm in diameter are up to 15 m^3 m^{-2} h^{-1}, that is up to 15 m^3 of liquid flowing per m^2 of bed area per hour; m h^{-1}. Flow rates are reduced when treating small solids such as viruses ($<< 1$ μm), where flow rates can be as low as 0.1–0.2 m h^{-1}.

Chemical coagulants are often added to the feed, see Chapter 5. These form gelatinous hydroxide precipitates which act as collectors for the very fine particles in the feed suspension. The probability of removal of the resulting larger agglomerate therefore increases. Polymer flocculants are also used to promote aggregation.

Capture efficiency within the filter varies with suspended particle size and flow rate, owing to the variety of mechanisms that exist within the bed. Suspended solid outlet (effluent) concentrations of < 0.1 mg l^{-1} are possible at low flows. It has been discovered [Yao et al, 1971] that filter efficiency can display a minimum at a particle size of 1μm, at constant flow rate, as the particles are too large to diffuse onto the bed solids and too small for the capture mechanisms of gravity and interception (Figure 6.2).

6.2.2 Cleaning

The filter is usually in service for 8–24 h between cleaning cycles. Automated cleaning is very common; the pressure drop across the bed is monitored until it exceeds a preset value, the filter is then taken off-line and the cleaning cycle initiated. After a preset cleaning time the filter is put back on-line or into standby. If a second filter is present the treatment of process water can remain uninterupted as one unit is on duty whilst the other is cleaning or on standby. It is also possible to control the cleaning cycle by monitoring filtrate quality, and to initiate cleaning when solids "breakthrough" into the filtrate occurs.

During cleaning the flow is reversed to wash off the deposited solids, this is called "back-flushing". This often fluidises the media, and frequently air scour is also used. The flow conditions are, therefore, very aggressive and turbulent, and particles attached to the media grains become dislodged and carried away by the back-flush water, which usually comes from the previously filtered process water. Typical back-flush rates are 36 m h^{-1}, for 3–8 min, using 1–5% of the filtrate. See Table 6.2 and the following text for a further discussion on operating data. The water used for cleaning has to be kept separate from the process fluid and, therefore, requires an alternative method of treatment to remove the suspended solids. Gravity settling after coagulation is one commonly employed technique.

One important consequence of cleaning is that if fluidisation is employed the smaller or less dense collection media grains will report to the top of the filter bed. The smaller grains possess a higher collection efficiency than the coarser ones. Thus if the normal mode of filtration is to use downflow operation the top of the filter will be much more efficient at removing suspended material than the rest of the filter. Also, the feed-suspended solid concentration will be greater than within the filter, so the top of the filter will become rapidly clogged by captured solids. This has led to some interesting bed designs where the normal mode of operation is to pass liquid slowly upwards; the coarsest grains prefiltering the feed suspension, prior to the finest grains polishing the liquid. This is discussed in greater detail in Section 6.2.3.

6.2.3 Design

The rational design of a deep-bed filter is based on an understanding of the processes present within the device. The collection mechanisms, see Section 6.1, are too complex to model collectively; but for practical purposes it is usual to require information on concentration of collected particles and pressure needed to effect a filtration. Both parameters are important in deciding how long a bed of certain size will remain in service between cleaning cycles. Collected particle solids retention and pressure drop are considered below.

Solids Retention. Mathematical modelling can be achieved by considering a mass balance of solids deposited within the bed (continuity) and a first order-type of rate equation, with respect to solid concentration and distance within the bed.

1) Mass balance

$$U_o C - \left(U_o C + \frac{\partial (U_o C)}{\partial y} dy \right) = \frac{\partial \sigma}{\partial t} dy$$

where, C is the volume fraction solids conc. of layer, y is the depth within filter, U_o is superficial velocity, and σ is "specific deposit" which is defined as:

$$\sigma = \frac{\text{Volume solids deposited within layer}}{\text{Volume of the filter layer}}$$

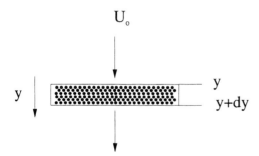

by mass balance on a differential layer
Mass input – Mass output = Accumulation

Cancelling terms within the mass balance provides an equation relating the suspended solid concentration gradient and the rate of change of specific deposit:

$$- U_o \frac{\partial C}{\partial y} = \frac{\partial \sigma}{\partial t} \tag{6.1}$$

2) Rate equation

In a uniform filter the filtration efficiency should also be uniform given a constant feed suspension. The removal efficiency is the fraction of particles removed per unit bed depth, thus:

$$\frac{\partial C}{\partial y} = - \lambda C \tag{6.2}$$

where λ is the filtration constant. Equation (6.2) is generally attributed to Iwasaki [1937]. The filtration constant is really a time-dependent variable and is only a true intrinsic property of the bed before filtration commences, i.e. at zero time. Under these conditions the constant is called the initial filtration constant (λ_o), and equation (6.2) can be integrated to give:

$$C = C_o \exp(- \lambda_o L) \tag{6.3}$$

where L is bed depth, or depth of the uniform section of bed if dealing with a mixed media system. Equation (6.3) is often used because of its inherent simplicity. A more complete solution to the variation of solids concentration and collection efficiency caused by solids deposition would have to consider how the filtration constant changes with time. This is illustrated in Figure 6.4.

The target which removes particles from the liquid stream is the surface area of the bed media. If solids are deposited this surface area increases and, therefore, the filtration constant should increase with time. However, pores within the bed will become increasingly clogged leading to straightening of flow channels with a consequent decrease

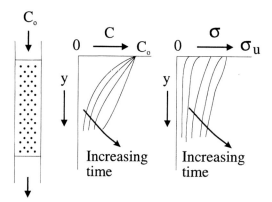

Figure 6.4 Variation of concentration and deposit within the filter bed

in the available bed surface area and an increase in the interstitial velocity. These two opposing factors can be incorporated into an equation describing how the filtration constant may vary with time and hence specific deposit:

$$\lambda = \lambda_o + b_1\sigma - \frac{b_2\sigma^2}{\varepsilon_o - \sigma} \tag{6.4}$$

where b_1 and b_2 are constants and ε_o is the original porosity of the bed. Equation (6.4) can be substituted into Equation (6.2) and a numerical solution to the concentration gradient may be attempted. However, it may be possible to simplify Equation (6.4) by approximating the following quadratic equation to the experimental data, i.e. to replace Equation (6.4) with:

$$\lambda = b_3 - b_4\sigma^2$$

Figure 6.5 illustrates some data in which it was acceptable to replace Equation (6.4) with the above simplification.

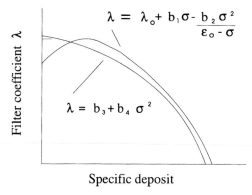

Figure 6.5 Comparison of filtration constant models

An alternative approach to applying the empirical quadratic equation is to consider a power function for each of the factors which affect the filtration constant [Ives & Pienvichitr,1965; Ives,1975]. The filtration constant is again related to the microstructure changes within the bed (deposit porosity, maximum specific deposit):

$$\frac{\lambda}{\lambda_o} = \left(1 + \frac{b\beta\sigma}{\varepsilon_o}\right)^{y}\left(1 - \frac{\beta\sigma}{\varepsilon_o}\right)^{z}\left(1 - \frac{\beta\sigma}{\sigma_u}\right)^{w} \qquad (6.5)$$

where,

$$\beta = \frac{1}{1 - \varepsilon_s}$$

and ε_s is porosity of the deposited solids, σ_u is the specific deposit when the bed is fully clogged (i.e. under conditions of minimum filtration efficiency), $y,z,w,$ and b are constants.

In Equation (6.5) the first term on the right-hand side represents the geometric change in the bed element due to solids deposition, the second the decrease in internal surface area (due to clogging) and the final term the flow through the bed at maximum loading. Equations (6.4) or (6.5) can be evaluated empirically and used with the mass balance and rate equations, Equation (6.1) and (6.2), to obtain deposited solids concentration as a function of position and time.

The simplified empirical model assumes that $y = w = 0$ and $z = 1$, which leaves fewer constants in Equation (6.5) to be determined experimentally. Under these circumstances there should be a linear relation between the filtration constant and the specific deposit. This is the required assumption behind the bed depth service time (BDST) approach.

Pressure drop. It was during the study of deep-bed filtration that Darcy discovered the law which now bears his name [Darcy, 1856], see Sections 2.3 and 2.4. Kozeny and Carman provided an explicit function for the bed permeability and their equation can be written as follows:

$$\frac{\Delta P}{L} = \mu\left[\frac{K(1 - \varepsilon)^2 S_v^2}{\varepsilon^3}\right]U_o \qquad (6.6)$$

(assuming streamline flow, see Section 6.1) and:

$$\Delta P = h\rho g$$

where h is the "head" of liquid. Then the "head loss" is

$$\frac{dh}{dy} = \mu\left[\frac{K(1 - \varepsilon)^2 S_v^2}{\varepsilon^3}\right]U_o\frac{1}{\rho g} \qquad (6.7)$$

note that $\int dy = L$, ΔP is pressure drop across the bed and K is the Kozeny constant (usually 5). Inches water gauge is a common industrial unit.

Equations (6.6) and (6.7) are useful in the prediction of the minimum pressure drop across the bed and, hence, the required pump size for a given a flow rate. As soon as deposition within the bed occurs, however, the pressure drop will rise. This can be predicted using an empirical equation for pressure correction:

$$\Delta P(h,t) = \Delta P(h,0) + k_p \, \beta \, U_o \, C_o t \tag{6.8}$$

where k_p is another constant and (h,t) refers to the pressure drop at a given height and time during the filtration, $(h,0)$ is the pressure drop at the same height but zero filtration time, i.e. the head loss due to clean media. The head loss due to Equation (6.8) H_d is illustrated in Figure 6.6, together with the initial head loss H_o and the head loss due to undesired surface clogging H_s. Surface clogging is to be avoided because it reduces the in-service time of the bed.

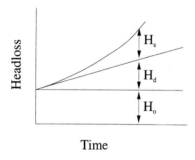

Figure 6.6 Head loss variation with time during filtration

The increasing head loss produced by the filter leads to two separate modes of operation: constant rate or declining rate. Under the former the filtration pressure must be increased to compensate for the head loss, until the maximum head loss permitted before cleaning is achieved. Declining rate filtration would result from the use of a constant filtration pressure; thus the superficial flow velocity and volumetric flow rate will decrease in accordance with Equation (6.7). In this instance the filter cleaning would be initiated when the filtration rate declined to an unacceptable value. In general declining rate filtration is preferred, as there is less likelihood of suspended solid concentrations in the effluent in excess of the permitted value under these conditions.

An optimum filter design would have been achieved if the time taken to reach the maximum permitted head loss, or to reach the minimum permitted flow rate, is the same as the time taken for solids to break through the filter bed. Solids break through in this context refers to the occasion when the effluent concentration exceeds some maximum permissible amount. Some solids leakage through the bed even when clean is inevitable.

Alternative Filter Types. A consequence of Equation (6.2), and its more complicated equivalent, is that the number of particles retained in an element within the bed is not uniform throughout. In a size-segregated bed, owing to the action of the back-wash, the finer particles which have a greater capture efficiency will report to the upper bed surface. This will result in short downflow filter runs with low utilisation of the lower bed region. Thus a uniform filter becomes clogged at the top layers whilst the lower layers are still serviceable, as discussed in Section 6.2.2. If the same filter is used in an upflow mode the effective deposition is similar, but the load is spread much more evenly throughout the bed, leading to longer bed service times. This principle is illustrated in Figure 6.7, where the two modes of operation are compared. Also included in this figure is a further alternative: multimedia filtration where the coarser particles have a lower density than the fine material. Thus during back-flushing the coarse but light particles report to the bed surface, permitting downflow filtration with more effective bed utilisation. Such beds are composed of garnet, sand and anthracite.

Another important development has been the introduction of continuous sand filtration, normally employing upflow mode of operation for the reasons detailed above. In this instance the filter remains on duty whilst simultaneously removing a portion of the bed for cleaning and reintroduction into the bed. The principal advantage of such a device is that only a single unit is required for uninterrupted process filtration, up to the limit of flow required from one vessel. This considerably reduces the capital cost, compared with purchasing and sequencing two or more beds to operate in duty, cleaning and standby modes. An example of a commercially available continuous upflow filter is illustrated in Figure 6.8.

The sand bed inside the DynaSand filter continually moves downwards, through the dirty sand hopper, until it reaches the bottom of an air lift pump, see Figure 6.8. Some previously deposited solids on the sand grains are removed in the turbulent air lift pump, additional sand cleaning takes place in the washbox, where some filtered water is admitted to wash the sand in counter-current flow. The clean sand returns to the top of the bed. The washwater and fine solids flow over a weir into the washwater outlet. Process water is fed into the filter through the inlet distributor, and flows counter-currently to the sand bed. Cleaned water emerges from the bed, flows over an outlet weir and leaves via the filtrate outlet.

Typical operating characteristics of the filters illustrated are provided in Table 6.2, for comparison.

The information provided in Table 6.2 refers to solids which are easily removed by back-flushing. In attempts to apply these techniques to the removal of specific chemical precipitates [Rushton & Spruce, 1982] strong particulate adhesion resulted in higher wash: filtrate volumetric ratios. Best washing conditions are obtained by maximising fluid shear near the particle surface; particle–particle contact is not believed to influence cleaning. Recent studies [Amirtharajah et al, 1991; Fitzpatrick, 1990] have identified the need for combined air scour and back-wash used at subfluidising rates in upflow. The latter apparently produces a "collapse pulsing" condition giving optimum cleaning. Field trials have further confirmed the rates of air/water needed. For successful cleaning [Chipps et al, 1993] combined rates in the range 38–63% of the minimum flow needed for fluidisation were reported for typical waterworks filters.

Biological filters do not normally remove solids from a process stream, their purpose is usually to provide a support medium for microorganisms which digest dissolved species, converting biodegradable solutes into biomass. However, since the 1980s the biological aerated filter has been used to reduce both the dissolved organic and solid

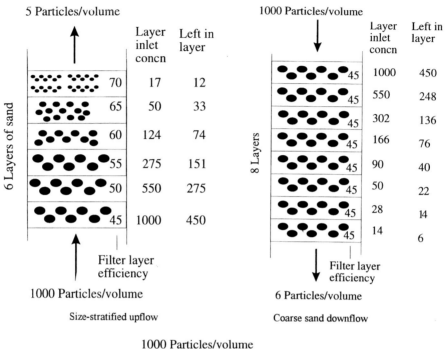

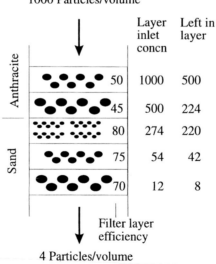

Figure 6.7 Size-stratified filter-different operating modes

Table 6.2 Characteristics of different types of deep bed filters

Characteristics	Continuous upflow	Coarse downflow	Multimedia
Direction of flow	up	down	down
Cleaning	continuous	sequenced	sequenced
Media size, mm	sand, 1–2	sand, 1–3	sand, 0.4–1 anthracite, 0.8–2
Sphericity of media	0.8	0.8–0.9	sand, 0.8 anthracite, 0.7
Media depth, m	1.2	1.5–2	sand, 0.5 anthracite, 0.3
Vessel depth, m	3.5–7	3–5	2–3.5
Surface loading, $m^3\, m^{-2}\, h^{-1}$	8–14	6–18	8–16
Hydraulic stability	low	high	high
Removal efficiency	moderate	low	sand, high anthracite, low

Back-wash cycle

water requirement, $m^3\, m^{-2}\, d^{-1}$	19	30	30
air requirement, $m^3\, m^{-2}$	10–12 at 7.5 bar	120 at 1 bar	50 at 0.5 bar
time	continuous	2 h/d	1 h/d
Relative capital cost	high	low	moderate
Relative running cost	high	low	low

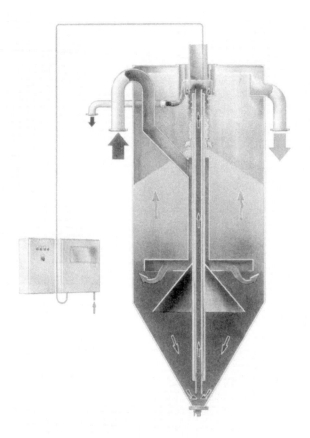

Figure 6.8 Continuous filtration: DynaSand (Courtesy of Nordic Water products AB)

loading of a process stream [Stephenson et al, 1993]. The hydraulic principle of operation is similar to that already described; upflow or downflow in use, with periodic back-flushing to remove entrapped solids and excess biomass. The filter is, therefore, a combined biological reactor and clarifier, as illustrated in Figure 6.9.

6.2.4 Laboratory Test Equipment

A comparative filterability test has been recommended [Ives, 1986] to assess water quality against standard filter media at standard rates and to test changes in rate, or filter media, using a constant feed. Such a test may be useful in preselecting filter media and flow conditions worthy of further investigation. Bed design data, however, need to be collected employing a pilot test filter.

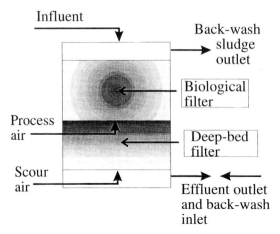

Influent

Back-wash
sludge
outlet

Biological
filter

Process
air

Deep-bed
filter

Scour
air

Effluent outlet
and back-wash
inlet

Figure 6.9 Schematic illustration of the downflow: biological aerated filter

Experimental filters can be built from Perspex, metals, etc. incorporating automatic back-flushing control or using manual monitoring of conditions and cleaning cycles. The recommended minimum diameter for a cylindrical test filter is 150 mm (6 in), the vertical height should be as close as possible to that of the full-scale unit. Different media and clogging rates are easily checked, and the ability to clean the filter should also be tested. The various empirical parameters required in Equations (6.1)–(6.8) can be obtained, providing reliable design data for the full-scale unit. A schematic illustration of the required laboratory test equipment is given in Figure 6.10.

A pilot column containing a set of manometers enables the pressure at known heights within the column to be monitored with time. In the absence of a bed the pressure profile would be a straight line corresponding to the hydrostatic head. However, the flow of liquid in the bed causes a dynamic pressure loss, due to fluid drag, hence the pressure profile shows a discontinuity at the bed surface in Figure 6.11. Below the bed surface the profile is again constant, with a clean bed or one at the start of filtration, as the dynamic pressure loss is uniform. However, as solids become deposited in layers the dynamic pressure loss across those layers increases, due to the increase in solid content and hence friction, leading to curved pressure profiles. Straight pressure profiles indicate insignificant deposition of solids within the bed.

6.2.5 Design Calculations

There are many factors to be considered in the design of a deep-bed facility, the initial design may be based on some preliminary testwork which needs to establish the efficiency of filtration of the medium i.e. the filtration constant λ and uses a knowledge of the physical properties of the medium.

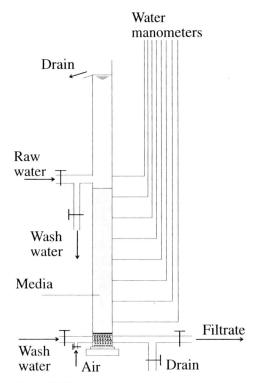

Figure 6.10 Laboratory test equipment

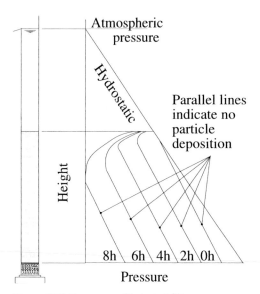

Figure 6.11 Pressure profile in test filter

Example: calculate the bed depths required to reduce the concentration of suspended matter from 100 to less than 10 mg l^{-1}, using a mixed medium of anthracite, sand and garnet operating in downflow, the specific gravities of these are 1.4, 2.65 and 3.83 respectively. First, the system must be designed to segregate during backflushing so that the coarser but lightest anthracite reports to the top of the bed and garnet to the bottom. For this calculation the Archimedes (*Ar*) and particle Reynolds (*Re*$_p$) number correlation may be used (note that the Archimedes number is also referred to as the Galileo number):

$$Re_p = \frac{xu\rho}{\mu}$$

where *u* is the particle settling velocity, and:

$$Ar = \frac{x^3\rho^2 g}{\mu^2}\left(\frac{\rho_s - \rho}{\rho}\right)$$

the correlation between these two is:

$$Re_p = [(14.42+1.827Ar^{0.5})^{0.5} - 3.798]^2$$

for 2<Re$_p$<20 000.

The medium particle size must be chosen so that the settling velocity of the components is in the order: anthracite, sand, garnet. Particle diameters of 1500, 750 and 550 μm respectively satisfy this requirement. The corresponding settling velocities are 0.083, 0.11 and 0.12 m s^{-1}, using the above correlation. Filtration tests on the medium at these sizes reveal the following values of the filtration constants: 0.5, 4 and 6 respectively. Equation (6.3) can then be used with selected bed depths to provide an acceptable solution, one solution would be:

Bed material, g l^{-1}	Bed depth, m	Concentration at exit, mg l^{-1}
Anthracite	0.2	90.5
Sand	0.4	18.3
Garnet	0.2	5.5

i.e. the final effluent concentration is 5.5 mg l^{-1}, which provides an adequate level of safety on the desired level of performance.

The hydraulic loading will be between 8–16 m^3 m^{-2} h^{-1}, see Table 6.2, if a 1 m^2 bed area is chosen the maximum head loss will be given by a flow of 16 m^3 m^{-2} h^{-1}. Equation (6.7) can be used to calculate the head loss given that the bed porosity will be approximately 50% and the specific surface area per unit volume S_V of the particles is:

$$\frac{6}{x\psi}$$

where ψ is particle sphericity, see Table 6.3. The head losses in each section of the multimedia bed are: 0.03, 0.18 and 0.22 m of water for the anthracite, sand and garnet parts respectively. It should be emphasised that this is the head loss when clean, the head loss will increase substantially over these values during service. This and other design parameters should be checked at the pilot scale testing stage, as described in Section 6.2.4.

Further information on the physical properties of media used in deep bed filtration is provided in Table 6.3.

Table 6.3 Physical properties of deep bed media

Property	Multilayer filter			Sand filter		
	Anthracite	Garnet	Magnetite	Crushed flint sand I	Crushed flint sand II	Quarry sand
Sieve size, mm	1.676–1.405	0.599–0.500	0.500–0.422	0.853–0.699	0.599–0.422	1.676–1.405
Sphericity:	0.745	0.865	0.90	0.78	0.82	0.765
Specific Gravity:	1.40	3.83	4.90	2.65	2.65	2.65
Porosity:	0.425	0.47	0.42	0.464	0.464	0.39

It is not possible to provide values for the filtration constants, as these are dependent on the material to be filtered, the flow conditions, filtration history, etc. They must be assessed by laboratory and pilot-scale tests.

6.3 Precoat Filtration

Filter aids have been considered already in Chapter 5, and filtration process equipment will be discussed in Chapter 11. Precoat filtration is the application of a filter aid on a mainly conventional filter under clean conditions; i.e. in the absence of suspended solids other than the filter aid, to form a highly porous and uniform filter cake. The suspension to be filtered is then introduced onto the filter. Clarification by filtration results by the action of the filter aid trapping the suspended solids within the filter aid cake, by the techniques described in Section 6.1. Only a thin layer of cake is usually considered to be important in this operation. If the filtration equipment is continuous by nature, e.g. a rotary vacuum filter, a thick filter aid cake may be employed as the top layer contaminated with the material filtered during the clarification can be scraped off and discarded. The layers below the surface are then exposed for further filtration and, therefore, economically viable rates of filtration ensue. Alternatively, if the filtration is conducted in a batch vessel thinner filter aid cakes are usually employed, with more frequent cleaning required.

The main difference between a drum filter employed for precoat filtration and one employed in conventional vacuum filtration is the level of submergence, which can be as high as 70% in the former–case. A filter aid cake may be 75 to 150 mm thick initially, and a knife is usually employed to shave off only the top layer of the contaminated cake. The rate of removal of the filter aid is controlled by the rate at which the knife is advanced towards the drum surface, and this depends on the concentration of suspended material to be clarified. A typical value would be 0.01–0.40 mm per revolution of the drum. Operation of the precoat filtration may last as long as 240 h before no filter aid is left on the drum, and the filtration must be stopped, the drum precoated again and filtration recommenced. The nature of the precoat cake is very important, it must be firm enough to be shaved with the filter discharge knife, adhere strongly to the filter cloth so as not to be easily knocked off, not prone to cracking or other cake inhomogeneities, provide minimal flow resistance to the clarified liquid and yet retain the solids to be clarified in a narrow depth of cake. The procedure of laying down the precoat with the intention of achieving these aims is very important. Precoating under the following conditions is recommended [Purchas, 1981]. High drum speed, dilute (1 to 2% by weight) precoat slurry, low (100 mm Hg) vacuum. During the precoating operation the vacuum level can slowly rise to the more normal value of 500 mm Hg, as required by the increasing flow resistance caused by the increasing cake height.

The rate at which the knife should be advanced and its dependency on the drum speed, cake resistance, etc. has been investigated [Hoflinger & Hacki, 1990], and Figure 6.12 shows an example of experimental data of the precoat blocking function against knife advance rate. The blocking function is a combination of the filtration resistance due to the deposited solids, the cake resistance due to the precoat and the knife advance rate itself as this influences the overall resistance to filtration. The optimum knife advancement rate is shown as a maximum on Figure 6.12, and is approximately 0.22 mm per drum revolution for this material.

Vertical filtration surfaces are preferred in batch pressure precoat filtration, e.g. tubular candles and filter leaves. When the precoat has become saturated with the solids removed from the clarified liquid, the cake is discharged by back-flushing off the support, or by means of water spray jets. A drainage port beneath the filter surface facilitates resuspended cake removal by gravity. A new precoat can then be applied to the filtration surface and the clarification recommenced. The pressure filter does not, therefore, need to be opened up for cleaning. One major disadvantage of a vertical surface is the possibility of forming a cake with poor homogeneity (greater solids depth at the bottom of the filter element, due to gravity). In some respects this is self-compensating because high clarification flows through the top of the filtration surface will result, leading to more rapid clogging of that region. Horizontal leaf filters which filter on the top surface of a plate only and are equipped with spray discharge to remove the filter cake can also be used for precoat filtration. They do not suffer from some of the disadvantages mentioned for vertical filtration surfaces, but give a lower filtration surface area per unit volume of space occupied.

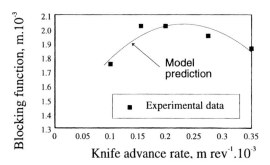

Figure 6.12 Precoat filtration blocking function with knife advance rate Hoflinger & Hacki, [1990]

6.4 Filter Candles

Filter candles for clarification generally rely on the use of a precoat to prevent media blinding. In many respects their operation is similar to that described for batch precoat pressure filtration. Cylindrical candle diameters of 10 mm or more are used, with many candles inside a pressure vessel. They usually operate vertically with the dirty precoat back-flushed off the support candle. Bottom discharge of the solids is then followed by coating the candles with precoat before recommencing clarification. Simple candles rely on the hydraulic pressure produced during the back-flush for solids removal, but this can lead to areas on the filtration surface which have not been effectively cleaned. On recommencement of filtration the only areas effectively used are those previously cleaned and hence the effective filtration area, and rate, will diminish. To overcome this problem candles that flex and vibrate during back-flushing have been devised. One example is the Culligan-type filter [Rushton, 1985] which has a cloth stretched over a spring-like support. During constant rate filtration the increasing pressure differential between the feed and filtrate compresses the spring until the end of the filtration cycle. To clean the filter the two pressures are equalised and the spring relaxes thus dislodging the filter cake. This filter is illustrated in Figure 6.13, together with a more conventional design employing simple back-flushing.

It is economically attractive to provide as many candles as possible inside a single pressure vessel. This will provide the greatest filtration area and hence product flow rate. However, one factor which considerably affects the ease of cake discharge is the occurrence of bridging between the filter candles. If no space exists for the cake to break

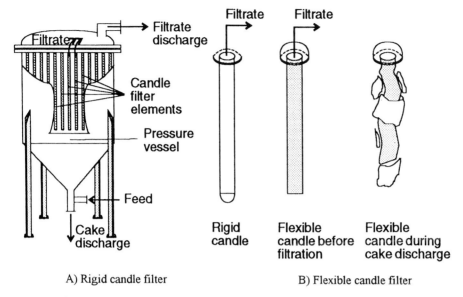

Figure 6.13 Candle filter designs

off the candles during cleaning, then back-flushing will be impossible. The minimum recommended distance between the cake on the candles prior to back-flushing is 20 mm. If the candle filter is used with a precoat for clarification then this distance should be fairly easy to maintain; the filter cake will not grow significantly during filtration as depth filtration in the precoat is more likely than cake filtration on the precoat surface. There are, however, some instances in clarification when additional cake may be formed and care in maintaining the clearance should be exercised.

In some of the more difficult clarifications, such as the filtration of yeasts, filter aids may need to be both precoated and body fed to ensure viable filtration rates. The mathematical analysis of constant-rate filtration on filter candles employing both methods of filter aid addition has been presented [Hermia, 1993]. During filtration on a filter candle the filter area increases during filtration, which is a significant advantage over a filter of planar geometry with constant area, such as a plate and frame filter press. Hence the increasing flow resistance due to deposition of solids is offset against the increase in filtration area. The modelling of the process has to be split into two stages and a mass balance employed for each:

1) During precoating

$$A_o a_p = Nh\pi((r_o + d_l)^2 - r_o^2)\rho_b$$

where A_o is the initial filter area, a_p is precoat dosage as mass per unit area, N is the number of candles, h is the height of the candles, r_o is the candle radius, d_l is the depth of precoat deposit and ρ_b is the bulk density of the precoat,

2) During body feed filtration

$$a_b V = N\pi h((r_o + d_1 + d_2)^2 - (r_o + d_1)^2)\rho_b$$

where a_b is body feed dosage as mass per unit volume of liquid, V is the filtrate volume and d_2 is the depth of body-fed cake on the candle.

If the presence of solids other than filter aid and the liquid retained in the cake are neglected the two mass balances can be combined with a differential form of Darcy's law and integrated to give:

$$\Delta P = \Delta P_o + \frac{\mu\alpha r_o \rho_b}{2}\left(\frac{Q}{A_o}\right)\ln\left[1 + \frac{2a_b}{r_o\rho_b + 2a_p}\left(\frac{Q}{A_o}\right)t\right] \tag{6.9}$$

where ΔP_o is the pressure drop at the start of the filtration (after precoating), α is the specific resistance of the filter aid and suspended solid mixture, Q/A_o is the constant filtration rate per unit area and t is the elapsed filtration time.

Worked example: [using the data from Hermia 1993], reproduced below, calculate the pressure as a function of time needed to maintain a constant flow rate of 1 m^3 m^{-2} h^{-1} on candle filters of radii 15 and 10 mm, and for a filter of planar geometry. Data: pressure at end of precoating 0.1 bar, beer viscosity 0.003 Pa s, body feed dosage 1 kg m^{-3}, precoat dosage 1.5 kg m^{-2}, specific resistance 1×10^{11} m kg^{-1} and filter aid bulk density 350 kg m^{-3}. The maximum pressure available from the pump is 3.5 bar. Applying Equation (6.9) at various times provides the results tabulated in Table 6.4.

Table 6.4 Pressure required to maintain constant rate filtration for various filter geometries

Filtration time	Pressure drop for constant rate on a filter of		
h	planar geometry bar	15 mm radius bar	10 mm radius bar
0	0.10	0.10	0.10
1	0.93	0.57	0.49
2	1.77	0.96	0.80
3	2.60	1.30	1.05
4	3.43	1.58	1.27
5	–	1.84	1.46
10	–	2.79	2.15
15	–	3.46	2.62
20	–	–	2.97

Table 6.4 is based on the same superficial velocity for all the filters considered. However, it is possible to obtain a greater filtration area per unit tank volume in the case of planar geometry filters, thus for the same volume of liquid produced the superficial velocity would be smaller for planar filters than candles.

6.5 Microstrainers

For the removal of low concentrations of coarse solids strainers may be employed in preference to sedimentation basins, as the amount of solids present may not justify the latter. Non-settling solids may also be removed in strainers. One simple strainer arrangement is to provide an in-line filter element to remove solids, as illustrated in Figure 6.14. The device illustrated is self-cleaning; the cleaning cycle is initiated automatically when the pressure drop reaches a preset value [de Vreede, 1987]. A typical application is in the sugar industry where the liquid viscosity is up to 0.040 Pa s, temperatures up to 95 ^{0}C, flow rates per filter area of up to 30 m^3 m^{-2} h$^{-1,}$ operating pressures up to 10 bar and screen sizes down to 50 µm.

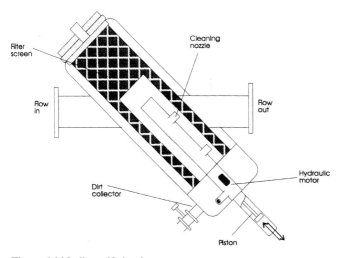

Figure 6.14 In-line self-cleaning screen

Microscreening is very common in the potable and wastewater industries where a woven metal mesh or fabric of 15–200 µm may be attached to the periphery of a rotating drum typically 1–3.3 m diameter and 0.6–5.1 m long. Figure 6.15 illustrates a typical microscreen. Flow enters in the centre and is radially filtered through the drum mesh. The drum rotates and the solids retained on the screen are removed in a section by back-flushing with the previously filtered water. A separate launder takes the back-flush suspension off for further processing. Rotation speed usually varies from 20 to 120 s, and flow rates of up to 3900 m^3 h^{-1} for a single unit are claimed [Anon, 1993].

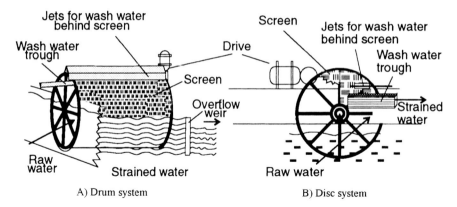

A) Drum system   B) Disc system

Figure 6.15 Microscreens with continual removal of solids

An alternative to the use of the drum periphery for screening is to employ a vertical disc Figure 6.15 (B), again with continual solid removal by back-flushing. In this instance it is possible to have a series of discs, one after another, with increasing fineness of screening media. Such an arrangement may prevent overloading of the finer screen if a high solid content is present, but pore plugging may become significant. Conventional filter media has been employed in the disc screen systems, e.g. felts and monofilament cloths. Removal of solids by back-flushing is effective at particle sizes which do not plug the media and for those which have insignificant surface forces. This limits screening to low concentrations of particles generally greater than 10 μm in diameter.

Six factors are considered important in the hydraulic design of a microscreen: maximum flow rate, allowable head loss, porosity of the medium, effective submerged surface area, drum speed and characteristics of the feed. These factors are numerically combined in Boucher's filterability index for water [Boucher, 1946–47]. The basis for this index is the empirical observation that the logarithm of the head loss H is linearly proportional to the volume filtered V, for constant-rate screening through a thin septum:

$$\ln H = IV + \ln H_o$$

where H_o is the head loss of the clean screen and I is a constant related to the water to be filtered (known as the filterability index). In viscous flow the head loss is proportional to flow rate and the above equation can be rearranged and expanded to provide one in the form:

$$H = \frac{bQC_h}{A}\exp(QtI) \tag{6.10}$$

where A is the submerged strainer area, C_h is the initial head loss under the standard filtration condition of a flow of 0.3 m s^{-1} flowing through the screen, this is called the clean filter factor, and b is a dimensional constant which is a function of temperature. Equation (6.10) is for batch straining, but it can be modified to provide an expression for continuous straining by considering the effective filtration area to be the product of the actual submerged area and the rotation speed of the strainer. This also introduces another dimensional constant into the expression.

6.6 Cartridge Filtration

Cartridge filters are a very important means of effecting a clarification and have found use in a wide variety of industries. Advances in membrane manufacture have enabled cartridge suppliers to use filtration media capable of retaining all particulates down to 0.1 μm in diameter, and with significant removal of colloidal material below this size. The market for such filters is extremely large, notably the pharmaceutical, electronic, automotive and other industries requiring the protection of machinery from suspended material. The specification for fineness of filtration varies from colloidal material up to tens of micrometres, depending on the industry and application. They have much in common with candle filters, but are often employed individually, replaced rather than cleaned, and may be of a type of construction not usually found in candle filtration. Cartridge prices range from under £ 100 to over £ 1000 per cartridge, depending on fineness, material and method of construction. A single element cartridge filter housing, and various types of filter geometries are illustrated in Figure 6.16.

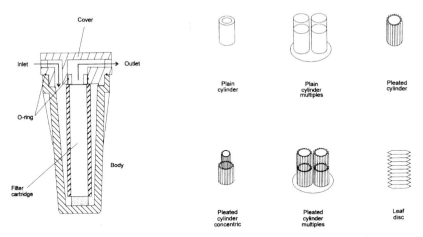

Figure 6.16 Cartridge filter housing and element geometries

They are, necessarily, dead-end filters and because of this have only limited capacity for the removal of suspended solids; a typical 10 in cartridge nominally rated at 10 μm may be able to remove up to 20 g of suspended material before the pressure drop across the filter becomes unacceptable. This limits their application to clarification, but it is common to find filters using many such cartridges in parallel which provides greater "dirt handling" capacity or longer service life. Such an arrangement is illustrated in Figure 6.17.

These filters are often referred to as "absolute" or "nominal" filters, the former implying complete removal of suspended material from a liquid stream above some rated size, the latter removing some high fraction (greater than 90% for example). It is the nature of filtration, however, that removal efficiency depends on the material being filtered and the flow conditions. Thus even with an absolute filter particles of diameters greater than the rating could report to the filtrate if they can deform, or have an aspect ratio that enables them to line up in a flow streamline to pass through the medium. The filter rating may also be optimistic and, of course, depends on the method used to investigate it.

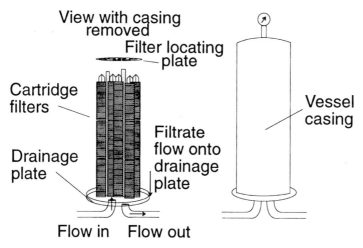

Figure 6.17 Multielement cartridge filter housing

Media Type and Arrangements. The following materials have found application in cartridge filters: glass fibre, Nylon 66, PTFE, polypropylene, PVDF, cellulose, sintered metal powder, sintered metal fibres and a variety of textiles and papers. When filtering water with hydrophobic membrane materials, such as PTFE and polypropylene, a low surface tension liquid miscible with water must be used to pre-wet the filter, or the pressure required to force water through the small membrane pores will be excessive. Cartridge filters can be classified into three groups: depth, surface and edge, with additional subdivision. However, it should be noted that there are many design similarities between filters of different classification. Furthermore, many so-called surface filters do, in fact, rely on depth filtration to achieve their nominal filter rating; the marketing of these filters is believed to be easier if surface filtration is implied. Absolute filtration by means of a barrier to particles above a specified size is more attractive than the somewhat statistical nature of depth filtration. For further discussion on the pore size distribution of a membrane, and its interpretation, see Section 6.6.1.

Depth Filters. All the mechanisms for particle collection provided in Section 6.1, except surface straining, are relevant in cartridge depth filtration. Metal filters are preferred to synthetic or natural fibres if the pressure drop over the filter might become very high, when filtering polymer melts for example. Element cleaning is not normally possible unless the suspended material can be dissolved without affecting the integrity of the cartridge filter. Depth filters require a reasonable thickness of filtering medium, by definition, so the cartridges tend to be straightforward in geometry, a plain cylinder for example, and with a consequent low filtration area compared with pleated surface filters of the same overall dimensions. Pleated, so-called membrane, pre-filters which rely on depth filtration mechanisms are also available.

Surface charged filters, negatively or positively, are also available to enhance particle capture during filtration. This is useful when filtering charged suspended material from liquids of low ionic strength or conductivity, such as deionised water.

1) Homogeneous type

Metal cartridge depth filters can be made from sintered metal powder, woven metal mesh, sintered metal fibres or any combination of these. So-called absolute filter ratings are 3–500 µm. The finer filters are all depth rather than surface filters. Thus the pores at the surface of the filter are much wider than the specification of the filter, but the flow path through the filter is tortuous and particles are intercepted by the metal media. The modern sintered metal fibre media produced by Bekaert provides a very high porosity (up to 87%) and, therefore high flow rate/low pressure drop, with particle retention down to 3 µm. Some degree of inhomogeneity is claimed for this material, but this is not great at the finer pore sizes. Metal filters are typically available in stainless steel 316L, Inconel 601, Hastelloy X, Fecralloy, Monel, nickel and brass.

Glass fibre and other polymer filters have a uniform felt type of internal structure. Modern PTFE filters also resemble a felt, with thin filaments stretching between nodes of PTFE, see Figure 6.18.

2) Inhomogeneous type

An alternative membrane arrangement to that shown above is possible by using a polymer filter medium which is more open on one surface than the other, such as the ASYPOR filter (Domnick Hunter). This is called an asymmetric membrane. The more open side faces the incoming flow, the finer side faces the outgoing flow, thus larger suspended particles are trapped prior to the finer membrane surface and the filter has its own pre-filtration stage built in. Superior flow characteristics are claimed for this design. The two membrane surfaces are shown in Figure 6.19.

A similar principle is used in the Pall Profile filters, which are made from Nylon or polypropylene. The pore size varies from 40 µm at the outer side down to 0.5 µm at the inside. This is achieved by varying the fibre diameter, whilst maintaining a uniform pore density which leads to a higher void volume and longer service life. This filter design is shown schematically in Figure 6.20.

Inhomogeneous filters using metal media are also available, the size of sintered bead changing within the depth of the filter to again provide a prefiltering effect for the finest filtration surface. Combinations of a very fine filtering surface formed from sintered

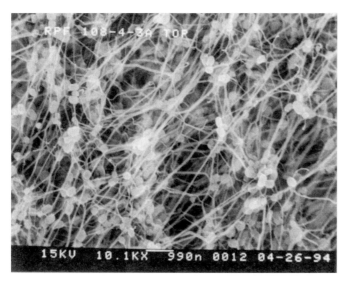

Figure 6.18 PTFE membrane filter (10 000×magnification-Courtesy of Pall Corporation)

beads on top of coarse fibres or a woven metal support are also possible, providing fine filtration with greater mechanical strength due to the support. Composite filters where the fine filtering surface may be a ceramic material bonded onto a metal support are also becoming available.

3) Bonded cartridge type

Synthetic or natural fibres are formed into a thick wall and then impregnated with resin to fix them in position. A light porous structure which is self-supporting and relatively cheap results. The pore size of these filters is restricted by both the method of construction and structural limitations.

4) Wound cartridge type

Winding a string of staple fibres on to a former or core provides a cartridge which can have a high dirt handling capacity provided the suspended solids are not removed by surface filtration. Figure 6.21 illustrates a yarn-wound cartridge.

The string may be formed from wool, cotton, glass fibre or synthetic materials which have been brushed to provide a fibrous surface for filtration. It is possible to provide some inhomogeneity by winding with different pitches on the same cartridge, but scope for this is limited. Wound cartridges with bound yarn are also available, these are self-supporting; there is no need for the former, and they are consequently cheaper. Structural strength and dirt handling capacity suffer, however.

Surface Filters. The mechanism of filtration of a true surface filter is surface straining only. Most commercial surface filter cartridges also filter by means of depth filtration

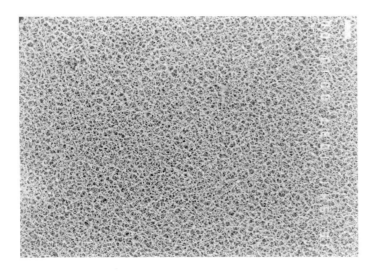

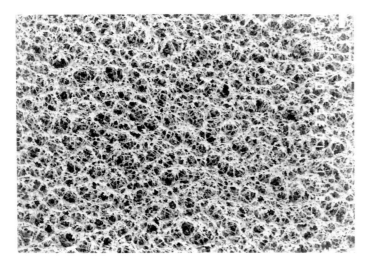

Figure 6.19 Asymmetric pore structure of ASYPOR membrane (Courtesy of Domnick Hunter Ltd)

mechanisms, however. Very low dirt handling capacity and limited filtration area would result from this mechanism if a cartridge of regular geometry was used. To overcome these disadvantages it is usual to pleat the surface filter medium as illustrated in Figure 6.22 or to use crossflow filtration, see Section 10.7.1.

When pleated, a 10 in cartridge with dimensions 230×34×16 mm (length×outside diameter×inside diameter) can have up to 0.5 m^2 of membrane surface area available for filtration. Filters made from a series of membrane discs are also available, with spacers

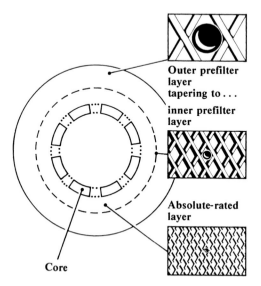

Outer prefilter
layer
tapering to . . .

inner prefilter
layer

Absolute-rated
layer

Core

Figure 6.20 Profile filter construction (Courtesy of Pall Corporation)

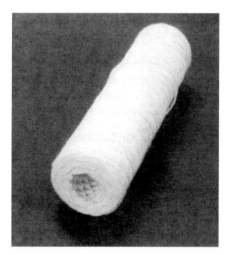

Figure 6.21 Yarn-wound cartridge

between each section. Pleated surface filters are generally made from flexible material such as polymers and are used as the final filter at the point of use. Absolute filter ratings down to 0.1 μm are available in microfiltration. Below this size ultrafiltration technology becomes applicable.

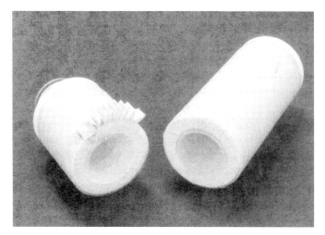

Figure 6.22 Cut-away section of pleated filter cartridge

The design of surface membrane filters can be very critical as they tend to be used as the last filter before an important point of use. They must be robust so that hydraulic shocks due to fluctuating flow does not cause particle shedding. They also require guaranteed seals, both the sealing rings required to push the cartridge into its housing and the membrane material bonded to the endcaps. This bonding must be robust enough to prevent collapse of the cartridge as the pressure differential increases during loading and when flow fluctuations occur.

Edge Filters. A stack of solid thin discs packed together by compression will provide a flow channel between each disc, in the absence of a sealing gasket or ring. This is the basis behind an edge filter. A flow channel towards the centre of the disc must be provided to allow the filtrate to leave the cartridge. Metal discs are usually employed, but polymer and treated paper discs are also available. Spring-like metal filters can be used, which may be cleaned by releasing the compressive force and back-flushing.

6.6.1 Tests and Characterisation

The wide range of media type, particle retention size and applications leads to a wide range of methods for testing and characterising cartridge filters. Many of the tests described below were designed for microfiltration membrane cartridges, as they tend to be applied to the higher value markets which require a means for absolute performance guarantee. The tests are also relevant to the chapter on microfiltration membrane technology, but will not be repeated there.

Bubble Point Test. This is a widely used test to establish the largest pore size available in the filter under test. It is also regarded as a standard integrity and quality control check on commercial membrane filters. It has its origin in the characterisation of fabrics [ASTM 316, BS3321]. Air is normally used in this test, the cartridge filter or membrane sample is wetted with whatever liquid readily soaks the membrane material. The air pressure is then steadily increased until bubbling through the membrane occurs. The first few bubbles are ignored as these rarely give reproducible results, and the test should be repeated a few times as a check. The required equipment and arrangement are shown in Figure 6.23.

The pore size can be related to the bubble point pressure as follows. The surface tension formula is :

$$\gamma = \frac{h\rho g r}{2\cos\phi\, B} \tag{6.11}$$

where γ is surface tension, h is height of rise, ρ is density of fluid, g is the acceleration due to gravity, r is the radius of a tube or pore, ϕ is the contact angle and B is a constant. Rearranging for diameter d and substituting the bubble point pressure ($\Delta P = h\rho g$)

$$d = \frac{4\cos\phi\, B\gamma}{\Delta P} \tag{6.12}$$

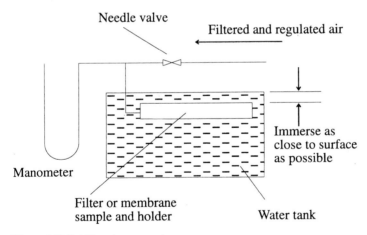

Figure 6.23 Bubble point test equipment

When the fluid fully wets the filter:

$$d = \frac{4B\gamma}{\Delta P} \tag{6.13}$$

For circular pores with pressure measured in Pascals, pore diameter in micrometres and surface tension in dyne/cm Equation (6.14) results:

$$d = \frac{2860\gamma}{\Delta P} \tag{6.14}$$

The range of surface tensions of common liquids is 72–22.3 dyne/cm (water to denatured alcohol).

Mean Pore Flow Test. Details of this test can also be found in ASTM F316. The mean pore flow test provides more information about the membrane under test than the simple bubble point value. The pore size distribution is determined by increasing the gas pressure and flow rate beyond the bubble point pressure. As the pressure increases then so does the number of pores passing the gas. This is because the increase in pressure blows out the liquid from the smaller pores of the membrane. The larger pores being the first to unblock and pass air. By comparing the flow rate of gas and the pressure required to effect that flow rate between the dry membrane and the wet membrane it is possible to deduce the relative proportion of each size of pores present. The mean flow pore pressure is calculated by drawing a graph with both the wet flow curve and the half-dry flow curve (i.e. 50% of the values read for the dry flow curve), and looking for the intersection. This is illustrated in Figure 6.24. The pore size frequency is calculated from:

$$Q_f = \left[\left(\frac{\text{Wet flow}_h}{\text{Dry flow}_h} \right) - \left(\frac{\text{Wet flow}_l}{\text{Dry flow}_l} \right) \right] \times 100\% \tag{6.15}$$

where Q_f is the percentage volume filter flow rate, l is the lower pressure of the division and h is the higher pressure. For example, on Figure 6.24 the pressures and equivalent pore sizes are marked at 2, 7, 15 and 21 l/min. Thus the percentage of the filter flow going through pores of size range 0.2–0.8 µm is:

$$Q_f = \left[\left(\frac{7}{21} \right) - \left(\frac{2}{15} \right) \right] \times 100\%$$

or 20%.

With this technique it is possible to determine the full pore size frequency for a membrane which is a good indication of the filtration performance of the filter. Clearly, it is most desirable to have a narrow pore size distribution. An automated version of this test is available commercially [Wenman & Miller, 1987].

Extractables and Testing. All filters release some material into the process fluid passing through. These are termed "extractables" and the amount depends on the type of filter medium, temperature, contact time and solvent. The amount of extractables can be measured by mechanically moving the test filter upwards and downwards whilst submerged under about 30 mm of liquid for 2 h [Howard & Nickolaus, 1986]. The filter should be reciprocated at 6–10 times per minute. The volume of liquid both before and after the test should be measured, and the remaining liquid is subjected to various other tests. First, a known volume is evaporated to dryness and the residue weighed. This provides the extractables on a mass basis (of nonvolatiles). Other tests include ultraviolet-, infrared-spectrometry and gas-liquid chromatography in an attempt to determine the nature of the material leaving the filter.

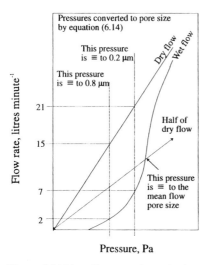

Figure 6.24 Mean flow pore determination and example of pore size frequency

Bacteriological Tests. Complete bacterial retention on filters is possible for a 0.2 μm filter. The usual test bacterium is Pseudomonas diminuta which is used in a filter challenge at a concentration of 10^7 organisms per cm^2. A filter passes this standard if no colonies can be detected downstream in the filtrate. The volume of filtrate taken and cultured is usually 200 ml, with around 6 samples taken over a 24 h period.

Grade Efficiency Tests. The term grade simply refers to a size range where the amount of particles upstream of the filter can be compared with the number of particles downstream, giving the efficiency of removal of particles in the grade by the filter. In this instance the reason for filtration is to remove particles from suspension, thus grade efficiency is defined in terms of the effectiveness of the filter in removing material from the flow. The most reliable method is to measure on-line the particle size distribution and concentration of suspended solids both upstream and downstream of the filter. This can be done with a variety of particle size analysis equipment including the Hiac Royco 346 using laser light scattering (0.5–28 μm), the Malvern using laser diffraction (0.3–1800 μm) and a multitude of other devices using light scattering or obscuration. The grade efficiency is:

$$\left(\frac{\text{Mass of particles upstream of filter} - \text{Mass of particles downstream of filter}}{\text{Mass of particles upstream of filter}} \right) \times 100\%$$

Particle sizing equipment that results in distributions in terms of number or area can also be used in the equation, substituting for mass, provided the solid density of the material in suspension is uniform across the size distribution.

Grade efficiency is usually time dependent because as the larger pores become clogged and the flow path becomes more tortuous the filtration efficiency often improves. Thus it is safer to measure grade efficiency with very dilute suspensions and new (clean) filters.

Dirt-Holding Capacity. As the filter becomes clogged with material the filtration efficiency, in terms of particulate retention capability, may increase. The filter rating, therefore, improves with time and use. The pressure drop required to maintain flow, however, also increases and will at some point become excessive. The filter requires changing at this point. The amount of material retained on the filter by mass is the filter dirt-holding capacity. A typical curve is illustrated in Figure 6.25.

The economic time to change the cartridge, indicated on Figure 6.25, would be about the knee of the curve. Clearly, the pump or prime mover must be capable of providing the pressure based on this partly clogged filter. If the filter is not changed in time the pressure will continue to rise (if the pump has spare capacity), with the serious risk of structural failure. Typical values for a 10 µm filter are 10–20 g of solids corresponding to a differential pressure of 200–300 mbar in water. All filter manufacturers specify the maximum differential pressure allowable with their filters, this is typically 350 mbar with polypropylene filters.

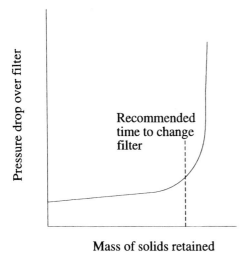

Recommended
time to change
filter

Mass of solids retained

Figure 6.25 Cartridge pressure drop with dirt-holding capacity

Details of the investigation of dirt-holding capacity, as a function of pressure drop, using a multipass filter test system can be found in ISO 4572.

6.6.2 Filter Sizing

Initial tests can be conducted with a small flat membrane or other media, usually 47 mm in diameter, of the same material as the proposed filter cartridge. Scale-up has been done by proportioning the sizes between the disc and cartridge, but with varying degrees of success. This is a consequence of the different flow patterns given by a filter cartridge, due to the need for drainage layers, construction, etc. An additional and ever-present problem is the need to obtain a representative sample from the process fluid to be filtered. After identifying a particular medium tests should be undertaken with a cartridge filter operating on a side stream from the process fluid. Only then can reliable scale-up and service life calculations, and hence economics, be performed.

Filter sizing based on flat disc tests usually consists of recording the filtration time, volume of suspension filtered and pressure drop across the filter. An estimate of the filter performance can be achieved by calculating the volume of filtrate produced per unit time, for constant-pressure filtration, or the volume of filtrate per unit pressure, for constant-rate filtration. Inspection of these rates will show if the membrane is acceptable, and the investigator can decide on the volume filtered before the filtration becomes either too slow or excessive in pressure drop. This provides a volume filtrate per unit area value which can be used to proportion the areas to arrive at the required filter area.

Finally, the process filtration will probably be performed over a different time from that of the laboratory test, so it is usual to calculate the flux obtained in the laboratory on the basis of volume per unit time and filter area and to check that this is considerably in excess of that required by the process after having applied safety factors to the process filtration area calculation (typically 1.5 or more).

6.6.3 Arrangements

The use of a number of cartridges operating in parallel within a single filter module has already been mentioned. In this respect the cartridge filters are operating as candle filters. In duties that require reliably clean liquids it is common to find a series of filter elements where the first filter in line acts as a prefilter for the subsequent ones. The finest filter is usually placed last as it is likely to be the one with the lowest dirt-holding capacity. Sometimes a train of filters of similar size are employed, the last filter(s) remaining clean during operation. This appears uneconomic, at first, but it should be remembered that the upstream filters will require changing and this is often done using a "duplex" system, during which particle shedding is inevitable. The final clean-up filters act to minimise this and to provide some defence against other actions which could cause shedding from the upstream filters. The additional capital and running costs of a final filter which remains unused are very small, and there are many instances when its inclusion has protected an extremely valuable process.

A duplex system uses two filter housings so that one filter is constantly on-line. When the pressure drop becomes excessive, and the filter requires changing, the flow is directed through the second filter usually by the action of a single handle or lever connected to two, or more, valves. The first filter can then be drained and replaced.

If laboratory tests using a disc of the filter membrane indicate that the filter will have insufficient throughput or life then prefiltration should be considered. Ideally both filters should come to the end of their service life together, i.e. the dirt holding capacity of both filters should be exhausted at the same time. This enables both filters to be replaced together. A duplex system switching over both prefilter and fine filter is available.

6.7 References

Amirtharajah, A. 1991. American Water Works Association, Research Foundation, Denver, USA.

Anon, 1993, Filtrat. and Separat., 30, p 289.

ASTM F316-86, 1987, Pore-size characteristics of membrane filters by bubble point and mean flow pore test, American Society for Testing Materials.

Boucher, P.L., 1946–47, J. Proc. Inst. (iv Eng), 27, pp 415–445.

BS 3321: 1969, The measurement of the equivalent pore size of fabrics, British Standards Institution, Milton Keynes.

Chipps, M.J., 1993, Filtech Conference,'93. Karlsruhe, Filtration Society, Horsham, p 63.

Darcy, H.P.G., 1856, Les Fountaines Publiques de la Ville de Dijon, Victor Delment, Paris.

de Vreede, F.G.A., 1987, Filtech Conference'87, Utrecht, Filtration Society, Horsham, p 422.

Fitzpatrick, C.S.B., 1990, Water Supply, 8, Jonkoping, p 177.

Hermia, J., 1993, Filtration in the beverage industry recent developments, Filtech Conference'93, Karlsruhe, Filtration Society, Horsham.

Hoflinger, W. and Hacki, A., 1990, Filtrat. and Separat., 27, pp 110–113.

Howard, G.W. and Nickolaus, N., 1986, Cartridge Filters, in Solid/Liquid separation equipment scale-up, D.B.Purchas and R.J.Wakeman, (Eds.) Uplands Press, Croydon.

ISO 4572.

Ives, K.J., 1986 in Solid/Liquid separation equipment scale-up, D.B. Purchas and R.J.Wakeman (Eds), Uplands Press, Croydon.

Ives, K.J., 1975, The Scientific Basis of Filtration, Noordoff, Leyden .

Ives, K.J. and Pienvichitr, V., 1965, Chem. Eng. Sci., 20, pp 965–973.

Iwasaki, T., 1937, J. Amer. Wat. Wks. Ass., 29, pp 1591–1602.

Purchas, D.B.,1881, Solid/Liquid separation technology, Uplands Press, Croydon.

Rushton, A., 1985, Filtrat. and Separat., 22, pp 388–391.

Rushton, A. and Spruce, F., 1982, Water Filtration Symposium, Antwerpen, Belgium, Koninklijke Vlaanse Ingenieurs p 233.

Shaw, D.J., 1989, Introduction to Colloid and Surface Chemistry, 3rd ed, Butterworths, London.

Stephenson, T., Mann, A. and Upton, J., 1993, Chemistry and Industry, pp 533–536.

Wenman, R.A. and Miller, B.V.,1987, A novel automatic instrument for pore size distribution, Particle Size Analysis 1985, P.J. Lloyd, (Ed.) Wiley, pp 583–590.

Yao, K.M., Habibian, M.T., O'Melia, C.R., 1976, Environ. Sci. Technol., 5, p 105.

6.8 Nomenclature

C	Solid concentration by volume fraction	–
D	Particle diffusion coefficient	$m^2\,s^{-1}$
d	Pipe diameter	m
g	Gravitational constant	$m\,s^{-2}$
K	Kozeny constant	–
L	Bed thickness	m
ΔP	Pressure Differential	$N\,m^{-2}$
Q	Volume flow rate	$m^3\,s^{-1}$
S_v	Specific surface area per unit volume	$m^2\,m^{-3}$
t	Time	s
u	Particle sedimentation velocity	$m\,s^{-1}$
U_o	Superficial liquid velocity	$m\,s^{-1}$
$\bar{v}$	Average fluid velocity	$m\,s^{-1}$
x	Particle diameter	m
y	Position in filter	m

Greek Symbols

ε	Porosity	–
γ	Surface tension	
λ	Empirical filtration constant	m^{-1}
μ	Liquid viscosity	$Pa\,s$
ρ	Fluid density	$kg\,m^{-3}$
ρ_s	Solid density	$kg\,m^{-3}$
σ	Ratio of solids deposited in filter to filter volume "specific deposit"	–
ψ	Particle sphericity	–

7 Sedimentation and Thickening

7.1 Batch Tests and Analysis

The most important task in specifying or designing a sedimentation process occurs at the outset and that is in recognising the form of settling which is taking place in the suspension under examination. Several parameters play significant parts in this behaviour. Particle size and concentration are the major factors, but the degree of colloid stability and the presence of particles of various densities are others.

Generally dilute suspensions of coarse minerals will settle as nonflocculated suspensions at rates determined mainly by the size and density of the particles and, at the other extreme, concentrated slurries of fine slimy materials will tend to flocculate and will settle as a bulky aggregate with its settling behaviour largely depending on the solids concentration.

Clarifier Design. Design methods for dilute sedimentation are based on the principle of providing enough residence time for the separation to take place. Estimates can be made using Stokes' law or Newton's law as described in Chapter 3, especially for well-defined homogeneous systems, but due allowance must be made for deleterious effects due to convection, air-induced surface waves, inlet and outlet turbulence etc.

Batch Settling: Dilute Systems. The residence time required for the batch settling of a suspension that is composed of a size distribution of particles of the same material can be calculated easily using the size of the smallest particle to estimate the time it will take to settle the full height of the vessel from surface to base.

However for many suspensions it is only economic to settle out the larger particles leaving a reduced concentration of smaller ones to be removed by other means such as clarification filtration. In this case the settled material will comprise two components in particle size terms; material of sizes that have been settled out totally and, secondly, material of sizes that have only been settled out partially.

Some of the particles of the smallest size to be settled out totally must have travelled the full height of the suspension and this condition defines that size. All sizes greater than this one must be separated. For smaller size particles the proportion that are separated is related directly to the proportion of the full height that they sediment in the time made available.

Example 7.1

A aqueous effluent containing mineral particles is to be treated prior to the discharge of the liquid to a watercourse. The first stage in the process is to pump the suspension into a 5 m deep vessel of area 10 m^2, and allow solids to settle for half an hour.

Calculate the solids concentration of the supernatant suspension and the mass (dry weight) of deposited material, if the initial concentration of solids in the effluent is 60 mg/l.

The size distribution of the mineral particles, measured by a sedimentation analysis method, is:

cumulative mass undersize, %	100	92	80	62	48	31	18	8	4	0
equivalent spherical diameter, μm		90	80	70	60	50	40	30	20	10	0

The densities of the particulate solids and the liquid are 2.6×10^3 kg/m^3 and 1.0×10^3 kg/m^3 respectively, and the liquid viscosity is 1.0×10^{-3} Ns/m^2.

Solution

Assume a well-mixed suspension, with the same composition at all parts initially. The size of the smallest particle that is settled out totally can be estimated from Stokes' law. This particle will have settled the full depth of the vessel, 5m, in 30 min.

Hence, from Equation 3.15

$$x_t = \left(\frac{18\mu u_t}{(\rho_s - \rho)g} \right)^{0.5} = \left(\frac{18 \times 1.0 \times 10^{-3} \times 5}{(2.6 - 1.0) \times 10^3 \times 9.81 \times 30 \times 60} \right)^{0.5} = 56.4 \ \mu m \qquad (7.1)$$

(Check that Stokes' law is valid by calculating the particle Reynolds number, which should be less than 0.2.

$$Re_p = \frac{xu_t\rho}{\mu} = \frac{56.4 \times 10^{-6} \times 5 \times 1.0 \times 10^3}{1.0 \times 10^{-3} \times 30 \times 60} = 0.157$$

If Re_p were to be greater than 0.2, Newton's law should be used—see Chapter 3). Thus all particles of sizes greater than or equal to 56.4 μm must settle.

In the next step the proportion of the particles of sizes smaller than 56.4 μm that settle is calculated. From Stokes' law the settling velocity is obtained for each of the top of the range sizes given in the size distribution and the distance settled in the available time (1800 s) is calculated. All particles that are initially within that distance from the base will be collected; those that are above will settle, but will not reach the base of the vessel. So the proportion settled is in the ratio of the distance settled to the overall depth of the vessel.

Create a table beginning with the cumulative undersize data provided, columns A and B, and include an interpolated point for 56.4 μm. Calculate the settling velocity, column C, the distance settled, column D, and the fraction settled, column E, for each of the sizes.

To obtain the fraction of the sub 56.4 μm material that settles, plot column E, the fraction of particles of a given size settling in 1800 s, against column B, the cumulative undersize fraction of material present, and integrate. The area under this curve is the fraction required. It is convenient to integrate numerically using the trapezium rule or similar method:

A	B	C	D	E
size, μm	cumulative fraction undersize	settling velocity m/s $(\times 10^{-3})$	distance settled in 1800 s, m	fraction settled
56.4	0.57	2.78	5.0	1.0
50	0.48	2.18	3.9	0.787 (= 3.9/5.0)
40	0.31	1.39	2.5	0.52
30	0.18	0.784	1.4	0.283
20	0.08	0.349	0.63	0.126
10	0.04	0.087	0.16	0.031

fraction undersize x	fraction settled y	$\frac{1}{2}(y_i + y_{i+1})\Delta x$
0.57	1	0.0842 {= 1/2 (1+ 0.787)(0.57−0.48)}
0.48	0.787	0.10957
0.31	0.502	0.05103
0.18	0.283	0.02045
0.08	0.126	0.00314
0.04	0.031	0.00062
		$\Sigma = 0.265$

Since all material above 56.4 μm settles, i.e. 0.43 of the original material, the total amount settled is 0.265 + 0.43 = 0.695 and the fraction remaining in suspension is 0.305. Thus the solids concentration of the supernatant is 60×0.305 = 18.3 mg/l

The mass of solids deposited = $5\times10^2\times(60-18.3)\times10^{-3}$ = 20.85 kg

Long Tube Test. When heterogeneous systems of particulates are considered, the exact values and ranges of particle sizes, shapes and densities may be ill-defined. In these cases it is necessary to conduct some experimental testing. In the long tube test the object is to achieve the required degree of separation in a tube having a height of a similar value to the depth of the basin necessary, allowing the particles to settle for a time equal to the pool detention time, t_d. This test is of particular importance when system behaviour involves some natural coagulation of the particles at a relatively slow rate which implies a dependence on detention time.

Combinations of pool depth and detention time must be discovered that will give the specified level of removal of solids and to this end a long tube settling column (up to 5 m length) is sampled at various depths H and times t_d, to give concentrations c' of solids. The solids concentrations may be corrected for unwanted colloidal material which is often present in such samples by subtracting the background concentration of unsettleable solids.

The apparatus may be used for a batch sedimentation by simply settling for prescribed periods of time and sampling at all the sample points. An arithmetic average concentration is computed for all the points above and including the one in question. In this way decantation of supernatant liquid above that point is simulated.

Batch Settling: Hindered Systems. In these more concentrated suspensions the proximity of the particles to each other is such that the settling rate is independent of the size and density of individual particles. Laboratory examination of the settling behaviour will show the development of a clear interface between the settling solids and an uppermost layer of clear liquid.

The behaviour of the settling system is characterised by the batch settling curve as described in Chapter 3 and design calculations are based on that data using the Kynch [1951] theory.

Example 7.2

The effluent from a factory consists of constant flow of 3 m³/h of a 3% by volume suspension of solids in water. Before the liquid effluent can be discharged to river it is necessary to filter out the solids. The filters available can handle 0.6 m³/hr of suspension and therefore the effluent must be concentrated in a large tank, effective dimensions 6×4×3m deep, which may be subdivided and used as a batch settler. Batch settling data for the effluent is as follows:

Height of Interface, cm	30	25	20	15	10	7.5	6	5	4	3	2.5
Time, min	0	9.5	18	27	37	44	48	52	57	66	80

Show how the tank may be arranged to allow continuous discharge of effluent from the factory and continuous feed of settled sludge to the filters. Calculate the time to fill settle and empty the batch settler and the average and minimum solids concentrations of the sludge discharged from the settler.

Solution

This problem suggests that there are three concurrent operations for the settler, i.e. filling with effluent, settling and discharging to the filters. Considering a 24 h cycle leads to an 8 h time for each operation with the tank divided into three compartments. A series of calculations must be made to see if this arrangement can be made to work.

Available tank volume must equal the effluent output per 24 h.

Tank Volume = 3×6×4 = 72 m^3

Effluent output/24 h = 3×24 = 72 m^3 ∴ O.K.

Settling time available is 8 hours. The sludge volume must be reduced to 0.6 m^3/h from 3 m^3/h. Inspection of the batch settling data shows a reduction from 30 to 6 cm in 48 min. Therefore the height of sludge interface would fall from 3 to 0.6 m in 480 min i.e. 8 h. So the settlement time available equals the settlement time required.

The average solids concentration of the sludge discharged is given by:

$$\frac{3\times 0.03}{0.6}\times 100 \ = \ 15\%$$

Height of interface cm

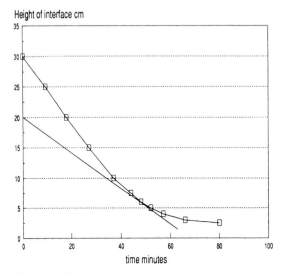

Figure 7.1 Batch settling curve

The minimum concentration is found by taking a tangent to the graph (Figure 7.1) at 6 cm, 48 min point and the required concentration is given by:

$$C_1 \ = \ \frac{C_0 h_0}{h_1} \ \text{(Equation (3.29)). } h_1 \text{ is 20 cm so}$$

$$C_1 = \frac{3 \times 30}{20} = 4.5\%$$

7.2 Design Methods for Continuous Settling

These processes can be described most simply by the Camp–Hazen ideal continuous sedimentation tank model which equates the residence times derived from the vertical settling motion of the particle and the horizontal plug flow movement brought about by the drag of the liquid. Thus for a residence time t the vertical equation is:

$$t = \frac{H}{u_t} \tag{7.2}$$

and the horizontal equation is:

$$t = \frac{AH}{Q} \tag{7.3}$$

where H is the tank depth, A its area, Q the volumetric flow rate of liquid through the tank and u_t the terminal settling velocity of the particle. Clearly the depth of the tank is eliminated from the equation and the relationship becomes:

$$A = \frac{Q}{u_t} \tag{7.4}$$

Since $u_t = f(x)$ the separation efficiencies can be estimated for a range of particle size values.

7.2.1. Continuous Settling: Dilute Systems. When applied to continuous sedimentation the long tube test is used to model the settling of particles in an elemental vertical volume of liquid moving through the length of the clarifier. The corresponding overflow rate V_o of a continuous ideal basin is given by H/t_d, where H is the sample point depth.

Since an overflow system is being designed the corrected concentrations must be averaged over the number of sample points above and including the one in question. For example, if the corrected concentrations measured at sample points 1, 2, 3, 4, and 5 (sample point 1 being uppermost) were c'_1, c'_2, c'_3, c'_4 and c'_5 the average concentrations at those points would be as follows:

Sample point number	1	2	3	4	5
Average concentration c'	c'_1	$\dfrac{c'_1+c'_2}{2}$	$\dfrac{c'_1+c'_2+c'_3}{3}$	$\dfrac{c'_1+c'_2+c'_3+c'_4}{4}$	$\dfrac{c'_1+c'_2+c'_3+c'_4+c'_5}{5}$

The results, plotted as log corrected concentration c' versus log overflow rate V_o for a range of detention times t_d, should appear similar to those shown in Figure 7.2.

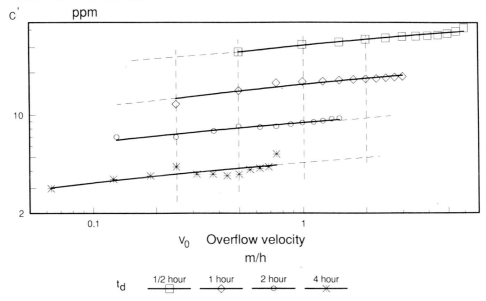

Figure 7.2 Plot of V_o vs c'. Squares ½ h; triangles 1 h; circles 2 h; crosses 4 h

Crossplots are then made of log c' vs log t_d for a range of values V_o. The points are taken from the intersections of the vertical dotted lines with the corrected concentration plots illustrated in Figure 7.2 and the crossplot is shown in Figure 7.3. From this plot for a chosen value of c', the desired clarity, a range of values of V_o and t_d can be obtained. Any of these combinations will give the required clarity in the effluent and suitable ones can be picked off another crossplot of log V_o vs log t_d (Figure 7.4).

Thus for a given throughput rate Q the required area A (= Q/V_o) and depth may be calculated. The non-idealities in the full-scale system are taken into account by using an area efficiency factor (commonly 65%) and a detention efficiency factor, which is a function of depth: diameter ratio and ranges from 20 to 70%, as shown in Figure 7.5.

For this empirical method to be successful it is necessary that the slopes of the plots of log c' vs log V_o have the same value, as shown in Figure 7.2, and the lines may have to be extrapolated to give the points required for the crossplots. Careful experimentation and interpretation of the raw data are vital.

Short Tube Test. If the separation is governed simply by overflow rate, i.e. by the settling rate of the particles, a simplified procedure using a short tube is employed, and this is especially suitable for classification and degritting applications. This test uses a plain transparent cylinder about 50 mm diameter and 1 m long with each end closed by a

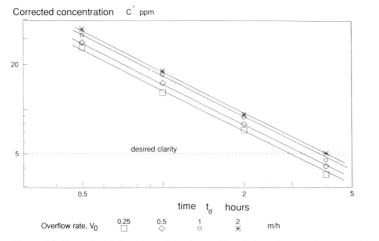

Figure 7.3 Crossplot giving c' *vs* t for values of V_o. Values for overflow rate V_0, m/h are as follows: squares, 0.25; triangles, 0.5; circles, 1; crosses, 2

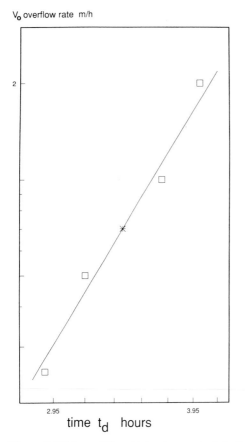

Figure 7.4 Values of V_o and t_d to give required clarity

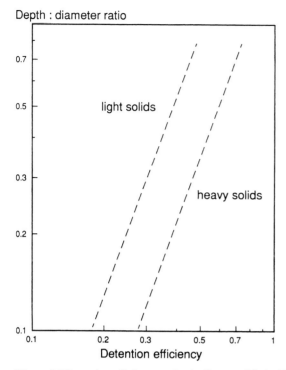

Figure 7.5 Detention efficiency *vs* depth: diameter [Fitch, 1968]

rubber stopper. A vacuum flask connected to a source of vacuum and fitted with a suction tube about 1.5 m long is also required. In each test two settling trials are required. In the first the suspension is prepared and mixed well by upending the tube several times. The tube is clamped in the vertical position, which must be exact, and the solids allowed to settle for the desired length of time. It is useful to have a scale on the side of the tube so that the original suspension height and the settled solids height can be recorded. At this point the tube is upended again and the solids remixed with the liquid. The tube is set vertically and the settling timed. At the set time the upper stopper is taken off and the supernatant liquid removed by the suction apparatus to the level of solids marked in the first settling procedure. Analysis of the supernatant and the original sample for the concentration and particle size distribution will enable the degree of separation to be established and the appropriate settling (or overflow rate) is given by the height settled (the distance between the two marks) divided by time of settling. It will be necessary to use a range of settling times in order to allow interpolation of results to give the desired information. These tests are described in full detail by Fitch [1986].

7.2.2. Continuous Settling: Hindered Systems/Thickening. This process is usually conducted in a raked cylindrical tank with a sloping conical base leading to a central outlet for the sludge, an overflow launder around the periphery of the tank for the clear liquid and a central feed point.

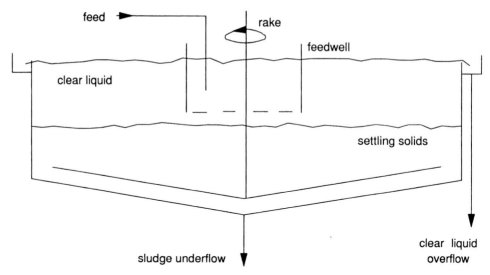

Figure 7.6 Continuous thickener

The major design calculation is to obtain a minimum area for the tank. This may be determined from the settling behaviour of a capacity limiting layer in the thickener whose concentration is not known. The downward settling velocity U is augmented in a continuous thickener by an additional velocity w due to the continuous operation. Thus for the capacity limiting layer, local concentration C and batch settling velocity U, the local flux G is given by:

$$G = C(U + w) \tag{7.5}$$

At constant operating conditions the flux must be the same at all levels so at the discharge concentration C_D, batch settling rate U_D, the flux is:

$$G = C_D(U_D + w) \tag{7.6}$$

and eliminating w from these equations gives:

$$G = \frac{(U - U_D)}{\left(\dfrac{1}{C} - \dfrac{1}{C_D} \right)} \tag{7.7}$$

Coe and Clevenger [1916] suggested finding the rate-determining flux, which has a minimum value, by calculation from Equation (7.7) using a range of values of U and C. If the volumetric flowrate of feed suspension is F and the concentration C_F then the appropriate values of thickener area are found from:

$$A = F \frac{C_F}{G} \qquad (7.8)$$

A graphical solution, the Yoshioka construction, is more appropriate when the batch settling velocity of the discharge concentration is low enough to be ignored, as is often the case. In this method the batch settling data is represented by a flux curve:

$$G_b = UC \qquad (7.9)$$

Material balances over the rate determining layer and the feed and discharge points produce:

$$FC_F = A(G_b + wC) = DC_D \qquad (7.10)$$

where D is the volumetric discharge rate of the concentrated sludge, given by:

$$D = A (U_D + w) \qquad (7.11)$$

Thus if U_D is zero, for unit area:

$$G_b + wC = \frac{FC_F}{A} = wC_D \qquad (7.12)$$

If the conditions in the thickener are at steady state, i.e. total flux constant then:

$$\frac{dG_b}{dC} = -w \qquad (7.13)$$

Since if U_D is zero the total flux is given by wC, then it may be found by taking a tangent to the batch flux curve from the known concentration C_D, as illustrated in Figure 7.7, and the required area found from Equation (7.8). These procedures rely absolutely on the Kynch postulate and thus only apply to conditions where the rate-determining step is the settling rate of some concentration in the bulk settling regime. Occasionally it may be noted that the results of the Talmage and Fitch [1955] procedure do not match the batch flux curves obtained from batch settling curves taken at a range of individual concentrations. In these cases one can suspect that the rate-limiting step to be within the compression stage and not in the sedimentation process.

Coe and Clevenger [1916] observed that for some mineral pulps the underflow sludge concentration, C_D attained in the compression zone is a function of time spent in that zone and it seems that in this regime the settling velocity is a complex function of the rate of change of concentration with height. Channelling often occurs in the compression zone which allows the liquid to find preferential paths so settling rates need not be limiting. By increasing detention time, rates can be increased to make fluxes non limiting.

A design method often favoured industrially is an empirical one due to Oltmann [Fitch, 1986]. To use this technique it is important to be able to identify the compression

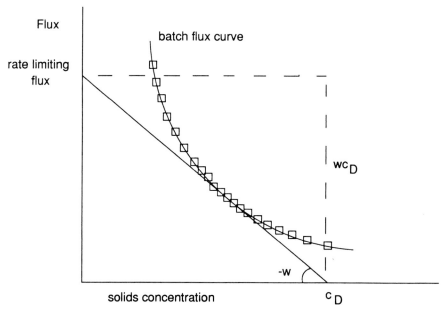

Figure 7.7 Yoshioka construction

point on the batch settling curve. If it is not obvious then a log–log plot may show up discontinuities or a plot of log $(H–H_\infty)$ *vs* linear time, where H_∞ is the height of the settled solids at infinite time, may enable the point to be identified. A batch settling test is conducted with the necessary dose of flocculant chemical and the compression point found as illustrated by point (c) on Figure 7.8.

The start of the constant-rate section is fixed at point (a) and a line drawn through point (c) to point (e). This is found at the intersection of a line drawn parallel to the time axis at a height such that:

$$H_D = \frac{C_o H_o}{C_D} \tag{7.14}$$

The detention time t_D is found from the times t_a and t_e corresponding to points (a) and (e) so that:

$$t_D = t_e - t_a \tag{7.15}$$

and the critical thickener flux is given by:

$$G = \frac{M}{1.2kt_D} \tag{7.16}$$

where M = mass of dry solids used in the test, k is the ratio of volume to height of the graduated cylinder used and the 1.2 constant incorporates a 20% safety factor.

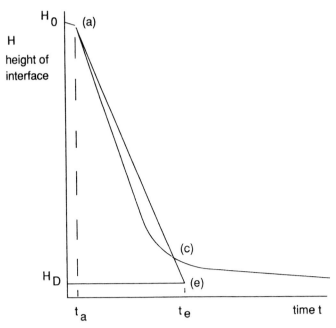

Figure 7.8 Oltmann construction

Example 7.3

From the single-batch settling test data given in the previous example construct a batch flux curve. Determine the required area for continuous thickening from 3 to 15% by (a) the Yoshioka Construction and (b) by the Coe–Clevenger procedure.

Solution

The batch curve is obtained by drawing tangents from calculated values of settling height to the single-batch settling curve. The value of H_l is calculated from appropriate values of concentration:

C, %	3	3.5	3.7	4	4.5	5	5.5	6	7
H_l, cm.	30	25.7	24.3	22.5	20	18	16.4	15	12.9
U	0.560	0.436	0.389	0.344	0.292	0.254	0.22	0.194	0.157
UC	1.682	1.525	1.438	1.374	1.314	1.268	1.21	1.169	1.10

The integer values of C were used for calculation first and the intermediate points calculated when the curvature in the plot became obvious. The column for $C = 3.7\%$ was the last to be obtained.

From Figure 7.9 a tangent to the curve from $C = 15\%$ cuts the flux axis at 1.88 cm/min %. The required area is calculated as:

$$\frac{3\times 3}{60\times 0.0188} = 7.98 \text{ m}^2$$

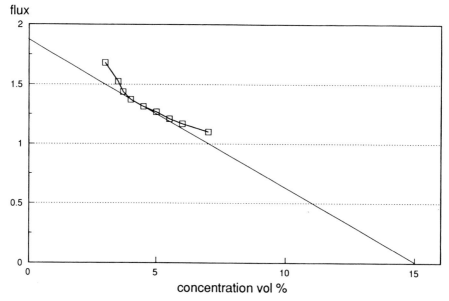

Figure 7.9 Batch flux curve

For a similar range of concentration values the Coe–Clevenger procedure produces, for:

$$U_D = 0, \quad C_D = 0.067$$

C	3.5	3.7	4	4.5	5	5.5
$\dfrac{1}{C}$	0.286	0.2703	0.25	0.222	0.20	0.182
U	0.436	0.389	0.344	0.292	0.254	0.220
$\dfrac{1}{C} - \dfrac{1}{C_D}$	0.219	0.203	0.183	0.155	0.133	0.115
$G = \dfrac{U}{\left(\dfrac{1}{C} - \dfrac{1}{C_D}\right)}$	1.945	1.913	1.88	1.88	1.909	1.913

and the minimum batch flux is 1.88 cm/min % which checks with the result of the Yoshioka construction.

7.2.3. Continuous Settling: Special Designs. In the usual design of a sedimentation process the chemical pretreatment stage is expected to take place separately before the settling step and in many cases this may be most appropriate. However there are applications where this approach is not particularly successful owing to low solids concentrations in the feed, long flocculation times and the formation of low-strength or small flocs.

In a floc blanket clarifier as used in the treatment of potable water with aluminium sulphate, a fluidised bed of aluminium hydroxide flocs is formed in the body of the vessel and the chemically treated feed water is led directly into this bed from below [Ives, 1968]. The effect of the floc blanket is twofold. First, the presence of a high concentration of flocs rapidly increases the rate of coagulation of the particles in the feed stream by orthokinetic coagulation and second there is a particle capture mechanism operating in which the smaller flocs, which might otherwise pass through the clarifier, collect on the larger ones.

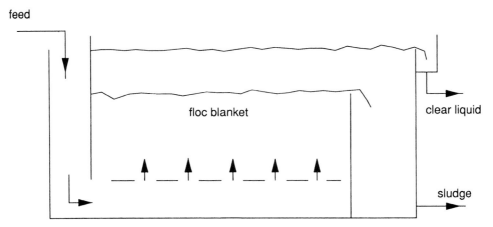

Figure 7.10 Floc blanket clarifier

The clear liquid passes upwards and out of the clarifier over weirs and the accumulating floc material is removed as a sidestream with a little of the clear liquid. With this type of clarifier much higher overflow velocities can be used than in the conventional design.

The high-rate thickener design is very similar to the conventional one but, differs in that it pays particular attention the careful addition of the polymeric flocculant in the feedwell and to the introduction of the flocculated suspension to the sludge bed without diluting or disturbing the settling suspension. Benefit is gained from having an increased concentration in that bigger flocs are produced faster and the time for floc development is reduced. It is claimed that up to ten times the loading rates of conventional designs can be achieved [Keleghan, 1980; Baczek, 1989].

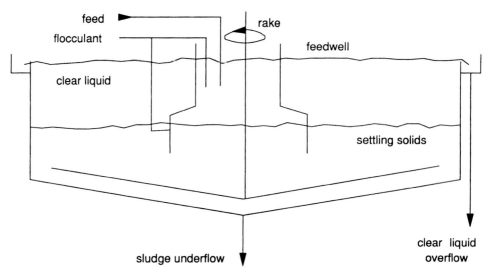

Figure 7.11 High-rate thickener

Table 7.1. Typical Performance Data for Conventional and High-Rate Clarifiers and Thickeners (Courtesy of Outokumpu Technology Pty Ltd., Australia)

High-Rate Thickeners and Clarifiers

Application	Feed (wt % solids)	Underflow (wt % solids)	Vol. Separation Rate (m^3/m^2 h)	Solids Rate (t/m^2 d)
Alumina Red Mud Washing	7-10	45	5.5-6.5	12
Coal Tailings	1-8	25-30	6-10	2-3.5
Copper Tailings	10-35	50-65	1-10	16-50
Kaolin Slurry	1-6	20-30	6-7.5	-
Magnesium Hydroxide from Seawater	2-6	10-25	2-10	1-10
Metal Hydroxide (ex. effluent treatment)	0.1-10	2-10	1.0-2.2	0.1-0.25
Molybdenum Tailings	20-30	50-70	5-10	30-50
FGD Scrubber Water	5-15	20-50	7-10	5-50
Zinc Concentrate	15-40	60-75	3-10	40-50

Conventional Thickeners

Activated Sludge	0.1-0.4	1-3	1.5-2.3	0.2
Cal. Phosphate effluent	0.006	4-7	1.2-1.5	0.02
Iron Oxide ex. S.R.	7-10	40-50	1.6-2.0	2-5
Lead Flotation conc'te	15-40	60-85	0.5-2.0	3-10
TiO2 Pigment clarif.	0.1-10	20-25	0.1-0.15	0.06
Zinc Concentrate	15-40	55-70	0.5-1.5	2-6

Some typical performance figures for high-rate and conventional thickeners are given in Table 7.1.

Deep-Cone Thickeners. Consolidation and compression occurs to some extent in all types of thickeners with most sludges in the sense that as solids reach the base of the vessel and the layers of particles build up, the underlying ones are constrained to pack more closely together thus achieving a higher concentration. Describing this process involves the use of soil mechanics and filtration models, which are difficult to develop because of the anisotropic nature of the solids system network.

No general solution to describe this behaviour is available presently. The constitutive equations have been set out and computer models developed for specific applications such as sewage sludge, which depend on experimentally measured parameters [Adorjan, 1986].

Some materials such as mine tailings can be concentrated to a significant degree by these mechanisms and a special apparatus, the deep cone thickener, has been developed for this purpose.

For these systems the unconfined compressive strength, σ can be related to the solids concentration, C by:

$$\sigma = k \left\{ \frac{C - C_g}{C_\infty - C} \right\}^m \tag{7.17}$$

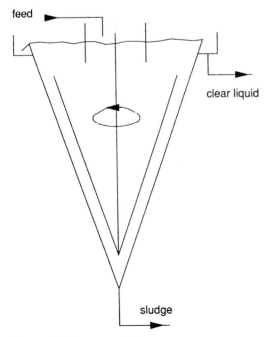

Figure 7.12 Deep-cone thickener

where C_g is the volume fraction at which solids just touch, C_∞ is the upper volume fraction where compressive strength is infinite and k and m are constants.

The depth H of the deep-cone thickener is then obtained from the equation:

$$H = k_2 \int_{C_g}^{C} \frac{(C - C_g)^{m-1}}{C(C_\infty - C)^{m+1}} \, dC \tag{7.18}$$

For colliery wastes $C_g = 0.17$, $k_2 = 0.03275$, $C_\infty = 0.33$ and $m = 0.607$ [Abbott, 1979].

7.3. Consolidation Tanks   In the municipal water industry it is usual to refer to thickeners as consolidation tanks. The tanks are designed to thicken a feed suspension from an average density of approximately 1020 kg/m^3 to values of between 1035 to 1090 kg/m^3. Water is the suspending liquid, therefore, the equivalent concentrations are about 2% w/w feed up to approximately 9% w/w thickened underflow. However, the concentration of the consolidated sediment depends upon the nature of the sludge being settled: primary sludges providing consolidated sediments of up to 9%, whereas activated sludges may only give a concentration of 3.5% w/w. The density of the settling solids may be taken to be 1050 kg/m^3 for the purpose of conversion of the concentrations by weight into those by volume employed elsewhere in this chapter, see Section 2.4.2.

The consolidation tank design procedure was investigated in the 1980's and early 1990's [Hoyland, 1986; Dee et al, 1994], and allows for compression effects within the design procedure. The mathematical analysis for the design is based on the equations briefly presented in Section 3.3. Laboratory test procedures are used to obtain the compression characteristics for the sludge and the resistance to consolidation. These are essentially similar to the parameters used in Equation (2.40) and the permeability expression in Equation (3.37). Hence it is possible to determine the solid concentration at any position within a batch tank using a similar process to that employed in Section 3.3. Also, the concentrations within a continuous consolidation tank can be deduced if the solids loading is known; this is the feed rate of solids per unit area. However, it is convenient to employ the inverse to solids loading: the specific plan area in the units of m^2 day/tSS, where tSS represents tonnes of settleable solids. Thus performance charts for each sludge undergoing sedimentation can be deduced from the numerical solution to the consolidation equations. An example performance chart is provided in Figure 7.13 [Dee et al, 1994].

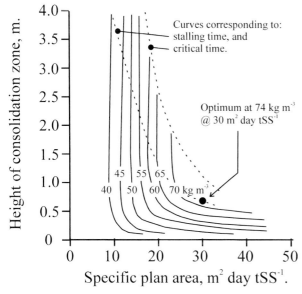

Figure 7.13 Example performance chart for sludge consolidation [Courtesy of WRc, Swindon, UK] (curves represent underflow concentrations resulting from given operating conditions)

The height of the consolidation zone refers to the region in which the solids are in network contact, thus transmitting the weight of the overlying solids to the base of the vessel. The total thickener height is obtained by adding further heights to allow for the clarification zone (approximately 1 m), transport zone at the base (approximately 1 m) and sludge blanket height adjustment zone (0.5 m). Thus a continuous thickener, or consolidation tank, height of approximately 3 to 5 m often results. However, when settling sludges of a biological origin consideration of the possibility of gas bubble attachment to the solids is required. If the solids remain in the tank for too long bubble formation is encouraged. Hence, there is a 'critical time' after which the presence of significant amounts of bubbles may cause the sludge blanket to rise, even in a batch vessel. Before the critical time is reached the presence of a lower quantity of bubbles will hinder the sedimentation, the time at which this becomes significant is called the stalling time. Hence, to obtain optimum consolidation or thickening of a sludge, a thickener should be operated with an average retention time of just less than the stalling time and with sufficient consolidation zone height to provide the required underflow concentration. For a given feed flow rate this operating condition can be read off a performance chart similar to the one provided in Figure 7.13. An example of a thickener designed to this procedure is shown in Figure 7.14.

Figure 7.14 Consolidation tank to WRc design with insert showing picket fence rakes

The picket fence rake is also visible on Figure 7.14, which is used to gently knock gas bubbles off the solids, thus enhancing sedimentation performance. These turn at a rate of 0.1 to 0.15 r.p.m. around the central axis.

The effect of increasing the consolidation zone height can be seen using Figure 7.13, by drawing a vertical line up from a fixed value of specific plan area. Thus a small change in height can have a significant change in the thickened sludge underflow concentration. This may have a deleterious effect on the pumping characteristics of the underflow. Hence an important control consideration is a knowledge, and the maintenance, of the sludge blanket height. An ultrasonic sludge blanket height detector is, therefore, usually employed and is positioned close to the wall of the vessel to prevent debris falling onto the sensor from the feed inlet and to avoid snagging on the picket fence.

7.4. Flotation. The use of air or gas bubbles to separate mineral particles by preferential attachment and subsequent flotation has been a feature of the mineral processing industry for many decades, but the application of the principle to solid–liquid separations is of more recent origin. The techniques have similarities but differences. In mineral flotation it is more common to use chemicals to enhance the collection of the particles and the stability of the froth and to produce the bubbles by beating air into the suspension, but in solid liquid separation it is more usual to dissolve air into the liquid under pressure, allowing the gas to expand out of solution under the lower pressure of the separation chamber, producing bubbles for the preflocculated particles to collect on without the need for further chemical addition.

Alternatives to pressure dissolution of air have been tried during the development of the process technique. These methods of bubble generation include:

1) Dispersed air flotation (DAF)—beating air into the suspension, using a high-speed agitator and an air entrainment system.

2) Electroflotation—passing an electric current between electrodes immersed in the suspension and producing bubbles by electrolysis of the liquid which is usually water.
3) Vacuum flotation—saturating the suspension with air at atmospheric pressure and transferring the suspension to an expansion chamber where the pressure is reduced to less than atmospheric; the dissolved air rapidly comes out of its supersaturated solution, producing bubbles.

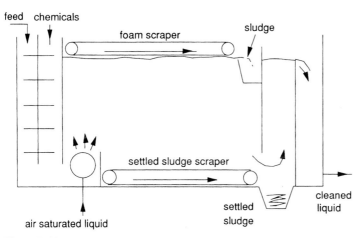

Figure 7.15 Conventional DAF unit with paddle flocculator [Degremont, 1979]

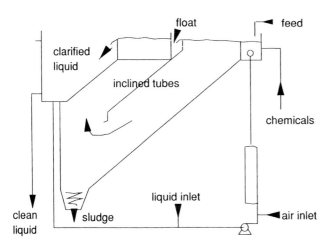

Figure 7.16 DAF unit with lamella separator [Ward, 1981]

Electroflotation is not economical in the treatment of wastewater and generally pressure dissolution of air (dissolved air flotation) is the most widely used method and it is this technique that is discussed in detail below. The objective of the process is to remove by flotation the fine-sized "difficult to settle" particles from a suspension. In wastewater treatment the suspension is usually dilute and the operation is clarification; the method can also be readily used to thicken and concentrate suspensions as in an activated sludge process.

Small bubbles usually <100 μm diameter are found to be most effective in intercepting the suspended particles. For most applications some pretreatment with coagulating chemicals or flocculating agents is necessary to produce a floc which will readily adhere to the bubbles. For aqueous suspensions conventional water treatment chemicals are used; preferably to give a small strong floc which is more likely to survive the shear forces in the liquid around the bubbles rather than larger deformable flocs.

The total incoming flow of feed can be treated with air under pressure, but it is more usual to introduce the air into a recycle stream of cleaned liquid and to allow the bubbles to develop in the flotation chamber at the point where the recycle stream enters. The feed stream is pretreated with chemicals, if necessary, prior to its introduction to the flotation chamber where it meets the bubble blanket. The fine particles in the suspension are captured by the bubbles and are floated up to the surface where they form a thick scum which is removed by a continuous mechanical scraper. Some solids will settle from most feed streams and it is usual to provide for the withdrawal of accumulated silt from the base of the chamber. Figure 7.15 shows a conventional diffused air flotation system housed in a rectangular tank and Figure 7.16 a DAF unit which incorporates an inclined plate settler to capture by sedimentation those particles that escape flotation.

Capture mechanism. The bubble rise velocity u can be described by Stokes' law:

$$u = \frac{x_b^2 \rho g}{18 \mu} \tag{3.15}$$

where x_b = bubble diameter.

The collection mechanisms may include interception, gravity settling and Brownian diffusion. If the overall collection efficiency for a single bubble is η_t defined as the ratio of the particle to bubble collision rate divided by the particle-to-bubble approach rate then η_t can be considered to comprise the sum of the efficiencies for the individual mechanisms so:

$$\eta_t = \eta_d + \eta_g + \eta_i \tag{7.19}$$

where the subscripts d, g and i denote diffusion, gravity and interception respectively.

Individual equations for these efficiencies have been developed for deep-bed filters and these can be written:

$$\eta = 0.9 \left(\frac{K_B T}{\mu x_b x_p u} \right)^{2/3} \tag{7.20}$$

$$\eta = (\rho_s - \rho) x_p^2 \frac{g}{18\mu u} \tag{7.21}$$

$$\eta = \frac{3}{2} \left(\frac{x_p}{x_b} \right)^2 \tag{7.22}$$

where x_p = particle diameter, K_B = Boltzmann's constant, T = absolute temperature, ρ_s = particle density.

Applying these equations to the bubble particle situation and substituting for the bubble rise velocity u gives new equations for η_g and η_d

$$\eta_d = 6.18 \left(\frac{K_B T}{g \rho x_p} \right)^{(2/3)} \left(\frac{1}{x_b} \right)^2 \tag{7.23}$$

$$\eta_g = \left(\frac{x_p}{x_b} \right)^2 \frac{(\rho_s - \rho)}{\rho} \tag{7.24}$$

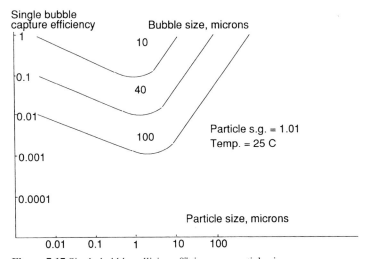

Figure 7.17 Single-bubble collision efficiency *vs* particle size

It can be seen that this single-bubble collision efficiency is related to particle and bubble size, particle and fluid density, and temperature. Edzwald et al [1991] investigated this relationship and their results show a minimum in the efficiency graph at a particle size of around 1 μm, irrespective of the bubble size, as shown in Figure

7.17. At sizes smaller than 1 μm, diffusion is the dominant mechanism; at larger particle sizes interception becomes progressively more important.

Extending this model to the assembly of bubbles and particles in the flotation chamber involves the introduction of the attachment efficiency η_a, which accounts for the proportion of successful particle to bubble collisions, the bubble volume concentration ϕ and the particle number concentration N, which leads to an equation for the particle removal rate:

$$\frac{dN}{dH} = -\left(\frac{3\,\eta_a\eta_t\phi N}{2x_b} \right)$$

(7.25)

If this rate equation is integrated over the tank depth H, from $N = N_o$ at the entry $H = 0$ to $N = N$ at the vessel exit $H = H$, an overall particle removal equation is produced:

$$N = N_o \exp -\left(\frac{3\eta_a\eta_t\,\phi H}{2x_b} \right)$$

(7.26)

The overall particle removal efficiency is given by:

$$\eta = 1 - \frac{N}{N_o} = 1 - \exp -\left(\frac{3\eta_a\eta_t\phi H}{2x_b} \right)$$

(7.27)

Clearly if efficiency is to be improved the possibilities are as follows:

Maximise the single bubble capture efficiency
Maximise the attachment efficiency
Increase the bubble volume concentration
Increase the tank depth
Minimise the bubble size

Capture and Attachment Efficiencies. Both these terms can be enhanced by using a suitable chemical pretreatment that will give flocs with a size of about 50–100 μm. The suspension should be thoroughly destabilised by the pretreatment because any unflocculated particles of small particle size, say less than 1μm, will be captured with minimum efficiency. Increasing the residence time for the feed suspension, for example by increasing the tank depth, will tend to improve the capture efficiency for the smaller particles because the diffusion mechanism will be more effective in those circumstances.

Surfactant Effects. The bubbles and the particles will both carry surface charges which can be measured as the zeta potential. Several authors, eg Deryagin et al [1959, 1964], have shown that theoretically the maximum flotation efficiency should occur when the product of the zeta potential of the bubbles and the zeta potential of the particles is zero. The presence of surfactants can have significant effects on zeta potentials even at low concentrations. Figure 7.18 shows the effect of hexatrimethylammoniumbromide on the zeta potential of gas bubbles and Figure 7.19 illustrates the effect of the same surfactant on the zeta potential of kaolin particles and the corresponding removal efficiency. Small amounts of surfactant can be beneficial but greater dosages can be detrimental.

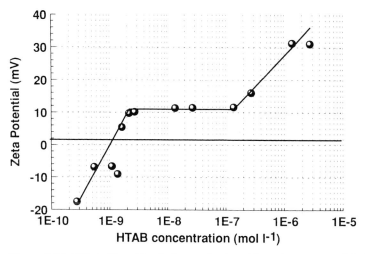

Figure 7. 18 Effect of hexatrimethylammoniumbromide on the zeta potential of gas bubbles [Ward et al, 1997]

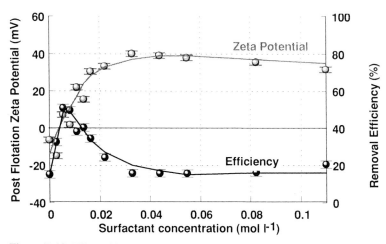

Figure 7. 19 Effect of hexatrimethylammoniumbromide on the flotation of a kaolin suspension [Ward et al, 1997]

As well as producing the surface charge effects, surfactants will alter the magnitude of the hydrophobic force existing between the bubbles and particles and can effectively change the nature of the force from repulsive to attractive. The DLVO theory has been applied to flotation and extended to include hydrophobic forces by Yoon and Mao [1996]. They showed that the hydrophobic forces between the bubbles and particles were greater than those between the particles and they established that the hydrophobic force plays the most important rôle in decreasing the energy barrier between the bubbles and particles. Such a reduction will lead to greater capture efficiency and a higher flotation efficiency.

Bubble Size and Volume Concentration. These parameters are governed by the design of the air–liquid saturator and diffuser system and by the recycle rate. The saturator may be a packed tower, spray tower or an ejector system; a major advantage of saturating a recycle stream of clean liquid is that more efficient saturator designs involving complex packings can be used with less fear of blockage by solids, which is likely to be the case if the feed stream were to be treated in this way. In any case it is a matter of detailed design to protect a complex packing saturator from gross invasion by excess solids.

The bubble size and size distribution is governed by the design and operating conditions of the injection system. Bubbles form and grow on nucleation sites within the surface roughness of the cavitation plate until buoyancy and drag forces are great enough to lift them into the liquid stream. The next stage of growth involves the precipitation of dissolved air onto the newly formed bubbles and the third and final stage of development is by coalescence. which increases bubble size and reduces numbers. A typical bubble size distribution is shown in Figure 7.20. It shows that the bulk of the bubbles is between 20 and 60 μm in diameter and that there is a "tail" of larger bubbles extending to 120 μm in size. Saturator pressure has very little effect on bubble size and distribution.

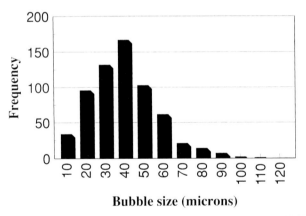

Bubble size (microns)

Figure 7. 20   Typical bubble size distribution [Ward et al. 1996]

Stevenson [1986] suggests that the minimum pressure for satisfactory bubble generation is about 400 kPa and that about 90% saturation is achieved at the working pressure. It is important to prevent release of air from the saturated liquid in the pipework leading to the injection point and to distribute the liquid widely into the flotation tank in order to promote the generation of fine bubbles, to minimise any tendency of bubbles to coalesce and to minimise turbulence which might otherwise cause break-up of flocs.

Edzwald et al [1991] suggest that treating the saturated water as a dilute solution of air in water, allows Henry's law to be used in the form:

$$c = f\frac{p}{K}$$ (7.28)

where c is the concentration of air in the saturated liquid, p is the absolute pressure, K is the Henry's law constant and f is an efficiency factor.

Values are given for f of about 0.7 for unpacked saturators and up to 0.9 for packed systems; for K at 0 and 25 °C figures are 2.72 and 4.53 kPa mg^{-1} l^{-1} respectively. So for 100% efficiency at 70 psig (482.3 kPa) the air concentration is 215 mg/l at 0 °C and 129 mg/l at 25 °C. Thus the concentration, c_r, of air released into the flotation vessel of a continuous system can be obtained from the following expression, assuming that the incoming water's air concentration is c_o,

$$c_r = \frac{(c_s - c_a)R - c_a + c_o}{(1 + R)}$$ (7.29)

where c_s is the concentration of air in the recycle stream, c_a is the air concentration that stays in solution at atmospheric pressure and R is the recycle ratio defined as the ratio of the recycle flow rate to the influent flow rate.

The bubble volume concentration ϕ is obtained by dividing the concentration by the density of saturated air; typical values for this are given as 1.29 and 1.17 g/l for 0 and 25 °C respectively. These authors give typical values for an 8% recycle at 70 psig and for 40 μm bubbles as 4600 ppm for the bubble volume concentration and 1.4×10^5 for the number of bubbles per millilitre. They suggest that for a typical application such as DAF treatment of water containing an algae this would represent about 1 bubble per particle of contaminant.

Most commercial designs of dissolved air flotation plants for use on aqueous systems are based on an empirical approach by Bratby and Marais [1974; 1977]. Their experiments used a pressure saturator in which a high air pressure is maintained in a cylindrical pressure vessel by means of a compressor whilst air is continuously recycled through the water by a circulating pump. Water is fed continuously and the rate controlled to keep a constant level in the vessel. No air is lost from the system.

Tests reported by these authors indicate that Henry's law is not strictly applicable and that the equation should be modified by an exponent on the pressure term:

$$c = \frac{p^m}{K}$$ (7.30)

Empirical values for saturator feed rates in the range 0–10 l/min at 20 °C were 2.45 for m and 370 for K (units for c and p were mg/l and kPa gauge respectively). These authors define a limiting downflow rate v_l in terms of the total hydraulic flow through the flotation vessel divided by its cross-sectional area. The limiting condition is considered to be that at which solids might just be drawn down with the effluent liquid. The limiting downflow rate is equivalent to the upward bubble velocity of the slowest moving bubble–particle agglomerate. Experimentally it was found that there was a relationship between v_l and a_s, the air: solids ratio defined as the mass of air precipitated per unit mass of solids:

$$v_l + v_s = k_1 a_s^{k_2} \tag{7.31}$$

where v_s is the settling velocity of an average particle in the absence of air and k_1 and k_2 are empirical constants. Stevenson [1986] gives values for these constants, for v_l in m/h and a_s in l/kg, as 275 m/h and 0.75 for an algal-laden effluent and 231 m/ h and 0.87 for activated sludge with $v_s = 1.5$ m/h. Generally the value of v_s will be negligible for most applications and the upward rise rate of the froth blanket will be the limiting factor.

Float Characteristics. The bubble-particle system that accumulates on the surface of the tank is termed the float. Prediction of the solids concentration of the float is a complex matter. Bratby and Marais [1977] indicate from their experimental work that generally the floating sludge solids concentration is directly proportional to the depth of the float and inversely related to the solids loading rate. They give equations which involve five empirical coefficients.

For activated sludge, for example, the floating sludge solids concentration, c_f (% w/v) is given by the equation:

$$c_f = 1.32 \, d_w^{0.2} \, \Psi^{-0.5} \tag{7.32}$$

where d_w is the height of the float above the liquid level (m) and Ψ is the solids loading rate (kg m^{-2} h^{-1}). Some of the float will be below the waterline and the depth d_b of this is related to d_w and the air: solids ratio by the equation:

$$\frac{d_b}{d_w} = \frac{0.76}{a_s^{0.45}} \tag{7.33}$$

Laboratory tests. It will be evident from the previous discussion that laboratory tests are necessary in order to establish the design parameters for a new process. Stevenson [1986] describes the equipment required including the Water Research Centre apparatus shown in Figure 7.21 and also explains in some detail how to conduct the experiments.

If flotation is not immediately successful. it will be necessary to undertake settling tests in order to establish the optimum chemical treatment for the effluent being examined. Pilot tests are also described by Stevenson and in some detail by Edzwald et al [1991] and Bratby and Marais [1974; 1977]. Both these references also report extensive experimentation to determine the appropriate pretreatment conditions for the flotation systems being investigated.

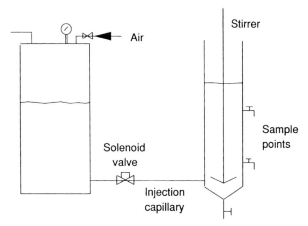

Figure 7.21 Test equipment

Example 7.4

Calculate the bubble volume concentration in a DAF that uses a recycle ratio of 8% with saturator pressure of 583.6 kPa abs. (70 psig) and a saturator efficiency of 70%. Assume an operating temperature of 25 °C and that the feed is saturated with air. Take the Henry's law constant to be 4.53 kPa mg^{-1} l^{-1} and the density of saturated air to be 1.17 g/l. If the bubbles are 40 μm in diameter, what is the number concentration of bubbles?

Solution

When the feed is saturated with air, Equation (7.29) reduces to:

$$c_r = \frac{(c_s - c_a)R}{(1 + R)}$$

The concentration of air in the stream from the saturator is found from Equation (7.28):

$$c_s = \frac{0.7 \times 583.6}{4.53} = 90.18 \text{ mg} / 1$$

The value of c_a is the saturation value at ambient conditions, i.e.:

$$c_a = \frac{101.53}{4.53} = 22.37$$

and hence the concentration in the tank of the unit is:

$$c_r = (90.18 - 22.37)\frac{0.08}{1.08} = 5.02 \text{ mg} / 1$$

The bubble volume concentration is thus:

$$\phi_b = \frac{c_r}{\rho_{sat}} = \frac{5.02}{1.17} \times 10^{-3} = 4.29 \times 10^{-3} \text{ vol fraction}$$

The volume of a 40 μm bubble is 3.35×10^{-14} m^3 or 3.35×10^{-11} l so the number of bubbles per litre is:

$$4.29 \times 10^{-3} / 3.35 \times 10^{-11} = 1.28 \times 10^{8}$$

Example 7.5

A 2.5 m deep DAF unit is being used to reduce the solids concentration in an effluent from 1000 to 10 ppm using an air: solids ratio of 0.3. What increase in air rate will be required to reduce further the output concentration to 4 ppm?

Solution

The air: solids ratio can be increased only by increasing the bubble volume concentration which is achieved by an increase in the recycle rate. So from Equation (7.26), assuming that the bubble size, tank depth and the efficiency terms stay the same:

$$\frac{3\eta_a\eta_t\phi_1 H}{2x_b} = -\ln\frac{10}{1000} \quad \text{and} \quad \frac{3\eta_a\eta_t\phi_2 H}{2x_b} = -\ln\frac{4}{1000}$$

Hence:

$$\phi_2 = 0.3 \times [-\ln(4/100)] / [-\ln(10/1000)] = 0.3 \times \frac{5.52146}{4.60517} = 0.36$$

Thus a 20% increase in air: solids ratio will be necessary to make the stated improvement.

7.5. Inclined Surface Equipment: Lamella Separators. The major advantages of settling under inclined surfaces have already been described in Chapter 3. These comprise a higher area for settling per unit volume of vessel than the conventional settling tank and secondly an enhanced settling rate due to a "convection" effect set up under the inclined surface. In comparison to conventional settlers lamella separators are smaller more compact units having a lower inventory of process materials which are cheaper to build and operate. Modular inclined plate or tube units can also be designed to be suspended within conventional gravity settling tanks to provide additional settling area.

A large number of successful applications of these settlers have been made to a wide range of sedimentation processes, but the design of such systems is not straightforward and in the past many such operations have not been as effective as their designers had expected them to be. Often the reason for this comparative failure was in the development of flow instabilities which interfered with the predicted settling behaviour of the particles, or to poor hydraulic design giving turbulent flow conditions in all or part of the separator.

It can be shown (see Example 7.6) that Q_f the volumetric feed rate for an inclined plate settler with a gross plate width W, plate length L, separation distance b, inclined at an angle α to the horizontal, feed concentration C_f producing a clear overflow and an underflow concentration C_u, may be described by the equation

$$Q_f = \frac{WLv\cos\alpha}{(1 - C_f / C_u)}\left(1 + \frac{L\sin\alpha\cos\alpha}{b}\right)$$

Dimensionless volumetric feed rate

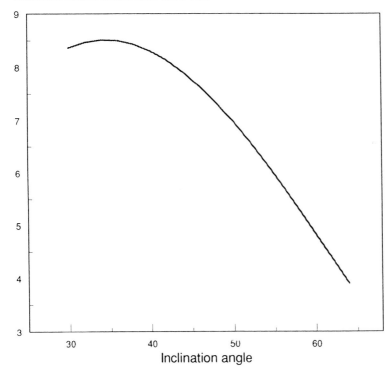

Figure 7.22 Optimum inclination angle

If Q_f is made dimensionless by dividing by the feed rate for the corresponding conventional settler, $\dfrac{WLv}{(1 - C_f / C_u)}$, to give

$$Q'_f = \cos\alpha + \frac{L\sin\alpha\cos^2\alpha}{b}$$

the expression shows the effect of inclination angle. Numerical differentiation produces a maximum value for Q'_f when α is about 35°. The change in Q'_f with α is described in Figure 7.22   for an aspect ratio of 20   which shows that the improvement falls significantly with increasing angle so that at 64° it is less than half the maximum theoretical value.

7.5.1. Dilute Systems. If fully developed laminar flow or uniform flow is assumed to exist between the lamellae, simple particle trajectory analysis may be attempted. For the co-ordinate system shown in Figure 7.23, a general trajectory equation of the following form may be developed:

$$\int \frac{u}{v_o} \, dY - \frac{v_s}{v_o} \, Y \sin\,\alpha + \frac{v_s}{v_o} X \cos\alpha \; = \; C \tag{7.34}$$

where u is the local flow velocity in the x direction, v_o is the average velocity of flow through the settler, v_s is the vertical settling velocity of the suspended particle,

$Y = y/b, X = x/b$

and C is an integration constant.

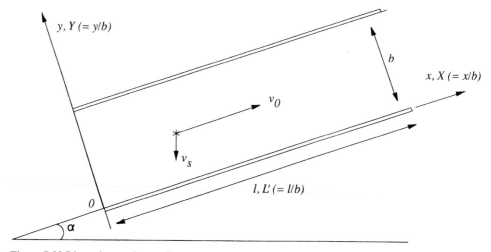

Figure 7.23 Dimensions and co-ordinates

The integral and constant in Equation (7.34) can be evaluated for a particular trajectory in a given lamella separator system. Thus for laminar flow in a circular tube:

$$\frac{u}{v_o} = 8\,(Y - Y^2) \tag{7.35}$$

which enables the integral to be evaluated and the constant is obtained from the boundary condition $X = L'$, $Y = 0$, where L' is the relative length of the settler–plate spacing distance b. Thus a general equation is obtained:

$$8\left[\frac{Y^2}{2} - \frac{Y^3}{3} \right] - \frac{v_s}{v_o}\,Y\sin\alpha + \frac{v_s}{v_o}\,(X - L')\cos\alpha = 0 \tag{7.36}$$

and a limiting trajectory for $X = 0$, $Y = 1$ gives:

$$\frac{v_s}{v_o}\,(\sin\alpha + L'\cos\alpha) = \frac{4}{3} \tag{7.37}$$

Similarly the limiting trajectory equations for laminar flow between parallel plates or in shallow trays and for uniform flow (i.e. $u/v_o = 1$) are of the form:

$$\frac{v_s}{v_o}\,(\sin\alpha + L'\cos\alpha) = 1 \tag{7.38}$$

and that for square conduits:

$$\frac{v_s}{v_o}\,(\sin\alpha + L'\cos\alpha) = \frac{11}{8} \tag{7.39}$$

If the constant on the R.H.S. of Equations (7.37)–(7.39), is given the symbol S and v_s ($= -dh/dt$) is substituted from Equation (3.40) a general trajectory equation is obtained for all three types:

$$\frac{Sv_o}{v} = (1 + kL'\sin\alpha\cos\alpha)(\sin\alpha + L'\cos\alpha) \tag{7.40}$$

An optimum value of α is obtained by differentiation of the R.H.S. of Equation (7.40) and setting the result equal to zero. This gives:

$$(L' + kL')\tan^3\alpha + (2kL'^2 - 1)\tan^2\alpha + (L' - 2kL')\tan\alpha - kL'^2 - 1 = 0 \tag{7.41}$$

Numerical solutions of Equation (7.41) for $k = 1$, $k = 4/\pi$ and $\sqrt{2}$ show the function to be insensitive to values of L' that are of practical interest, producing optimum values of α close to $35°$ for $L' > 1$.

The effect of increasing L' is most easily seen by considering the ratio of vertical tube particle settling velocity v to the average axial velocity v_o as a function of L'. The equation is:

$$\frac{v}{v_o} = \frac{S}{(1 + kL' \sin \alpha \cos \alpha)(\sin \alpha + L' \cos \alpha)} \qquad (7.42)$$

and Figure 7.24 is a plot of v/v_o vs L' for $\alpha = 35°$. Little difference is apparent in the results for circular tubes, square tubes and parallel plates. In each case the value of the ratio falls rapidly with increase in L' and it is clear that there is little to be gained in using values of L' greater than about 15.

The volumetric throughput Q of a given upflow separator of width w and having n spaces between its plates (or in the case of tubes, with n separating compartments) can be obtained by simply substituting

$\dfrac{Q}{nwb}$ for v_o in Equation (7.42) to give:

$$\frac{Q}{v} = (1 + kL' \sin \alpha)(\sin \alpha + L' \cos \alpha) \frac{nwb}{S} \qquad (7.43)$$

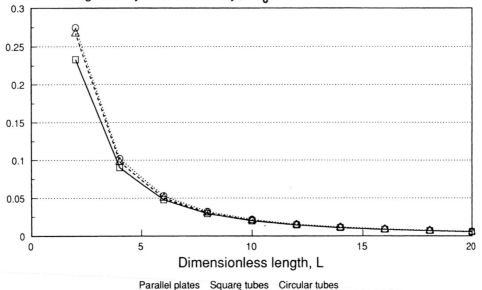

Figure 7.24 Ratio of velocities as a function of dimensionless length. Squares, parallel plates; triangles, square tubes; circles, circular tubes

This equation has the dimensions of area and represents the settling capacity of a given separator. Substitution into Equation (7.43) of the appropriate values for k and S with $L' = 10$ and $\alpha = 35^0$ gives, for parallel plates:

$$\frac{Q}{v} = 49.96 \, nwb \tag{7.44}$$

for square tubes:

$$\frac{Q}{v} = 48.73 \, nwb \tag{7.45}$$

and for circular tubes:

$$\frac{Q}{v} = 45.90 \, nwb \tag{7.46}$$

Hence it might be expected that a parallel plate configuration would require the minimum area to achieve a given duty.

In the trajectory analysis, fully developed laminar flow is assumed to exist which may be justified if Reynolds numbers are kept low. The entry length required for the development of laminar flow is well known to be given by:

$$L'' = k_1 \, Re$$

where L'' is an entry length ratio to plate spacing b and k_1 is a constant equal to 0.04 for flat parallel plates or 0.058 for circular tubes.

Therefore if the separator feed is to be taken from a pipe with the liquid in turbulent flow then some sort of calming section will always be required. In most cases chemical treatment or conditioning to achieve coagulation or flocculation will be used prior to settlement and the design of this section can be used to obtain quiescent flow conditions at the entry to the lamellae. If the desired Re value is 500 then the entry length ratio will be of the order of 20–30 thus making the total length ratio of the order of 35–45 (for $L' = 15$) and with a plate spacing of 5 cm, an excessively high actual plate length of 1.75–2.25 m is obtained in order to ensure that the sedimentation takes place under fully developed laminar flow conditions.

However the transition from flow to laminar flow can be expected to proceed via uniform conditions which will provide similar trajectory equations to those for laminar flow so in practical terms it is only necessary to ensure the elimination of turbulent flow at the entry to the separation section.

Practical experience and experimental comparison shows that the trajectory analysis over-predicts the performance of the settler and it is therefore prudent to allow a 20% increase in the calculated length of plate required. The choice of plate type is resolved principally in terms of ease of construction. Circular tubes do not pack easily, leaving dead spaces in between and are therefore rejected. Parallel plates can be obtained cheaply in a variety of materials and may need to be strengthened in large installations in order to support the settled sludge adequately. There is also the possibility of flow instabilities developing in wide channels which may thus require the use of longitudinal separator strips to maintain a width: depth ratio of about 5 [Yao, 1970]. The square section conduit may be packed very effectively in a given space, can form a very strong structure and although it has marginally less theoretical capacity than the parallel plate configuration the liquid flow pattern is likely to remain stable.

Plate inclination angle. Although it has been shown that an optimum angle of 35° exists, an overriding consideration in the choice of plate inclination concerns the ability of the settled sludge to discharge under the influence of gravity and the surface drag force of the overflowing liquid. Lamella separators may be conceived to operate in a variety of configurations, shown in Figure 7.25, which may include co-current or counter-current flow between the two phases. Clearly with the co-current downward flow of liquid the two forces reinforce and with upward counter-current flow they oppose.

Studies on dispersed systems in batch mode identified four types of movement [Ward and Brown,1996]. Layer movement entails the sliding of the top layers of solids over a relatively thin and almost stationary underlying layer which may be typically 1-2 particle diameters in thickness. The gross behaviour is analogous to the flow of a viscous fluid. and the movement of the overlying layers appears to have a flat velocity profile.

Heap flow occurs when the flow consists of the motion of small aggregates over the entire inclined surface.

In bulk movement the complete thickness of the sludge moves en masse; this behaviour resembles the motion of a block down an inclined plane.

In fracture flow extensive breaks appear within the sludge layer; below these the solids flow in the heap or layer modes of movement. The fractures are formed continuously above the initial shear region. This mode resembles the flow of dry powders.

In most practical applications the sludge layer is usually of the order of a few particle diameters in thickness. Layer movement will be the particle motion to occur at low inclination angles but as the angle increases the movement changes to the heap or bulk type. Particle size and shape, and the nature of the supporting inclined surface are important factors. For glass beads, ground glass, crushed limestone and zircon powders with mean particle sizes above 100 microns layer movement occurred at an angle of 35° or less. This was independent of the nature of the inclined surface. However greater angles were necessary when mean particle sizes were less than 100 microns. Layer flow leaves an underlying coat of particles on the surface; if true plug flow, described above as bulk flow, is required then a higher inclination angle is necessary.

Low bulk density sludges such as metal hydroxides are better handled in a co-current arrangement which allows higher surface loading at closer plate spacing than does counter-current flow. Generally 30–40° is a suitable inclination when working co-

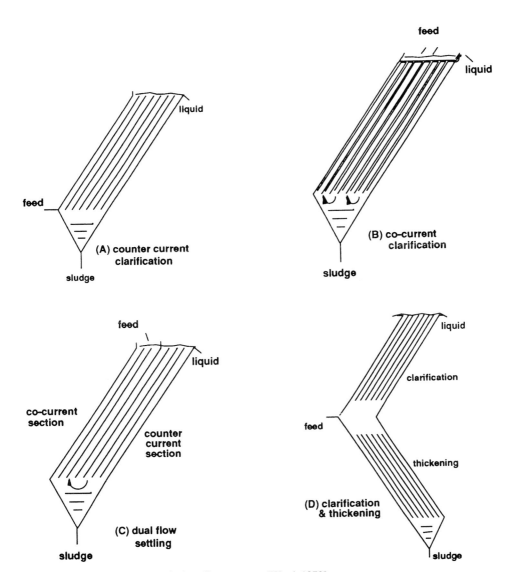

Figure 7.25 Flow arrangements for lamella separators [Ward, 1979]

currently whereas an increase up to 55-60° may be necessary with the counter current arrangement. However in co-current flow a lower limit of throughput is met in the maximum volumetric sludge flowrate that can be handled; which tends to restrict applications to those having small sludge volume fraction. Since counter-current flow implies a simpler design of separating plate this mode of operation is preferred for solids separation. In many commercial designs the inclined plates are ridged to provide shallow triangular-shaped channels in which the settled sludge accumulates to aid discharge.

Plate spacing. In order to minimise the physical size of the separator and thus achieve the most economical design the plate spacing should be as small as possible. Blockage of the conduits by accumulated sludge or tramp material at their base may impose a lower limit on plate spacing and values of 50 mm tend to be common in commercial designs.

7.5.2. Hindered systems. The design of lamella separators for bulk settling or thickening duties involves using the bulk settling rate for the suspension as measured in a vertical sedimentation vessel and modifying that value using the Nakamura–Kuroda model under conditions where that model may be applied. If such conditions cannot pertain, extra settling area must be provided because the modified settling rate will not be achieved.

As discussed in Chapter 3, two dimensionless numbers are used to characterise the settling behaviour; G_R, a sedimentation Grashof number and R, a sedimentation Reynolds number defined by the equations:

$$G_R = \frac{h^3 g \rho (\rho_s - \rho) C_f}{\mu^2}$$

(3.38)

and

$$R = \frac{h v p}{\mu}$$

(3.39)

The ratio of G_R: R is Λ given by:

$$\Lambda = \frac{h^2 g (\rho_s - \rho) C_f}{\mu v}$$

(3.40)

If the Nakamura and Kuroda equation is to predict adequately the separating capability of both batch and continuous lamella separators, then Λ should be in the range 10^4–10^7 and R in the range 0–10 [Ward and Poh, 1994]. Steady state conditions must be maintained and separator dimensions chosen so that flow instabilities are avoided. Typical values for Λ and R are as follows:

Material	Conc, v/v	Settling rate, m/s	Λ	R	h, m
Cement/water	0.173	3.03×10^{-5}	2.3×10^{10}	15.1	0.5
Iron ore/water	0.019	2.39×10^{-4}	8.5×10^{8}	120	0.5
Glass beads/ plasticiser	0.010	2.55×10^{-4}	7.6×10^{4}	0.70	1.13
Limestone (fine)/water	0.005	2.29×10^{-5}	9.2×10^{8}	11.6	0.5
Limestone (coarse)/water	0.005	1.87×10^{-3}	1.1×10^{7}	936	0.5
Activated sludge	0.0011	6.35×10^{-4}	4.3×10^{4}	318	0.5
	0.0013	3.18×10^{-4}	8.5×10^{4}	191	0.5
	0.0017	2.95×10^{-4}	1.4×10^{5}	147	0.5
	0.0018	1.52×10^{-4}	3.0×10^{5}	76	0.5
	0.0021	8.14×10^{-5}	6.2×10^{5}	41	0.5

The table gives some values for Λ and R for a number of common settling systems. Most of the values for Λ are within the desired range, but those for R are usually too great which will indicate that some instability may be present. The effect of this can be countered by providing additional settling area as shown in the example below.

For a given settler there may be two distinct operating modes which have different velocity profiles. The subcritical mode is one in which the clear liquid layer thickness is less that half the channel spacing at the top of the settling channel and decreasing gradually to a minimum at the base of the channel, whereas the supercritical mode is one where the clear liquid layer thickness is greater than half at the top and increasing gradually to a maximum at the base (see Figure 7.26). The flows may be arranged in three distinct ways:

1) Counter-current flow in which the feed and sludge streams move in opposite directions.
2) Co-current flow in which the feed and sludge streams are in the same direction
3) Cross-current flow in which the direction of the feed stream is perpendicular to the sludge stream

Generally counter-current flow systems are easier to design and build.

The aspect ratio h/b for the separator is significant in that there exists an optimum ratio beyond which the design becomes uneconomic owing to the development of flow instabilities. The value of this optimum decreases with increasing feed particle concentration as shown in the table below. Such considerations lead to a design strategy to provide shorter and broader settling channels when treating suspensions of higher concentrations.

Feed volume concentration C_f % vol.	h/b max
0.5	25–30
2.0	15–22.5

As a design guideline the constraint on the value of b, the channel spacing, is given by:

$$b^3 > \frac{192 \tan(90 - \alpha)v\mu x}{g(\rho_s - \rho)C_f} \qquad (7.48)$$

where α is the angle of the inclined surface to the horizontal and x is the distance along the inclined plate surface. The plate separation b, has a minimum value to allow for potential clogging problems due to overfeeding.

Generally the co-current supercritical mode of operation is superior to both the subcritical mode and counter-current flow. It is an inherently more stable system and this reduces the problems of particle re-entrainment so that maximum achievable overflow rates may be attained. Sludge thickening performance is better because greater consistency in the underflow solids is achieved and relatively high underflow concentrations are obtained at shallower angles of inclination, thus giving much higher overall separator throughputs.

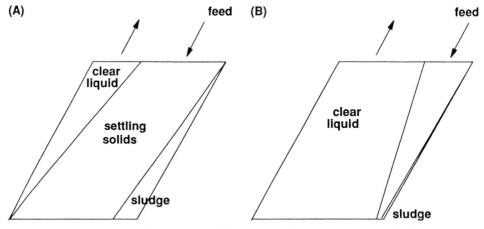

Figure 7.26 Operating modes [Probstein et al, 1977]. A) Subcritical mode; B) Supercritical mode

Example 7.6

Design an inclined plate separator to handle 15 m^{3}/h of an aqueous suspension containing 0.118% vol. of copper oxide. Laboratory tests have shown that the solids settling rate in a vertical vessel is 3.54×10^{-4} m/s and that the settled solids will transport satisfactorily on a 45 degree inclined plane. Assuming that the solids settle to a concentration of 0.4% vol. develop an outline design for a bottom-fed counter-current separator. Density of copper oxide is 5.6×10^{3} kg/m^{3}.

Solution

The core of this problem concerns the specification of an acceptable plate arrangement, i.e. length, width, spacing and number. The calculations are easier if they are done on a spreadsheet so that the interdependence of the variables can be seen instantly. The flowrates and concentrations are related by mass balance equations:

Volume flow: $\quad Q_f = Q_u + Q_o$

Solids flow: $\quad Q_f C_f = Q_u C_u$

hence:

$$Q_f = Q_o / (1 - C_f/C_u),$$

where Q indicates flow rate, C indicates concentration and the subscripts f, u and o indicate feed, underflow and overflow.

Assume plain flat inclined plates. The area is given by $W L \cos \alpha$ where W is the total width of all the plates, i.e. $W = n w$ for n plates each w in width, L is the plate length and α is the inclination angle.

The overflow rate, Q_o is related to the area via the apparent settling rate, v_s so that $Q_o = v_s W L \cos \alpha$. The settling rate is obtained from the Nakamura–Kuroda equation (3.40), with $k = 1$

$$v_s = v \left(1 + \frac{h \cos \alpha}{b} \right)$$

so

$$Q_f = \frac{WLv \cos \alpha}{(1 - C_f / C_u)} \left(1 + \frac{h \cos \alpha}{b} \right)$$

Criteria:

Liquid Reynolds number:

$$Re = \frac{2Q_o \rho}{\mu(W + nb)} = \frac{2Q_f \rho\left(1 - \dfrac{C_f}{C_u}\right)}{\mu(W + nb)}$$

must be less than 500.

Sedimentation Reynolds number:

$$R = \frac{hv\rho}{\mu}$$

should be ideally less than 10.

Ratio of Grashof number to sedimentation Reynolds number:

$$\Lambda = \frac{h^2 g(\rho_s - \rho)C_f}{\mu v}$$

should be in the range 10^4–10^7.

Set up these equations on a spreadsheet and use an iterative approach.

First iteration. Given $\alpha = 45^0$. Take plate spacing b as 40 mm, plate length L as 0.4 m and plate width w as 0.4 m. W is 4.89 and n is 12.

This gives Re as 1092 (too high), Λ as 1.20×10^7 (acceptable) and R as 100 (too high). Clearly, it is necessary to reduce Re and R. It is vital to reduce Re, R can only be reduced by using a smaller plate length. To get R down to 10 requires a plate length of 0.05 m – not practical. Hence accept a higher value for R and compensate by a 40% increase on W.

Second iteration. Multiply W by 1.4 to give 6.85 m. Using same plate dimensions gives n as 17, but Re as 780, which is still too high.

Third iteration. Increase spacing to 60 mm. Plate length and width of 0.4 m gives W as 9.48 m, $n = 24$ and $Re = 539$. R and Λ remain the same.

Fourth iteration. Hydraulic performance would be better if plates were longer than they are wide; there are less likely to be any short-circuit flows developing. Increase spacing to 80 mm, plate length to 0.45 m and reduce width to 0.33 m. W becomes 9.58 m, $n = 29$, $Re = 493$, $\Lambda = 1.52 \times 10^7$ and $R = 112$, which is acceptable in view of the 40% increase in W already included.

7.6 References

Abbott J., 1979, Filtrat. Separat. 16, pp 376–380.

Adorjan L.A., 1976, Trans.I.M.M. 85, p 157.

Baczek, F., 1989, in Flocculation, Sedimentation & Consolidation, Proc. Engineering Science Foundation Conf., 1985, United Engineering Trustees, Inc. pp 261–272.

Bratby J., and Marais G.v.R., 1974, Dissolved air flotation, Filtrat. Separat. 11, pp 614-624.

idem, 1977, Flotation, Chapter 5 in Solid-liquid separation equipment scale-up, D.B. Purchas, (ed.) Uplands Press, London.

Coe, H.S, Clevenger, G.H., 1916, Trans. A.I.M.E. 55, p 356.

Dee, A., Day, M. and Chambers, U.C., 1994, Guidelines for the Design and Operation of Sewage Sludge Consolidation tanks, WRc Publications, Swindon, U.K.

Degremont, 1979, Water treatment handbook, 5th ed. Wiley, pp 191–202.

Deryagin, B.V., Dukhin S.S., and Lisichenko V.A., 1959, Russian Journal of Physical Chemistry, 33, pp 389-393; 1964, *ibid.*, 34, pp 248-251.

Edzwald, J.K., Malley J.P. and Yu C., 1991, Water Supply 9, pp 141–150.

Fitch, E.B., 1986, Chapter 4 in Solid-Liquid Separation Equipment Scale-up, D.B. Purchas and R.J. Wakeman (eds.), 2nd ed., Uplands Press and Filtration Specialists, London.

Hoyland, G., 1986, Proc. Symp. on Effluent Treatment and Disposal, Bradford, April 1986, I.Chem.E. London.

Ives, K.J., 1968 Proc. Inst. Civ. Engrs., 39, p 243.

Keleghan W., 1980, Filtrat. Separat. 17, pp 534–538.

Kynch, G.J., 1951, Trans. Faraday Soc., 48, p 166.

Nakamura, H. and Kuroda, K., 1937, Keijo J. Med., (8), pp 256–296.

Probstein, R.F., Yung, D. and Hicks, R.E., 1977, Engineering Foundation Conference Asilomar.

Stevenson, D.G., 1986, Flotation Chapter 5 in Solid-liquid separation equipment scale-up 2nd ed., Purchas, D.B., Wakeman, R.J. (eds.) Uplands Press/Filtration Specialists.

Talmage, W.P., Fitch, E.B., 1955, Ind. Eng. Chem. 47, p 38.

Ward, A.S., 1979, Filtrat. Separat., 16, pp 477–492.

Ward, A.S., 1981, I.Chem.E. Symp. Series No. 67, pp. 71–85.

Ward, A.S. and Poh, P.H., 1994, First Int. Particle Tech. Forum, Denver, USA, AIChE, New York, Part III, pp 367–372.

Ward, A.S. and Brown, D.J., 1996, Proc 5th World Congress of Chemical Engineering, AIChE, New York, Vol. V, pp 437-442.

Ward, A.S., Stenhouse, J.I.T., Jefferson, B. and Ponting, J., 1996, Proc. IChemE Res. Event, IChemE London, pp 286-288.

Ward, A.S., Stenhouse, J.I.T., Jefferson, B. and Petıraksakul, A., 1997, Proc. IChemE Jubilee Res. Event, IChemE London, pp 457-459.

Yao, K.W., 1970, Intl. Water Poll. Contr. Fed. 42, pp 218–228.

Yoon, R-H., and Mao, L., 1996, J. Colloid Interface Sci., 181, pp 613-626.

7.7 Nomenclature

A	Area	m^2
b	Plate spacing	m
C	Solid concentration by volume fraction	–
c'	Suspended solids concentration	$kg\ m^{-3}$
D	Underflow discharge rate	$m^3\ s^{-1}$
F	Feed suspension flow rate	$m^3\ s^{-1}$
g	Gravitational constant	$m\ s^{-2}$
G	Solids settling flux	$m\ s^{-1}$
H	Height of settling vessel or suspension	m
h	Suspension height	m
L	Plate length	m
L'	Dimensionless directional plate length	–
N	Number concentration	m^{-3}
n	Exponent in Richardson & Zaki equation (3.21)	
Q	Flow rate	$m^3\ s^{-1}$
t	Time	s
U	Settling velocity of suspension	$m\ s^{-1}$
u_t	Terminal settling velocity	$m\ s^{-1}$
U_o	Superficial liquid velocity and sedimentation interface velocity	$m\ s^{-1}$
v	Characteristic settling rate	$m\ s^{-1}$
$V_o, ,v_o$	Overflow rate	$m\ s^{-1}$
v_s	Apparent settling rate	$m\ s^{-1}$
w	Velocity in thickener due to underflow withdrawal	$m\ s^{-1}$
X,Y	Dimensionless directional co-ordinates	–
x,y	Directional co-ordinates	m
x_b	Bubble diameter	m
x_p	Particle diameter	m

Greek Symbols

ϕ	Volume concentration of bubbles	–
η	Collection efficiency	–
μ	Liquid viscosity	Pa s
ρ	Fluid density	$kg\ m^{-3}$
ρ_s	Solid density	$kg\ m^{-3}$

8 Centrifugal Separation

Centrifugal separation makes use of an enhanced field force over that provided by gravity to cause particle or liquid motion, and can be used for liquid–liquid separation as well as liquid–solid separation. Centrifuges may be divided into two main categories: firstly those that separate using a sedimentation principle and employing a solid, or imperforate, bowl; and secondly those that use a filtration principle employing a perforated bowl. In the filtering centrifuge, the g-force provides the pressure to force the mother liquor through the cake. Hydrocyclones have much in common with the sedimenting centrifuge, but the energy needed to cause the liquid to rotate comes from the liquid rather than an external mechanical drive. The two types of centrifuge and the hydrocyclone are dealt with in separate sections following a discussion of basic particle mechanics in a centrifugal field, which is relevant to all three types of process.

8.1 Fundamentals

Section 8.1.1 is designed to provide a background to the physical basis of centrifugal separation, explaining the origin of the centrifugal acceleration and how it is related to the centripetal force. Section 8.1.2 takes the result of the centrifugal acceleration and arrives at an expression for particle velocity in a centrifugal field.

8.1.1 Angular Velocity and Acceleration

Consider a restrained body in circular motion around a point O with angular velocity ω, as described in Figure 8.1.

The velocity change, i.e. acceleration, from point A to point B is:

$$v_B - v_A = v \, \frac{\delta \theta}{\delta t}$$

where v_A and v_B are the vector quantity velocity vectors at point A and B respectively, v is the speed of travel of the point or object and δ represents a very small change in either the angle θ or time t. If the vectors are summed by means of a vector plot and $\delta 0 \rightarrow 0$ then the side in the vector plot between $v_B - v_A$ points towards the centre O of the circle. This leads to the conclusion that there is a velocity change, i.e. acceleration, and therefore a force acting *towards* the centre of the circle, which is known as centripetal force.

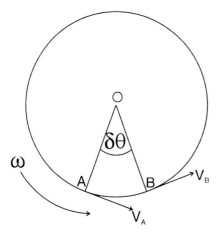

Figure 8.1 Illustration of angular velocity during rotation

When approaching the limit $\delta t \rightarrow 0$:

$$\frac{\delta \theta}{\delta t} = \frac{d\theta}{dt} = \omega$$

the speed of an object is constant but its direction changes, therefore, the acceleration a is:

$$a = v\frac{d\theta}{dt} = v\omega$$

Also, using the result that:

$$v = r\omega$$

the centrifugal acceleration is:

$$a = r\omega^2$$

which acts towards the centre of a circle for a restrained object.

A particle in suspension is an unrestrained object which will be free to move tangentially away from the orbit described by Figure 8.1. If the fluid is rotating at the same rate as the bounding surfaces and if it is assumed that an unrestrained particle accelerates to the speed of the fluid at each radial position from the origin, due to the viscous nature of the supporting fluid, it will receive an impulse from each layer of liquid that it passes, i.e. the trajectory followed by the particle will not be tangential to only one orbit, but tangential to a series of orbits, as illustrated in Figure 8.2.

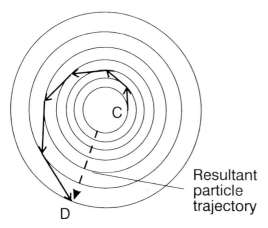

Figure 8.2 Particle trajectory crossing several orbits in a centrifuge

The net effect is that the particle appears to move directly outwards from the centre of rotation with a centrifugal acceleration equal and opposite to that causing the centripetal force.

8.1.2 Particle Velocity in a Centrifugal Field Force

In Chapter 3 the Stokes settling velocity of a small particle was derived, Equation (3.15), by neglecting all forces other than the gravitational field force and the liquid drag. Stokes' law was used for the drag expression. It is possible to derive an analogous equation for velocity in a centrifugal field force by following the same procedure, i.e. equating the product of mass and acceleration (centrifugal field force) with the liquid drag and assuming the particle to be spherical:

$$\frac{\pi}{6} x^3 (\rho_s - \rho) r \omega^2 = 3\pi\mu x \frac{dr}{dt}$$

where x is the particle diameter, ρs is the solid density, ρ is the liquid density, μ is the fluid viscosity and particle mass has been replaced by the product of volume and density, accounting also for the buoyancy effect. The particle velocity outwards from the centre of rotation is, however, dependent on the radial distance from that centre; the term r appears in the acceleration expression, unlike in gravity settling where the acceleration is constant. Hence the velocity must be written in the differential form rather than a constant. This equation can be rearranged to provide the following equation analogous to gravity settling:

$$\frac{dr}{dt} = \frac{x^2}{18\mu} (\rho_s - \rho) r \omega^2 \qquad (8.1)$$

which will be used extensively in the following sections.

Comparison of Equation (3.15) with Equation (8.1) provides a value for the commonly quoted *g* factor:

$$g\text{-factor} = \frac{r\omega^2}{g}$$

Thus the *g*-force of a machine will be the *g*-factor multiplied by the acceleration due to gravity and a mass. The mass will, however, be constant regardless of the acceleration.

8.2 Centrifugal Sedimentation

In centrifugal sedimentation the *g*-force encourages rapid settlement of solids towards the containing bowl. The removal of solids from industrial centrifuges can be divided into batch, continuous and semicontinuous. The machines relevant to these classifications are shown in Figure 8.3, and their designs are considered in the following sections.

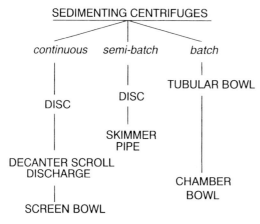

Figure 8.3 Industrial sedimenting centrifuges

8.2.1 Designs

There are a variety of designs available, each with a particular feature which enhances performance for a specific application. For example, the tubular bowl centrifuge has no internal structure, unlike the scroll discharge centrifuge. Therefore, solids discharge is limited to intermittent manual cleaning, but much higher rotational speeds can be used because of the simpler design. Thus the tubular bowl is a better machine to use for polishing or clarification of very fine suspended material, whereas the scroll discharge decanter is a better machine when dealing with more concentrated suspensions. Table 8.1 lists some of the more common designs and their operating characteristics.

Table 8.1 Sedimenting centrifuge types

Machine	Axis	Solids removal	*g* force	Internals/geometry
Disc stack	vertical	none, intermittent or continuous	6000 – 12 000	50–200 thin conical discs
Scroll discharge decanter	mainly horizontal	via a continuous, archimedean screw	2000 – 5000	cylindrical and conical sections
Screen bowl	horizontal	continuous archimedean screw	2000 – 3000	as above with bar screen beach
Skimmer pipe	vertical	intermittent pipe	1000	a cream like slurry discharged through pipe
Tubular bowl	mainly vertical	none	10 000 – 16 000	simple design
Chamber bowl	mainly vertical	continuous, one– chamber discharges	12 000– 18 000	solids flow through concentric chambers

High rotational speeds are generally found on machines operating on immiscible liquid separation rather than solid–liquid separation; the chamber bowl centrifuge is a good example of this. Some of the above characteristics can be seen in Figure 8.4, which illustrates the internals of three of the most common sedimenting centrifuges. These will be discussed in greater detail in later sections.

8.2.2 Simple Sigma Theory

This theory concerns clarification in cylindrical bowl centrifuges, such as the tubular bowl machine. It was first published by Hebb and Smith [1948], and is generally known as Ambler's sigma theory [Ambler, 1952].

a

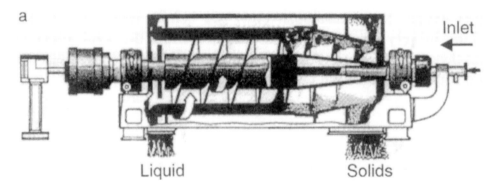

Liquid   Solids

b

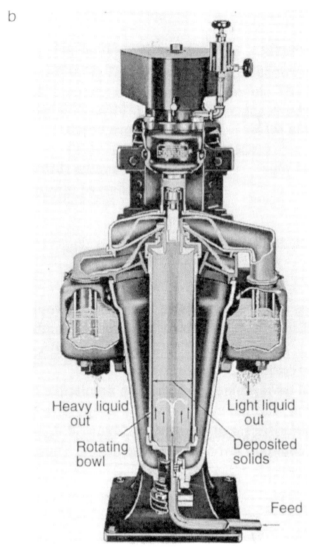

Heavy liquid   Light liquid
out   out

Rotating   Deposited
bowl   solids

Feed

c

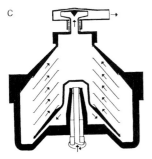

Figure 8.4 Three common sedimenting centrifuge designs. A) Scroll discharge decanter; B) Tubular bowl; C) Disc stack, no solids discharge [Courtesy of Alfa Laval Separation Ltd.]

The theory concerns the settlement of a small spherical particle and equates the time required for settling to the bowl wall to the time required for the element of liquid, in which the particle settles, to travel from the point of entry to the discharge. It is analogous to the settlement of solids in a clarification basin by the Camp–Hazen model (see Chapter 7). Plug flow of the liquid down the axis of the machine is inherently implied. It is also assumed that small particles reach their terminal settling velocity virtually as they enter the centrifuge pond, and that the Stokes settling velocity modified for centrifugal settling (Equation 8.1) is valid. This last assumption could be checked by means of the particle Reynolds number. Under these conditions the particle which is just captured within the centrifuge will travel with the trajectory shown in Figure 8.5. It is assumed that particles are removed from the system when they reach the wall of the centrifuge. Particles that do not reach the wall will be swept out of the machine in the centrate. A particle would have to follow a trajectory above that shown in Figure 8.5 for this to occur, this being a consequence of a smaller diameter in accordance with Equation (8.1).

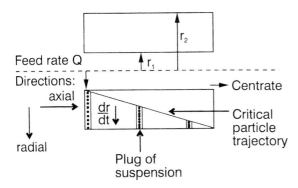

Figure 8.5 Critical particle trajectory for capture

The critical particle trajectory is the particle diameter where the trajectory goes from the top surface (inner radius) of the centrifuge bowl to the bottom surface (outer radius) in the residence time within the machine. Particles of a similar diameter entering the machine between r_1 and r_2 will not present a problem as they will follow a parallel trajectory to the critical one shown, and hence intercept the wall before the end of the machine.

The residence time of the particle in the axial direction (from left to right) will be:

$$t = \frac{V_c}{Q} \tag{8.3}$$

where V_c is the volume of the suspension in the centrifuge and Q is the volume flow rate of material fed to the machine. The radial velocity (up and down the page) is given by Equation (8.1). Integrating this equation using the limits of $r=r_1$ at $t=0$ to $r=r_2$ at $t=t$, and rearranging for residence time gives:

$$t = \frac{18\mu\ln(r_2/r_1)}{x^2(\rho_s - \rho)\omega^2} \tag{8.4}$$

Combining the two equations for residence time (Equations 8.3 and 8.4):

$$\frac{Q}{V_c} = \left(\frac{x^2}{18\mu}(\rho_s - \rho)g \right) \frac{\omega^2}{g\ln(r_2/r_1)}$$

where the first bracketed term is the Stokes settling velocity u_t, Equation (3.15). Substituting this term:

$$\frac{Q}{u_t} = \frac{V_c\omega^2}{g\ln(r_2/r_1)} \tag{8.5}$$

Equation (8.5) has the dimensions of area, and is equivalent to the required plan area of a settling tank operating under ideal conditions needed to perform the same clarification duty as the centrifuge (see Equation 7.4).

The right-hand side of Equation (8.5) is called the *machine parameters*, and is the theoretical area value at 100% efficiency. The left-hand side is called the *process parameters*. Both are given the symbol Σ. The machine sigma term is comprised solely of physical characteristics of the centrifuge, being a measure of the clarification ability of the machine. The settling velocity and volume flow rate to give the desired clarification duty are functions of the process.

The foregoing sigma theory is derived on the assumption of 100% collection of a particle of a critical diameter. It is common to find machines characterised on the basis of only 50% efficiency of collection, i.e. 50% of particles captured within the machine and 50% allowed to enter the centrate. The sigma theory can be modified to take this into account as follows.

It is assumed that the suspension entering the centrifuge is homogeneously mixed, thus to remove only 50% of the particles of a given diameter only 50% of the suspension needs to be processed completely. The critical particle trajectory is therefore altered to that shown in Figure 8.6, where the critical particle enters the machine at a radial position somewhere between that of r_1 and r_2.

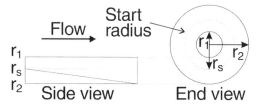

Figure 8.6 50% capture efficiency particle trajectory

Note that this is not halfway between these radii as the model is based on the premise that 50% of the volume of the suspension, or machine, will be completely processed and the machine is circular in cross section; thus half the volume in the machine will lie at a radial position nearer r_2 than r_1. The radius at which the critical particle starts its journey is called the start radius (r_s). The residence time of the particle in the axial direction will still be:

$$t = \frac{V_c}{Q}$$

The start radius will now enter into the radial residence time instead of r_1

$$t = \frac{18\mu \ln(r_2 / r_s)}{x^2 (\rho_s - \rho)\omega^2}$$

The knowledge that equal volumes of the machine must exist between the sections $r_1 - r_s$ and $r_s - r_2$ can now be used to substitute for r_s:

$$\frac{1}{2} = \frac{\pi L(r_2^2 - r_s^2)}{\pi L(r_2^2 - r_1^2)} \tag{8.6}$$

where L is centrifuge length, or:

$$(r_2^2 - r_s^2) = (r_s^2 - r_1^2)$$

hence:

$$r_s = ((r_2^2 + r_1^2)/2)^{1/2} \tag{8.7}$$

Substituting Equation (8.7) into the expression for the radial residence time provides:

$$t = \frac{g \ln [2r_2^2 / (r_2^2 + r_1^2)]}{2u_t \omega^2} \tag{8.8}$$

Finally, combining the equations for residence time gives:

$$\frac{Q}{u_t} = \frac{2V_c\omega^2}{g \ln [2r_2^2 / (r_2^2 + r_1^2)]} \qquad (8.9)$$

The machine parameter, the term on the right-hand side of Equation (8.9), has increased over the value given in Equation (8.5), the process parameter remains unchanged. For a machine of a given size the theoretical equivalent area of a settling basin will be bigger when capturing only half the particles compared with 100% collection efficiency, for the same throughput. Note that the difference between these two machine parameters is not, however, simply a factor of two. This is a consequence of the fact the r_s is not half of r_2 and r_1, as explained when considering the volume ratios.

The comparison with a gravity settler at 50% cut-off gives:

$$\frac{Q}{2u_t} = \Sigma_{machine}$$

where

$$\Sigma_{machine} = \frac{V_c\omega^2}{g \ln[2r_2^2 / (r_2^2 + r_1^2)]} \qquad (8.10)$$

The effective volume of a tubular machine will be:

$$V_c = \pi L(r_2^2 - r_1^2)$$

It is possible to represent a natural logarithmic function by a series, if only the first term in the series solution is considered:

$$\ln f(r) \cong 2\left(\frac{f(r) - 1}{f(r) + 1} \right)$$

where:

$$f(r) = \frac{2r_2^2}{r_2^2 + r_1^2}$$

Combining the equations for machine volume and the series approximation with Equation (8.10) provides:

$$\Sigma_{machine} \cong \frac{\omega^2}{2g} \pi L(3r_2^2 + r_1^2) \qquad (8.11)$$

Sigma can be simplified further if only a shallow pool is used as $r_1 \rightarrow r_2$, hence:

$$\Sigma_{machine} \cong \frac{2\omega^2}{g} \pi L r_2^2 \tag{8.12}$$

The concept of sigma being equivalent to the plan area of a settling tank at 100 or 50% collection efficiency is useful for comparison purposes, but can lead to erroneous conclusions. The diffusional forces on very small particles are much less relevant in a centrifuge, hence sedimenting centrifuges can be used to capture particles which would never settle in gravity sedimentation basins. Particles smaller than 2 μm in diameter are, in general, poorly separated by gravity sedimentation, but may be adequately processed in a sedimenting centrifuge.

8.2.3 Particle Collection Efficiency

The concept of the start radius and its relation to the volume of suspension processed is a very important one in relation to understanding the collection efficiency of a centrifuge. For example, it is possible to rearrange Equation (8.9), using Stokes' law for u_t to provide an equation for the particle size when the collection efficiency is 50%, thus:

$$x = \left(\frac{9 Q \mu \ln [2 r_2^2 / (r_2^2 + r_1^2)]}{V_c \omega^2} \right)^{1/2} \tag{8.13}$$

Considering efficiencies other than 50% Equation (8.6) can be modified to give:

$$p = \frac{\pi L (r_2^2 - r_s^2)}{\pi L (r_2^2 - r_1^2)} \tag{8.14}$$

where p is the fractional proportion of particles of a given diameter that are collected. Thus the start radius, which is a function of particle size, becomes

$$r_s = (r_2^2 - p(r_2^2 - r_1^2))^{1/2} \tag{8.15}$$

which can be used instead of Equation (8.7) in the mathematical development which resulted in Equation (8.9). Rearrangement of such an equation to provide an expression for "cut" size gives:

$$p = \frac{r_2^2}{r_2^2 - r_1^2} \left(1 - \exp\left(\frac{-2 u_t V_c \omega^2}{Q g} \right) \right) \tag{8.16}$$

Alternatively, the start radius associated with a given particle size can be calculated by combining the equations for radial and axial residence times and rearrangement to give:

$$r_s = \frac{r_2}{\exp\left(u_t V_c \omega^2 / gQ\right)}$$

the fraction of particles retained then comes from Equation (8.14). Note that Equation (8.16) is only valid whilst r_s lies between r_1 and r_2. An example of a collection efficiency graph for a tubular centrifuge is provided in Figure 8.7. The curve illustrated in Figure 8.7 does not have the characteristic S shape expected of a "grade efficiency" curve; the top of the curve bends only slightly away from the grade efficiency axis. This is one of the problems with Equation (8.16). It is generally found to overestimate the achieved grade efficiency of a machine by a factor of about 40%. The error is often interpreted as being due to turbulences, end effects, and some of the other assumptions mentioned earlier being inapplicable. In order to overcome this an "efficiency" factor can be applied to the machine parameter, of approximately 60%. A more accurate figure could be obtained from process tests enabling comparison of the process and machine parameters, leading to the fraction required to balance Equation (8.5). It should be noted, however, that if the feed solids concentration is sufficiently high (above 2% by weight) some degree of hindered settling is inevitable and Stokes' law will no longer be valid.

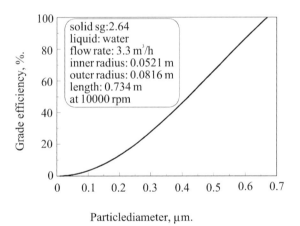

Figure 8.7 Particle collection efficiency with size by Equation (8.16)

8.2.4 Hindered Settling in a Centrifuge

The relation between concentration of the discharged centrifuge solids and feed flow rate, via residence time within the machine, can be illustrated by considering the effect of hindered settling on the suspension in a bowl-type centrifuge.

Under hindered settling conditions the profile of the top interface between the clean centrate and the centrifuged suspension will be as illustrated in Figure 8.8. A weir set a radial distance r_3 will retain the solids from the centrate and some form of continual solids discharge (not shown) must be employed. Consider a layer of suspension settling towards the bowl wall. If forces due to inertia and the solids stress gradient are neglected then the remaining forces are effective solids weight (in a centrifugal field) and liquid drag:

$$0 = C(\rho_s - \rho)r\omega^2 - \frac{\mu dr}{k dt}$$

assuming insignificant net water movement outwards, where k is the permeability of the layer of suspension to the flow of centrate.

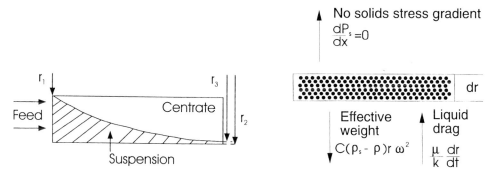

Figure 8.8 Hindered settling. A) Profile of suspension in centrifuge; B) Layer of centrifuged suspension

The residence time for the top of the suspension to travel from r_1 to r_3 is the same as the time required to travel axially down the machine, assuming plug flow. The axial residence time is again given by Equation (8.3), the radial residence time is:

$$\int_0^t dt = \frac{\mu}{(\rho_s - \rho)\omega^2} \int_{r_1}^{r_3} \frac{dr}{Crk}$$

Many expressions exist for permeability, one of the most suitable for settling suspensions can be derived from Equation (3.21), noting that $\varepsilon = 1 - C$ thus:

$$k = \frac{x^2(1 - C)^n}{18C}$$

and

$$\int_0^t dt = \frac{18\mu}{x^2(\rho_s - \rho)\omega^2} \int_{r_1}^{r_3} \frac{dr}{r(1 - C)^n}$$

where x is a particle diameter representative of the settling velocity of the size-distributed material. If the solid concentration of the suspension is assumed to be constant at the final discharge value, which will provide the slowest settling velocity and an over design, then the radial residence time is:

$$t = \frac{g}{u_t \omega^2} \frac{1}{(1 - C)^n} \int_{r_1}^{r_3} \frac{dr}{r}$$

(8.17)

where the Stokes settling velocity has been introduced to replace several constants. Note that when $C \rightarrow 0$ and $r_3 \rightarrow r_2$ Equation (8.17) reduces to the expression for radial residence time in free settling.

A mass balance on the solids provides:

$$\pi L(r_2^2 - r_3^2)C = \pi L C_o (r_2^2 - r_1^2)$$

where C_o is the feed concentration by volume fraction, thus:

$$r_3 = \left(r_2^2 - \frac{C_o}{C}(r_2^2 - r_1^2) \right)^{1/2}$$

(8.18)

Conditions employed in Figure 8.9 are: angular velocity 524 s^{-1}, inner and outer bowl radii 52.1 and 81.6 mm, centrifuge length 0.734 m, particle diameter 2 μm, liquid viscosity 0.001 Pa s, inlet concentration 4% by weight, solid and liquid density 2640 and 1000 kg m^{-3}.

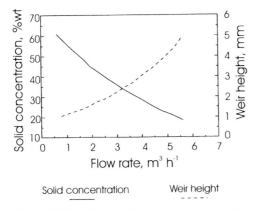

Figure 8.9 Centrifuged solids concentration variation with flow rate. Full curve, solid concentration; broken curve, weir height

It is possible to model the effect of flow rate on centrifuged solids content by solving Equations (8.3), (8.17) and (8.18). Figure 8.9 provides an example of this, and the required weir height (r_2-r_3) to prevent discharge of solids in the centrate. The weir height is the minimum value given ideal separating conditions; it is likely that a weir height in excess of that provided in Figure 8.9 will be required to ensure centrate clarity.

8.2.5 Decanter Scroll Discharge Machine

A simplified diagram of a scroll discharge machine was provided in Figure 8.4. A cut-away section of a commercial machine is provided in Figure 8.10.

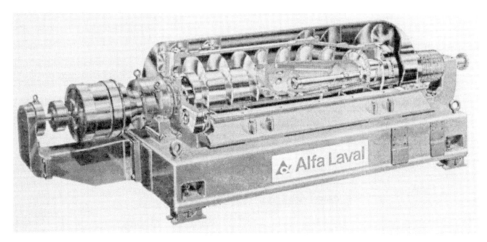

Figure 8.10 The Alfa-Laval Super-D-Canter centrifuge [Courtesy of Alfa Laval Separation Ltd.]

The archimedean screw inside the centrifuge bowl rotates at a slightly slower speed than the bowl, continuously conveying the solids out of the pool of liquid and onto the beach of the machine. Liquid overflows from the pool by means of a weir at the opposite end of the centrifuge. The beach is usually imperforate, i.e. solid similar to the centrifuge bowl, but the "screen bowl" decanter centrifuge has been designed to use a perforated beach to enhance the dewatering of the centrifuged solids. This development of the decanter centrifuge can be used only with large particles.

The detailed internal designs of decanters tend to be an industrial secret and, therefore, little is published on the more advanced technology of these machines. Advances in terms of materials of construction, manufacturing techniques and process applications are also being made. This enables larger diameter machines to spin faster, and improved means of controlling the centrifugation also lead to optimised performance. Some companies now provide a remote telemetry service, where the centrifuge is controlled from the manufacturers' offices, which may be hundreds of kilometres from the process application.

One of the main reasons why decanter centrifuges are becoming more popular for applications which were not believed practicable until recently is the development of shear-resistant flocculants. This has opened up the market of thickening biological sludges, which without these special flocculants would lead to very finely broken up particles of material only slightly heavier than water and, therefore, very difficult to dewater. Other applications are discussed below.

8.2.5.1 Applications

The scroll discharge decanter centrifuge is a ubiquitous machine for the thickening and dewatering of mineral and chemical slurries, digested sludges, activated sludges, raw and mixed sludges, and other biological sludges. The major disadvantage of the machine is the wear that the internal surfaces may be subject to. This has limited their use to slurries of low abrasion. Manufacturers have, however, made some advances in the use of modern high-wear resistant material for the flights on the conveyer, etc. Some applications are provided in Table 8.2.

Table 8.2 Some applications of decanter centrifuge operation

Aplication	Feed conc % wt	Cake conc % wt	Centrate clarity, ppm	Polymer dosage, kg t^{-1}	Capacity, m^3 h^{-1}
FGD - gypsum	35	60–70	<3% solids	0	4–24
Anode slimes	15	75	few ppm	0–1	4–20
Anodising effluent (metal hydroxides)	2–5	15–22	<0.1%	1–2	5–40
Food industry waste	2–4	16–20	400–800	3–6	3–35
Water treatment sludge	1–6	10–35	<1000	1–3	10–20
Waste activated sludge	0.5–1.5	6–8	<1000	1–3	16–180
Digested sludge	2–4	8–20	<1000	1–5	17–160
Paper industry:					
De-inking (TiO$_2$/fibre)	2–4	25–35	<4000		15–20
Caustic effluent	20–40	60–70	<100		20–30
General effluent	2–10	25–40	<100		9–14

The capacity varies with the size of the machine employed; the values given in Table 8.2 represent the range in feed flow rates that may be expected from commercial machines.

Some modern decanter centrifuge designs are claimed to eliminate the need for polymer flocculant, notably in the water treatment industries, but centrate quality usually suffers. During flocculation the very fine particles can become attached on the surface of the larger agglomerates, enhancing centrate quality. Some polymers marketed for use in centrifuges claim to increase the solid content of the cake.

For most neutral and acidic slurries, including municipal sewage, cationic polyelectrolytes are the most suitable flocculant. For waterworks sludges, inorganic sludges and those of neutral to high pH anionic polyelectrolytes tend to be more applicable. Thus anionic and neutral polymers are generally used for mineral slurries where doses are generally below 0.25 kg t^{-1}. In municipal wastes doses may be as high as 10 kg t^{-1}. It is often an important cost in the process and precoagulants, such as lime, alum or iron salts, may be added prior to polymer dosing.

8.2.5.2 Sigma Theory for Scroll Discharge Decanters

Expressions for the sigma factor at 50% collection efficiency have been derived for the scroll discharge decanter centrifuge. The geometry of these machines is more complex than for the simple bowl centrifuge and the formulae for the sigma factor are likewise more complex. Two basic geometries are considered: a machine consisting of a conical bowl with scroll, and a machine that has both cylindrical and conical sections and a scroll. The latter is more common. For a conical bowl alone the following expression can be used for sigma [Purchas, 1981]:

$$\Sigma = \frac{2\pi\omega^2 L_1}{g}\left(\frac{r_2^2 + 3r_2 r_1 + 4r_1^2}{8}\right) \tag{8.19}$$

where L_1 is the length between where the feed enters the centrifuge and the liquid overflows the discharge weir. For a cylindrical and conical bowl-type centrifuge the sigma factor is:

$$\Sigma = \frac{2\pi\omega^2 L_2}{g}\left(\frac{3r_2^2 + r_1^2}{4}\right) + \frac{2\pi\omega^2 L_3}{g}\left(\frac{r_2^2 + 3r_2 r_1 + 4r_1^2}{8}\right) \tag{8.20}$$

where L_2 is the length of the cylindrical part of the centrifuge and L_3 is the length between where the feed enters the centrifuge and the cylindrical part.

8.2.5.3 Power and Efficiency

The power required to run a centrifuge is estimated from three separate power requirements. These are, the power necessary to accelerate the process streams to their discharge radii P_p, the power necessary for scrolling P_s and the power necessary to overcome windage (entrained air) and friction P_{wf} [Records, 1990]:

$$P_p = 7.615 \times 10^{-16} S^2 D^2 M \quad \text{kW} \tag{8.21}$$

$$P_s = 1.047 \times 10^{-4} ST \quad \text{kW} \tag{8.22}$$

where S is the bowl speed in rpm, D is the discharge diameter in mm, M is the process mass flow rate in kg h^{-1} and T is the gearbox pinion torque in N m. The windage and friction component is a complex function of bowl speed and bowl speed squared. A manufacturer can usually provide a windage and friction graph (against speed) for the machine supplied.

Efficiency can be split into two parts: that due to scrolling and that of dewatering. Poor scrolling efficiency will result in solids not being conveyed onto and up the beach for discharge. When the solids emerge onto the beach they will lose the buoyancy effect of the liquid, and they may fall back into the pond. Scroll and beach friction and beach angle are important variables in determining this effect. Some machines are designed to work with submerged beaches to lower the beach friction and to assist cake discharge using the liquid buoyancy. The scrolling mechanism has been described in detail in the literature [Lavanchy and Keith, 1964]. It is often found in processes that the conveying torque developed in a decanter is closely related to the dryness of the cake produced. This is most useful in the process control of this type of equipment.

The cake dryness is generally dependent on the g-Force and residence time inside the centrifuge, as described by Equations (8.3), (8.17) and (8.18). In a decanter the residence time on the beach will have a major effect on the cake dryness. Scrolling efficiency has an important influence on this residence time and, therefore, cake dryness. It has been proposed [Records, 1990] that the scrolling rate is used in the scale-up of decanter centrifuges, a similar rate providing similar cake dryness.

8.2.6 Disc Stack Machine

The disc machine is able to produce 6000–12 000 g and is widely used for clarification duties and oil–water separation. It is a high–capacity machine, but because of its nature the degree of thickening it can achieve is limited. There are several types of disc machine, including the opening bowl, the nozzle discharge, the nozzle discharge with recycle and various nozzle discharge actuated by timers or sludge density. These designs are employed in an effort to remove the thickened suspension from the machine continuously or intermittently. These designs are illustrated in Figure 8.11.

These centrifuges are also classified according to their duty. The "separator" or "solids-retaining" disc centrifuge is for removal of small amounts of sludge and has no cleaning mechanism other than stopping the machine and manual removal of solids. The "purifier" is designed for separation of two liquid phases with, possibly, some solids present. The emphasis in this machine is on production of a clean light liquid phase, e.g. a water-free oil phase. The heavier liquid phase can contain appreciable amounts of the lighter phase and the solids. The "concentrator" is designed to remove small amounts of light liquid-phase from a heavier phase, e.g. oil from water, and the "clarifier" is designed to remove solids from a single liquid phase only. The purifier, concentrator and clarifier can be either of the solids-retaining variety or employ intermittent or continuous removal of the separated phases.

If the solids are retained in the machine the sludge capacity becomes important; this varies from 0.8 to 23 l for bowl diameters of between 0.24 and 0.72 m. The corresponding liquid throughput varies from 0.05 to 50 m^3 h^{-1}, dependent on liquid properties as well as machine size. In general, the ease of separation and volume throughput vary in the order: cylinder oil, grinding oil, heavy diesel to fuel oil (easiest last).

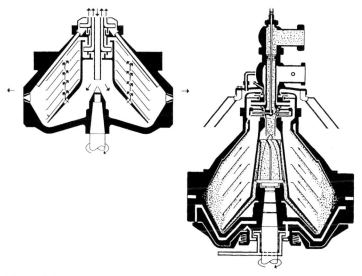

Figure 8.11 Various designs for solid discharge from a disc stack centrifuge [Courtesy of Alfa Laval Separation Ltd.]

8.2.6.1 Modified Sigma Theory

The liquor to be clarified flows between closely spaced conical discs from the outside to the axis. A parabolic velocity distribution exists between the discs and the particles settle to the underside of the discs by virtue of the radial centrifugal force, settling from the high-velocity zone to where the velocity is zero. Thus the settled solids slide down the underside of the discs and into the chamber outside the disc. If a_s is the distance between the discs and n is the number of discs:

$$\frac{dx}{dt} = \frac{dr}{dt \sin \theta} = \frac{Q}{2\pi a_s nr}$$

and

$$\frac{dy}{dt} = \frac{dr}{dt} \cos\theta = \frac{x^2}{18\mu} (\rho_s - \rho) \, r \, \omega^2 \cos \theta$$

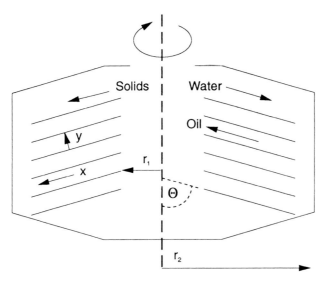

Figure 8.12 Schematic diagram of the disc stack

These two expressions for residence time, after integration, can be equated in a critical particle trajectory analysis similar to that described in Section 8.2.2. When considering only 50% capture of particles the volumes processed now have to be related to the area of a frustum (a truncated cone), the start radius being that value that has equal volumes of the section between each disc above and below it. The resulting expression for sigma (machine parameter) is:

$$\frac{2\pi\omega^2 (r_2^3 - r_1^3)n}{3g \tan \theta} \tag{8.23}$$

The above theoretical developments have been reviewed recently [Leung, 1998] along with a description of modern developments and applications in a wide range of solid bowl centrifuge machines.

8.2.7 Design Calculation Examples

1) A continuous tube centrifuge with a bowl 1.5 m long and 0.75 m diameter operating with a pool depth of 0.1 m at 1800 rpm, is clarifying an aqueous suspension at a rate of 5.4 m³ min⁻¹. All particles of diameter greater than 10 μm are being removed. Calculate the efficiency of this machine and the estimated grade efficiency curve. The solid and liquid specific gravities are 2.8 and 1.0, and the liquid viscosity is 0.001 Pa s.

Equation (8.5) can be used to provide values for the sigma machine and process parameters. Thus:

$$\Sigma_{process} = \frac{Q}{u_t} = \frac{5.4}{60} \frac{18(0.001)}{(1 \times 10^{-5})^2 (2800 - 1000)9.81}$$

or 917 m^2, this is the plan area of a settling tank which the centrifuge replaces or is equivalent to, i.e. a settling tank of this plan area would do the same clarification job as this centrifuge, neglecting any effects due to diffusion or local convection within the tank. The theoretical equivalent plan area of the settling tank doing the same duty as the centrifuge, calculated from the centrifuge dimensions is:

$$\Sigma_{machine} = \frac{V_c \omega^2}{g \ln (r_2 / r_1)} = \frac{\pi (r_2^2 - r_1^2) L \omega^2}{g \ln (r_2 / r_1)}$$

where the outer radius is 0.75/2 m, the inner radius is 0.75/2–0.1 m (i.e. the pool depth). This calculation provides a value of 3580 m^2. So, at 100% efficiency the plan area of an equivalent settling tank is 3580 m^2, actual plan area is, however, only 917 m^2 hence the efficiency is 917/3580 or 25%. If we assume that this efficiency is valid for all particle sizes then this factor can be included in a revised Equation (8.16):

$$p = \frac{r_2^2}{r_2^2 - r_1^2} \left[1 - \exp\left(\frac{-2u_t E_A V_c \omega^2}{Qg} \right) \right]$$

where E_A is the efficiency as a fraction. The following table compares the results with and without the correction for centrifuge efficiency for this example and at various particle sizes.

	Grade or collection efficiency (%)							
Particle size μm:	0.2	1	2	4	5	6.6	8	10
Centrifuge efficiency %								
25:	0.1	1.3	5.2	19.9	40.3	50	69.4	100
100:	0.2	5.2	19.9	69.4	100	100	100	100

8.3 Hydrocyclones

A hydrocyclone is a device employing centrifugal separation without the need for mechanically moving parts, other than a pump. They are cheap, compact and versatile as a means of solid–liquid separation. It is similar in operation to a centrifuge, but with much larger values of g-force (ranging from 800 g in a 300 mm diameter cyclone to 50 000 g in a 10 mm diameter cyclone). This force is, however, applied over a much shorter residence time. The most significant difference in the fluid mechanics between the centrifuge and the hydrocyclone is that the liquid in the former rotates as a solid body with constant angular velocity, a forced vortex, whereas the hydrocyclone approaches constant angular momentum conditions, a free vortex. The former is akin to a gramophone record, the latter analogous to an ice dancer who changes speed of rotation by arm movement. A density difference between the dispersed phase and liquid is an essential requirement for both hydrocyclone and centrifuge. As this density difference reduces so does the effectiveness of the centrifugal separator. The residence time inside a centrifuge could be extended to compensate for a low density difference and even batch centrifugation is possible. A similar option does not exist with hydrocyclones. Subject to a significant density difference hydrocyclones can be effective in separating particles down to 2 µm in diameter, below this size the efficiency is lowered by the complex flow patterns and turbulence inside the cyclone. The principal features and flow patterns are shown schematically in Figure 8.13.

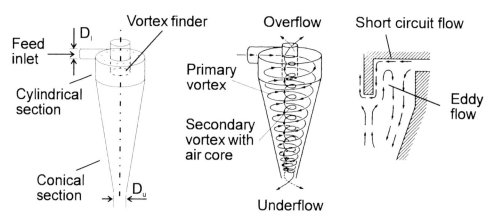

Figure 8.13 Principal features and flows inside a hydrocyclone

The most noticeable flows in the hydrocyclone are the primary and secondary vortices. The primary vortex lies outside the secondary one, and carries suspended material down the axis of the hydrocyclone.

The secondary vortex carries material up the axis and into the overflow vortex finder. There is also an air core of a few millimetres diameter at the centre of the vortices. Some of the suspended material "short-circuits" the vortices by leaking around the top of the hydrocyclone and into the overflow. The vortex finder design is important in reducing this loss of unclassified material. Hydrocyclones can act as classifiers and/or thickeners; the overflow being a dilute suspension of fine solids, the underflow is a concentrated suspension of more coarse solids. The two designs which are used to enhance either the thickening or classifying action of the hydrocyclone are shown in Figure 8.14. The long cone provides thicker underflow concentrations, but poorer sharpness of separation than the long cylinder.

8.3.1 Cut Point and Fractionation

It is usual to describe the efficiency of hydrocyclone performance in terms of x_{50}; where particle separation is 50% efficient, i.e. the particle has a 50% chance of entering the overflow or underflow from the cyclone. Figure 8.15 shows an idealised size distribution of a cyclone feed, and how it might be split into the cyclone overflow (OF) and underflow (UF). Figure 8.16 shows the grade efficiency curve for this cyclone which is defined as the fraction of material, usually by mass, of a particular size range (grade) appearing in the underflow from the cyclone (E_i):

$$E_i = \frac{\text{Mass in size grade in UF}}{\text{Mass in size grade in feed}} \tag{8.24}$$

where the subscript i represents the grade under consideration. The mass flow rate may also be used instead of mass in Equation (8.24) for each grade. Grade efficiency is sometimes called the solids recovery and the recovery curve illustrated in Figure 8.16 is sometimes known as the Tromp curve. It is worth noting that the term "grade" simply implies size range and, clearly, the efficiency of a device depends upon its application. Thus grade efficiency curves that are the mirror image of Figure 8.16 are also possible, where the efficiency is defined in terms of the recovery of solids reporting to the fine cut of a classifier, the overflow in the case of the hydrocyclone.

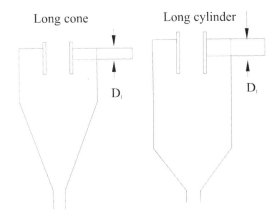

Long cone D_1 <Long cylinder D_1

Figure 8.14 Hydrocyclone designs

Further consideration of hydrocyclone types is given in Section 8.3.9.

An ideal classifier would split the particle size distribution fed to it into two fractions, one of a size below the cut point the other greater than the cut point. In

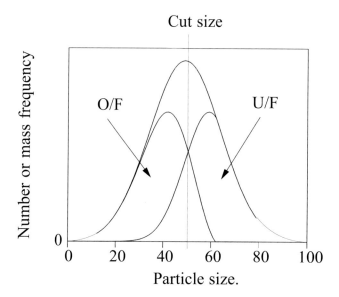

Figure 8.15 Size distributions

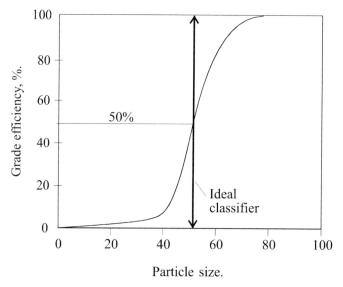

Figure 8.16 Grade efficiency curve

practice this degree of classification can be achieved only by hand picking particles into two piles using a pair of tweezers. All mechanical separators will provide undersized material in the coarse fraction, and very often oversize in the fine cut. A factor for the sharpness of separation can be defined by ratioing the particle diameters corresponding to the 75 and 25% value of efficiency, i.e.:

$$\text{Sharpness of separation} = \frac{x_{25\%}}{x_{75\%}} \tag{8.25}$$

Thus the ideal classifier will have a sharpness of separation equal to unity. For non-ideal classifiers the sharpness of separation will be fractional, and the lower the value the poorer the separation.

If the full size distribution of a material is known then the grade efficiency based on particle number, length, area, mass or volume will be the same as any factor required to convert between these quantities will cancel. Very often in practice, however, the full size distribution is not accurately known, there may well be a large amount of the distribution in the finest channel of the equipment used for particle size analysis. It is therefore advisable to work in terms of a mass distribution, and to provide a grade efficiency based on this. The total mass fraction of feed solids reporting to the underflow is termed the gross efficiency [Bradley, 1965], or total solids recovery.

8.3.2 Reduced Grade Efficiency

The hydrocyclone, in common with other wet classifiers, provides an underflow which could contain solids by means of entrainment and short-circuiting from the feed, rather than due to the act of classification. Thus the action of simply splitting the flow into two equal fractions will effect a grade efficiency of 50%, according to the earlier definition. This efficiency can be achieved without any grading according to particle size. Furthermore, a grade efficiency of 100% could be obtained by simply blocking off the overflow, i.e. whilst achieving nothing. The concept of reduced grade efficiency is used to overcome the effect due to flow split. It is assumed that the amount of material entering the underflow without experiencing classification is proportional to the volume flow split going to the underflow. For example if the solids feed to a hydrocyclone of a particular size (grade) is 10 kg h^{-1} and 6 kg h^{-1} enters the underflow the grade efficiency is:

$$\frac{6}{10} \times 100\% = 60\%$$

If 20% of the total volume of the feed reports to the underflow after classification the reduced grade efficiency is:

$$\left(\frac{6}{10} - 0.2 \right) \times 100\% = 40\%$$

i.e. only 4 kg h^{-1} of the material is entering the underflow as a consequence of the classifying action of the hydrocyclone.

In general, the reduced grade efficiency E^*_i is therefore:

$$E^*_i = E_i - R_f \tag{8.26}$$

where R_f is the volumetric flow split of the underflow relative to the feed. It is likely that Equation (8.26) will produce the correct limit at the lowest particle size, i.e. zero, but does not provide 100% efficiency at the largest size. Some consideration of grade efficiency has been made [Bradley, 1965] and the following equation proposed to overcome this effect:

$$E_i^* = \frac{E_i - R_f}{1 - R_f} \qquad (8.27)$$

The different grade efficiency curves are illustrated in Table 8.3 and Figure 8.17.

Table 8.3 Example of a grade efficiency calculation

Particle diameter, μm	Mass in range inlet, %	Mass in range overflow, %	Mass flow in range inlet, g s^{-1}	Mass flow in range overflow, g s^{-1}	Grade efficiency Eq. (8.24), %	Grade efficiency Eq. (8.26), %	Grade efficiency Eq. (8.27), %
118.4	0	0	0	0			
88.1	0.6	0	2.64×10^{-2}	0	100	89.6	100
65.6	1.6	0	6.56×10^{-2}	0	100	89.6	100
48.8	3.2	0	0.131	0	100	89.6	100
36.3	5.1	0	0.209	0	100	89.6	100
27	6.7	0.4	0.275	5.5×10^{-3}	98.0	87.6	97.8
20.1	8.0	1.1	0.328	1.51×10^{-2}	95.4	85.0	94.9
15	10.7	4.3	0.439	5.92×10^{-2}	86.5	76.1	85.0
11.1	12.5	9.9	0.513	0.136	73.4	63.0	70.3
8.3	16.0	18.7	0.656	0.257	60.8	50.4	56.2
6.2	15.8	24.6	0.648	0.338	47.8	37.4	41.7
4.6	10.3	19.9	0.422	0.274	35.2	24.8	27.7
3.4	5.0	10.3	0.205	0.142	30.9	20.5	22.9
2.6	2.3	5.1	9.43×10^{-2}	7.02×10^{-2}	25.6	15.2	17.0
1.9	0.6	1.5	2.46×10^{-2}	2.06×10^{-2}	16.1	5.7	16.4
1.4	0.5	1.3	2.05×10^{-2}	1.79×10^{-2}	12.8	2.4	2.7
1.2	1.1	2.9	4.51×10^{-2}	3.99×10^{-2}	11.6	1.2	1.3
0	0	0	0	0		0	0

Other conditions are: inlet and overflow concentrations, 20.3 and 7.6 kg m^{-3}, and rates 2.02×10^{-4} and 1.81×10^{-4} m^3 s^{-1} respectively. Therefore, mass flow rate inlet and overflow was 4.10×10^{-3} and 1.38×10^{-3} kg s^{-1} respectively, and flow split was 0.104.

The three grade efficiencies are plotted against the midpoint of the size grade in Figure 8.17 .

The cut point is the most important part of the grade efficiency curve and it is notable that this value is not altered greatly by the choice of definition. It is nowadays conventional to employ Equation (8.27) for calculations of reduced grade efficiency, as it provides the correct values of 0 and 100% for the limits. Another notable observation from Figure 8.17 is that the accuracy of the curve is poorer towards the origin. Very small values of percentage mass in the feed or overflow distribution can substantially alter the position of the curve. Even the use of three significant figures precision provides "step" changes in the curve in this region. The problem is accentuated when using the underflow distribution size analysis data to calculate the grade efficiency, as this stream has only very small amounts of material in the fine grades.

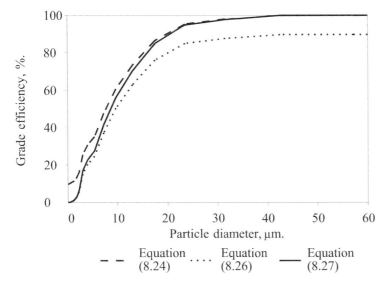

Figure 8.17 Examples of grade efficiency curves. Curve (a). Eq. (8.24); curve (b), Eq. (8.27); curve (c), Eq. (8.26)

An alternative graphical means of determining the reduced grade efficiency curve has been proposed [Svarovsky, 1984], which is claimed to be more accurate than considering incremental particle sizes, and plotting midpoints. It does, however, require precise and accurate values for the cumulative mass undersize of the overflow and underflow, and employs graphical differentiation of a plot of these parameters. These requirements and procedures are, regrettably, seldom successfully achieved in practice. The use of the volumetric flow split to provide a numerical value for the bypassing of solids from the classifier has also been questioned [Austin and Klimpel, 1981] and an alternative proposed.

8.3.3 Velocities

Hydrocyclones are separators which rely on non–axisymmetric flow to provide the means of separation, i.e. the feed is introduced off centre and in one or possibly two positions only. To understand how a hydrocyclone functions it is, therefore, important to consider the three types of velocities present within the device, and to bear in mind its asymmetric nature. The lack of symmetry is most important when considering numerical solutions to the flow inside a hydrocyclone; any solution based on assuming symmetry provides only a trivial solution to the continuity (mass balance) equation and provides no information about the liquid flow. Some distinction between the solid and liquid velocity has to be drawn in the following discussion; clearly the liquid must have a tendency to concentrate in the overflow, the solids in the underflow. Thus the solid and liquid velocities must differ in at least one direction.

8.3.3.1 Tangential Velocity (Fig. 8.18)

The tangential velocity inside the hydrocyclone is very important, it is the means by which a suspended particle following the liquid flow path because of drag will experience a centrifugal force. The tangential velocity of the solid will be similar to the liquid on entry into the hydrocyclone and it is assumed that this is also the case at any instant at radii less than that of the entry.

The linear velocity of the feed at the inlet v_f is related to volume throughput Q and area of inlet nozzle A_I as follows:

$$v_f = \frac{Q}{A_I} \tag{8.28}$$

Equation (8.28) provides a value for the tangential velocity at the outer radius of the hydrocyclone. Tangential velocities at radii less than that of the hydrocyclone can be estimated by means of the principle of conservation of angular momentum: under frictionless conditions:

$v_i r_i = \text{constant}$

where v_i is tangential velocity at the radius of rotation r_i. In real systems energy is lost and it is argued that the angular momentum will be less than that given by the above equation. Equation (8.29) is often used to account for this:

$$v_i r_i^n = \text{constant} \tag{8.29}$$

note that if n ≠ 1 the constant has the SI units $m^{(1+n)} \, s^{-1}$ and not the units of angular momentum. In fact, if the values of *n* and *r* are fractional then the product of velocity and radius will be greater than that when *n* is unity, i.e. under these conditions it is no longer possible to call Equation (8.29) an expression for angular momentum as the momentum will be greater than under frictionless conditions! Use of this equation will be illustrated later, see Equation (8.35). Empirical values for n are as follows [Kelsall, 1952]: 0.7 for water and approaching 0.5 for 15–20% w/w slurries. Experimental measurements of the tangential velocity show that it varies in a way similar to that shown in Figure 8.18. The velocity increasing in accordance with Equation (8.29) until close to the inner air core, after which some decline is observed.

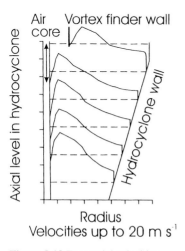

Figure 8.18 Tangential velocities

8.3.3.2 Radial Velocity (Fig. 8.19)

The solid and liquid radial velocities must differ substantially for there to be an increase in the liquid: solid ratio in the overflow, and vice versa in the underflow, relative to the feed stream.

The net liquid velocity inwards can be estimated from a knowledge of the net flow into the overflow, assuming uniform flow inwards through a surface of known dimensions, see Section 8.3.5. The net solid flow outwards could be estimated at each point in the hydrocyclone by means of a force balance between the centrifugal field and liquid drag forces. Such an approach is used to determine the equilibrium orbit which will be discussed in detail later. The maximum values of both the liquid and solid velocities are found at the wall of the hydrocyclone, diminishing to zero at the air core. Experimental measurements suggest that the radial liquid velocity is negligible in the cylindrical section of the hydrocyclone.

8.3.3.3 Axial Velocity (Fig. 8.20)

The net solid and liquid flows are in the same direction, unlike radial flow, but there are two distinct regions inside the hydrocyclone with net velocities in different directions. The secondary vortex spins into the vortex finder and, therefore, takes material into the overflow. Thus net flow in the secondary vortex is upwards. In the primary vortex net flow is downwards towards the underflow.

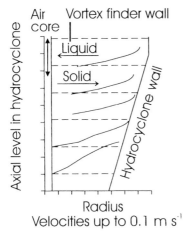

Radius
Velocities up to 0.1 m s^{-1}

Figure 8.19 Radial velocities

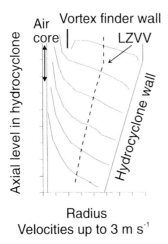

Radius
Velocities up to 3 m s^{-1}

Figure 8.20 Axial velocities

One important consequence of the existence of two regions of velocities in opposing directions is that there must exist a position of no net vertical movement which acts as the boundary between these two regions. This boundary is three-dimensional extending down through the cylindrical and conical parts of the hydrocyclone and should be rotated around the central axis. Such a rotation would provide a surface or locus of zero vertical velocity (LZVV). The position where the locus changes from being a cylinder to a cone is subject to some debate. Superficially, the change may be thought of as following the geometry of the hydrocyclone, but there is experimental evidence to suggest that the cylindrical section extends into the cone as discussed below.

8.3.4 Locus of Zero Vertical Velocity and Mantle

Experimental evidence suggests that there is a region at the top of the locus of zero vertical velocity (LZVV) in which the radial velocity is also zero [Bradley and Pulling, 1959]. Dye tracer experiments have shown that dye collects in this region, which forms a hollow cylinder around the vortex finder. This stagnant surface of liquid, and consequently particles trapped within the surface, is called the mantle. The LZVV with radial flow has been found to be conical in shape, starting from a height where the diameter of the conical section is 0.7 of the diameter of the cyclone, and extending down to the bottom of the cyclone [Kelsall, 1952]. There is, however, little evidence to suggest that the mantle will extend into the conical section of the hydrocyclone to the value of $0.7\,D_c$ for all geometries.

8.3.5 Equilibrium Orbit Theory

The principle behind the equilibrium orbit theory is to equate the liquid drag on a particle, caused by radial flow of liquid into the centre of the hydrocyclone with the centrifugal field force on the particle. If the two forces are in balance the particle will not move inwards or outwards. It will orbit with the appropriate tangential velocity for its radial position. If the radial position corresponds to the LZVV then the particle will not show any preference for transfer into the overflow or underflow, i.e. the diameter of a particle orbiting at the radius corresponding to the LZVV will be the hydrocyclone separation size x_{50}. In order to use the equilibrium orbit theory the radius of the LZVV must be fixed, this is usually achieved by considering the volume flow split going into the overflow, and assuming that this is equal to the volume flow split within the hydrocyclone. The radial liquid flow can then be calculated based on the volume flow rate entering the overflow and the surface area of the LZVV. If only that part of the LZVV which lies below the mantle is used then detailed knowledge of the mantle is required. Alternatively, the full surface area of the LZVV may be used in which case the theory will predict a finer cut size than is measured. Stokes' law is often used to equate the liquid drag through this plane with the centrifugal field force. Refinements to this theory have included the use of alternative liquid drag equations and the addition of hindrance terms to account for hindered settling effects, as demonstrated in Section

8.2.4. The development shown below will use the full LVZZ for the purpose of illustration.

Consider the hydrocyclone to have a LZVV consisting of a cylinder and cone as shown in Figure 8.21. The volume of a cylinder and cone are $\pi r^2 l$ and $1/3\ \pi r^2 l$ and the flow split to the overflow is $1-R_f$, hence:

$$1 - R_f \;=\; \frac{\pi R^2 l_1 + 1/3\pi R^2 l_2}{\pi r_o^2 l_1 + 1/3\pi r_o^2 l_2} \;=\; \left(\frac{R}{r_o}\right)^2 \tag{8.30}$$

where R and r_o are equilibrium orbit and hydrocyclone radii respectively.

The centrifugal acceleration of a particle in the hydrocyclone is:

$$a = r_i \omega^2$$

where ω is the angular velocity which can be replaced by the ratio of the tangential velocity to the radius:

$$\omega \;=\; \frac{v_i}{r_i}$$

Hence:

$$a \;=\; \frac{v_i^2}{r_i} \tag{8.31}$$

now:

Centrifugal force = Stokes drag

thus:

$$\frac{\pi}{6} x^3 (\rho_s - \rho) \frac{v_i^2}{r_i} \;=\; 3\pi\mu x u$$

which can be rearranged to give:

$$x \;=\; \left(\frac{18\mu u R}{v_i^2 (\rho_s - \rho)}\right)^{0.5} \tag{8.32}$$

where the equilibrium orbit R is used for r_i.

The liquid velocity u is obtained from:

$$u \;=\; \frac{Q_{OF}}{2\pi R l_1 + \pi R l_2} \tag{8.33}$$

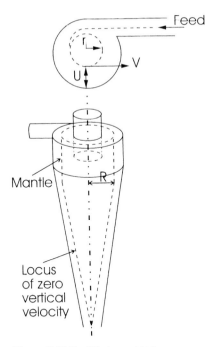

Figure 8.21 Equilibrium orbit theory

or if no radial liquid flow over the mantle is assumed:

$$u = \frac{Q_{OF}}{\pi R l_2} \tag{8.34}$$

i.e. the overflow volume flow rate divided by the surface area of the cylinder and/or cone. The tangential velocity at the equilibrium orbit v_R comes from the Equation (8.29)

$$v_R = v_f \left(\frac{r_o}{R} \right)^n \tag{8.35}$$

where v_f is the tangential velocity at the inlet, which is simply the inlet volume flow rate divided by the cross-sectional area of the inlet pipe. Thus Equation (8.30) can be rearranged to provide a value of R, if the flow split and hydrocyclone geometry are known. Equation (8.33) or (8.34) provide a value of u and Equation (8.35) a value of v_R if the volume flow rates and n are known, and Equation (8.32) predicts the cut size if the physical properties of the solid and liquid are known. Figure 8.22 illustrates how equilibrium orbit theory can be used to predict how the cut size will vary with feed flow rate for the operating conditions described below. Also included on the figure are experimentally measured points and the measured pressure drop.

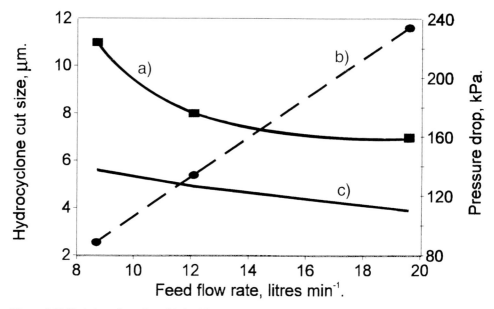

Figure 8.22 Variation of cut size with feed flow rate: measured and equilibrium orbit theory. Hydrocyclone dimensions: total length 273 mm; conical length 221 mm; diameter 42 mm; inlet diameter 5 mm. Solid and liquid densities 2710 and 1000 kg m^{-3}; viscosity 0.001 Pa s. Flow splits in order 8.7, 12.1 and 19.6 l min^{-1}: 0.138, 0.104, and 0.082
Curve (a), cut size; curve (b), pressure drop; curve (c), equilibrium orbit with it n = 0.7

Equilibrium orbit theory is a useful means of correlating and explaining the relation between flow rate and hydrocyclone cut size. Its use as a predictive tool is limited, however, as tests must be conducted to determine several parameters required in the model; notably the flow split and the exponent on the radius in the cyclone. It does not provide any information on the pressure drop required to perform the separation, or on the sharpness of the cut.

8.3.6 Residence Time Model

The original residence time model [Rietema, 1961] was developed on the assumption that for 50% collection efficiency a particle must travel from the centre of the inlet pipe to the wall of the hydrocyclone, i.e. a distance of half of the inlet diameter, in the time that the particle is present within the hydrocyclone. The concept that only half of the uniformly suspended solids in the feed is processed giving the particle diameter at 50% collection efficiency was used in sigma analysis, see Section 8.2.2.
Equation (8.32) can be rewritten as:

$$\frac{dr}{dt} = \frac{x_{50}^2(\rho_s - \rho)v_i^2}{18\mu r_i}$$

where dr/dt represents the solid particle radial velocity in an effectively stationary liquid. Now resolving the velocity vectors, assuming that the local vectors are identical to the overall dimensions of the hydrocyclone:

$$\frac{dz}{dr} = \frac{L}{r_o} = \frac{dz}{dt}\frac{dt}{dr}$$

by the chain rule. Thus:

$$\frac{dz}{dt} = \frac{L}{r_o}\frac{x_{50}^2(\rho_s - \rho)v_i^2}{18\mu r_i}$$

The centrifugal head is:

$$\frac{dP}{dr} = \rho\frac{v_i^2}{r_i}$$

and:

$$\frac{dz}{dt} = v_z$$

therefore:

$$v_z = \frac{L}{r_o}\frac{x_{50}^2(\rho_s - \rho)}{18\mu}\frac{1}{\rho}\frac{dP}{dr}$$

both sides can be divided by expressions for the volume flow rate to give:

$$\frac{4v_z}{\pi d_I^2 v_f} = \frac{L}{r_o}\frac{x_{50}^2(\rho_s - \rho)}{18\mu Q}\frac{1}{\rho}\frac{dP}{dr}$$

where d_I is the diameter of the feed pipe to the hydrocyclone. The distance boundary condition for the integration is from the centre of the feed pipe to the wall, i.e. from 0 to $1/2\ d_I$, thus:

$$\frac{x_{50}^2(\rho_s - \rho)L\Delta P}{\mu\rho Q} = \frac{36v_z r_o}{\pi v_f d_I} \tag{8.36}$$

which provides the result that $\Delta P = f(x_{50}^{-2})$.

Rietema went on to speculate that for an efficient separation the left-hand side of Equation (8.36) should be as small as possible, and that the terms on the right-hand side are effectively constant (the ratio of axial to inlet velocity being a constant), thus there should be a hydrocyclone design which provides the optimum design, i.e. the lowest value in Equation (8.36). The right side of Equation (8.36) is known as the characteristic cyclone number and experimental work yielded a minimum value of 3.5 for the designs that Rietema tested. These have become a set of standard or optimum hydrocyclone designs.

8.3.7 Dimensionless Group Model

The use of a set of dimensionless correlations to relate cut size to pressure drop and flow rate of the hydrocyclone has been proposed [Svarovsky, 1984]. Tables of constants for various commercial hydrocyclones are provided enabling prediction of cut size and pressure drop without the need for any laboratory test. This approach does, therefore, provide more information than the equilibrium orbit theory and also has the advantage of giving a result without preliminary testing. The dimensionless group model can, of course, be refined if testing is conducted in order to fix the scale-up constants accurately. The model was developed for slurries of chalk and alumina hydrate, which had densities of 2 780 and 2 420 kg m^{-3}, respectively, in water with 0.1% Calgon added as a dispersing agent. The feed volume was varied from 1 to 10% by volume. The correlations involve the use of the following relations:

$$Stk_{50} = \frac{x_{50}^2(\rho_s - \rho)v_z}{18\mu d_c} \qquad (8.37)$$

where Stk_{50} is defined as the Stokes-50 number at the cut size, d_c is the cyclone diameter and v_z is the characteristic liquid velocity inside the hydrocyclone which is defined as:

$$v_z = \frac{4Q}{\pi d_c^2} \qquad (8.38)$$

i.e. equivalent to the axial liquid velocity. The Reynolds number is defined as:

$$Re = \frac{v_z d_c \rho}{\mu} \qquad (8.39)$$

and the Euler number as:

$$Eu = \frac{\Delta P}{\rho v_z^2 / 2} \qquad (8.40)$$

$$\frac{x_{50}^2 (\rho_s - \rho) L \Delta P}{\mu \rho Q} = 3.5$$

Equations (8.37)–(8.40) can be combined with this to give:

$$Stk_{50} Eu \frac{2L18}{\pi d_c} = 3.5$$

Now, Rietema's optimum design used a L/d_c ratio of 5, hence:

$$Stk_{50} Eu = \frac{3.5\pi}{180} = 0.061 \tag{8.41}$$

and the relation between Euler and Reynolds numbers for Rietema's cyclone geometry was empirically found to be:

$$Eu = 24.38 \, Re^{0.3748} \tag{8.42}$$

Thus if the feed flow rate to such a cyclone is altered the new characteristic velocity can be calculated from Equation (8.38), Reynolds number from (8.39), Euler number from (8.42) and pressure drop from (8.40). The new cut size can be calculated from Equation (8.37) after using (8.41) to give the new Stokes-50 number. However, Equations (8.41) and (8.42) are valid only for cyclones of Rietema's optimum geometry.

Svarovsky proposed that there is a general relation between the Stokes and Euler numbers for all geometries, thus:

$$Stk_{50} Eu = \text{constant} \tag{8.43}$$

and the general relation between the Reynolds and Euler numbers:

$$Eu = KpRe^{n_p} \tag{8.44}$$

where K_p and n_p are also empirically derived constants. He went on to provide a table of these constants for some more common hydrocyclones, see Table 8.4.

The factor in the final column in Table 8.4 is called the running cost criterion where:

$$\Delta P \approx Stk_{50}^{4/3} Eu \tag{8.45}$$

the power required to effect a separation being directly proportional to the pressure drop.

It can be seen that the general trends shown in Figure 8.23, cut size and pressure drop with flow rate, are similar to those displayed in Figure 8.22 for the measured data. It is

Table 8.4 Dimensionless scale-up constants [Svarovsky, 1984]

Cyclone type and diameter	$Stk_{50}Eu$	K_p	n_p	$Stk_{50}^{4/3}Eu$
Rietema d_c=0.075 m	0.0611	316	0.134	2.12
Bradley d_c=0.038m	0.1111	447	0.323	2.17
Mozley d_c=0.022 m	0.1203	6381	0	3.20
Mozley d_c=0.044 m	0.1508	4451	0	4.88
Warman model R d_c=0.076 m	0.1709	2.618	0.8	2.07
RW 2515 (AKW) d_c=0.125 m	0.1642	2458	0	6.66

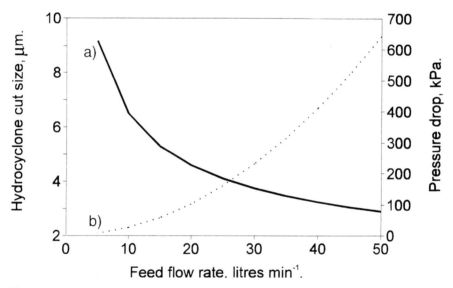

Figure 8.23 Cut size and pressure drop using Svarovsky's scale-up constants.
Hydrocyclone diameter 0.04445m; solid and liquid densities 7000 and 1000 kg m^{-3}; liquid viscosity 0.001
Pa s; scale-up constants: $Stk_{50}^{4/3}Eu$ 0.1508, K_p 4451 and n 0. Curve (a) cut size; curve (b) pressure drop

also possible to adjust the "constants" used in the correlations to provide a good fit to
the data if some existing data can be used to obtain values for this "curve fitting"
exercise. Should the operating conditions change in future such data can be used to
predict, or optimise, hydrocyclone performance. Some tests to obtain empirical values
for the constants used in the performance and scale-up relations are recommended, as it
is well known that suspension concentration affects hydrocyclone performance for a
number of reasons: hindered settling, viscosity modification, etc.

8.3.8 Numerical Solutions of Continuity and Flow

The hydrocyclone is discretised with many different grid points representing positions within it and the governing differential equations are solved by a finite difference or element models. Continuity equations are assigned to the fluid, solid and momentum whereas the fluid motion is described by the Navier–Stokes equation for incompressible flow. The commercially available computational fluid dynamics software package called PHOENICS has been used in this way [Rhodes et al, 1989] as a means of predicting the performance of the cyclone. The input parameters required are geometry of the cyclone and feed inflow conditions. The model can be used to predict the pressure drop, cut size, grade efficiency, flows into each discharge port and is very useful when investigating either an alteration in inlet conditions or the effect of variable cyclone geometry (design). It is very common for numerical modellers of hydrocyclones to assume symmetry around the axis, which considerably simplifies the computational requirement. A true representation of the hydrocyclone would have to permit variation in all three velocities within space and time, to permit turbulent liquid flow inside the device and to accommodate a large number of solid phases to represent a size-distributed particulate material. The computational requirement for all these variables is enormous, and some numerical studies have constrained the flow to be laminar, axis symmetric and even with only an overflow [Bloor et al, 1989]. The greater the number of assumptions or constraints the less reliable the results will be.

The potential for the information provided by this technique in design and optimisation is substantial, but considerable care must be taken in exercising the numerical model, which often includes various damping or "upwinding" factors to assist in the mathematical solutions. This may cause concern over the validity or general applicability of the results, especially when the assumptions made in formulating the model may also be physically unrealistic. Thus practical testing should accompany even this technique for the purposes of model validation. The results of one such study are illustrated in Figure 8.24.

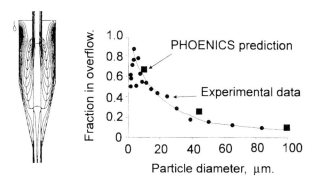

Figure 8.24 Numerically modelled flow streamlines by computational fluid dynamics [Rhodes et al, 1989] and model prediction. Large squares, fluid dynamics predictions; small circles, experimental data; curve, best fit to experimental data

8.3.9 General Relations

There have been numerous studies of hydrocyclone performance relating cut size, pressure drop and flow rate; many involve the application of empirical constants in the theoretical models already discussed. The constants are strictly valid only for the hydrocyclone geometry tested and operating conditions used, notably feed concentration. The situation is further complicated when the proposed equations are dimensionally inconsistent in the absence of empirical dimensioned constants. Some of the more well known equations are due to Bradley [1965], Holland–Batt [1982], Plitt & Kawatra [1979], Rietema [1961] and Trawinski [1969]. These have been reviewed and compared in the book by Svarovsky [1984].

A simple and reliable set of equations to predict cyclone performance from a known size distribution, cyclone geometry and flow rate does not exist. Laboratory testing must be used to ensure that a desired separation can be accomplished. It is useful, therefore, to have some notion of how the important variables are related, in order to minimise the required amount of test work to optimise performance. The important design and process variables are cut size, cyclone diameter, flow rate and pressure drop.

$$x_{50} = a_1 d_C^{\gamma} \qquad 1.36 < \gamma < 1.52$$

$$\Delta P = a_2 d_C^{\beta} \qquad -3.6 < \beta < -4.1$$

Note as cyclone diameter tends to zero so does the cut size, and pressure drop tends to infinity. Also:

$$x_{50} = a_3 Q^{-\chi} \qquad 0.53 < \chi < 0.64$$

$$\Delta P = a_4 Q^{\phi} \qquad 2 < \phi < 2.6$$

where subscripted a represents a constant.

Many other equations are available for predicting, for example, the cut size as a function of the ratio of underflow diameter to cyclone diameter, or as a function of the ratio of overflow diameter to cyclone diameter, etc. In most installations, however, it is usual to have some means of throttling the underflow diameter to provide some fine tuning in situ to establish the desired operating cut size.

8.3.10 Arrangements, Types and Designs

It is apparent that high retention efficiency of solids, i.e. a low cut size, is given by a hydrocyclone of low diameter. This imposes limitations on the throughput of the device. It is, therefore, common to encounter multiple installations of hydrocyclones with many units operating in parallel. Furthermore, consideration of the grade efficiency curve which is usually fairly shallow in slope leads to the conclusion that to obtain a "sharp" cut several hydrocyclones need to be employed in series, or in conjunction with other classifiers. Both of these arrangements can be economically viable as hydrocyclones are not expensive. A multiple hydrocyclone arrangement is shown in Figure 8.25.

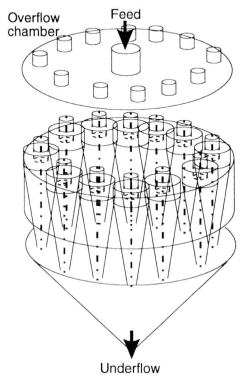

Figure 8.25 Multiple hydrocyclone arrangement

Two different types of hydrocyclones were illustrated in Figure 8.14; the design favouring thicker underflow having a much longer cone than that favouring a sharper cut. Hydrocyclones have also been employed for the removal of oil dispersed in water, or vice versa. The designs for these applications differ substantially from those discussed above, and are shown in Figure 8.26.

The design illustrated in Figure 8.26A, would be used for the removal of oil from water, that in 8.26B would be employed for the removal of water from oil. In use on the North Sea oilfields the first design is capable of cleaning a feed concentration of 1000 ppm dispersed oil to an overflow concentration below 40 ppm of oil. The development work leading to these hydrocyclone designs involved the use of computational fluid dynamics and has been well documented [Colman and Thew, 1983].

Another interesting design for oil–water separation and mineral flotation involves the use of a porous wall through which air passes into the hydrocyclone body. This is described as the air-sparged hydrocyclone [Beeby and Nicol, 1993] and is illustrated in Figure 8.27.

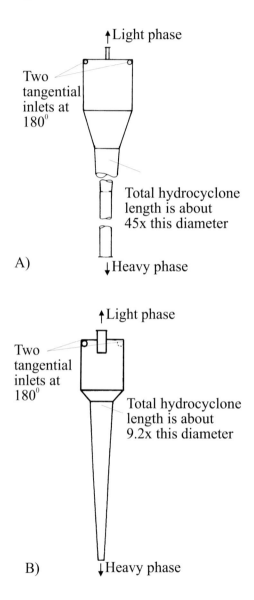

Figure 8.26 Hydrocyclones for liquid–liquid separation. A) Light dispersed phase; B) Heavy dispersed phase

Hydrocyclones have been tested for many applications including removal of suspended material from domestic washing machines and the lysing and separation of biological cell material. The optimum design for each application is likely to be different from the conventionally available units, leading to slight variations in design.

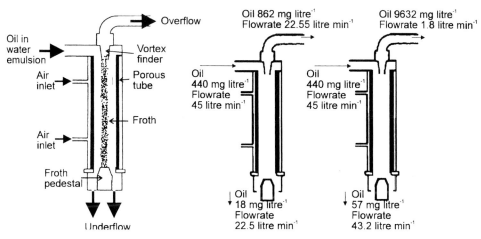

Figure 8.27 Air-sparged hydrocyclones

8.3.11 Applications

The dual function of the hydrocyclone as a classifier and a thickener, together with its relatively low cost and simplicity make it very valuable for many applications. The poor sharpness of separation can be overcome by employing hydrocyclones in series, one such arrangement is given in Figure 8.28. The hydrocyclone's ability to thicken the coarser material is particularly useful as a pretreatment prior to filtration. The underflow can be used as a precoat for a belt filter, the overflow being added further down the filter thereby protecting the filter medium from the fines and blinding. Hydrocyclones may also be used to assist if a solid–liquid separation device can not adequately treat the throughput required of it. It may be put in parallel with a thickener or filter taking a side stream from those devices. It may also act as a thickening device before a filter, as filter throughput increases with increasing slurry concentration, see Chapter 11. Under such circumstances it is usual to recycle the overflow into a process and a build-up of fine material within the recycle loop may result [Svarovsky, [1990].

One very common use is in wet grinding circuits, where the product from a grinding mill is classified through a hydrocyclone with the underflow recycled to the mill, i.e. closed-circuit grinding. In mineral processing the shearing action inside the hydrocyclones may also be used to provide de-sliming; removal of the fine particles loosely attached to larger ones. Many other applications may be found in the field of

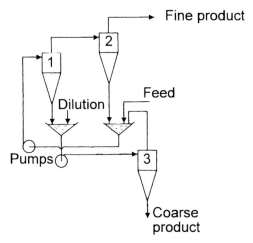

Figure 8.28 Arrangement for hydrocyclones used in series to sharpen the separation

mineral processing, including as a separator using an artificially enhanced continuous phase density (dense medium separation, often employing magnetite suspensions) in which other suspended mineral material is separated by means of density and size. The hydrocyclone originated in the mineral processing industries: kaolin production, metal ore processing, coal dressing, etc., but is becoming established in many other industries because of its lack of moving parts and relative cheapness.

8.4 Centrifugal Filtration

In centrifugal filters, separation is effected by directing the solid–liquid suspension on to the inner surface of a perforated, rotating bowl. In fine separations, the inner surface of the bowl will be lined with an appropriate filter medium.

During the filter cake formation period, filtrate passes radially outwards through the filter medium and the bowl. Solids removal may take place continuously or batch-wise; in the latter case slurry feed to the centrifuge is terminated during cake removal.

8.4.1 Batch Discharge Centrifuges

These units find wide application in processes involving cake formation and dewatering; they are also used where cake washing is required. The machines can be classified on the basis used for the discharge of the cake. The latter can be: manual (machine stationary); automatic (machine rotating at reduced speed or maintained at full speed).

A typical example of a manual discharge centrifuge is shown in Figure 8.29 which depicts the so-called basket or three-column centrifuge. Such machines can be over- or under-driven. The over-driven centrifuge is applied to free filtering materials such as sugar crystals. The times required for loading, spin drying and washing will be determined by the filtration characteristics of the particles; theoretical aspects of the various rates involved in the three operations are considered below.

The "peeler" centrifuge shown in Figure 8.30 is typical of the automatic batch-operated machine; when fitted with the plough-type knife for cake discharge, deceleration of the bowl is effected before operation of the discharge mechanism. Horizontal machines fitted with reciprocating thin-bladed peelers can be operated at full speed during cake discharge. Here obvious gains in productivity have to be weighed against the onset of crystal breakage.

Increased capacity of peeler centrifuges has also been obtained by use of a siphonic action to enhance the effective pressure differentials, as shown in Figure 8.31.

8.4.2 Batch Discharge Centrifuge Capacity

In the estimation of the capacity of a filtering centrifuge it is necessary to have data on the times required to: a) accelerate the machine up to a speed suitable for feeding; b) feed a prescribed quantity of slurry; c) accelerate to a higher speed for deliquoring and maintain that speed for a time; d) apply a quantity of wash liquor; e) spin dry; f) decelerate; and g) unload.

Typical information taken from the technical literature [Ambler, 1952] for a 1.2×0.74 m multispeed basket centrifugal separating an organic solid is reported in Table 8.5.

Table 8.5 Batch centrifuge process data

Process	Time, s
Accelerate from 50 to 500 rpm	40
Load at 500 rpm	277
Accelerate to 1050	90
Spin dry at 1050 rpm	119
Wash at 1050 rpm	10
Spin dry at 1050 rpm	236
Decelerate to 50 rpm	90
Unload at 50 rpm	15

Total cycle time 877 s
Basket load per cycle, dry solids 140 kg
Capacity per hour 575 kg

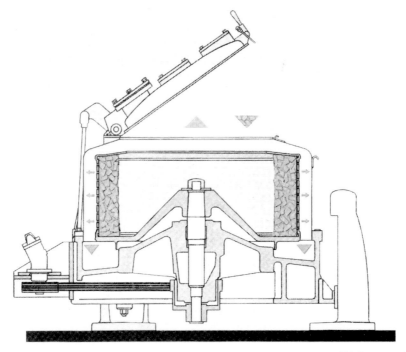

Figure 8.29 Manual discharge 3 column centrifuge [Courtesy of Krauss Maffei Company, Germany]

Figure 8.30 Peeler centrifuge [Courtesy of Buchau-Wulff, Krauss Maffei Company, Germany]

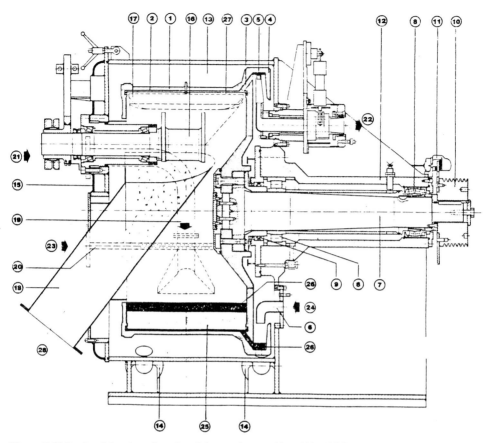

Figure 8.31 Sectional drawing of rotating siphon peeler centrifuge [Alt, 1984]

1) Siphon basket;
2) filter medium;
3) siphon holes;
4) siphon ring cup;
5) filtrate peeling tube;
6) back-wash pipe;
7) basket shaft;
    25) cake;
8) shaft bearing;
9) shaft seal;

10) V-belt pulley;
11) disc brake;
12) bearing stand with casing;
13) casing;
14) liquid run-off connecting piece;
15) casing door;

17) peeling knife;
18) solids chute;

19) feed pipe;
20) wash pipe;
21) suspension inlet;
22) filtrate outlet;
23) wash liquid inlet;
24) filtrate recycle;
16) solids knife discharge;

26) liquid layer;
27) residual cake layer;
28) solids discharge

The information in Table 8.5 was obtained in practical evaluations in the separation of the solids which consisted of cubic crystals, mean size 120 µm; the latter had been obtained by careful study of crystallization conditions in which an uncontrolled precipitation (mean size 30 µm) gave material unsuitable for centrifugal filtration.

The information presented points above to the importance of load, spin and wash times in overall cycles and, in view of the wide–spread use of centrifugal methods and the associated energy consumption, considerable attention has been given in research centres to the problem of relating particulate and fluid properties to centrifugal separation conditions [Zeitsch, 1978].

8.4.3 Continuous Discharge Machines

Typical examples of continuous solids discharge machines are shown in Figures 8.32 and 8.33 which depict the pusher and oscillatory and tumbler screen centrifuges. In the former unit the slurry is directed towards the back of the rotating bowl which contains a rotating, reciprocating pusher plate. During the back stroke of the latter, filtration ensues in the space created behind the filter cake; the forward stroke then causes the cake to be pushed towards the outer, lipless edge of the bowl. Feed rates have to be controlled in order to prevent flow of unfiltered slurry over the surface of the filter cake, in which event the surface of the cake becomes scored with deep rivulets.

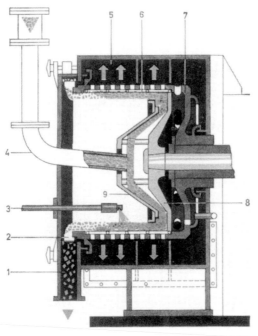

Figure 8.32 Schematic illustration of pusher centrifuge [Courtesy of Krauss Maffei Company, Germany]

1) Solids collector;	4) feed pipe;	6) slotted screen;	8) pusher plate;
2) scraper;	5) filtrate collector;	7) basket;	9) feed conc
3) wash pipe;			

8.4.4 Selection of Filtering Centrifuges

Factors which influence centrifuge selection include [Quilter, 1982]:

1) Type of separation required: liquid–solid, liquid–liquid–solid
2) Particle size, shape and density
3) Solids concentration
4) Relative densities of feed
5) Required flow

Figure 8.34 contains guidelines on the effect of particle size and solids concentration on the selection of centrifuges, including solid-bowl devices. Hultsch & Wilkesmann [1977] discuss, in great detail, the best procedures associated with the selection and sizing of centrifugal filters.

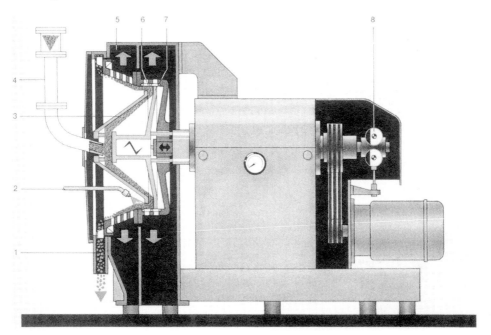

Figure 8.33 Schematic illustration of an oscillating centrifuge [Courtesy of Krauss Maffei Company, Germany]

1) Solids collector;	3) feed cone;	5) filtrate collector;	7) pusher plate;
2) wash pipe;	4) feed pipe;	6) cylindro-conical basket;	8) mechanical oscillation device

Zeitsch, [1977] has provided the following guidelines for centrifuge selection, using the concept of intrinsic permeability k', where:

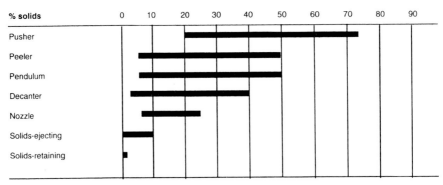

Figure 8.34 Particle Size and Concentration Ranges for Centrifugal Separators [Quilter, 1982]. A) Particle size; B) Feed solids content

$$k' = k / \mu$$

and k is the permeability of the filter cake defined by Darcy's law:

$$k = Q \mu L / \Delta P A$$

Intrinsic permeability k', m^4 N^{-1} s^{-1}	Equivalent Cake resistance* $\bar{\alpha}$, m kg^{-1}	Machine selected
> 20×10^{-10}	5×10^8	Continuous pusher
20×10^{-10} – 1×10^{-10}	5×10^8–8×10^9	Peeler
1×10^{-10}–0.02×10^{-10}	8×10^9 –4×10^{11}	Three-column
< 0.02×10^{-10}	>×10^{11}	Solid-bowl
centrifuge		

* Estimated from assumed porosity ε= 0.5 and solids density 2500 kg m^{-3}

The specific resistance is calculated from: $\bar{\alpha} = 1 / (k (1 - \varepsilon) \rho_s)$ where k is estimated from k' using $\mu = 0.001$ N s m^{-2}.

8.4.5 Centrifuge Productivities

Purchas [1981] lists typical productivities for different centrifuges:

Type of machine	Particle size range, μm	Cake conc'n, wt%	Production t/h
Vibratory/Oscillatory Screen	500–10 000	40–80	5–150
Pusher	40–5000	15–75	0.5–50
Peeler	20–1000	0–30	0.1–5
Three-column	10–1000	2–10	0.1–1

The latter publication gives a comprehensive description of many centrifuge types and data on performance.

8.4.6 Filtration and Permeation in Centrifugation

The early work [Maloney, 1946; Hassett, 1956; Storrow & Burak, 1950; Grace, 1953; Storrow & Haruni, 1952] resulted in equations which could be used to estimate permeation (wash) times (Equation 8.46) and filtration time (Equation 8.47):

$$Q = \frac{dV}{dt} = \frac{4\pi^3 N^2 \rho h k (r_o^2 - r_f^2)}{\mu \ln(r_o / r_c)} \tag{8.46}$$

where the various radii are shown in Figure 8.35, in which h is the bowl depth and $N = \omega/2\pi$. This equation was developed in terms of the permeability of the filter cake; the work ignored the presence of the filter medium which can present serious flow resistance, particularly for woven, multifilament media or where a "heel" of filter cake is left in the machine. The "heel" is that part of the deposited solids missed by the removing mechanism (plough or knife).

Grace included the medium resistance in his work to produce:

$$Q = \frac{dV}{dt} = \frac{4\pi^3 N^2 \rho h (r_o^2 - r_f^2)}{\mu (\bar{\alpha} \rho_s (1 - \varepsilon) \ln (r_o / r_c) + R_m / r_o)} \tag{8.47}$$

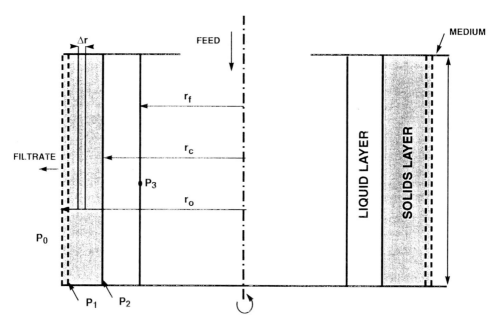

Figure 8.35 Centrifugal filter section

a result which followed the assumption that both $\bar{\alpha}$ the cake average resistance and ε the porosity were invariant with radius; in addition the effect of centrifugal body forces, created by spinning the solids themselves, is ignored in this analysis.

A serious limitation in the data collected by these researchers was the fact that $\bar{\alpha}$ was measured in *permeation* rather than in *filtration* conditions; the packing (porosity) and the resistance of newly formed filter cake under the dynamic conditions obtaining during filtration may be expected (in general) to be different from those applying after prolonged rotation. The latter results in the densest packing and highest α values.

Grace's data were obtained in simulating filtration by successive applications of several increments of solid, which were permeated to evaluate $\bar{\alpha}$ Storrow's data all pertain to permeation, not filtration, and will therefore apply to the washing stage only. Furthermore, the materials studied were relatively incompressible. Given that realistic values of $\bar{\alpha}$ and R_m are available from test work, the course of filtration may be followed by a modification of Equation (8.47) to yield an equation quite similar to the conventional "constant-pressure" design equation used in vacuum and pressure work for batch filtration:

$$Q = \frac{\Delta P}{\mu\left(\dfrac{\bar{\alpha}w}{A_{lm}} + \dfrac{R_m}{A_o}\right)} \qquad \text{where } A_0 \text{ and } A_{lm} \text{ are defined in Section 8.4.10} \quad (8.48)$$

A full derivation of this basic relationship is presented in Section 8.4.11.

The term w may be evaluated from the slurry concentration and $w = cV/A_{av}$. The pressure drop is calculated from the hydrostatic pressure created by centrifugal force to yield:

$$\Delta P = \rho\, \omega^2\, (r_o^2 - r_f^2) / 2 \tag{8.49}$$

where ρ is the liquid density.

Given that data are available on the effect of pressure on a filter cake, it has been proposed that the average resistance $\overline{\alpha}$ in Equation (8.48) can be calculated from:

$$\overline{\alpha} = \frac{\Delta P_c}{(1-\varepsilon)\displaystyle\int_o^{\Delta P_c} \mathrm{d}P/(1-\varepsilon)} \tag{8.50}$$

where ΔP_c is the pressure drop causing flow in the filter cake. Claims were made [Grace, 1953; Vallery & Maloney, 1960] that (α, ε) data taken from compression cell (CP) measurements could be used for such evaluations; centrifugally measured values (actually permeation values) within 20% of CP results were recorded. Such information was contradicted by Oyima and Sumikawa [1954] who obtained quite different results between cell and centrifuge; in the latter, using diatomaceous earth and powdered silica, $\overline{\alpha}$ values were generally much higher than obtained in static cell tests. On the other hand, it is probable that for low density solids of an incompressible nature, Grace's approach could find use in process estimations.

Careful experimentation [Spear & Rushton, 1975] provided evidence on the true value of $\overline{\alpha}$ during filtration and the resistance of the cake in subsequent permeation. Figure 8.36 depicts the expected result that $\overline{\alpha}$ (filtration) is generally lower than $\overline{\alpha}$ (permeation). Use of the latter to calculate cake formation or load time will lead to overestimation, which is on the conservative side in process terms.

In centrifugal filtration, the drag force created by filtrate flow over the surface of the particles (as in pressure filtration) is enhanced by the effect of rotation on the particles themselves.

Solids density effects were observed by studies in which the pressure drop due to fluid friction ΔP_f was maintained constant, whilst variation in the "solids pressure" ΔP_s caused by the rotation was increased. This was effected by varying N and $(r_o^2 - r_f^2)$ to produce a constant ΔP despite the use of various bowl speeds.

The "body force" term is calculable from:

$$\mathrm{d}P_s = (\rho_s - \rho)\,(1 - \varepsilon)\,\omega^2\, r\, \mathrm{d}r \tag{8.51}$$

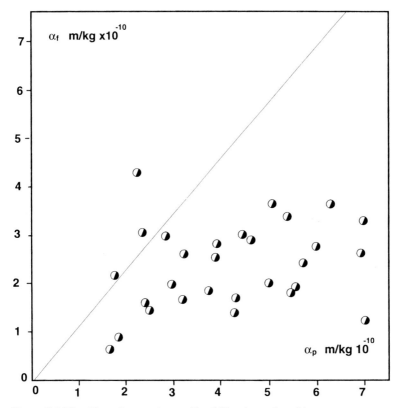

Figure 8.36 Specific resistances in centrifugal filtration and washing

and Figure 8.37 shows the effect of increase speed on $\bar{\alpha}$ despite the same ΔP value. This result points to the need for evaluation of $\bar{\alpha}$ or load times at various speeds so that the effect of body force can be evaluated — a feature of importance if scale-up of the data to large radii is envisaged.

Other work [Spear & Rushton, 1970] reported the effect of rotational speed on R_m, the medium resistance. Using clean multifilament media, a gradual increase in R_m up to an asymptotic maximum was noticed, probably due to yarn compression. The same cloths used in solids separation demonstrated increased R_m values following partial blinding by solids; this phenomenon has also been reported for vacuum and pressure separations with (R_m used/ R_m clean) being measured in the range 2–30.

It may be observed in such calculations that where $\bar{\alpha}$ is high the effect of the medium resistance is negligible whereas for low $\bar{\alpha}$ ignoring the presence of the medium can produce up to 10% errors in the drainage rate. The filter cloth resistance in the used condition should be included in such calculations and determined practically by permeating the cloth after repeated use, when (R_m used) has reached an equilibrium value.

8.4.7 Wash Time

The time required for washing will be decided by the amount of wash required and the effective wash rate. If the bowl speed remains unaltered after first drainage, the wash rate will equal the drainage rate as calculated above if it can be assumed that application of washwater does not cause an increase in $\bar{\alpha}$ or R_m (cake disturbance, particle migration, prolonged spinning, etc.). These practicalities have received much recent interest in the filtration literature [Leung, 1998].

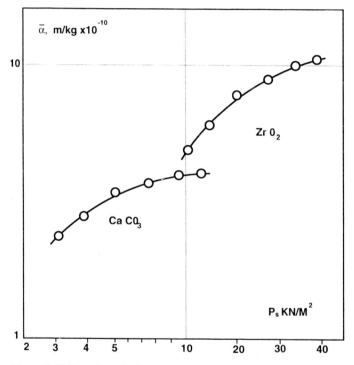

Figure 8.37 Plot of α *vs* body pressure

The quantity of wash required will be determined by the state of flow in the filter cake, the probability of pockets of undrained first liquors, etc. A washing curve of the type shown in Figure 8.38 may be obtained practically to determine the wash required. Wakeman [1980] provides a basis for calculating the wash volume in conditions where the cake remains flooded after filtration, or is partially drained by spinning. In the example in Table 8.5 the wash time constituted a small proportion of the overall cycle and as such must have constituted a relatively easy wash; in such non-stringent cases delivery of, e.g. two void volumes of wash (void volume = cake porosity×cake volume) can reduce the residual solute to acceptable limits. This of course, is not always the case and in certain circumstances — where slow diffusional processes are necessary for solute removal — considerable wash volume and time may ensue. Washing efficiency

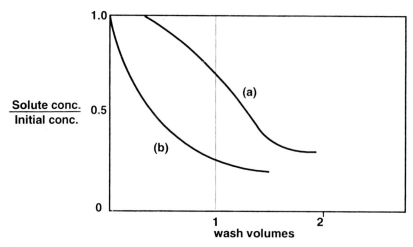

Figure 8.38 Typical washing curve. Curve (a), flooded cake; curve (b), dewatered cake

in centrifugal filters ranges from fair (50% removal of solute) to excellent (95% removal). In view of the short time available, high recoveries, particularly where mass transfer processes are involved, were seldom achieved [Ambler, 1979] in former centrifugal units. Recent developments permit re-slurry washing in situ on the spinning bowl, with high wash efficiencies [Leung, 1998].

8.4.8 Spin Dry

The dewatering process kinetics and the equilibrium saturation S_∞ (fraction of liquor remaining in cake after prolonged spinning) will be determined by the relative value of the centrifugal drainage force and the resistance to drainage offered by the capillaries in the cake. It will be realised that since the cake structure is created during the filtration period, a strong connection exists between $\bar{\alpha}$ and the spin dry time. This is particularly true for incompressible cakes which, upon deposition, quickly reach equilibrium porosity, etc.

Domkrowski and Brownell [1954] proposed a two-regime model for the residual saturation including bulk drainage during complete flooding of the filter cake and a slower film drainage; the latter obtained when the radius of the surface of the drainage liquid exceeds the cake radius. The draining force available increases in the radial direction, a fact which distinguishes this drainage mode from gravity and vacuum systems.

Ambler [1979] reports the equations below for the equilibrium liquor contents S_∞ and the instantaneous value S_t, obtaining at any time t during the spin; the constants k^1 and k^2 depend on the material processed:

$$S_\infty = k^1 \bar{\alpha}^{-0.5} (\omega^2 r / g)^{-0.5} \rho^{-0.25} \tag{8.52}$$

$$S_t = \frac{k^2}{\overline{d}}\left(\frac{\mu g}{\rho\omega^2 r}\right)\frac{(r_0 - r_c)^m}{t^b} + S_\infty \tag{8.53}$$

The exponent m on the cake thickness is reported in the range $0.5 < m < 1.0$ for relatively incompressible cakes; exponent b on the spin time takes values in the range $0.3 < b < 0.5$.

Other work [Wakeman, 1976] points to a general relationship for the residual saturation of coarse granular particles:

$$S_\infty = f(\rho, k, (r\omega^2), \sigma \cos \theta) \tag{8.54}$$

This approach by combination of dimensional analysis and the permeability–particle size relationship may be used to yield a capillary number N_c useful for the correlation of S_∞ with drainage conditions; experimental data are presented in Figure 8.39 where $\overline{d}$ is the average particle size:

$$S_\infty = \text{constant}\left[\frac{\varepsilon^3 \overline{d}^2 \rho N^2 r}{(1-\varepsilon)\sigma \cos \theta}\right]^z \tag{8.55}$$

i.e.

$$S_\infty = f[N_c]^z \tag{8.56}$$

Such data should be integrated over the cake thickness for accurate calculations; in Figure 8.39 the average permanent saturation is recorded with r defined at the midpoint of the cake.

Later work [Wakeman, 1979] demonstrated that the above information may be represented by equations of the form:

$$S_\infty = 0.0524 \, N_c^{-0.19}; \ 10^{-5} < N_c < 0.14 \tag{8.57}$$

$$S_\infty = 0.0139 \, N_c^{-0.88}; \ 0.14 < N_c < 10 \tag{8.58}$$

A sharp break in the drainage curve is noticeable as the drainage force increases (increase in $N_c > 0.14$); this may be due to breakdown of retention forces leading to a pendular state in the liquor between particles.

Using the relationship [Wakeman, 1979]:

$$\overline{d}1 = 13.4 \, [(1 - \varepsilon) / \overline{\alpha} \, \rho_s \varepsilon^3]^{0.5} \tag{8.59}$$

the expressions for S_∞ may be approximated in terms of cake resistance $\overline{\alpha}$ and solids specific gravity s_g, for low values of the contact angle θ.

$$N'_c = \frac{180 \, N^2 r}{\overline{\alpha} s_g (1 - \varepsilon)\sigma} \tag{8.60}$$

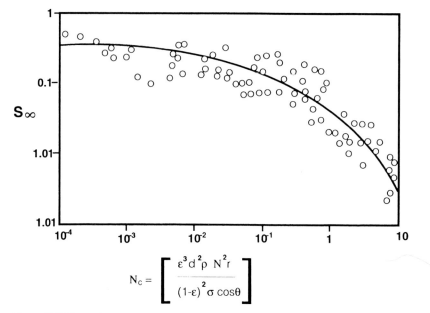

$$N_c = \left[\frac{\varepsilon^3 d^2 \rho\, N^2 r}{(1-\varepsilon)^2 \sigma \cos\theta} \right]$$

Figure 8.39 Saturation *vs* N_c [Wakeman, 1979]

Equation 8.60 along with the data reported in Figure 8.39 has been used to produce Figure 8.40 which relates the equilibrium saturation S_∞ with bowl speed and cake resistance. As expected, an increase in $\overline{\alpha}$ above 1×10^{11} m/kg leads to high equilibrium saturations.

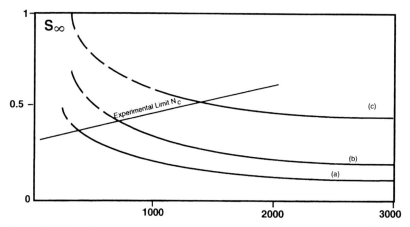

Figure 8.40 S_∞ *vs* Speed and cake resistance. Cake resistance values: curve (a), 10^9; curve (b) 10^{10}; curve (c) 10^{11} m/kg

8.4.9 Practical Equilibrium Saturation Studies

Experimental values of S_∞ in centrifugal conditions have been reported [Rushton & Arab, 1986; Daneshpoor, 1984]. Deviations between measured and predicted saturation levels, as a function of bowl speed, were recorded for materials such as calcium and

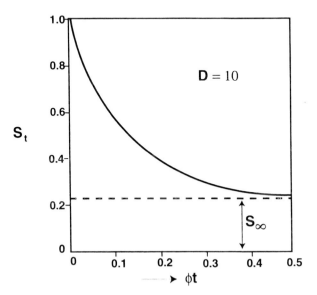

Figure 8.41 Theoretical drainage curve [Zeitsch, 1978]

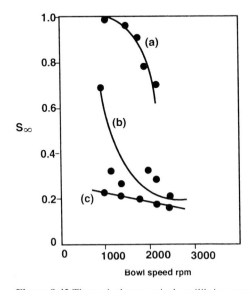

Figure 8.42 Theoretical *vs* practical equilibrium saturation at various bowl speeds. Curve (a), Equation (8.64); curve (b), experimental values; curve (c), Equation (8.57) and (8.58) [Daneshpoor, 1984]

magnesium carbonate filter cakes. In general, the capillary model discussed above tended to underestimate the dewatering effect of bowl rotation, Figure 8.42. On the other hand the empirical relations, Equations 8.57 and 8.58, over-predict the drainage. It will be noted that the latter equations were based on studies involving sand particles, glass beads, etc. It is suggested that the constants and exponents in Equations 8.57 and 8.58 obtain only for the materials used in the original studies. This does not detract from the usefulness of the proposed dimensionless groups in correlating drainage data.

8.4.10 Drainage Kinetics

Zeitsch [1977] has presented an analysis of capillary drainage involving a Boltzman-type distribution of pore diameters to produce a method for the calculation of drainage times for filter cakes of known permeability k or specific resistance. The equations take the form reported below in which l is the cake thickness:

$$S_t = S_\infty + \frac{1}{D^2\left(\frac{1}{D}+\phi t\right)}\left[\frac{1}{\frac{1}{D}+\phi t}\exp-\left(\frac{1}{D}+\phi t\right)-\frac{\sqrt{\pi}\left[1-erf\left(\frac{1}{D}+\phi t\right)^{0.5}\right]}{2\left(\frac{1}{D}+\phi t\right)^{0.5}}\right] \quad (8.61)$$

in which:

$$D = \frac{kr_o^2\omega^4\rho^2 l^2}{\varepsilon\sigma^2\cos^2\theta}, \text{ Zeitsch drainage number.} \quad (8.62)$$

$$\phi = \frac{\sigma^2\cos^2\theta}{2\mu l^3\rho r_o\omega^2}, \text{ Zeitsch drainage rate constant, where } r_o \text{ is the bowl radius} \quad (8.63)$$

$$S_\infty = 1-\exp-(1/D)+\frac{\sqrt{\pi}\sqrt{D}}{2D}\left[1-erf\left(\frac{1}{\sqrt{D}}\right)\right] \quad (8.64)$$

Again S_t is the fraction of pores remaining filled with liquid at time t and a typical curve showing the effect of prolonged spinning is presented in Figure 8.41. Equation 8.64 may be used to calculate experimental D values from measurements of S_∞ [Rushton & Arab, 1986]:

D	0.5	1.0	10	50	100
S_∞	0.92	0.77	0.28	0.13	0.09

A drainage problem involving Zeitsch's analysis is presented later. Wakeman (1979) provides an analysis of filter cake drainage kinetics. This is reported in detail in the next chapter.

8.4.11 Theoretical Filtration Rates in Centrifuges

After the first acceleration period, which is usually the fixed time required to bring the empty machine up to full operational level, a further period of process time is needed to separate a prescribed quantity of feed. The feed rate has to be considered carefully in order to ensure even cake distribution over the filter surface. In the case where the volume of feed is smaller than the volume of the basket the feed rate can be as high as permitted by the feeding mechanism and connections. A common problem involves the processing of large volumes of feed, so that repeated filling and emptying of the bowl have to be effected. Process information is thus necessary on the rate of cake build-up, liquor drainage requirements, etc. so that optimal and safe operation of the machine is ensured. In practice, pilot trials are conducted on small-scale machines to measure overall productivities, examine the influence of filter media, etc. Scale-up to full productivity in larger, similar machines is usually successful [Alt, 1984], except where principal changes in the filterability of the suspension have occurred. The effect of major process variables on the separation can be ascertained by inspection of the mathematical model for centrifugal filtration outlined below. The situation obtained during the filtration of a slurry in a centrifuge is shown above in Figure 8.35 where at a particular time t the inner radii of the slurry, filter cake and filter medium (fixed) are, respectively r_f, r_c and r_o . Consider a differential cylindrical element of filter cake between radii r and $r + dr$. The mass of solids contained in this element is:

$$dW = \rho_s (1- \varepsilon)\, 2\pi h r dr \tag{8.65}$$

The area available for filtrate flow is:

$$A = 2\, \pi r h \tag{8.66}$$

Pressure drop in the fluid, caused by the volumetric flow of liquid Q at time t, through the element is given by Darcy's expression:

$$d\, P_f = Q\, \mu\alpha\, dW / A^2 \tag{8.67}$$

From 8.65 and 8.66 this gives:

$$d\, P_f / dr = (Q\mu\alpha\, (1- \varepsilon)\rho_s / 2\pi r h) \tag{8.68}$$

The actual pressure differential over the element of cake is given by the difference between this frictional effect and the pressure developed radially in the fluid by rotation:

$$-d\, P = d\, P_f - d\, P_r \tag{8.69}$$

where the centrifugal pressure effect has been shown to be:

$$dP_r = \rho \omega^2 r\, dr \tag{8.70}$$

Therefore, over the filter cake from radius r_c to r_o:

$$-\int_{P2}^{P1} dp = \left[\frac{Q\mu\rho_s\alpha(1-\varepsilon)}{2\pi h} \right] \int_{r_c}^{r_o} \frac{dr}{r} - \rho\omega^2 \int_{r_c}^{r_o} r dr \qquad (8.71)$$

Here it has been assumed that both α and ε are invariant with radius; this should be reasonably true for crystalline, inorganic precipitates.

Thus:

$$-(P_1 - P_2) = \left(\frac{Q\mu\rho_s\alpha(1-\varepsilon)}{2\pi h} \right) \ln\left(\frac{r_o}{r_c} \right) - \frac{\rho\omega^2(r_o^2 - r_c^2)}{2} \qquad (8.72)$$

In Figure 8.35, assuming that $P_o = P_3 = 0$:

$$P_1 - P_o = P_1 = (Q \mu R_m) / (2 \pi r_o h) \qquad (8.73)$$

$$P_2 - P_3 = \rho\omega^2 \int_{r_f}^{r_c} r dr = \rho\omega^2(r_c^2 - r_f^2)/2 \qquad (8.74)$$

Equation 8.73 accounts for the pressure drop caused by flow through the filter medium and Equation 8.74 relates to the development of hydraulic pressure over the supernatant liquid (slurry) layer to the inner cake surface. Hence:

$$-(P_1 - P_2) = \left(\frac{\rho\omega^2(r_c^2 - r_f^2)}{2} \right) - \left(\frac{Q\mu R_m}{2\pi r_o h} \right) \qquad (8.75)$$

and in Equation (8.72)

$$\frac{\rho\omega^2(r_o^2 - r_c^2)}{2} + \frac{\rho\omega^2(r_c^2 - r_f^2)}{2} = \frac{Q\mu\rho_s\alpha(1-\varepsilon)}{2\pi h} \ln\left(\frac{r_o}{r_c} \right) + \frac{Q\mu R_m}{2\pi r_o h}$$

which reduces to:

$$Q = \frac{(\rho\omega^2/2)[r_o^2 - r_f^2]}{\frac{\mu\rho_s\alpha(1-\varepsilon)}{2\pi h} \ln\left(\frac{r_o}{r_c} \right) + \frac{\mu R_m}{2\pi r_o h}} \qquad (8.76)$$

which may be rewritten in terms of the mass (or volume) filtered as follows: the total mass of solids present at time t is given by:

$$\int dW = \rho_s(1-\varepsilon)2\pi h \int_{r_c}^{r_o} r dr$$

Hence:

$$W = \rho_s (1-\varepsilon) \pi h (r_o{}^2 - r_c{}^2)$$

Thus:

$$\rho_s (1-\varepsilon) = W / \pi h (r_o{}^2 - r_c{}^2)$$

Finally:

$$Q = \frac{(\rho \omega^2 / 2)(r_o^2 - r_f^2)}{\left[\dfrac{\alpha \mu W \ln(r_o / r_c)}{2\pi h[\pi h(r_o^2 - r_c^2)]} + \dfrac{\mu R_m}{2\pi r_o h} \right]} \tag{8.77}$$

Note that:

$$r_o{}^2 - r_c{}^2 = (r_o + r_c)(r_o - r_c)$$

Hence:

$$2\pi h(\pi h(r_o^2 - r_c^2)) / \ln(r_o / r_c) = \frac{(2\pi h(r_o + r_c))}{2} \frac{(2\pi h(r_o - r_c))}{\ln(r_o / r_c)}$$
$$= A_{av} / A_{lm}$$

Where:

A_{av} = average area of flow
A_{lm} = logarithmic mean area of flow

Also:

A_o = bowl or medium area = $2 \pi r_o h$

Substituting in Equation (8.77):

$$Q = \Delta P \bigg/ \left[\frac{\mu \overline{\alpha} W}{A_{lm} A_{av}} + \frac{\mu R_m}{A_o} \right] \tag{8.78}$$

The latter equation may be compared with Equation (8.48) in which $w = W/A_{av}$ is the mass deposited per unit area.

This analysis leads to the suggestion that plots of t / V versus W (or V) can be used in estimating cake and medium resistance as in pressure and vacuum filtration.

8.4.12 Centrifugal Cake Thickness Dynamics

The cake radius-time relationship for peeler centrifuge dynamics may be developed from the volumetric balance [Alt, 1984]:

Increase in volume of slurry = volumetric–volumetric
in machine   feed rate   filtrate rate

$$2\pi r h \, dr/dt \qquad = Q_o - Q \tag{8.79}$$

The volume of solids in the cake and hence the cake radius is given by

$$\pi (1-\varepsilon)(r_o^2 - r_c^2)\, h = Q\, t\, [C_f/(1-C_f)]$$

$$= V[C_f/(1-C_f)] \tag{8.80}$$

where C_f is the volumetric concentration of the solids and V is the volume filtrate at time t. The cake radius is:

$$r_c = \left(r_o^2 - \frac{C_f V}{\pi h (1-\varepsilon)(1-C_f)} \right)^{0.5} \tag{8.81}$$

The volumetric filtrate rate (Equation 8.76) may be written:

$$Q = \frac{\rho \omega^2 \pi h (r_o^2 - r_f^2)}{\mu(\alpha^* \ln (r_o/r_c) + R_m/r_o)} \tag{8.82}$$

where α^* is defined in terms of the cake thickness l:

Cake resistance $= \alpha^* \, l \;\; (\mathrm{m^{-1}})$,

Note: $\alpha^* = \alpha \, \rho_s (1-\varepsilon)\;(\mathrm{m^{-2}})$

From Equations (8.79)–(8.81) the change in the radius of the slurry surface with time is given by:

$$\frac{dr}{dt} = \frac{1}{2\pi r h} \left(Q_o - \frac{\rho \omega^2 \pi h (r_o^2 - r^2)}{\mu(\alpha^* \ln(r_o/r_c) + R_m/r_o)} \right) \tag{8.83}$$

Equation (8.83) can be used to calculate the movement of the slurry and filter cake surface for various process conditions, e.g. where the bowl has been filled with slurry to a certain level and then Q_o reduced to zero (Figure 8.43). Similar changes are shown in Figure 8.44 where Q_0 exceeds the filtration rate.

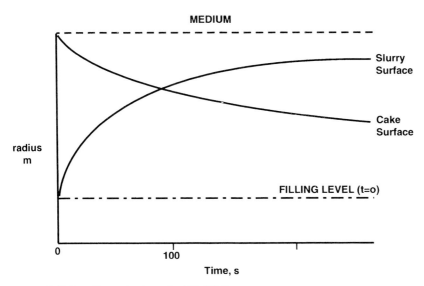

Figure 8.43 Centrifuge cake dynamics [Alt,1984]

The r_f, r_c, t relationship is obtained from Equation (8.84) below which follows from the substitution for r_c in Equation (8.83):

$$\frac{dr}{dt} = \frac{1}{2\pi r h} \left[Q_o - \frac{\rho \omega^2 (r_o^2 - r^2)\pi h}{\mu \left\{ \alpha * \ln \left[\frac{r_o^2}{r_o^2 - \dfrac{C_f Q t}{(1-\varepsilon)(1-C_f)\pi h}} \right]^{0.5} + \dfrac{R_m}{r_o} \right\}} \right] \tag{8.84}$$

The volume of filtrate V is related to the volume of feed suspension V_o by:

$$V = V_o - \pi\, h\, (r_o^2 - r^2)$$

in which case:

$$r = (r_o^2 - (V_o - V) / \pi\, h)^{0.5}$$

The filter cake resistance $\alpha*$ may depend on the operational speed of the machine and the number of peelings, as shown in Figure 8.45. Similar theoretical analyses are available on the performance characteristics of pusher centrifuges [Alt, 1984; Hallitt, 1975; Zeitsch, 1977].

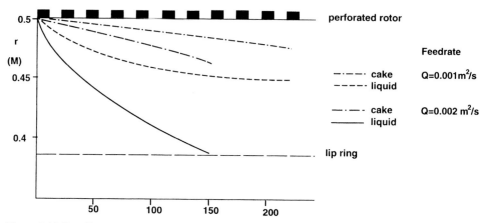

Figure 8.44 Centrifuge cake dynamics. Liquid and cake profiles for two feed rates, where $Q_f < 2$ litres per second [Alt, 1984]

Design Problem Examples

(A) Calculate the washing rate of a filter cake 0.025 m thick deposited on a centrifuge basket (0.635 m diameter; 0.254 m height) rotating at 20 rps.

Assume the cake is incompressible, has a porosity of 0.53 and a specific resistance of 6×10^9 m/kg. A medium with a resistance of $R_m = 1 \times 10^8$ m^{-1} is used to line the perforate basket. What time is required for the passage of two void volumes of wash?

Assume no supernatant liquid layer over the cake.

Solids density: 2000 kg/m^3
Liquid density: 1000 kg/m^3
Liquid viscosity: 0.001 N s m^2

Answer

460 l/h; 85 s

(B) A filter cake of specific resistance $\alpha = 1.85 \times 10^{10}$ m/kg is formed in a 0.5 m diameter bowl, operating at 250 rps. Use the Zeitsch method to calculate the relationship between the cake saturation and time, assuming the data below.

Solids density: 2400 kg/m^3
Porosity: 0.53
Cake thickness: 0.05 m
Liquid viscosity: 0.001 N s m^{-2}
Surface tension: 0.07 N m^{-1}
Contact angle: 0 Degrees

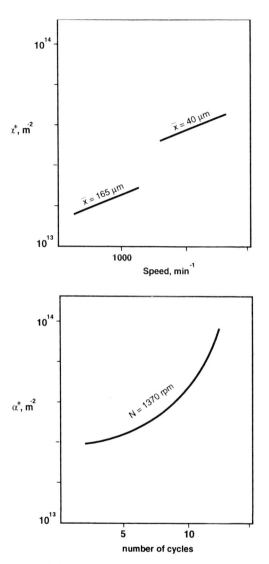

Figure 8.45 Effect of operating speed and cycles on specific resistance

Step 1:   Calculate cake permeability $k = 4.25 \times 10^{-14} \, \mathrm{m}^2$
Step 2:   Calculate Zeitsch drainage number $D = 10$
Step 3:   Estimate $S_\infty = 0.28$
Step 4:   Estimate drainage factor $\phi = 1.254 \times 10^{-3} \, \mathrm{s}^{-1}$

Assume various levels of ϕ_t; estimate S_t from Equation (8.61).

Answer

ϕ_t	S_t	t, s
0.1	0.56	80.0
0.2	0.34	160.0
0.3	0.31	239.0
0.4	0.30	319.0
0.5	0.29	400.0
∞	0.28	∞

It should be observed that Figure 8.41 applies for $D = 10$ only and S_T values at other drainage numbers must be calculated from Equation (8.61). For convenience approximate values of $erf(1/\sqrt{D})$ are recorded below in the range $0 < D < 100$.

D	1	2	3	4	5	10	25	100
$erf(1/\sqrt{D})$	0.843	0.683	0.586	0.520	0.473	0.345	0.223	0.112

Figure 8.46 H.F. Inverting Filter Centrifuge [Heinkel GmbH & Co., Germany]

8.4.13 Centrifugal Filter Developments

In common with developments reported in Chapter 11 on combined vacuum-pressure filters for continuous operation, centrifuges also have been modified to allow pressurisation up to 6 bars during centrifugation [Meyer, 1995]. A unit incorporating this principle is shown in Figure 8.46 where the gas-pressurised inverting filter centrifuge finds increasing applications in the filtration and dewatering of fine, sticky particulates. The latter would be difficult to separate in normal centrifugal condition.

The inverting filter receives attention, along with all other batch and continuous centrifugal units, in a comprehensive technical review by Leung [1998]. In the latter text, which is completely devoted to solid and perforated bowl centrifuge technology, recent developments are reported which have facilitated the extended application of these machines in many industries. Applications of fully automated devices are listed in the bulk chemicals, food processing, petrochemical, speciality chemicals and pharmaceutical industries.

8.5 References

Alt, C., 1984 Centrifugal Separation, in Mathematical Methods in SLS. A. Rushton (Ed.), Nato ASI Series; Appl. Sci. No. 88. Kluwer, Dordrecht.

Ambler, C.A., 1979, Sect. 4.5, Handbook of Separation Techniques, P.A. Schweitzer (Ed.), McGraw Hill.

Ambler, C.A., 1952 Chem. Eng. Prog. 48, pp 150–58.

Austin, L.G., and Klimpel, R., 1981 Powder Technology 29, pp 277–281.

Beeby, J.P. and Nicol, S.K., 1993, Filtrat. Separat., 30, pp 141–146.

Bloor, M., et al., 1989, Solid-liquid Separation III, I.Chem.E. Symp. Ser., 113, pp 1–16.

Bradley, D., 1965, The Hydrocyclone, Pergamon, Oxford.

Bradley, D. and Pulling, D.J., 1959, Trans. I.Chem.E., 37, pp 34–35.

Coleman and Thew, 1983, Chem. Eng. Res.Des., pp 233–239.

Daneshpoor, S., 1984, M.Sc. Thesis, Manchester University.

Dombrowski, H. and Brownell, 1954, I.E.C., 46, p 1201.

Grace, H.P., 1953, Chem. Eng. Prog., 49, p 154.

Hallitt, J., 1975, Filtrat. Separat., 12, (6), p 675.

Hasset, N.T., 1956, Chem. & Ind. (Rev), 35, p 917.

Hebb, M.H. and Smith, F.H., 1948, Encyclopedia of Chemical Technology, vol. 3, 1st Edn, Kirk, R.E. and Othmer, D.F., (Eds.) Intersciences, New York, pp 501–521.

Holland-Batt., A.B., 1982, Trans. Inst. Min. Metall. Section C., pp C21–C25.

Hultsch G. and Wilkesmann, H., 1977, Chap. 12. Solid Liquid Separation Equipment Scale-Up, D. Purchas and R. Wakeman (Eds), Uplands Press, Croydon.

Kelsall, D.F., 1952, Trans. I.Chem.E. 30, pp 87–104.

Lavanchy, A.C. and Keith, F.W. 1964, Encyclopedia of Chemical Technology vol. 4, 2nd Edn. Kirk, R.E. and Othmer, D.F. (Eds) Intersciences, New York.

Leung, W.W., 1998, Industrial Centrifugation Technology, McGraw-Hill, New York.
Maloney, J.D., 1946, I.E.C., 38, p 24.
Meyer, G., 1955, U.K. Filtration Soc. Symp., N.E.C. Birmingham, 7th June.
Oyima Y. and Sumikawa S., 1954, Kogaku Kagatu, 18, p 593.
Plitt, L.R. and Kawatra, S.K., 1979, Int. J. Min. Proc., 5, pp 369–378.
Purchas, D.B., 1981, Solid–Liquid Separartion Technology, Uplands Press. U.K., 241.
Quilter, J.A., I.Ch.E., N. W. Branch Symp., No. 3, Section 7.
Records, F.A., 1990, Centrifugal Sedimentation Loughborough University Short Course Notes.
Rhodes, N. et al, 1989, Solid Liquid Flow, 1, pp 34–41.
Rietema, K., 1961, Chem. Eng. Sci., 15, pp 298–325.
Rushton, A. and Arab, M., 1986, 4th World Filtration Congress, Ostend, Flemish Filtration Society, KVIV Institut. Engrs. Antwerp, Belgium.
Storrow, J.A. and Burak, N., 1950, J.Soc.Chem.Ind., 69, p 8.
Storrow, J.A. and Haruni, M., 1952, Chem. Eng. Sci., 1, p 154.
Spear, M. and Rushton, A., 1975, Filtrat. Separat., 12(3), p 254.
Spear, M. and Rushton, A., 1970, ibid, 3, p 236.
Svarovsky, L., 1984, Hydrocyclones, Holt, Rinehart & Winston, London.
Svarovsky, L., 1990, Solid-Liquid Separation, 3rd Edn., Butterworths, London.
The Institution of Chemical Engineers, 1987, User Guide for the Safe Operation of Centrifuges 2nd Edn., I.Chem.E., Rugby.
Trawinski, H.F., 1969, Filtrat. Separat. 6.
Vallery,V. and Malony, J.D., 1960, A.I.Ch.E.T., 6, p 382.
Wakeman, R.J., 1976, Chemical Engineers, 314, p 668.
Wakeman, R.J., 1979, Filtrat. Separat. 16(6), p 655.
Wakeman, R.J., 1980, ibid, 17(1), p 67.
Zeitsch, K., 1977, Chap. 14, Solid-Liquid Separation, L. Svarovsky (Ed), Butterworths, London.
Zeitsch, K., 1978, Int. Symp., Soc, Belge. Filtn., Antwerp, June.

8.6 Nomenclature

A	Area	m^2
a	Centrifugal acceleration	$m\,s^{-2}$
a_s	Distance between centrifuge discs	m
C	Solid concentration by volume fraction	–
C_o	Solid feed concentration by volume fraction	–
D	Drain number, Equation (8.62)	
d_c	Hydrocyclone diameter	m
$\overline{d}$	Mean particle diameter	m
E	Particle grade efficiency	–
g	Gravitational constant	$m\,s^{-2}$
h	Filter cake thickness; centrifuged cake or bowl depth	m

k	Filter cake permeability	m^2
k'	Intrinsic permeability - ratio of permeability to viscosity	$m^4\,N^{-1}\,s^{-1}$
L	Separator length	m
N	Rotation speed	$rev\,s^{-1}$
N_c	Capillary number, Equation (8.55)	–
ΔP	Pressure differential	$N\,m^{-2}$
ΔP_c	Pressure differential over filter cake	$N\,m^{-2}$
ΔP_s	Compressive pressure differential due to solids density	$N\,m^{-2}$
p	Proportion of particles of given size collected in a centrifuge	
Q	Volumetric flow rate	$m^3\,s^{-1}$
Q_o	Volumetric feed rate	$m^3\,s^{-1}$
q	Filtrate flow rate	$m^3\,s^{-1}$
R	Equilibrium orbit radius in hydrocyclone	m
R_f	Volumetric flow split of hydrocyclone underflow relative to feed	
r	Radial position	m
R_m	Medium Resistance	m^{-1}
S	Cake saturation	
S_∞	Saturation at infinite drainage time	
t	Time	s
u	Liquid velocity over equilibrium orbit radius	$m\,s^{-1}$
u_t	Terminal settling velocity	$m\,s^{-1}$
V	Volume of filtrate	m^3
V_c	Volume of centrifuge	m^3
v	Tangential velocity	$m\,s^{-1}$
v_z	Axial velocity in a hydrocyclone	$m\,s^{-1}$
W	Mass of solids	kg
w	Mass of solids deposited per unit area	$kg\,m^{-2}$
x	Particle diameter	m
x_{50}	Hydrocyclone cut size	m
z	Axial position within a hydrocyclone	m

Greek Symbols

α	Specific cake resistance	$m\,kg^{-1}$
ε	Local porosity	–
μ	Liquid viscosity	$Pa\,s$
ρ	Fluid density	$kg\,m^{-3}$
ρ_s	Solid density	$kg\,m^{-3}$
Σ	Ambler's sigma value for a centrifuge	m^2
ω	Angular velocity	s^{-1}

9 Post-Treatment Processes

9.1 Introduction

The term "post-treatment" refers to processes which are used after the principal solid–liquid separation has been achieved. The final step in the latter operation is usually some type of cake filtration which will leave a cake that may be fully saturated with liquid or partially drained and it is extremely unlikely that the solids are in an acceptable condition for any subsequent operation such as thermal drying. Thus the post treatment processes are washing and deliquoring which are concerned respectively with removing the soluble solids from the liquid remaining in the cake and then purging the major proportion of the liquid from the cake pores.

9.2 Washing

Many solid–liquid separation systems must recover a valuable solid product from a suspension which has undesirable constituents such as soluble dissolved solids in its liquid component. Crystallisation and precipitation are obvious examples where a solid phase has to be produced in a pure state from a contaminant liquor. In such a situation the cake must be washed with a clean wash liquor to reduce the contaminants to an acceptable level.

Reslurry washing where the cake is repulped with clean wash liquor and refiltered, often in several stages, is an obvious way to obtain the desired outcome, but it is expensive in terms of capital equipment and running costs. It is much more economical to conduct the washing operation on the filter if the specification can be achieved. Thus there is considerable interest in filters where effective washing can be undertaken and in calculation techniques that are able to predict wash rates, wash times and solute concentrations. For suspensions where the particles are relatively coarse in size and the solid material is not porous a quick wash may be all that is required to give the necessary effect and this can be undertaken on a wide variety of filters. In that situation the choice of filter will be decided by factors concerned with other aspects rather than the washing. In contrast, for systems where the particles are small and the solids material is intrinsically porous a significant proportion of the residual filtrate may be trapped between the particles and in the pores of particles themselves. In this case the removal rate of the final traces of solutes is determined by a diffusion-controlled transport mechanism and a significantly long residence time will be necessary for the wash to be effective. The range of filters suitable for this duty is more limited and is likely to be governed by the washing requirements rather than those for filtration.

The physical picture of the washing of a saturated cake of small porous particles is shown diagrammatically in Figure 9.1 It involves an initial displacement by the wash liquor, Figure 9.1 A and B, with longitudinal mixing and diffusion following, Figure 9.1 C and D. In the later stages mixing and diffusion from the boundary layers on the particle surfaces and from the interparticle voids are important steps and finally diffusion and mixing from the particles pores become the rate-determining stages. In practice cakes are often drained prior to washing and the washing curve for a typical drained cake will be as shown in Figure 9.2. Note the difference in shape when compared with the curve for the saturated cake; this is due to the absence of a displacement stage. In practice most cakes will be partially drained and the washing curve will fall between the two shown in the figure. Rhodes [1934] proposed an exponential decay equation to describe the way in which the concentration of solute material in the discharged wash liquor changes with time. This model assumes that c_s the concentration of solute in the wash liquor issuing from the cake is directly proportional to the concentration of solute in the cake and that changes in the two are equal and opposite.

Hence if the cake volume is V_c and m the mass of solute in the cake:

$$c_s = \frac{k'm}{V_c}$$

and if c_s increases to $c_s + \mathrm{d}c_s$ and m decreases to $m-\mathrm{d}m$:

$$(c_s + \mathrm{d}c_s) = \frac{k'm}{V_c}(m - \mathrm{d}m)$$

so:

$$\mathrm{d}c_s = -\frac{k'}{V_c}\mathrm{d}m$$

The amount $\mathrm{d}m$ removed in a small element of time $\mathrm{d}t$ is also given by:

$$\mathrm{d}m = conc \times \frac{wash\ rate}{area} \times area \times \mathrm{d}t$$

If F is the volumetric rate of wash liquid per unit area and A is the cake area then:

$$\mathrm{d}m = c_s FA\mathrm{d}t$$

and:

$$-\frac{V_c}{k'}\mathrm{d}c_s = c_s FA\mathrm{d}t$$

The cake thickness L is given by V_c/A so:

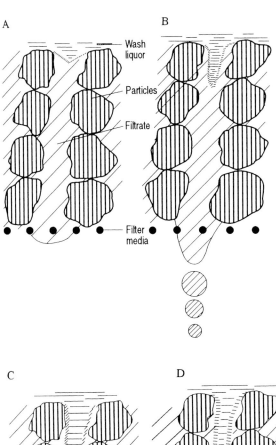

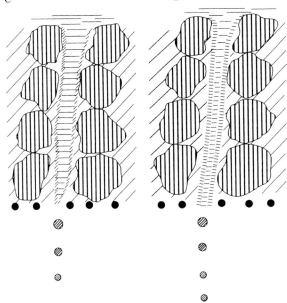

Figure 9.1 Washing of a saturated cake of small porous particles.

A) Start of displacement stage;   C) Breakthrough of wash liquor, start of diffusion/dispersion stage;

B) Displacement washing;   D) Diffusion/dispersion washing

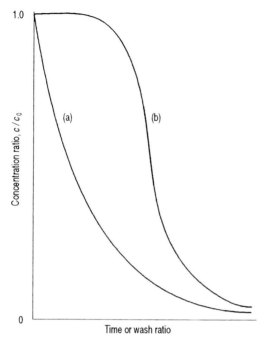

Figure 9.2 Concentration ration *vs* time or wash ratio. Curve (a), drained cake; curve (b), saturated cake

$$\frac{dc_s}{c_s} = -\frac{Fk'}{L}dt$$

which integrates between the limits $c_s = c_o$ at $t = 0$ and $c_s = c_s$ at $t = t$ to give:

$$c_s = c_o \exp\left(-\frac{k'Ft}{L}\right) \tag{9.1}$$

where k' is an empirical constant to be determined experimentally. This model is only accurate for short wash times since it does not take into account diffusional processes. However it has been applied to longer washes by establishing different values of the constant for consecutive portions of the wash curve. Marecek and Novotny [1980] used this approach to simulate successfully the removal by washing of ferrous sulphate from hydrated titanium dioxide in plant-scale leaf filters with wash ratio values of up to 6.

9.2.1 Wash Ratio

The wash ratio n is the ratio of the volume of wash liquor to the volume of filtrate remaining in the saturated cake after filtration. The factor is often included in the mathematical descriptions of short washes because the analysis allows the washing behaviour to be related to the filtration behaviour. Choudhury and Dahlstrom [1957] were interested in washing on continuous vacuum filters and developed Rhodes' Equation (9.1) using material balances to give:

$$R = \left(1 - \frac{E}{100}\right)^n \tag{9.2}$$

where R is the mass fraction of the original soluble material remaining in the cake after washing and E is the washing efficiency expressed as a percentage. In an extensive study experimental values of efficiency were found in the range 35–86% with the lower values applying to fast-washing cakes such as the ones composed of long fibres often found in paper manufacture. The washing efficiency was lowered by 10% for full-scale applications to take account of uneven cake thicknesses and wash liquor maldistribution and a figure of 70% assumed for general application. Good agreement of experiment with theory was found for wash ratios less than 2.1.

For constant-pressure filtration and washing, such as occur on vacuum filters, the filtration time and washing time can be related through the wash ratio if the wash liquor and filtrate are assumed to have similar physical properties. From the equation for incompressible filtration at constant pressure expressed in terms of α, specific resistance:

$$\Delta P = \frac{\mu c \alpha V^2}{2 A^2 t} + \frac{\mu R_m V}{A t} \tag{2.23}$$

and considering conditions at the end of the filtration operation, the medium resistance may be neglected leading to a simplified and rearranged equation:

$$\frac{V_f}{A} = \left(\frac{2 \Delta P t_f}{\mu c \alpha}\right)^{1/2} \tag{9.3}$$

where V_f is the volume of filtrate produced in time t_f, the cake formation time.

If the washing liquor is applied under the same pressure difference and the washing process assumed to proceed in the same way as the filtration, the washing rate can be written as:

$$\left(\frac{1}{A} \frac{dV}{dt}\right)_w = \frac{A \Delta P}{\mu c \alpha V_f} \tag{9.4}$$

If V_w is the volume of wash liquid then the amount applied per unit area can be expressed as:

$$\frac{V_w}{A} = t_w \left(\frac{1 dV}{A dt} \right)_w = \frac{A \Delta P t_w}{\mu c \alpha V_f} \tag{9.5}$$

Substituting for A/V_f from Equation 9.3 and rearranging gives:

$$\frac{V_w}{A} = \left(\frac{\Delta P}{2 \mu c \alpha t_f} \right)^{1/2} t_w \tag{9.6}$$

The volume of residual filtrate in the saturated cake V_m is proportional to the quantity of cake and proportional to the volume of filtrate so from Equation 9.3:

$$\frac{V_m}{A} = k'' \left(\frac{2 \Delta P t_f}{\mu c \alpha} \right)^{1/2} \tag{9.7}$$

and the ratio of wash liquid volume to residual filtrate volume, i.e. the wash ratio n, is given by:

$$\frac{V_w}{V_m} = \left(\frac{\Delta P}{2 \mu c \alpha t_f} \right)^{1/2} \frac{t_w}{k''} \left(\frac{\mu c \alpha}{2 \Delta P t_f} \right)^{1/2} \tag{9.8}$$

Hence:

$$\frac{V_w}{V_m} = \frac{t_w}{2 k'' t_f} = n \tag{9.9}$$

Experimental plots of the wash time t_w versus the wash ratio n give straight lines for different values of the cake formation time t_f (Figure 9.3).

Example 9.1

An aqueous suspension containing 10 wt% of solids is filtered at a constant pressure difference of 4 bar on a 0.4 m^2 area filter. The volume filtered is 0.02 m^3. The filter cake is then given a wash with water at the same pressure for 30 s. Calculate the washing rate and the wash ratio.

Data:

Cake specific resistance $\alpha = 1.24 \times 10^{12}$ m/kg
Cloth resistance $R_m = 4 \times 10^{10}$ m^{-1}
Cake porosity at 4 bar, $\varepsilon = 0.45$

Solids density $\rho_s = 2 \times 10^3 \, kg/m^3$
Density of water $\rho = 10^3 \, kg/m^3$
Viscosity of water $\mu = 10^{-3} \, Nsm^{-2}$

Solution

Calculation of cake and filtrate volumes

It is necessary to change the weight percent concentration to volume percent in order to calculate the cake and filtrate volumes. One kilogram of suspension at 10 wt% solids contains 0.1 kg of solids and 0.9 kg of liquid. The volume of the suspension is:

$$\frac{0.1}{2.0 \times 10^3} \, (\text{solids}) + \frac{0.9}{1.0 \times 10^3} \, (\text{liquid}) = 5 \times 10^{-5} + 9 \times 10^{-4} = 9.5 \times 10^{-4} \, m^3$$

Therefore volume % of solids is:

$$\frac{5 \times 10^{-5} \times 100}{9.5 \times 10^{-4}} = 5.26\%$$

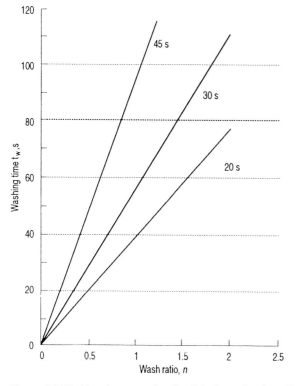

Figure 9.3 Washing time *vs* wash ratio. Cake formation time 45, 30 and 20 s

Hence 0.02 m^3 of suspension consists of 5.26/100×0.02=10.52×10^{-4}m^3 solids associated with (0.02−10.52×10^{-4}) = 18.948×10^{-3} m^3 liquid.
Liquid volume in cake.
At a porosity of 0.45, the volume of liquid remaining in the cake after filtration is:

$$0.45/0.55 \times (10.52 \times 10^{-4}) = 8.61 \times 10^{-4} \text{ m}^3.$$

$$\text{Total cake volume} = (10.52 + 8.61) \ 10^{-4} = 19.13 \times 10^{-4} \text{ m}^3$$

It follows that the cake thickness L = (cake volume)/area = 0.00478 m.

Filtrate rate and wash ratio
    The final filtration rate after the passage of 0.0181 m^3 of filtrate is given by a differential form of Equation (2.23):

$$q = \frac{dV}{dt} = \frac{\Delta PA}{\mu \left(\dfrac{\alpha c V}{A} + R_m \right)}$$

where:

$$c = \left(\frac{s\rho}{1 - sm} \right)$$

$$m = \frac{\text{mass of wet cake}}{\text{mass of dry cake}} = \frac{10.52 \times 10^{-4} \times 2 \times 10^3 + 8.61 \times 10^{-4} \times 10^3}{10.52 \times 10^{-4} \times 2 \times 10^3} = 1.41$$

s = mass fraction of solids in suspension = 0.1 so

$$c = (0.1 \times 10^3)/(1 - 0.1 \times 1.41) = 116.4$$

Hence:

$$q = \frac{4 \times 10^5 \times 0.4}{10^{-3} \left(\dfrac{1.24 \times 10^{12} \times 116.4 \times 0.0181}{0.4} + 4 \times 10^{10} \right)} = 2.44 \times 10^{-5} \text{ m}^3/\text{s}$$

Total volume of wash water used in 30 s = 7.31×10^{-4} m^3.
Void volume in filtercake = 8.61×10^{-4} m^3.
Wash ratio = (7.31×10^{-4})/(8.61×10^{-4}) = 0.85

9.2.2 Longitudinal Dispersion, Mixing and Diffusion

As the wash ratio increases the mechanism being used to remove the solute changes from displacement and other effects such as dispersion, mixing and diffusion become significant. A great deal of theoretical work has been done to describe these effects in capillary tubes and many workers have applied the same arguments to the problem as it occurs in porous media. Brenner [1962], for example, discusses the behaviour of an initially sharp interface between two miscible liquids, having identical dynamical and kinematical properties, in terms of the concentration c_s of solute at a distance x from the point of introduction of the wash liquid at time t. The process is described by the equation:

$$\frac{\partial c_s}{\partial t} + u \frac{\partial c_s}{\partial x} = D \frac{\partial^2 c_s}{\partial x^2} \tag{9.10}$$

where u is the average interstitial velocity and D is an axial dispersion coefficient. Other studies have recognised that the adsorption of solute by the particles and the subsequent desorption of solute into the wash liquor is a significant factor in many applications.

Lapidus and Amundson [1952] provided a modification to Equation (9.10) which takes adsorption effects into account:

$$\frac{\partial c_s}{\partial t} + u \frac{\partial c_s}{\partial x} + \left(\frac{1-\varepsilon}{\varepsilon} \right) \frac{\partial \eta}{\partial t} = D \frac{\partial^2 c_s}{\partial x^2} \tag{9.11}$$

where ε is the cake porosity and η is the amount of solute adsorbed per unit of solid material. The relationship between η and c_s is important and these workers took a simple proportionality where:

$$\eta = K c_s$$

and

$$\frac{\partial \eta}{\partial t} = K \frac{\partial c_s}{\partial t}$$

so Equation (9.11) becomes:

$$\frac{1}{\lambda} \frac{\partial c_s}{\partial t} + u \frac{\partial c_s}{\partial x} = D \frac{\partial^2 c_s}{\partial x^2} \tag{9.12}$$

with:

$$\frac{1}{\lambda} = 1 - K + \frac{K}{\varepsilon}$$

Taking the case of a step input gives a solution of the form [Michaels et al, 1967]:

$$\frac{c_s}{c_o} = 1 - \frac{1}{2}\left\{ erfc\left[\left(\frac{uL}{4Dn\lambda}\right)^{\frac{1}{2}}(1-n\lambda)\right] + \exp\left(\frac{uL}{D}\right)erfc\left[\left(\frac{uL}{4Dn\lambda}\right)^{\frac{1}{2}}(1+n\lambda)\right]\right\} \quad (9.13)$$

where L is the cake thickness and n, the wash ratio, is equal to ut/L. The boundary conditions for this solution are:

$$c_s(x,0) = c_o$$

$$c_s(0,t > 0) = 0$$

$c_s(x \rightarrow \infty, t)$ is finite.

For the perfect displacement situation when:

$$\left(\frac{uL}{D}\right) \rightarrow \infty$$

the second term becomes negligible leaving:

$$\frac{c_s}{c_o} = 1 - \frac{1}{2}erfc\left[\left(\frac{uL}{4Dn\lambda}\right)^{\frac{1}{2}}(1-n\lambda)\right] \quad (9.14)$$

For the case of perfect mixing when the wash liquor is introduced to the cake i.e. when

$$\left(\frac{uL}{D}\right) \rightarrow 0$$

and in the absence of sorption, Brenner [1962] shows that:

$$\frac{c_s}{c_o} = \exp(-n) \quad (9.15)$$

which is a similar equation to the Rhodes model, Equation (9.1).
    Note: The complimentary error function $erfc(z)$ is defined as:

$$erfc(z) = 1 - erf(z) = \frac{2}{\sqrt{\pi}}\int_z^\infty \exp(-y^2)dy$$

and has the following values:

z	0.05	0.1	0.2	0.3	0.4	0.5	0.6	0.7	0.8	0.9
$erfc\ (z)$	0.944	0.888	0.777	0.671	0.572	0.480	0.396	0.322	0.276	0.203

z	1.0	1.1	1.2	1.3	1.5	1.7	2.0	2.9	3.5
$erfc\ (z)$	0.157	0.120	0.090	0.066	0.034	0.016	0.005	4×10^{-5}	7×10^{-7}

The proper application of the dispersion model is bedevilled by difficulties with the initial boundary condition and in obtaining accurate values for the axial dispersion coefficient D. Sherman [1964] realised that expressing the initial condition by the step function, although convenient for the mathematical analysis, was not easy to do in practice especially with thin cakes and he contrived to measure the solute concentration at the inflow face of his beds. The initial boundary condition was then represented by a power–exponential function with six experimental constants:

$$c_s = c_o(k_o + k_1 n + k_2 n^2 + k_3 n^3 + k_4 n^4)\exp(-\gamma n) \tag{9.16}$$

A solution of Equation (9.11) using Laplace transforms was obtained which enabled Sherman to represent accurately the results from laboratory washing experiments with particle and fibre beds.

The system is characterised generally by the dispersion parameter $D_n = uL/D$ which includes the bed depth L as the characteristic dimension, the average interstitial velocity u and the axial dispersion coefficient D. Aris [1956] showed from fundamental considerations that the latter is related to the molecular diffusion coefficient D_m by the relationship:

$$D = D_m + \frac{\kappa x_p^2 u^2}{192 D_m} \tag{9.17}$$

where x_p represents the mean pore diameter and κ is a pore shape factor.

Later workers correlated their data in a similar form using a particle Péclet number, with $u\bar{x}/D_m$, with $\bar{x}$ representing a mean *particle* size, in the second term [Greenkorn,1982], i.e.:

$$\frac{D}{D_m} = A + B\left(\frac{u\bar{x}}{D_m}\right)^b$$

Taking account of the tortuosity of the pore paths through the cake established a value for constant A as $1/\sqrt{2} = 0.707$, but the values of the constants in the second term seem to be related to cake depth. Some studies of deep beds of coarse regular particles gave $B = 1.75$ and $b = 1$; however Wakeman and Attwood [1988], from an extensive experimental study of relatively thin cakes of fine particles, obtained:

$$\frac{D}{D_m} = 0.707 + 55.5 \left(\frac{\bar{x}u}{D_m} \right)^{0.96} \tag{9.18}$$

The first term in the equation represents the effect of molecular diffusion and that controls up to:

$$\frac{u\bar{x}}{D_m} \approx 5 \times 10^{-3}$$

The second term represents the effect of convective dispersion and that dominates at the higher Péclet numbers.

These authors continued Sherman's approach in a computer model using finite difference equations. They used a simpler initial boundary condition:

$$c_s = c_o k_o \exp(-\gamma n)$$

Values of k_0 and γ were assumed in order to fit the numerical solution to their experimental results. For $k_0 = 0.15$ and $\gamma = 7.5$ a distinct "tail" was obtained in the predicted washing curve which then fitted the experimental results. This tail is indicative of a wash liquor distribution effect on the cake surface affecting about 2.5% of the cake depth.

General solutions to the partial differential equation for the non-sorption case are presented by Wakeman and Attwood in the form of charts, see Figures 9.4 and 9.5, and tables giving the relationship of c_s/c_0 and R to the wash ratio n for values of D_n in the range 0.01–500. The parameter R is the fraction of solute retained in the cake.

Although these results make it relatively easy to make washing estimates using the dispersion model great care must be taken in its application. Cracking in the filter cake, channelling and initial maldistribution of the wash liquor can each produce major error effects. Machine designers can take steps to eliminate cracking in some filter configurations and can arrange to introduce the wash liquor evenly. However, channelling is known to be a problem in thin cakes. Crozier and Brownell [1952] observed significant channelling in beds of thicknesses, $L \leq 200\bar{x}$. Wakeman [1986] examined this problem statistically and found that a bed thickness equal to approximately 10 000 particle diameters was necessary to eliminate channelling. Other factors which produce anisotropic effects are variations in particle size and particle shape. Washing of flocculated cakes can be difficult due to hold-up of solute in the flocs; however if the flocs break down during fitration and washing to give cakes of porosity < 0.86 reproducible behaviour can be expected [Michaels et al,1967].

9.2.3 Diffusion Coefficient

Calculations using the dispersion model usually take literature values of the binary molecular diffusion coefficient for the solute as a starting point in estimating the value of the axial dispersion coefficient. However the presence of other dissolved species can have a significant effect on molecular diffusion coefficients and the use of literature values for pure solutions could give rise to serious errors in the cases where several soluble components are present. Relevant examples of systems exhibiting these interactions are given by Cussler [1976]. These include systems where:

1) Components show strong thermodynamic interactions such as large salting in or out, large electrostatic effects or reversible reactions
2) Components have very different molecular weights
3) Solutions are not dilute
4) One solute gradient is much larger than that of a second solute

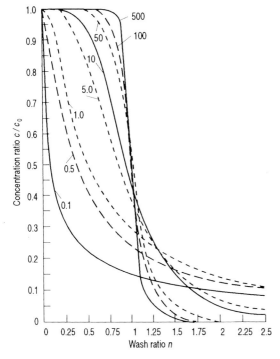

Figure 9.4 Concentration ratio *vs* wash ratio. Range of *uL/D* 0.1–500 [Wakeman and Attwood, 1988]

In electrolyte systems the components diffuse as ionic species and the effective ionic diffusivities are related to the nature and concentration of the other ions present. Taylor and Krishna [1993] give as an example the diffusion of an aqueous solution of HCl and $BaCl_2$. The infinite dilution diffusion coefficients for the individual ions in water are:

$$\begin{array}{lll} H^{+} & D_{H} & = 9.3\times10^{-9}\ m^{2}/s \\ Cl^{-} & D_{Cl} & = 2.0\times10^{-9}\ m^{2}/s \\ Ba^{2+} & D_{Ba} & = 0.85\times10^{-9}\ m^{2}/s \end{array}$$

and the calculated effective ionic diffusivities at a concentration of 1 kmol/m^3 of each electrolyte become:

$$\begin{array}{lll} H^{+} & D_{H} & = 6.81\times10^{-9}\ m^{2}/s \\ Cl^{-} & D_{Cl} & = 2.54\times10^{-9}\ m^{2}/s \\ Ba^{2+} & D_{Ba} & = 0.395\times10^{-9}\ m^{2}/s \end{array}$$

which are matched by experimental results. Note that the chlorine ion diffusivity is increased, the hydrogen ion diffusivity reduced and the barium ion diffusivity decreased from an already low value. The differences in comparative values of the effective diffusion coefficients increase with increase of concentration.

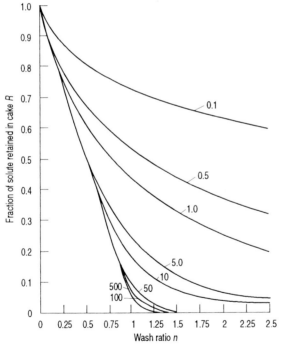

Figure 9.5 Fraction of solute retained *vs* wash ratio. Range of *uL/D* 0.1–500 [Wakeman and Attwood, 1988]

9.2.4 Washing Efficiency

Application of the models to washing on full-scale equipment raises the question of washing efficiency. One approach involves the use of a straightforward efficiency factor which is related principally to the type of filter to be used and to the application, as mentioned above. Purchas and Wakeman [1986] noted the relationship between the washing efficiency and the type of system found by Choudhury and Dahlstrom for rotary drum vacuum filters and proposed a correction of the dispersion parameter rather than the use of an efficiency factor. This correction may be expressed as:

$$\left(\frac{uL}{D}\right)_F = 0.49 + 1.348 \ln\left(\frac{uL}{D}\right)_C \qquad (9.19)$$

where $(uL/D)_C$ is the calculated value and $(uL/D)_F$ is the corrected value to be used in the estimation of the washing behaviour on the filter.

9.2.5 Drained Cakes

Allowing cakes to drain partially prior to washing is usually beneficial in that the amount of solute to be removed by washing is reduced. The washing rate and the value of the final residual solute are not usually affected unless the cakes are fully drained or deliquored to such an extent that resaturating the cake with wash liquor becomes a rate-determining operation. Such a situation is likely to occur on a belt filter for example if the cake is drained under vacuum until air breaks through.

9.2.6 "Stop-start" Washing

If long washing periods are found to be necessary to achieve the desired solute levels, some advantage might be found in "stop-start" washing. In this type of operation the cake is flooded and washed to the extent of one or two wash ratios and the washing stopped for a period of time to allow the diffusion process to take place. A further displacement washing is then performed and the sequence repeated as required. Stop-start washing can reduce the amount of wash liquor used in situations where cake cracking cannot be prevented, where the solid particles adsorb significant quantities of the solute, or where maldistribution of the wash liquid may be unavoidable, such as in some filterpresses.

Example 9.2

Using the information, data and results of Example 9.1 estimate the recovery of the solute if the molecular diffusion coefficient for the solute is 2×10^{-12} m^2/s.

Solution

The mean particle size may be related to the specific resistance, α and the porosity, ε of the cake by the equation:

$$\bar{x}^2 = \frac{180(1-\varepsilon)}{\alpha\rho_s\varepsilon^3} \tag{9.20}$$

So:

$$\bar{x} = 13.4\sqrt{\frac{0.55}{1.24\times10^{12}\times2\times10^3\times(0.45)^3}} = 6.62\times10^{-7}\,\text{m}$$

The washing superficial velocity q/A is 6.1×10^{-5} m/s so the average interstitial velocity is:

$$u = \frac{q}{\varepsilon A} = \frac{6.1\times10^{-5}}{0.45} = 1.36\times10^{-4}\,\text{m/s}$$

The particle Péclet number is:

$$\frac{u\bar{x}}{D_m} = \frac{1.36\times10^{-4}\times6.62\times10^{-7}}{2\times10^{-12}} = 44.86$$

The axial dispersion coefficient D is then calculated from the correlation Equation (9.18):

$$D = 0.707\times2\times10^{-12} +55.5(44.86)^{0.96}\times2\times10^{-12} = 4.28\times10^{-9}\,\text{m}^2/\text{s}$$

The cake thickness is:

$$L = \frac{\text{cake volume}}{\text{area}} = \frac{19.13\times10^{-4}}{0.4} = 4.78\times10^{-3}\,\text{m}$$

and thus the calculated dispersion parameter is:

$$\left(\frac{uL}{D}\right)_c = \frac{1.36\times10^{-4}\times4.78\times10^{-3}}{4.28\times10^{-9}} = 151.5$$

From the graph the value of R is estimated for $n = 0.85$ and $uL/D = 152$ at 0.18 so the amount of solute removed is $(1-R)$ expressed as a percentage, i.e. 82%. This calculation assumes 100% washing efficiency, which is unlikely even on a pressure filter. The actual washing efficiency is probably about 50% when bypassing and leakages are taken into account so the effective wash ratio would be 0.43 and the value of R falls to 0.59 and the value of solute removed becomes 41%.

Example 9.3

Use the information, data and results of Examples 9.1 and 9.2 to estimate the wash ratio and wash time required to achieve a solute reduction of 95% if the washing efficiency is 50%.

Solution

Using the value of uL/D calculated in Example 9.2 of 152 and an R value of 0.05 the wash ratio can be estimated from the graph as about 1.1. For a washing efficiency of 50% the wash ratio must be doubled to give 2.2.

From Example 9.1, the cake void volume is 8.61×10^{-4} m^3 and the wash rate is 2.44×10^{-5} m^3/s so the wash liquid volume is $2.2 \times 8.61 \times 10^{-4} = 1.89 \times 10^{-3}$ m^3 and the wash time is $1.89 \times 10^{-3}/2.44 \times 10^{-5} = 77$ s.

Obviously these calculations only estimate the amount of wash liquor necessary for the *actual washing operation*; the amounts of wash liquid required to fill pipework, feed vessels, pumps and receivers should not be forgotten.

Although the calculation method is based on some rigorous mathematics, the approximate nature and fragility of the technique must be emphasised. In addition to the assumptions already mentioned it must be noted that the physical state of the filter cake is represented by an assembly of uniformly sized spheres with the liquid flowing through uniformly sized pores in a manner that may be described by the Kozeny–Carman expression.

9.2.7 Other Mathematical Models

Alternative approaches that have received attention include the *uniform film* and *blind-side channel* models. In the uniform film model [Kuo, 1960] it is assumed that the initial displacement stage leaves a channel for wash liquor with a stagnant film of solute-bearing liquid on the pore surface. The subsequent removal of solute is determined by mass transfer across the stagnant film boundary. Various assumptions are made such as plug flow in the channel, no sorption by the particles and no axial dispersion.

The blind-side channel model [Han, 1967; Han & Bixler, 1967; Wakeman, 1971] is claimed to deal readily with drained cakes. It assumes that the residual filtrate is contained in a system of blind channels in the cake and that the flow of wash liquid is along straight channels without any axial mixing. The soluble material in the trapped filtrate is transferred by a diffusion process and the solute is transported through the cake in plug flow. The blind-side channels are assumed to be wedge-shaped and the straight channels are assumed to be empty of filtrate so that diffusion of soluble material from the side channel starts when the wash liquid reaches the point of emergence of the side channel into the main channel.

Comparison of these models with experimental results has shown a match over part of the washing curve on some occasions; however both are difficult to apply and for these reasons are not preferred to the dispersion model.

9.2.8 Reslurry Washing

If satisfactory washing with the cake in situ cannot be achieved, reslurry washing may be attempted. In this process the cake is discharged from the filter, mixed with the wash liquor and refiltered. It can be shown that it is more efficient to divide the wash liquid into a number of equal amounts and to conduct the operation in several stages.

Let a volume V_w of wash liquor containing a concentration c_w of the solute be used to wash a cake with a volume V_m of liquor holding a concentration c_o of solute. If the concentration of the resulting liquid mixture of filtrate and wash liquor is c_s then, for a single stage, by a material balance on the solute:

$$V_m c_o + V_w c_w = (V_m + V_w) c_s$$

which transposes to:

$$\frac{c_s - c_w}{c_o - c_w} = \frac{1}{1 + (V_w / V_m)} \qquad (9.21)$$

If the wash liquid is divided into z aliquots and the wash performed in z stages, with the output concentration in the wash liquor from the final stage now being c_s and the output concentration from the intermediate stages bearing the stage number as a subscript, it follows that:

Stage 1

$$\frac{c_{s1} - c_w}{c_o - c_w} = \frac{1}{1 + (V_w / z V_m)}$$

Stage i

$$\frac{c_{si} - c_w}{c_{si-1} - c_w} = \frac{1}{1 + (V_w / zV_m)}$$

Stage z

$$\frac{c_s - c_w}{c_{sz-1} - c_w} = \frac{1}{1 + (V_w / zV_m)}$$

The overall effect is obtained by multiplying the stage effects:

$$\frac{c_s - c_w}{c_o - c_w} = \frac{c_{s1} - c_w}{c_o - c_w} \ldots \times \frac{c_{si} - c_w}{c_{si-1} - c_w} \ldots \times \frac{c_s - c_w}{c_{sz-1} - c_w}$$

which gives:

$$\frac{c_s - c_w}{c_o - c_w} = \frac{1}{\left[1 + (V_w / zV_m)\right]^z} \tag{9.22}$$

Thus for example if the wash liquid is three times the volume of the residual filtrate in the cake, corresponding to a wash ratio of 3, and c_w is zero then c_s/c_o reduces rapidly as the number of stages increases:

Stage (z)	1	2	3	4	6	8
c_s/c_o	0.25	0.16	0.125	0.107	0.088	0.078

9.3 Deliquoring

The filtration and washing processes leave the filter cake substantially saturated with liquid. The presence of liquid is usually a disadvantage in subsequent applications of the solids so there is some processing benefit in removing as much of the liquid as possible whilst the cake is on the filter. This objective may be achieved either by blowing with air or inert gas or on some filters by mechanically squeezing the cake using an inflatable impermeable membrane.

The gas-blowing operation can be conducted on most types of pressure and vacuum filters and is commonly used to follow filtration or washing stages. The parameters to be determined are the pressure difference, the gas flow rate and the saturation, which is the proportion of the voids occupied by liquid. The initial mechanism is displacement leading at breakthrough to two-phase laminar flow and after an extended period to two phase transition range flow. Long periods of blowing after breakthrough may begin to remove the remaining liquid by a drying mechanism.

A plot of pressure difference versus saturation is called a *capillary pressure curve* and has the typical shape shown in Figure 9.6. Note that there is a minimum pressure required to achieve an initial reduction in saturation; this is the threshold pressure, P_b. There is a final finite saturation value beyond which no further reduction in liquid content is possible; this is the irreducible saturation, S_∞. The irreducible or infinite saturation is taken as a limit in many theoretical approaches and a reduced saturation is used and defined as:

$$S_R = \frac{S - S_\infty}{1 - S_\infty}$$

(9.23)

where S is the achieved saturation measured.

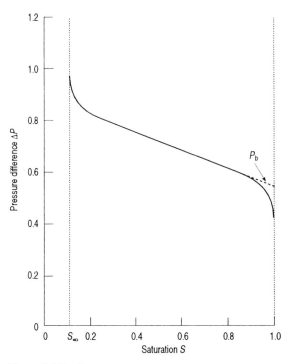

Figure 9.6 Capillary pressure curve

The calculation method due to Brownell et al [1947; 1949] is based on an analogy between pipe flow and flow in porous media. For laminar flow a Reynolds number and a friction factor are defined which include the sphericity and diameter of the particles and the porosity of the filter cake. These are defined as:

$$Re = \frac{\bar{x}u\rho}{\mu\varepsilon^{m}} \qquad (9.24)$$

and

$$f = \frac{2g\bar{x}\Delta P\varepsilon^{n}}{Lu\rho}$$

where the exponents m and n are functions of the ratio of sphericity to porosity. From their experimental data Brownell et al produced design charts to give values to these exponents.

The concept was extended to the two-phase flow situation by applying the relationship to each of the phases.

The effective saturation, defined as:

$$S_{E} = \frac{S - S_{R}}{1 - 2S_{R} + SS_{R}} \qquad (9.25)$$

is included in the equations for the wetting phase, raised to a power y which is another function of bed porosity. The flow of the nonwetting phase is described by a similar relation which incorporates the wetted porosity and the wetted sphericity which are defined respectively as:

$$\varepsilon' = \frac{(1 - S)\varepsilon}{(1 - S_{R})} \qquad (9.26)$$

and

$$\Psi' = \frac{\text{surface area of a sphere of the same volume as the wetted particle}}{\text{surface area of the wetted particle}} \qquad (9.27)$$

Whilst wetted porosity is a function of saturation, these authors failed to establish a relation between wetted sphericity and saturation. For the nonwetting (gas) phase the experiments are used to determine the wetted porosity and the wetted sphericity. Fuller details of the design method and copies of the charts can be found in the original papers and in Brown [1950].

The use of this approach in the calculation of gas flow in the deliquoring of filter cakes on rotary drum vacuum filters is described by Brownell and Gudz[1949]. However when applied to air blowing in a pressure filter the calculated gas flow under estimates the experimental flow by a considerable margin and the prediction of the liquid flowrate compares poorly with the desaturation rate actually obtained. Extrapolation of this calculation method is not successful due to the difficulty of evaluating the required exponents outside the porosity limits set by the authors.

9.3.1 Relative Permeability

Lloyd and Dodds [1972] calculated the flow of liquid from the cake by application of Darcy's law to elemental sections of the cake using a relative permeability model. Wakeman [1979] extended this approach by incorporating a pore size distribution function in the relationship between saturation and capillary pressure. The model involves a computation of liquid and gas flow by a simultaneous solutions of Darcy's law. The relative permeability values used are obtained from a pore size distribution index λ which relates the average reduced saturation of the cake to the ratio of the modified threshold pressure P_b to the capillary pressure P_c. The equations are:

$$P_b = \frac{4.6(1-\varepsilon)\sigma}{\varepsilon \bar{x}} \tag{9.28}$$

where σ is the surface tension of the liquid:

$$S_R = \left(\frac{P_b}{P_c}\right)^{\lambda} \tag{9.29}$$

$$k_{rL} = S_R^{\frac{2+3\lambda}{\lambda}} \qquad \text{and}$$

$$k_{ra} = (1-S_R)^2(1-S_R)^{(2+3\lambda)/\lambda} \tag{9.30}$$

where k_{rL} and k_{ra} are the liquid and gas relative permeability respectively. Experimental measurements by the capillary pressure method are necessary in order to obtain values for λ and S_R which are required for the calculation.

The irreducible saturation correlates with the capillary number N_c by the equation:

$$S_\infty = 0.155(1 + 0.031N_c^{-0.49}) \tag{9.31}$$

where:

$$N_c = \frac{\varepsilon^3 \bar{x}^2(\rho g L + \Delta P)}{(1-\varepsilon)^2 L\sigma} \tag{9.32}$$

for such material as coal fines and coarse crystallised materials [Wakeman,1986].

The rate of deliquoring is estimated from charts (Figure 9.7) which correlate the reduced saturation with the dimensionless time defined as:

$$\theta = \frac{kP_b t}{\mu L^2(1-S_\infty)\varepsilon} \tag{9.33}$$

and with dimensionless pressure difference:

$$\Delta P_a = \left(\frac{P_a}{P_b} \right)_{inlet} - \left(\frac{P_a}{P_b} \right)_{outlet} = P_{ai}^* - P_{ao}^* \qquad (9.34)$$

where P_a is the gas pressure and k is the permeability of the cake to the liquid at complete saturation.

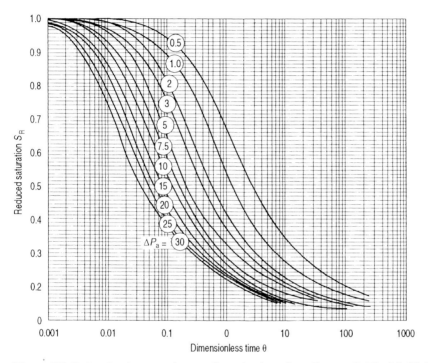

Figure 9.7 Reduced cake saturation *vs* dimensionsless time. Range of ΔP_a 0.5–30 [Purchas and Wakeman, 1986]

Example 9.4

The washed filter cake described in Examples 9.1 and 9.2 is blown with air at 7 bar abs for 150 s. Calculate the final moisture content. The surface tension of the filtrate is 0.075 N/m and the air pressure below the cake is 1.013×10^5 N/m^2.

Solution

Calculation of capillary number N_c from Equation 9.32 gives:

$$N_c = \frac{(0.45)^3 \times (6.61 \times 10^{-7})^2 (1000 \times 9.81 \times 4.78 \times 10^{-3} + 6 \times 1.013 \times 10^5)}{(1 - 0.45)^2 \times 4.78 \times 10^{-3} \times 0.075} = 2.23 \times 10^{-4}$$

The irreducible saturation is obtained from Equation (9.31):

$$S_\infty = 0.155(1 + 0.031 N_c^{-0.49})$$

as:

$$S_\infty = 0.155(1 + 0.031 \times (2.23 \times 10^{-4})^{-0.49}) = 0.45$$

The threshold pressure is calculated from Equation (9.28):

$$P_b = \frac{4.6 \times (1 - 0.45) \times 0.075}{0.45 \times 6.61 \times 10^{-7}} = 6.38 \times 10^5 \, N/m^2$$

and the dimensionless pressure difference

$$\Delta P_a = \left(\frac{7 \times 1.013 \times 10^5}{6.38 \times 10^5} \right) - \left(\frac{1.013 \times 10^5}{6.38 \times 10^5} \right) = 0.952$$

Cake permeability, k is derived from the specific resistance:

$$k = \frac{1}{\alpha(1 - \varepsilon)\rho_s} = \frac{1}{1.24 \times 10^{12} \times 0.55 \times 2000} = 7.33 \times 10^{-16} \, m^2$$

and the dimensionless time:

$$\theta = \frac{7.33 \times 10^{-16} \times 6.38 \times 10^5 \times 150}{10^{-3} \times (4.78 \times 10^{-3})^2 \times (1 - 0.45) \times 0.45} = 12.42$$

Inspection of the chart (Figure 9.7) at the relevant values of ΔP_a and θ gives S_R as 0.18. Hence the final saturation is

$$S = S_R (1 - S_\infty) + S_\infty = 0.18 \times (1 - 0.45) + 0.45 = 0.55$$

Although the reduced saturation is reasonably low at 0.18 the high value of the irreducible saturation, 0.45 means that gas blowing this cake will have a limited effect. Increasing the air pressure to 10 bar abs for the same blowing time gives a capillary number of 3.35×10^{-4}, an irreducible saturation of 0.40, a dimensionless pressure difference of 1.43 and a dimensionless time of 11.32, which lead to a reduced saturation of 0.15 and a final saturation of 0.49. Not a great improvement on the earlier value, because the major factor is the small particle size which appears in the expressions for the capillary number and the threshold pressure.

The averaged dimensionless gas rate $v_a{}^*$, defined as:

$$v_a^* = \frac{v_a \mu_a L}{k P_b} \tag{9.35}$$

where v_a is the actual averaged gas rate and μ_a is the gas viscosity, is obtained from a correlation with dimensionless time θ, see Figure 9.8, which arises from the relative permeability model [Wakeman, 1979]. Since the basis for the design charts is:

$$\left(\frac{P_a}{P_b} \right)_{inlet} = 100$$

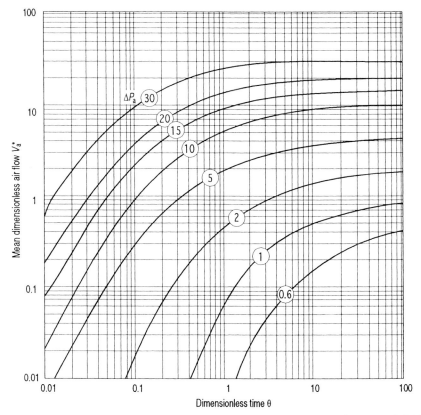

Figure 9.8 Mean air (gas) flow rate *vs* dimensionsless time. Range of $\Delta P_a = 0.6–30$ [Purchas and Wakeman,1986]

the dimensionless air rate has to be corrected for this and for the actual pressure used in the deliquoring operation. Assuming the gas behaviour is ideal over the conditions encountered the correction to be applied to the chart value of $v_a{}^*$ is:

$$\frac{100 - \Delta P_a}{P_{ao}^*} \left(\frac{P_{ao}^{*2} - P_{ai}^{*2}}{(100 - \Delta P_a)^2 - 10^4} \right) \tag{9.36}$$

Example 9.5

Using an air viscosity of 1.7×10^{-5} N s/m^2 calculate the air rate required to perform the blowing operation discussed in Example 9.4.

Solution

From the Figure 9.8, a value of the averaged dimensionless air rate v_a^* of 0.50 can be obtained for the values of θ and ΔP_a calculated in Example 9.4. The dimensionless inlet and outlet pressures are calculated as:

$$P_{ai}^* = \left(\frac{7 \times 1.013 \times 10^5}{6.38 \times 10^5} \right) = 1.112$$

and:

$$P_{ao}^* = \left(\frac{1.013 \times 10^5}{6.38 \times 10^5} \right) = 0.159$$

and the pressure correction is, from equation 9.36:

$$\frac{100 - 0.952}{0.159} \times \left(\frac{0.159^2 - 1.112^2}{(100 - 0.952)^2 - 10^4} \right) = 3.980$$

Thus the corrected averaged dimensionless air rate is 0.5×3.98 = 1.99. The actual averaged air rate is obtained from equation 9.35 as:

$$v_a = \frac{v_a^* k P_b}{\mu_a L} = \frac{1.99 \times 7.33 \times 10^{-16} \times 6.38 \times 10^5}{4.78 \times 10^{-3} \times 1.7 \times 10^{-5}} = 0.0115 \text{ m/s}$$

and the volumetric flowrate on 0.4 m^2 is 4.6×10^{-3} m^3/s at 7 bar absolute pressure. This method only gives values for the air or gas required in the deliquoring step. Additional volumes will be required to fill and pressurise vessels and pipework.

All the calculation techniques presented and discussed here are approximate and set great store on the solid–fluid systems being uniform and representative. Clearly an application to systems having wide distribution of particle sizes, densities and shapes or with complex mixtures of solutes will involve an greater degree of approximation.

9.4 Equipment for Washing and Deliquoring

Washing of filter cakes in situ can be carried out on most but not all types of filters. The two principal types which do not allow for washing are disc filters and recessed-plate filter presses, although some manufacturers would claim that a crude form of washing can be arranged on the latter. It will become clear that successful washing requires the wash liquor to be supplied and distributed evenly across a filter cake that is of uniform thickness and structure. These requirements make major demands on the filter design.

Plate and frame filter presses, when arranged for washing, introduce the wash liquor behind the filter media at the plate face and allow the liquid to traverse the full thickness of the cake. Only alternate plates are needed for the introduction of the wash liquor and these are usually identified separately to distinguish them from the nonwashing plates. Obviously only alternate drains are used when washing and the used wash liquid may be collected independently from the filtrate. It is important that the design of the drainage channels on the plate surface and the choice of filter media both allow a good distribution of wash liquid to develop across the cake surface. Similarly the development of anisotropy in the cake during the initial stages of filtration must be avoided by ensuring that the feed suspension is homogeneous and that sedimentation in the filter chambers is avoided by feeding from the bottom if necessary.

Washing can be achieved successfully on most batch pressure filters. In this type of equipment the principal potential problem is sluicing the filter cake from the plates with the wash liquid, which could occur if the cake were sloppy and poorly formed.

On rotary drum vacuum filters the extent of washing is limited to a segment of the drum so that in some cases only a short wash may be possible. It can be difficult to arrange for the successful introduction of the wash liquor onto the upper curved surface of the drum and a high proportion may be lost. However cracking on the cakes can be removed prior to washing by using a roller to slightly compress and consolidate the cake. If extensive washing is required on a vacuum filter a belt type is usually a better choice because the washing and dewatering sections can be designed independently from the filtration stage.

Blowing with air or inert gas can be arranged on many types of filters. Obviously pressure differences are limited on vacuum filters and lower saturations may be obtainable in pressure filters. This operation can be difficult to arrange on filterpresses because of problems in achieving good sealing.

Filter cakes should be uniform in porosity and thicknesss, and should have no cracks. A support medium with a fine pore structure will help to obtain a good gas distribution at the cost of some pressure loss.

More details about equipment and information on the analysis of the filtration, washing and deliquoring cycle will be found in Chapter 11.

9.5 References

Aris, R.,1956, Proc. Roy. Soc. A235, p 67.

Brenner, H., 1962, Chem. Eng. Sci. 17, p 229.

Choudhury, A.P.R. and Dahlstrom, D.A., 1957, AIChE J. 3, p 433.

Brown, G.G. et al, 1950, Unit Operations, Wiley, New York.

Brownell, L.E. and Gudz, G.B., 1949, Chem. Eng. 56, p 112.

Brownell, L.E. and Katz, D.L., 1947, Chem.Eng. Prog. 43, p 537.

Crozier, H.E. and Brownell, L.E., 1952, Ind. Eng. Chem. 44, p 631.

Cussler, E.L., 1976, Multicomponent Diffusion, Elsevier, Amsterdam.

Greenkorn, R.A., 1982, Flow Phenomena in Porous Media, Dekker, New York.

Han, C.D., 1967, Chem. Eng. Sci. 22, p 837.

Han, C.D.and Bixler, H.J., 1967, AIChE J. 13, p 1058.

Kuo, M.T., 1960, AIChE J. 16, p 566.

Lapidus, L. and Amundson, N.R., 1952, J. Phys. Chem. 56, p 984.

Lloyd, P.J. and Dodds, J.A., 1972, Filtrat. Separat. 9, p 91.

Marecek, J. and Novotny, P., 1980, Filtrat. Separat. 17, p 34.

Michaels, A.S., Baker, W.E., Bixler, H.J. and Vieth, W.R., 1967, IEC Fund. 6, p 33.

Purchas, D.B., Wakeman, R.J., 1986, Chapter 13 in Solid/Liquid Separation Scale-up 2nd ed, Purchas, D.B and Wakeman, R.J. (eds.) Uplands Press/Filtration Specialists, London.

Rhodes, F.H., 1934, Ind. Eng. Chem. 26, p 1331.

Sherman, W.R., 1964, AIChE J. 10, p 855.

Taylor, R. and Krishna R., 1993, Multicomponent Mass Transfer, Wiley, New York.

Wakeman, R.J., 1971, MSc Thesis, Manchester University.

Wakeman, R.J., 1979, Filtrat. Separat. 16, p 655.

Wakeman, R.J. and Attwood, G.J., 1988, Filtrat. Separat. 25, p 272.

Wakeman, R.J., 1986, Chem. Eng. Res. Des. 64, p 308.

9.6 Nomenclature

A	Area	m^2
c	Dry mass of solids per unit filtrate volume	$kg\ m^{-3}$
c_s	Solute concentration in wash water	$kg\ m^3$
D	Axial dispersion or diffusion coefficient	$m^2\ s^{-1}$
E	Washing efficiency	$\%$
F	Volumetric flow rate of wash liquid per unit filter area	$m\ s^{-1}$
f	Friction factor	$-$
g	Gravitational constant	$m\ s^{-2}$
h	Mass transfer coefficient	$m\ s^{-1}$
k	Permeability	m^2

k'	Empirical constant	–
L	Cake thickness	m
m	Mass solute in cake	kg
N_c	Capillary number	–
n	Wash ratio: volume wash liquor/pore volume in cake	–
P_b	Threshold pressure	$N\ m^{-2}$
P_c	Capillary pressure	$N\ m^{-2}$
ΔP	Pressure Differential	$N\ m^{-2}$
q	Filtrate flowrate	$m^3\ s^{-1}$
R	Mass fraction of solute remaining in the cake	–
R_m	Medium Resistance	m^{-1}
s	Mass fraction of solids in feed mixture	–
S	Saturation: volume of liquid in cake/volume of cake pores	–
t	Time	s
t_f	Cake formation time	s
t_w	Time during washing	s
u	Interstitial velocity of wash liquid	$m\ s^{-1}$
V	Filtrate volume	m^3
V_c	Cake volume	m^3
V_f	Filtrate volume produced in time t_f	m^3
V_m	Volume of residual liquor in saturated cake	m^3
V_w	Volume of wash liquid used	m^3
$\bar{x}$	Mean particle size	m
x_p	Pore diameter	m

Greek Symbols

α	Specific cake resistance	$m\ kg^{-1}$
ε	Filter cake porosity	–
η	Amount of solute absorbed per unit volume	$kg\ m^{-3}$
λ	Pore size distribution index	–
μ	Filtrate viscosity	Pa s
ρ	Liquid density	$kg\ m^{-3}$
ρ_s	Solid density	$kg\ m^{-3}$
σ	Liquid surface tension	$J\ m^{-2}$

10 Membrane Technology

Appropriate membrane technologies in the context of solid–liquid separation are microfiltration and the use of the more open membranes in ultrafiltration. Other membrane processes, including the pressure-driven processes of hyperfiltration and reverse osmosis, are concerned primarily with the removal of dissolved species from a solvent and shall not be considered. The boundary between the finer end of microfiltration and the coarser end of ultrafiltration is not sharp, and ultrafiltration is used for fine colloid–liquid separation. The start of the regions of microfiltration, ultrafiltration and hyperfiltration occurs, approximately, with the filtration of particles of diameter 10, 0.1 and 0.005 μm, respectively.

Further distinction has to be made between conventional filtration of fine particles, less than 10 μm in diameter, and microfiltration. It would be unusual for the filtration of such particles on a conventional filter cloth to be described as microfiltration. Thus microfiltration is constituted by the filtration of small particles *and* by the medium which is used for the filtration. Conventional filtration is undertaken on filter cloths with a very open structure, see Chapter 4, whereas membrane filtration is usually concerned with filtration employing membrane media: where the "equivalent" pore size is of the order of 10 μm, or less. These definitions are, however, becoming less distinct as it is now possible to obtain conventional filtration equipment employing membrane-type filter media, as discussed in Chapter 4, and crossflow microfilters employing conventional filter cloth.

The use of membranes in cartridge microfiltration has been discussed already in Chapter 6, Section 6.6. That section also contained a number of test procedures employed for membrane characterisation which will not be repeated here. This chapter provides details of membrane configuration other than cartridges, mathematical models to assist in the understanding and control of the processes. Industrial applications or investigations of microfiltration and to a lesser extent ultrafiltration are also discussed.

Filtration membranes are made from a variety of polymers, including: cellulose acetate, polyamides, polyethers, polycarbonates, polyesters, regenerated cellulose, polyvinyl chloride, polyvinylidene fluoride (PVDF), PTFE, acrylonitrile copolymers and polysulphones. Surface sulphonation can be used to form hydrophilic membranes which would otherwise be naturally hydrophobic in nature. When filtering aqueous suspensions the untreated hydrophobic membranes may require wetting with a solvent miscible with water, depending on the pore size and pressures used, in order to overcome the capillary pressure resisting water ingress, see Equation (6.14).

Inorganic membranes are becoming increasingly popular because of their chemical inertness, strength and hydrophilicity. Thus strong chemical cleaners and aggressive temperatures can be used for cleaning. Transmembrane pressures of 10–20 bar are possible, although not necessarily leading to increased flux rates, and there is no problem with wetting the membrane before or during filtration. They are, however, usually very brittle and more expensive than polymers.

The membrane material is formed into structures by a variety of processes that are adequately described elsewhere [Gutman, 1987]. The membrane structure depends upon many factors: ease of manufacture, desired retention size, mechanical integrity, chemical and physical resistance, etc. Three basic structures are commonly used: homogeneous, asymmetric and composite; these are illustrated in Figure 10.1.

Homogeneous: uniform pore profile through filter

Asymmetric: finer filtering surface faces feed suspension

Composite: two types of materials used

Figure 10.1 Membrane structural and filtering surface arrangements

The simplest structure is the homogeneous type which has no significant variation in pore channel diameter from the filtering surface through to the other side. The asymmetric membrane has a much smaller pore diameter on the filtering surface compared with the rest of the filter depth; the latter acts as a mechanical support for the filtering surface. This arrangement is the reverse of the asymmetric membrane used in cartridge filtration, see Figure 6.20. In dead-end filtration the more open pore diameter protects the finer pores by removing some of the coarser particles, i.e. prefiltration. However, in many membrane filtrations alternative strategies, such as crossflow filtration, are employed to remove the coarse particles. The prime requirement is, therefore, to prevent media blockage or to facilitate the ease of media cleaning. The asymmetric membrane in the crossflow orientation is less likely to retain particles internally. The composite membrane is similar to the asymmetric in having a thin layer of small pores on top of a coarser pore-sized material acting as a mechanical support but, in this instance, the filtering layer is made from a different material, or polymorph, from the support. In the ultrafiltration of molecules a composite membrane is, strictly speaking, a composite of two materials used to impart some physical–chemical effect, such as rejection of specific molecules. However, in microfiltration such specific rejection of material is not found.

Most membranes, regardless of their structure, have an appreciable pore size distribution and an irregular arrangement of pores on the membrane surface. Figure 10.2 is a photograph taken under a scanning electron microscope (SEM) of a typical microfiltration membrane with a nominal pore size of 0.2 µm; included on the figure is a scale bar representing 5 µm.

Membrane filtration has many similarities to conventional filtration, and the mathematical description of the process uses many concepts already introduced in Chapter 2. However, there are significant differences in the terminology employed: the filtrate is referred to as the "permeate", the residual slurry or suspension from the filtration is called the "retentate" and the permeate filtration rate is the "flux rate", which in microfiltration is conventionally reported in the units of litres per square metre of membrane area per hour ($1 \text{ m}^{-2} \text{ h}^{-1}$). This rate is equivalent to the superficial liquid velocity through the membrane. In nearly all the instances of constant-pressure membrane filtration the flux rate declines in a manner similar to that illustrated in

Figure 10.3. If some method is employed to restrict or halt the depth of the deposit on the membrane surface the flux rate may become roughly constant at a "pseudo-equilibrium" value. If the deposit is not restricted, the flux rate will decay to an imperceptible amount, in a similar manner to constant-pressure cake filtration.

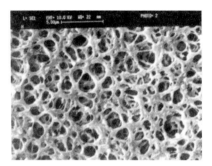

Figure 10.2 A microfiltration polymer membrane surface

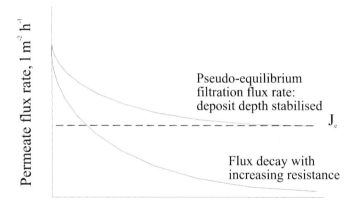

Permeate flux rate, $1 \, m^{-2} \, h^{-1}$

Pseudo-equilibrium filtration flux rate: deposit depth stabilised

J_e

Flux decay with increasing resistance

Filtration time, s

Figure 10.3 Permeate flux decline during membrane filtration

The efforts that can be made to restrict the amount of flux decline, i.e. to establish an equilibrium flux rate, can be split into two categories, dependent largely on the scale of operation: stirred cells and crossflow filtration. The former is of use in the laboratory for small-scale separations, the latter is more appropriate for process applications. Both stirring and crossflow employ the same principle: high shear at the surface of the deposit. Further techniques to restrict deposit thickness, minimise fouling or regenerate flux are discussed in Section 10.7. All the main membrane arrangements are illustrated in Figure 10.4.

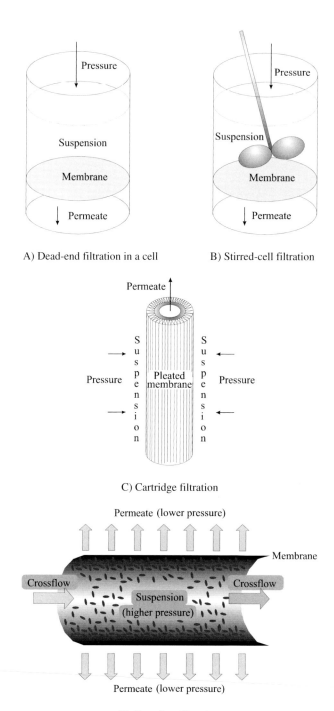

A) Dead-end filtration in a cell   B) Stirred-cell filtration

C) Cartridge filtration

D) Crossflow filtration

Figure 10.4 Membrane filter arrangements

This chapter is concerned primarily with process-scale membrane filtration, and phenomena or effects that are relevant to such filtration. Cartridge filtration has already been discussed in Chapter 6, hence most of the following work refers to membrane filtration under crossflow conditions. This technique is applicable to both microfiltration and ultrafiltration, and the results from either type of filtration are often similar, though often due to different phenomena. The flux rate declines to an equilibrium value as illustrated in Figure 10.3, and that value is often pressure dependent only over a restricted pressure range as shown in Figure 10.5.

One important consequence of the effect illustrated in Figure 10.5 is that increasing the filtration pressure and, therefore, the operating costs may lead

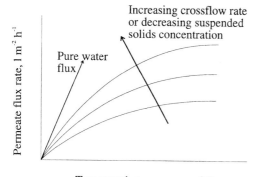

Figure 10.5 Effect of pressure during filtration

to no improvement in permeate, or production, rate. It is, therefore, important to have some understanding of the crossflow filtration process before attempting to modify it.

The driving force for the filtration is the "transmembrane" pressure, which is the average pressure difference between the inside of the crossflow filter and the permeate pressure. It is often assumed to be equal to the mean average of the feed and retentate pressures minus the pressure in the permeate line. It is usual, however, to employ pressure measurement upstream and downstream of the filter in different diameter tubes, hence the transmembrane pressure calculated in this way is only very approximate. Under these circumstances, a better representation of the transmembrane pressure can be obtained from a consideration of an energy balance on the liquid, such as the simplified version of Bernoulli's equation:

$$\frac{P_1}{\rho} + \frac{\bar{v}_1^{\,2}}{2} = \frac{P_2}{\rho} + \frac{\bar{v}_2^{\,2}}{2}$$

where P_1 and P_2 are the pressure before the filter tube and in the filter tube, ρ is the liquid density, and $\bar{v}$ represents the mean suspension crossflow velocity at the same position as the pressure readings.

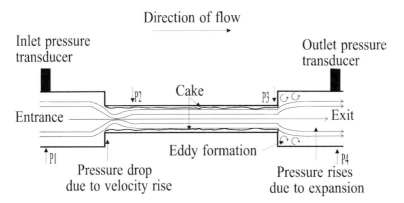

Figure 10.6 Pressure with position during membrane filtration

Considering Figure 10.6, the velocity increases as the suspension enters the filter tube, hence the pressure will fall in accordance with the energy balance given above. The true transmembrane pressure is really the pressure difference between the mean value of P_2 and P_3, marked on the figure, and the pressure of the permeate. However, it is common to use the measured values represented by P_1 and P_4 in the diagram. Both P_2 and P_3 will be below the measured values of P_1 and P_4. It is possible to incorporate a correction into the equation for transmembrane pressure using Bernoulli's equation, which for a Newtonian liquid gives:

$$\text{Transmembrane pressure} = \frac{P_2 + P_3}{2} = \frac{P_1 + P_4}{2} - \frac{\rho\bar{v}_1^{\,2}}{2}\left[\left(\frac{d_1}{d_2}\right)^4 - 1\right] \qquad (10.1)$$

where $\bar{v}_1$ is the mean suspension velocity in the filter manifold d_1 and d_2 are the internal diameters of the inlet manifold and filter tube respectively. If there is significant pressure in the permeate line, the transmembrane pressure is reduced further; i.e. the pressures in Equation (10.1) are values measured in excess of the permeate line pressure. If the suspension density is significantly greater than that of the liquid, then it should be used instead of the liquid density in Equation (10.1). In many practical microfiltration instances the manifold before the filter is a pipe 10 to 50 mm in diameter and the filter tube, or capillary, may be 2 mm in diameter; hence the pressure correction can be significant. An illustrative example is given in Figure 10.7, for a filter tube 3 mm in diameter positioned after a manifold 8 mm in diameter.

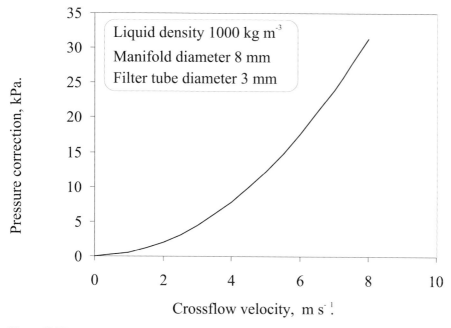

Liquid density 1000 kg m^{-3}

Manifold diameter 8 mm

Filter tube diameter 3 mm

Figure 10.7 Example of pressure correction to mean value during tubular membrane filtration

In microfiltration transmembrane pressure of less than 1 bar and crossflow velocities up to 8 m s^{-1} (defined as velocity within the filter tube) are often reported. At 8 m s^{-1} the pressure correction illustrated in Figure 10.7 is approximately 0.3 bar, hence the true transmembrane pressure of such a system would be 0.3 bar **less** than that deduced from the uncorrected inlet and outlet pressure measurements. Transmembrane pressure correction is complicated further by two occurrences:
- the deposition of a filter cake, thus d_2 becomes indeterminate, and
- at higher solid content the rheological properties often become non-Newtonian.

A further discussion of rheology is included in Appendix B, and a method to determine the transmembrane pressure drop under both of these conditions has been published by Cumming *et al* [1999]. The error is not so significant in ultrafiltration because the operating pressures are appreciably higher and deposit thickness is much lower.

There are many suppliers of membranes and the reader is referred to trade directories and magazines which provide regularly updated listings of these [Anon, 1994]. There is also a well respected South African World Wide Web site providing details of current research, software, applications and up-to-date listing of suppliers [http://www.ccwr.ac.za/emily/ or the European mirror site at http://www.isah.uni-hannover.de/emily/index.html].

10.1 Microfiltration

The most popular nominal pore size in microfiltration is 0.2 μm, with many polymer membranes and some ceramic types available at this size. These filters are believed to provide a sterile liquid because bacteria are retained in the retentate and the permeate is bacteria free. In order to ensure sterile conditions a 0.1 μm membrane is sometimes used. This is also, therefore, a popular commercial microfiltration membrane pore size. For details of the standard biological test see Section 6.6.1. Other cellular material of a biological origin, such as beer and wine yeasts, can be filtered by similar pore size membranes. Coarser pore sizes are available: 0.45, 0.8, 1.0, 1.2, 2.0, 3.0, 5.0 and 10 μm, for other commercial duties, such as the recycling of paint pigments in paint baths, removal of suspended particles in effluents, prefiltration, laboratory analysis, and similar clarification duties.

Popular composite and asymmetric ceramic membranes are formed from a monolithic support consisting of a very porous network, typically of equivalent pore size of 8 μm or more, on which is superimposed a thin skin of much finer pore size to act as the filtering membrane. These membranes are available in tubular and capillary form. The membrane and support are made from various types of alumina, which are chemically very stable and can withstand substantial mechanical pressures, but are susceptible to brittle fracture. The monoliths, or filter elements, come in a variety of dimensions and channel shapes, including circular, square and even star shaped cross-section. The common channel diameters range from 2.7 to 6 mm internal diameter, with typically 7–19 channels in parallel within a monolith. Figure 10.8 illustrates some of the ceramic filter types available.

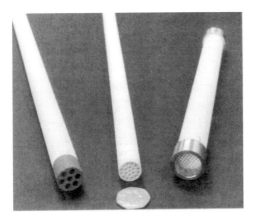

Figure 10.8 End view of some ceramic filter microfiltration elements

Metal microfiltration membranes are also available. These can provide very high mechanical, thermal and chemical stability and are less prone to brittle fracture than the ceramics. These membranes are usually homogeneous in membrane design and are made by sintering together small spheres of metal to provide a rigid porous network on which to filter. Typical porosity of such a filter would be close to that of randomly packed spheres, approximately 50%. Metal fibres can also be used as a filtering medium, and this has the advantage that surface porosity can be increased up to 80%. Thus filtration flux rates are greater with fibres than with sintered spheres. Membrane cleaning is often easier with the fibres, both because of the higher porosity and the internal structure of the filter. Metal filters made from spheres and fibres are available in stainless steel 316L, Inconel 601, Hastelloy, brass, etc. Equivalent pore diameters are much coarser than with the polymer or ceramic membranes. A pore opening of approximately 3 μm is the finest commercially available metal filter, without using an excessive filter thickness; i.e. over-reliance on depth filtration mechanisms, with a consequent inability to produce adequate cleaning of the filter for reuse.

In membrane technology some care has to be exercised in interpreting the pore opening data supplied by manufacturers, and from the type of equipment described in Section 6.6.1 for all types of membrane. For example, Figure 10.9 shows the pore size distribution of a microfiltration material, and Figure 10.10 is a picture taken under SEM of the surface of the same filter material.

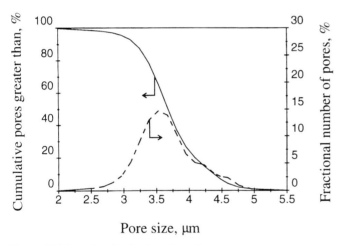

Figure 10.9 Pore size distribution of membrane

Figure 10.10 Photograph of membrane surface

The data presented in Figure 10.9 suggest that this filtration material contains pore openings in the range of 2.2–5.2 µm, with a median pore size of 3.6 µm. Figure 10.10 illustrates that this material has actual pore openings up to at least an order of magnitude greater than this. The equivalent pore opening is the size of opening equivalent to a smooth parallel channel of the same flow resistance or fluid displacement properties as the real flow channel which is highly tortuous. Thus, the apparent pore diameter is considerably less than the surface opening, as can be seen in Figure 10.10. The material illustrated does, in fact, filter non-deforming particles of 3 µm diameter from suspension with very high efficiency, and can provide efficiencies greater than 99% even when filtering particles less than 1 µm diameter from a suspension of low solid concentration. To some extent, therefore, microfiltration often includes a depth filtration effect similar to that described in Section 6.1.

At high solid concentration, greater than approximately 1% by weight, a fouling layer of solids may form on the surface of the filter if the true pore opening is small enough or, alternatively, it may form a combined fouling and filter layer within the membrane surface. Further filtration will then take place on the newly formed fouling layer and subsequent filtration behaviour may depend only on this fouling layer. This has been called a "dynamic" or "secondary" membrane.

In membrane filtration, however, the term dynamic is also used to represent a filter with moving surfaces. In this chapter the term dynamic membrane will not be used for membrane filtration on a fouling layer formed in-situ on a secondary membrane support material. When filtering under such circumstances the original membrane may play only a minor role in the remaining filtration. This effect is illustrated in Figure 10.11 [Raasch, 1987] where a finely ground limestone suspension of 2% by volume solids, mean particle size 4 µm, was filtered on a conventional 0.2 µm polymer membrane and two sieve plates of pore size 25 and 42 µm.

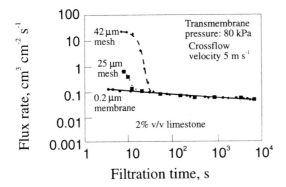

Figure 10.11 Filtration of limestone on different media [Raasch, 1987]

When using the sieve plates the initial permeate was cloudy, but after 40 s there was no perceptible difference between the clarity of the filtrate and the flux rates for all the filter media employed. The coarse mesh had a pore size that is approximately an order of magnitude above that of the particle diameter, yet it appears that the concentration of 2% by volume is sufficient to form a particle bridge over the mesh pores and cause a secondary membrane to form. This is a striking illustration of the effect of the secondary membrane and how it can dominate the filtration performance during crossflow filtration. Particle segregation, however, is a well-known effect during crossflow filtration [Tarleton and Wakeman, 1993]. An increase in crossflow velocity may lead to more of the coarser particles being removed from the deposit with a thinner cake of finer particles being left on the membrane. During secondary membrane filtration such segregation could lead to the reduction in concentration of particles, and size of deposited particles necessary to form a secondary membrane. Hence the bridging over coarse pores needed to form a secondary membrane may also be a function of the crossflow velocity as well as the membrane support and particle size distribution of the suspended material.

In some instances, however, the use of a finer pore-sized membrane provides higher stable fluxes over longer periods than is found with coarser membranes. This is particularly true when the suspended solids particle size is close enough to the membrane pore size for internal filter clogging to occur. Figure 10.12 illustrates an experiment conducted with two polymer membranes and a very low concentration latex suspension, with particle size in the range 0.2–2 μm.

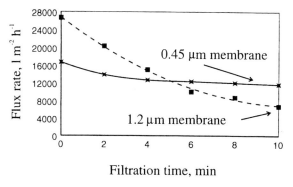

Figure 10.12 Filtration of latex suspension by polymer membranes

During the filtrations illustrated on Figure 10.12 all the operating conditions were identical except the membrane pore size. The latex was suspended in very pure water and very high filtration fluxes resulted. However, the filtration resistance when using the 1.2 μm membrane increased to more than that of the 0.45 μm membrane during the experiment owing to the greater ability of the particles to deposit inside the coarser membrane and the flux rates, therefore, crossed over. The finer membrane appeared better able to resist internal particle deposition, and the flux rate remained fairly constant throughout. Latex particles are a straightforward material to filter and the effect illustrated by Figure 10.12 is even more common with the highly size distributed material encountered in industry. Thus when selecting a membrane, consideration has to be given to the finer end of the suspended particle distribution and the coarser end of the membrane pore size distribution; if these overlap low fluxes may result [Tarleton and Wakeman, 1994 a,b]. At high suspended solids concentration, however, the internal clogging effect is not so prevalent as solids can bridge over the filter medium to form a protective layer, in a similar fashion to a filter cake bridging over filter cloths during dead-end filtration.

In several instances, in order to avoid membrane clogging and to guarantee filtrate quality, ultrafiltration membranes are used for what would normally be regarded as a microfiltration duty [Gutman and Knibbs, 1989].

10.2 Ultrafiltration (UF)

Most ultrafiltration membranes are made from PVDF or polysulphone and are asymmetric. However, two common inorganic UF membranes are made from a composite of carbon (support) with zirconia (skin) at molecular weight cut-off (MWCO) of 20 000–50 000, and an alumina composite with alpha (support) and gamma (skin) at 40 nm. Instead of MWCO the Dalton unit is now in common use (1 Dalton is equal to 1 on a molecular weight scale).

The simple interpretation of cut-off is to regard the membrane as a sieve which will not pass molecules above a threshold value, e.g. a 20 000 Dalton UF membrane would filter out molecules of a molecular weight in excess of this value, but let molecules significantly smaller pass. Molecules are not, however, easy to characterise in such a simple fashion, and filtration can rarely be regarded as simple sieving. Hence a MWCO is usually defined as being the size at which 90% of some test molecule, normally a globular molecule, is retained. Ultrafiltration is applied to the separation or concentration of macromolecules or polymers that vary from molecular weights of 300 to 2 000 000, and may also be applied to colloid filtration. In some instances, however, an UF membrane is used in preference to a microfiltration membrane, see the discussion around Figure 10.12. Table 10.1 provides some indication of molecule size of various materials and colloids, and an inexact relation between MWCO and equivalent pore size has been employed to provide some indication of size.

Transmembrane pressures during ultrafiltration are generally higher than for microfiltration, but pressures in excess of 6 bar are rarely used. Equilibrium flux rates become pressure independent as illustrated in Figure 10.5, often because of a "gel layer" which is described in greater detail in Section 10.4. Permeate flux rates from ultrafiltration membranes are usually less than $50 \, l \, m^2 \, h^{-1}$.

10.3 Module Design

Module designs for both types of filtration have grown in variety according to the requirements of the user. The simplest membrane design uses flat sheets which may be supplied up to 1 m in width. The design of the module in which to mount the flat-sheet

Table 10.1 Suspended material: molecular weight and equivalent pore size

Suspended material	Molecular Weight (nominal average)	Equivalent Pore Size (µm)
Sucrose	342	0.0013
Raffinose	504	0.0016
Bacitracin	1400	0.0023
Peptones	2000–5000	0.0027–0.0038
Insulin	6000	0.0040
Albumin	67 000	0.0101
Gamma Globulin	160 000–900 000	0.0141–0.0273
Dextran T-500	500 000	0.0218
Blue Dextran	2 000 000	0.0371
Colloidal silica		0.1–1
Viruses		≤ 0.1
Clays		≤ 2
Individual bacteria		0.2–10

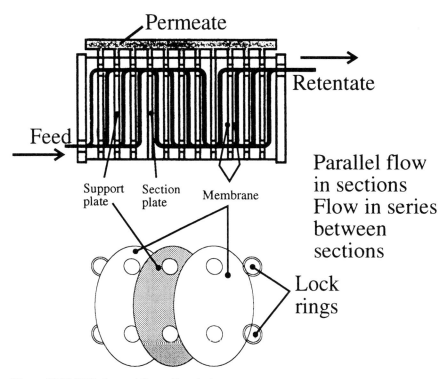

Figure 10.13 DDS plate and frame filter design

membrane and provide adequate flow distribution is, however, much more complex. Nevertheless a variety of laboratory and process scale UF and MF are available using flat-sheet membrane. Very high values of surface area of membrane per unit volume of space occupied can be provided by this design. The major disadvantage is the increased difficulty of membrane cleaning, lack of simple module construction which complicates membrane replacement and, often, poor flow distribution within the unit if parallel flow channels are employed. One of the more common plate and frame membrane filters that successfully employs parallel flow channels, with serial flow between internal sections, is illustrated in Figure 10.13.

The simplest module design is one which contains capillary or tubular UF or MF elements, of internal diameters ranging from 1–20 mm, such as that shown in Figure 10.14.

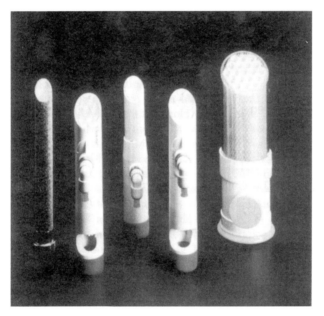

Figure 10.14 Tubular and capillary crossflow filter elements [Courtesy of Koch Membrane Systems Inc.]

It is usual to filter on the inside surface of the tubular or capillary elements and to remove the permeate from the outside. Module construction is simply a means of providing adequate inlet and outlet manifolds to prevent the process fluid mixing with the permeate. This is the usual cause of high filtration flux rates and poor rejection performance, and any filtration which does not provide a flux response such as that described by Figure 10.3 should be regarded with suspicion. The amount of membrane area that can be packed into a given volume is relatively low when using this configuration.

Hollow-fibre membrane modules are similar to the capillary type described above, but with fibres of outside diameters ranging from 80 to 500 μm. It is usual to pack a hollow-fibre module with many hundreds or thousands of these fibres, thus membrane area per unit volume is extremely high. It should be apparent that filtration using hollow-fibre modules is only realistic with process fluids prefiltered to prevent fibre blockage; this limits the technology and it is applied mainly in UF. Also used in ultrafiltration is a spiral-wound membrane module which is often compared to a Swiss roll. The membrane and a spacer are wound round a former, with an appropriate permeate spacer; flow is introduced and removed from the ends. This module design is not appropriate for solid–liquid separation, even when filtering colloids, because of the possibility of flow channel blockage and so it will not be discussed any further.

Membrane surface area to volume of space occupied varies from around 8000 $m^2\,m^{-3}$ for hollow fibres, down to a few hundred for tubular systems with plate and frame systems providing an intermediate membrane packing density.

10.4 Filtration Resistances

There are several resistances to permeate flow during membrane filtration [van den Berg and Smolders,1988]. These are illustrated in Figure 10.15.

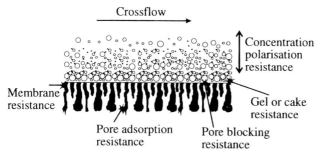

Figure 10.15 Filtration resistances during microfiltration and ultrafiltration

The most well known resistance is that of "concentration polarisation". If the membrane is successful in retaining the suspended material at its surface there must be a concentration gradient between that surface and the bulk flow of feed suspension. Resistance to permeate flow increases with suspended solids content, as implied by permeability expressions, see Section 2.4. Hence resistance to permeate flow increases with any effect that increases the concentration at the membrane surface. The concentration at the membrane surface may reach a maximum or limiting value as transmembrane pressure is increased. Any further increase in pressure does not increase the concentration but may increase the depth of the deposit at the membrane surface. Hence, there is an additional resistance to concentration polarisation owing to the constant-concentration layer at the membrane surface, but with the possibility of variable deposit depth. This resistance is often called that of the "gel layer". This term is more applicable to ultrafiltration rather than microfiltration as polymers and very fine species are more prone to formation of deposits of roughly constant concentration at low solid concentrations. However, suspended solids do form filter cakes that may have similar properties to the gel layer resistance.

The membrane, even in the absence of any suspended material, will have a natural flow resistance that may be determined during a clean liquid flow test. During filtration suspended matter might become attached to the pore channel of the membrane thereby reducing the flow channel dimension, or pores may become blocked off altogether. These last two effects lead to resistance terms that are due to adsorption and pore blocking. It is not usual to be able to independently quantify these two resistances when filtering suspensions which also provide a substantial resistance due to deposit formation. If the membrane acts as a true surface filter then the membrane resistance will remain constant but, usually, some penetration of the membrane takes place and the effective membrane resistance has to be deduced in situ. This is a situation similar to that of classical cake filtration, as discussed in Chapter 2. The effective membrane resistance, therefore, includes contributions due to adsorption and pore blockage. In membrane filtration modelling it is also usual to consider only a single deposit resistance term: the effective deposit resistance. Therefore, this resistance has contributions from concentration polarisation and gel (or cake) layer formation, as well as membrane and deposit interactions such as pore adsorption and blocking.

10.5 Equipment Scale-up and Modelling

Process scale microfiltration usually employs several membrane modules operating in parallel in order to achieve the desired permeate flow. Preliminary testing with a single module is undertaken easily and scale-up by means of proportioning the areas and flows may be sufficient for process design. There are, however, important choices regarding the optimum method of operation:

1) Single pass through a cascade
2) Multiple pass with suspension thickening between batches
3) Multiple pass with "feed and bleed"

The schematic arrangements for these are illustrated in Figure 10.16. Single pass requires a large membrane surface area to achieve an adequate degree of suspension thickening. The concentrate, or retentate, from one set of modules operating in parallel forms the feed to the next set of modules. The number of modules in parallel reduces between each set as the removal of the permeate reduces the volume of suspension to be treated, thus reducing the number of modules in parallel maintains the crossflow velocity, or can even be designed to increase it. Multiple-pass suspension thickening is essentially a batch operation in which the feed concentration to the membrane modules increases during filtration as the permeate is removed from the system. In process-scale

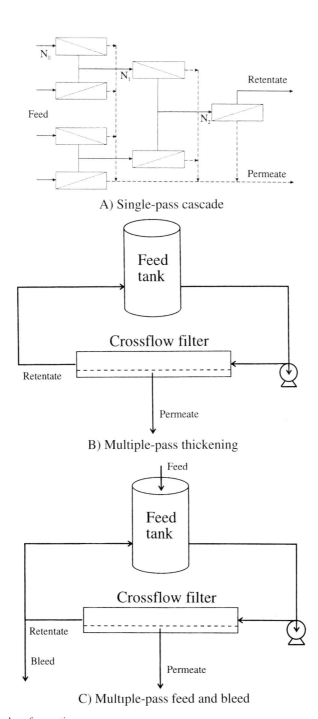

A) Single-pass cascade

B) Multiple-pass thickening

C) Multiple-pass feed and bleed

Figure 10.16 Modes of operation

multiple-pass thickening it is common to have suspension fed into the system at the same rate as the permeate leaves the system. In laboratories, or for small-scale operation, the continuous feed may be absent. In multiple-pass feed and bleed the feed concentration to the membrane is maintained constant, usually at a value that provides appreciable flux rates. The rate at which permeate and retentate bleed is removed is matched exactly by the introduction of fresh suspension to be treated. The first mode of operation is more energy efficient as pumping costs are minimised, but to minimise the membrane cost the flux rates must be high. In order to simplify the preliminary testing it is usual for the test to be undertaken in multiple pass thickening mode in the absence of continuous feed. Often a single test is performed and the flux rate is plotted against feed suspension concentration, as illustrated in Figure 10.17. However, this procedure may lead to an overestimation of the flux rate as the instantaneous flux rate during a thickening operation is usually greater than the sustained flux rate at the same concentration. The deposited cake requires some time to adjust to the prevailing flow conditions, including suspension concentration. During this period of adjustment the equilibrium flux rate often declines in a similar fashion to that illustrated in Figure 10.3. This may be a problem if the test data is used to predict the performance of single-pass or multiple-pass feed and bleed operation. Further testing, allowing the equilibrium flux rate to develop, may be required for these modes of operation.

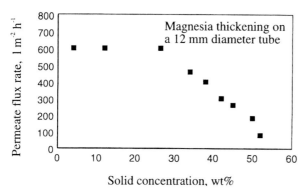

Figure 10.17 Flux decay with magnesia suspension thickening

During single-pass operation the membrane area required at each stage can be calculated from the flux and feed rates via a mass balance. In multiple-pass thickening mode, with continuous feed, the filtration loop can be considered as being a well mixed system and the following mass balance results:

$$N_o JA = V_o \frac{dN}{dt}$$

where J is the flux rate ($m^3\ m^{-2}\ s^{-1}$ in SI units), A is the membrane area, V_o is the tank or system volume, N and N_o are the system and feed concentrations (any consistent concentration units may be used but mass per unit volume is conventional). Integration under the appropriate boundary conditions gives:

$$N = N_o + \frac{N_0 A}{V_o} \int_o^t J\mathrm{d}t \qquad (10.2)$$

where N is the solid concentration at time t. If the flux rate is a constant with respect to concentration and consequently time, then

$$N = N_o + \frac{N_0 A J t}{V_o} \qquad (10.3)$$

If the flux is a function of concentration, and measurable, the function can be incorporated into the mass balance. For a linear function, such as that valid for concentrations above 30% on Figure 10.17, the resulting equation for concentration with time is:

$$N = \frac{k'}{m}\left[\left(1 + \frac{mN_T}{k'}\right)\exp\left(\frac{mN_0 At}{V_o}\right) - 1 \right] \qquad (10.4)$$

where N_T is the concentration in the tank, or system, prior to the period in which flux declines with increasing concentration, m and k' are the gradient and intercept on the flux *vs* concentration plot. Equations (10.3) and (10.4) can be solved for various combinations of membrane area and tank volume to provide the desired degree of thickening in the required time available.

Worked example. Calculate the time required to thicken a magnesia suspension to 50% solids by weight in a 100 litre tank using 1 m^2 membrane area. The feed concentration and initial concentration in the tank is 3% solids and the flux with concentration data is provided in Figure 10.17.

As can be seen on Figure 10.17, the concentration at which the flux is no longer constant with concentration, i.e. N_T, is 30%. Hence Equation (10.3) can be rearranged for the time taken to thicken to 30% solids, this is 90 minutes. Equation (10.4) can then be used to investigate the concentration with time relation for the rest of the filtration, using: $k' = 1200\ \mathrm{l\ m^{-2}\ h^{-1}}$, $m = -20\ \mathrm{l\ m^2\ h^{-1}\ \%wt^{-1}}$ and $N_o = 3\%$ wt%. The time taken for the second stage 30–50% solids is 110 min, making a total time of 200 min. The resulting graph over both stages is given in Figure 10.18. The volume of permeate produced during the 200 min can be obtained by an overall mass balance and is 1570 litres.

During-multiple pass feed and bleed operation the flux rate is likely to stabilise, after an initial period of decay, and at equilibrium the concentration in the system is related to the permeate flux rate, membrane area, bleed rate and feed concentration only:

$$N = N_0 \left(1 - \frac{JA}{Q} \right)^{-1}$$

(10.5)

where Q is the bleed rate in units consistent with the product of the flux and membrane area. Equation (10.5) can be used to investigate various combinations of bleed rate and membrane area for a given concentration.

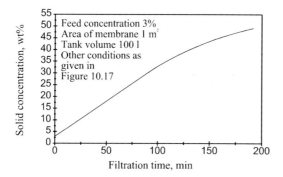

Figure 10.18 Concentration during multiple pass thickening

The mass balances required for process-scale modelling and design are very dependent upon the permeate flux rate, and the variation of the flux rate with operating conditions. Hence, there has been some considerable research into factors affecting flux rates and means to minimise flux decay. If a complete mathematical model of crossflow filtration is available then scale-up can be achieved reliably from the appropriate overall mass balance together with the model and, if necessary, with experimental test work to discover the empirical components of the model. Such a complete model is rarely achieved in microfiltration, but may be found in some ultrafiltrations.

10.5.1 Models of Flux Decline

The flux decline period has already been illustrated in Figure 10.3. Empirical mathematical models of a power law type have been fitted to this curve. These models do not provide an insight into the effects taking place during flux decline, but may be useful if information for process-scale modelling is required over such a period. Models which relate the flux decline to physical phenomena usually make use of the filter models already described in Section 2.7, these are: cake filtration [Belfort & Mark, 1979] intermediate blocking [Bhattacharya et al, 1979], complete blocking [Gutman,

1977; Belfort et al, 1993a; Granger et al, 1985,], and standard filtration [Belfort et al, 1993b]. In microfiltration, the cake filtration model is often appropriate for concentrated systems and the standard filtration model is more applicable for dilute ones undergoing flux decline. The cake filtration model will be described in greater detail in the context of equilibrium flux models in the next section.

The standard filtration model can be represented as:

$$\frac{t}{V} = \alpha_1 t + \beta_1 \qquad (10.6)$$

where α_1 and β_1 are constants. If the pores are assumed to be circular, of equal length and radius, and the volume of the major flow path reduces in direct proportion with the volume of filtrate passing, then a mass balance and Poiseuille's law can be used to expand the equation constants to:

$$\alpha_1 = \frac{4}{\pi h_m n_p d_p^2 A} \frac{N}{\rho_s (1- \varepsilon)} \qquad (10.7)$$

and

$$\beta_1 = \frac{h_m}{N} \frac{128 \mu}{\pi d_p^4 A \Delta P_{TMP}} \qquad (10.8)$$

where N is the suspended solid concentration entering into the pores in mass per unit volume terms, h_m is the depth of the flow channel, n_p the number of pores, d_p the internal diameter of the pores before filtration, ε the porosity of the deposit formed on the surface of the pore and ΔP_{TMP} the transmembrane pressure. Note that the suspended solid concentration entering the pores may not be the same as that in the bulk crossflow, depending on the effectiveness of the filtration, i.e. the rejection.

Equation (10.6) can be rearranged and differentiated with respect to time to give:

$$\frac{dV}{dt} = \frac{\beta_1}{(\alpha_1 t + \beta_1)^2}$$

Hence, filtration flux rate is:

$$J = \frac{1}{A} \frac{dV}{dt} = \frac{\beta_1}{(\alpha_1 t + \beta_1)^2} \frac{1}{A} \qquad (10.9)$$

which provides an equation where flux rate is a function of filtration time.

The blocking models are not so common in microfiltration as ultrafiltration, probably because of the structure of most microfiltration membranes. Two membranes were illustrated in Figures 10.2 and 10.10. In both pictures the tortuous structure below the surface pores is evident. If a particle becomes lodged within the porous structure of the membrane, fluid flow is still possible around the particle via the interconnected porous network. Thus a deposited particle reduces the effective flow path for the fluid, or increases the flow resistance, but does not block off the flow. Increasing resistance due to a decreasing effective flow path is the model on which the standard filtration law was based, hence the standard law appears to be followed more closely than the blocking models. Membranes which do exhibit uniform pore openings and little interconnected flow channels throughout the depth, such as the Nuclepore membrane, may be modelled more accurately by the blocking models. However, most microfiltration membranes are of the interconnected variety.

The main difference between crossflow filtration and conventional filtration is the restriction of cake depth, or deposit resistance, due to the crossflow. Field et al [1995] have modelled the flux decline during crossflow filtration using the filtration laws provided in Table 2.3 and included a term representing the reduction in deposit resistance due to the crossflow. This is easily illustrated by a consideration of the cake filtration model. The conventional cake filtration equation, Equation (2.26), is:

$$\frac{dt}{dV} = \frac{\mu\alpha c}{A^2 \Delta P_{TMP}} V - \frac{\mu R_m}{A \Delta P_{TMP}}$$

where

$$J = \frac{dV}{dt}\frac{1}{A} \qquad \text{and when } V=0, \text{ then:} \qquad -\frac{\mu R_m}{\Delta P_{TMP}} = \frac{1}{J_0}$$

Hence, the cake filtration equation can be written as:

$$\frac{1}{J} - \frac{1}{J_0} = \frac{\mu\alpha c}{\Delta P_{TMP}}\frac{V}{A} \tag{10.10}$$

where V is cumulative permeate volume, as is the case in dead end filtration. The integral of the filtration flux with respect to time is the cumulative filtrate volume over the area, hence:

$$\frac{1}{J} - \frac{1}{J_0} = \frac{\mu\alpha c}{\Delta P_{TMP}}\int_0^t J dt$$

An erosion factor S, SI units of kg m^{-2} s^{-1}, can be added to the above equation to represent the rate of erosion per unit cake area due to the crossflow:

$$\frac{1}{J} - \frac{1}{J_0} = \frac{\mu\alpha c}{\Delta P_{TMP}}\int_0^t J dt - \frac{\mu\alpha S}{\Delta P_{TMP}}t \tag{10.11}$$

assuming that the erosion factor is not a function of filtration time. Differentiating with respect to time and rearranging provides the following equation:

$$-\frac{1}{J^2}\frac{\mathrm{d}J}{\mathrm{d}t} = \frac{\mu\alpha c}{\Delta P_{TMP}}\left(J - \frac{S}{c}\right)$$

and calling $S/c=J^*$, then:

$$-\frac{1}{J^2}\frac{\mathrm{d}J}{\mathrm{d}t} = \frac{\mu\alpha c}{\Delta P_{TMP}}(J - J^*)$$ (10.12)

i.e. the term J^* represents the rate of cake erosion per unit area divided by the mass of cake deposited per unit permeate volume, and connects the rate of erosion with the permeate volume. Hence, when $J=J^*$ the rate of cake erosion is equal to the rate of cake deposition and $\mathrm{d}J/\mathrm{d}t$ is zero. Thus J^* can be regarded as a 'critical flux' [Field et al, 1995], when $J>J^*$ then the flux rate will decline but with $J<J^*$ the permeate flux rate will be stable. Clearly, the value of S, and hence J^*, will be dependent upon the prevailing flow conditions: shear, etc. The existence of a critical flux has been reported on many occasions during crossflow microfiltration [Raasch, 1987], and is illustrated in Figure 10.19. The filtration conditions used in obtaining the data for Figure 10.19 were: PVC particles of median diameter 100 µm filtered from water on a surface filter with pore diameter of 18 µm and surface porosity of 1%. The suspension concentration was 1000 ppm and the filter was a tube of 14 mm internal diameter, 370 mm long. A fuller description of surface filters is provided in Section 10.7.1.

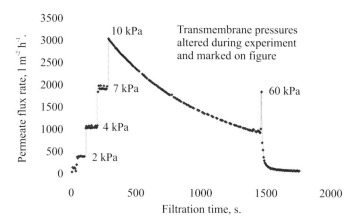

Figure 10.19 Critical flux during the crossflow filtration of PVC particles on a surface filter

At low transmembrane pressures the permeate flux is below the critical flux and there is no evidence of a deposit restricting the permeate flow. Above a threshold value of pressure the flux declined, in a manner similar to Figure 10.3. The threshold pressure, and its consequent flux rate, appears to be between 7 and 10 kPa; corresponding to a critical flux of between 2000 and 3000 $1\ m^{-2}\ h^{-1}$. However, the process is irreversible: when the flux rate drops below the critical value it continues to decline to a value similar to one provided at a much lower transmembrane pressure before operating under critical conditions. A further increase in operating pressure (to 60 kPa) resulted in a very short lived increase in flux, followed by a rapid reduction to an even lower value. The effect demonstrated in Figure 10.19 is probably exaggerated because of the large particle diameter, uniform and unconnected filter pores and narrow particle size distribution, but the principle has been shown to exist during crossflow filtration of yeast and similarly sized particles on conventional microfilters [Field et al, 1995].

A similar analysis to that shown in the derivation of Equation (10.12) can be performed for the other filtration laws represented in Table 2.3, the resulting general filtration equation for the crossflow filtration under the condition of constant pressure, i.e. the analogue to Equation (2.54) for dead end filtration is:

$$-\frac{dJ}{dt} J^{n-2} = k'(J - J^*)$$ (10.13)

where   n = 0 for cake filtration,
         n = 1 for intermediate blocking,
         n = 2 for complete blocking, and
         k' is a constant dependent upon the appropriate filtration mechanism [Field et al, 1995].

In the application of the cake filtration model to crossflow filtration, Equations (10.10) to (10.12), the concepts of specific resistance and dry cake mass per unit volume of permeate, familiar from Chapter 2, have been employed. However, segregation of the cake deposited during crossflow filtration by particle size has already been mentioned, giving a deposit with a larger proportion of fines than a cake formed under dead end filtration conditions, or when compared to the proportion of fines in the feed suspension [Baker et al, 1985]. Hence, under conditions of equivalent cake forming pressure, the specific resistance of a cake formed in crossflow filtration is normally greater than that provided during dead end filtration. However, there are also some recorded occurrences of crossflow filter cakes of lower resistance than those formed in dead end filtration [Yazhen Xu-Jiang et al, 1995].

10.5.2 Equilibrium Flux Models

Equilibrium flux models are concerned with the period of the flux–time curve when the pseudo–equilibrium period has been established, see Figure 10.3. In general, the more concentrated the suspension the more rapidly this is established. The equilibrium value is of great importance as it is the flux rate that will pertain during long process scale filtrations and the filtration may, or may not, be economically viable because of it. Strategies to improve the flux rate are discussed in Section 10.7. This section details some of the more commonly accepted models for predicting or correlating the equilibrium flux rate, and experimental measurements that need to be performed, with a consequent discussion on the important principles that the models illustrate.

Correlating data during ultrafiltration or reverse osmosis is in general more reliable than during microfiltration. The behaviour of solutes during mass transfer is well known and has been investigated over a long period of time. Solutes also behave in a more uniform way; molecules of a species have a unique molecular weight and it is possible to deduce a reliable diffusion coefficient. Suspensions of particles usually have a size distribution with no obvious single size representative of the whole distribution. During crossflow the size distribution of the deposit on the membrane is a function of the crossflow rate [Tarleton and Wakeman, 1994a], the coarser particles may be scoured off the membrane surface, hence the behaviour of the particles is a function of the system. Thus there is the need for empirical tests to provide data for all of the models which will be discussed. In all the models the deposit is assumed to control the filtration rate; however, in some cases if particle penetration of the membrane has occurred this may result in a very high apparent membrane resistance which controls filtration performance. Each of the filtration models has been found to be appropriate under certain conditions tested. It is extremely unlikely that a single model will ever be found to correlate all data, the most appropriate model to the suspension treated under the prevailing operating conditions should be used.

The form of Darcy's law which is usually applied in membrane filtration is:

$$J = \frac{\Delta P_{TMP}}{\mu(R_m + R_c)} \tag{10.14}$$

where R_m and R_c are the membrane and cake resistances respectively. The membrane resistance includes all those resistances attributable to the support mentioned in Section 10.4, the cake resistance includes concentration polarisation and gel or cake layers.

Microfiltration usually provides complete retention of the particle phase on, or in, the membrane but for ultrafiltration this may not be the case and the retention (R_t) on the membrane is defined as the fraction of retained species by the membrane:

$$R_t = 1 - \frac{N_p}{N_b}$$

where N_p is solute concentration in the permeate and N_b is solute concentration in the bulk. Another important distinction between microfiltration and ultrafiltration is the shape of the concentration profile of the retained species with distance towards the membrane. With particles a discontinuity in solid concentration at the surface of the deposit is assumed, whereas with solutes a more, gentle and continuous concentration profile exists. Both profiles are illustrated in Figure 10.20.

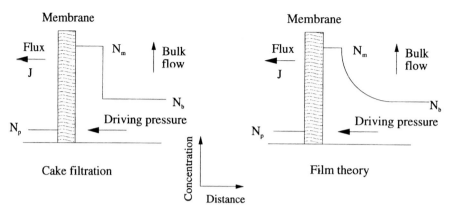

Figure 10.20 Concentration profile of retained species at the membrane surface

Cake Filtration model. This model treats both the membrane resistance and deposit resistances in a very similar way to that described in Chapter 2. Equation (10.14) is the starting point for further development which is concerned initially with the increase in cake resistance to some equilibrium value. This is analogous to accounting for the increasing cake depth in conventional dead-end filtration, see Section 2.5. In membrane filtration a mass balance over the deposit layer is performed:

$$N_b R_t V = \delta A N_m \tag{10.15}$$

where V is the cumulative volume of permeate produced and δ is the depth of the deposit. Initially, all the material convectively moving towards the membrane will deposit there; however, after some time the crossflow will shear off deposited material and Equation (10.15) will no longer be valid, as discussed in the previous section. The deposit resistance R_c is the depth of the deposit divided by its permeability. Using the cake filtration model for an equilibrium permeate flux requires a knowledge of how the values for filtration resistance varies with crossflow filtration operating conditions. These are best obtained by experiment. Equation (10.10) is valid at the start of a crossflow membrane filtration; i.e. the inverse flux rate is proportional to the cumulative filtrate volume, in much the same way as described in Section 2.6.1. The intercept of such a plot can be used to provide an in situ value for the membrane resistance. This resistance is usually much greater than the clean water permeation test value for the same membrane. This is due to the effect of the interaction of the initial layers of deposit within the membrane structure. These layers add substantially to the effective membrane resistance, i.e. additional resistances due to adsorption and blocking, see Section 10.4. This situation is again very similar to that for conventional filtration, where the filter medium resistance increases at the start of the filtration. If the membrane was a true surface filter this would not happen, but almost all membrane filters do permit some initial penetration of particulates.

Figure 10.21 illustrates the use of the equations to determine the membrane resistance in situ, and includes the membrane resistance obtained during clean water permeation tests. There are two orders of magnitude difference between the two values. The gradient can be used to provide a value for the increase in cake resistance with permeate volume, in the manner that is valid for conventional cake filtration:

$$R = R_m + \alpha c \frac{V}{A} \tag{10.16}$$

where R is the total resistance to filtration for cake filtration in both dead end and crossflow modes of operation.

If the crossflow is successful in limiting the build up of the deposit then the deposit resistance will become constant when the equilibrium deposit thickness, and deposit composition, is reached. Hence a curve such as that illustrated in Figure 10.21 will become a line parallel to the filtrate volume axis. Under such circumstances Equation (10.14) can be written:

$$J = \frac{\Delta P_{TMP}}{\mu(R_m + R_c - R_s)} \tag{10.17}$$

where R_s is the reduction in the deposit resistance due to crossflow shear [Fane, 1986].

The cake filtration model is applied as an equilibrium flux model by determining the net value of deposit resistance $(R_c\text{-}R_s)$ under different operating conditions. This entails correlating the equilibrium flux rate with the transmembrane pressure in accordance with Equation (10.17).

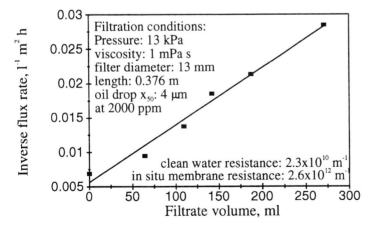

Figure 10.21 Inverse flux rate with filtrate volume for oil emulsion

Film Theory Model. There will be convective mass transfer of solute to the wall of the membrane, which is balanced by diffusional transport away from the membrane owing to a concentration gradient. At steady state, assuming the diffusion coefficient D to be constant:

$$J = \left(\frac{D}{\delta} \right) \ln \left[\frac{N_m - N_p}{N_b - N_p} \right]$$

where δ is the distance over which the diffusion occurs (film thickness). It is possible to replace the diffusion coefficient and the distance over which a species diffuses by the mass transfer coefficient K. Furthermore, if the membrane is 100% efficient:

$$J = K \ln \left(\frac{N_M}{N_B} \right) \tag{10.18}$$

The membrane concentration cannot continue to increase ad infinitum and so a "gel" concentration is postulated N_g, which is a constant and replaces membrane concentration in Equation (10.18). The existence of this gel concentration is used to explain why an increase in pressure during UF does not always increase the flux rate. The increase in pressure simply increases the depth of the gel layer at the surface of the membrane, the concentrations remain constant and, therefore, so does the driving force for back diffusion, Equation (10.18), hence flux rate becomes independent of pressure.

The usefulness of film theory becomes apparent when the effect of an alteration of a process parameter, such as bulk flow velocity, is to be investigated [Blatt et al, 1970]. Using the Leveque solution for transfer coefficient, the following equation correlates flux rate with shear rate at the wall of the membrane (γ_W), where L is the channel length and B is a constant:

$$J = B \left(\ln \frac{N_g}{N_b} \right) \left(\gamma_W \frac{D^2}{L} \right)^{1/3} \tag{10.19}$$

The following relations are applicable for wall shear rate for Newtonian liquids:

Channel shape	Linear dimension of system (h)	Wall shear rate
Rectangular slit	Half channel height	$\dfrac{3\,\bar{v}}{h}$
Circular tube	Tube radius	$\dfrac{4\,\bar{v}}{h}$

In the table above $\bar{v}$ is the mean tangential velocity of flow over the membrane surface. Equation (10.19) is valid for laminar flow where the concentration profile is still developing. Under conditions of turbulent flow the following correlation is often used:

$$\frac{Kd_h}{D} = B\,Re^a\,Sc^b \tag{10.20}$$

where d_h is equivalent diameter of a channel, and equal to 4 times the cross-sectional area of the channel divided by the wetted perimeter. The Reynolds number Re is defined as:

$$Re = \frac{d_h \bar{v} \rho}{\mu}$$

where ρ is fluid density and μ is viscosity. The Schmidt number Sc is defined as:

$$Sc = \frac{\mu}{\rho D}$$

and B, a and b are constants. Table 10.2 contains values for the empirical constants used in the correlations.

Table 10.2 Empirical constants used in Equation (10.20)

Correlation	B	a	b
Calderbank–Young	0.082	0.69	0.33
Chilton–Colburn (1)	0.023	0.80	0.33
Chilton–Colburn (2)	0.040	0.75	0.33

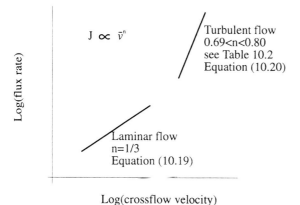

Figure 10.22 Equilibrium flux with crossflow velocity

In ultrafiltration, or microfiltration of very small particles, the equilibrium permeate flux rate is often found to be a function of the crossflow rate, in accordance with Equations (10.19) and (10.20). This is illustrated in Figure 10.22.

Film theory has had less success in microfiltration than ultrafiltration for predicting or correlating experimental data. Diffusion coefficients for molecules can be successfully calculated from the Stokes–Einstein equation, whereas those calculated for particles are extremely small. The predicted flux rates, from Equation (10.18) are much smaller than those measured experimentally. Hence additional forces which act to limit the deposit resistance have been postulated. Film theory is often modified by an additional term assisting the back-diffusion of the particles to the bulk of the suspension flow, this is called the "pinch" effect. Equations exist for the magnitude of the reduction in particle polarisation due to this effect [Rautenbach and Albrecht, 1989].

In most instances it is a simple matter to investigate empirically the mass transfer coefficient and exponents; hence Equations (10.18) and (10.20) can be used in certain microfiltration equipment design, performance monitoring and control by applying the empirically determined values. This model tends to work best for very fine particles or biological materials.

Shear–Drag Force Models. There are two directions in which a particle deposited on a membrane surface may be dragged: the crossflow induces a motion parallel to the membrane surface and the permeate drag on the particle acts towards the membrane. For the particle to be dragged, or sheared, away from the deposit the crossflow shear force must be greater than the permeate drag force. This governing equation is analogous to the coefficient of friction that objects exhibit when in contact with flat surfaces. Thus the permeate flux correlates with the fluid Reynolds number by an exponent that depends upon the flow regime and where the particle centre line is. This could be the viscous sublayer, buffer layer or turbulent boundary layer, the exponent in the correlation takes the values: 1.75, 1.26 or 1.0, respectively [Rautenbach and Albrecht, 1989]. The permeate flux is also related to the size of the particle deposited, the channel diameter, the fluid viscosity and the fluid density. For turbulent flow the correlation is:

$$J = BRe^{1.26}\left(\frac{\mu}{\rho d}\right)\left(\frac{\bar{x}}{d}\right)^{0.44} \tag{10.21}$$

where B is a constant. However, the slurry viscosity could be significantly greater than that of the permeate and the correlation can be modified as follows for this:

$$J = BRe^{1.26}\left(\frac{\mu}{\rho d}\right)\left(\frac{\bar{x}\mu}{d\mu_s}\right)^{0.44} \tag{10.22}$$

where μ_s is the slurry viscosity, $\bar{x}$ and d are the mean particle and channel diameters.

Both equations have had some success correlating data from crossflow microfiltration of clay and sand particles in water at low to medium solid concentration, volume concentrations of 1–15%.

As the concentration of the dispersed phase increases it is usual for the suspension to exhibit non-Newtonian flow characteristics. This can be included in a more general model applying the shear principle. The shear force parallel to the membrane surface is proportional to the wall shear stress τ_w, thus the permeate flux rate should be proportional to the wall shear stress:

$$J = k_1 \bar{x}\tau_w + k_0 \tag{10.23}$$

where k_0 and k_1 are empirical constants [Blake et al, 1992]. The wall shear stress is easily obtained from the pressure drop down the filter tube due to the crossflow (*not* the transmembrane pressure), which for a tube of circular geometry is:

$$\tau_w = \frac{\Delta P d}{4 L} \tag{10.24}$$

where ΔP is the difference between the inlet and outlet pressures on the filtering side of the tube and L is the filter length. Alternatively, from the definition of the Fanning friction factor f the wall shear stress is related to the flow rate Q by:

$$\tau_w = \frac{f}{2} \rho_m \left(\frac{4Q}{\pi d^2} \right)^2 \tag{10.25}$$

where ρ_m is the density of the suspension. For both Newtonian and non-Newtonian suspensions well-known expressions exist linking pressure drop with flow rate or friction factor to Reynolds number, and these can be used to provide values of wall shear stress and flux rate when the empirical constants in Equation (10.23) are known.

It is the diameter open to flow which should be employed in Equation (10.25), or expressions linking flow rate and pressure drop, and this is not always easy to assess. The deposit builds up on the tube wall, as illustrated in Figure 10.23. So, for the purposes of predictive crossflow filtration modelling or simulation the deposit thickness must be known before the shear stress and hence flux rate can be calculated. This can be achieved by combining Equation (10.23) with the various flow equations as follows. Darcy's law in cylindrical coordinates is:

$$\frac{dP}{dr} = \frac{\mu}{k} \frac{Q_p}{2\pi r L} \tag{10.26}$$

where Q_p is the permeate flow rate. The pressure drop over the cake due to permeate flow is, after integration:

$$\Delta P_c = \frac{\mu}{k} \frac{Q_p}{2\mu L} [\ln r]_{r_i}^{r_o} = \frac{\mu}{k} \frac{Q_p}{2\pi L} \ln \left(\frac{d_o}{d_f} \right) \tag{10.27}$$

using the dimensions shown in Figure 10.23.

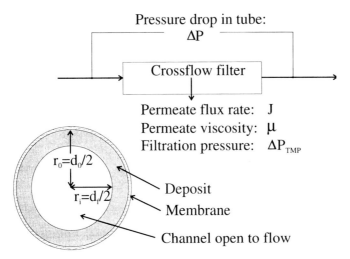

Figure 10.23 Filter section with deposit

Darcy's law can also be applied over the membrane:

$$\Delta P_m = \frac{\mu}{k_m} L_m \frac{Q_p}{\pi L d_0} = \mu R_m \frac{Q_p}{\pi L d_0} \qquad (10.28)$$

where L_m is the constant thickness of the membrane, hence R_m is the constant membrane resistance. Assuming pressure drops are additive, i.e.:

$$\Delta P_{TMP} = \Delta P_c + \Delta P_m$$

then:

$$\Delta P_{TMP} = \frac{Q_p}{\pi d_f L} \frac{\mu}{2} \left[\frac{d_f}{k} \ln\left(\frac{d_0}{d_f} \right) + 2 R_m \frac{d_f}{d_0} \right] \qquad (10.29)$$

Note that:

$$J = \frac{Q_p}{\pi d_f L}$$

Hence:

$$\Delta P_{TMP} = J \frac{\mu}{2} \left[\frac{d_f}{k} \ln\left(\frac{d_0}{d_f} \right) + 2 R_m \frac{d_f}{d_0} \right] \qquad (10.30)$$

Now, assuming $k_0 = 0$, Equations (10.23), (10.25) and (10.30) can be combined to give:

$$\Delta P_{TMP} = k_1 \bar{x} \frac{f}{2} \rho_m \left(\frac{4Q}{\pi d_f^2} \right)^2 \frac{\mu}{2} \left[\frac{d_f}{k} \ln \left(\frac{d_0}{d_f} \right) + 2R_m \frac{d_f}{d_0} \right] \tag{10.31}$$

Rearranging for diameter open to flow gives:

$$d_f^3 = \frac{k_1 \bar{x} f \rho_m 4 Q^2 \mu}{\pi^2 \Delta P_{TMP}} \left[\frac{\ln(d_0/d_f)}{k} + \frac{2R_m}{d_0} \right] \tag{10.32}$$

which can be iteratively solved if a value of friction factor can be determined. Equation (10.32) is a general equation valid for both Newtonian and non-Newtonian flow. In the case of Newtonian laminar flow the resulting expression for diameter is:

$$d_f^2 = \frac{k_1 \bar{x} 16 Q \mu_s \mu}{2 \Delta P_{TMP}} \left[\frac{\ln(d_0/d_f)}{k} + \frac{2R_m}{d_0} \right] \tag{10.33}$$

where μ_s is the suspension viscosity. In the case of non-Newtonian laminar suspension flow the resulting expression is:

$$d_f^{3n-1} = \frac{k_1 \bar{x} \eta \mu}{2 \Delta P_{TMP}} \left[\frac{4Q}{\pi} (6 + 2/n) \right]^n \left[\frac{\ln(d_0/d_f)}{k} + \frac{2R_m}{d_0} \right] \tag{10.34}$$

where η is the consistency coefficient and n is the flow index, see Appendix B.

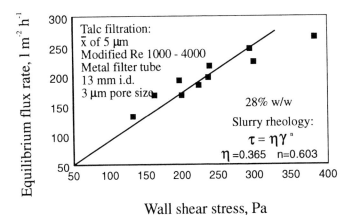

Figure 10.24 Permeate flux with wall shear

Figures 10.24 and 10.25 show the correlation of the equilibrium flux rate with wall shear stress for a suspension of talc at 28% by weight, and the deposit thickness predicted by Equation (10.34) as a function of transmembrane pressure at three crossflow rates. Further experimental details are provided on the figures.

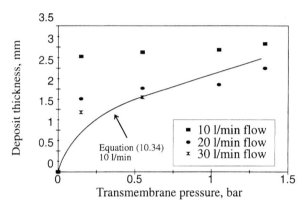

Figure 10.25 Talc deposit depth with pressure

The equilibrium flux rate correlates very well with wall shear stress, Figure 10.24, under laminar and turbulent flow conditions. However, the experimentally determined values of deposit depth are appreciably greater than those predicted by solving Equation (10.34) at low transmembrane pressures, but the deposition model becomes acceptable at transmembrane pressures greater than 1 bar.

The easiest way to apply the flux with the shear model is to use Equation (10.23), where the wall shear stress can be obtained from Equation (10.25). This model has been used to correlate data in both laminar and turbulent crossflows. If the deposit depth is significant, when filtering more concentrated suspensions, the channel diameter open to flow must be deduced. This can be achieved but requires a more detailed knowledge of the suspension rheology and an iterative solution to Equation (10.32), or its analogue for filters of different geometry [Cumming et al, 1999].

10.6 Diafiltration

Diafiltration is a process for washing a solute out of a suspension. During crossflow filtration the permeate rate, or volume if run in batch mode, is matched by the addition of fresh wash liquor, or buffer solution. The solids are retained at the membrane surface, but the solute, i.e. the dissolved species, can pass into the permeate. It is illustrated schematically in Figure 10.26.

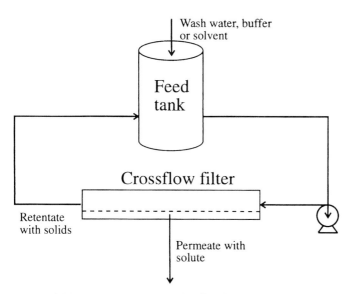

Figure 10.26 Schematic illustration of diafiltration

If the crossflow filtration system is well mixed, the concentration of solute in the system and the permeate c_p are equal at any instant in time and are given by a stirred tank model:

$$c_p = c_w \left[1 - \left(1 - \frac{c_o}{c_w} \right) \exp\left(\frac{-JAt}{V_o} \right) \right]$$

(10.35)

where V_o is the system volume, A is the membrane area, t is the filtration time, c_o and c_w are the initial and wash liquor solute concentrations, respectively. If the wash liquor does not contain any solute Equation (10.36) results:

$$c_p = c_o \left[\exp\left(\frac{-JAt}{V_o} \right) \right]$$

(10.36)

The mathematical modelling of the process is based on displacement washing only and is not valid for solutes which adsorb onto the surface of a solid. In order to adapt it for displacement and diffusional washing detailed information on the diffusion, or dispersion, coefficients of a solute would be required.

Diafiltration is used in some biotechnological industries to remove dissolved species from a biotransformation, and can be applied in any other situation where conventional washing techniques, see Chapter 9, may be inappropriate [Holdich et al, 1994].

10.7 Permeate Flux Maintenance and Regeneration

It is apparent that shear on the membrane surface plays an important part in minimising fouling; the higher the crossflow rate the greater the resulting shear will be. However, higher crossflow rates lead to greater pumping costs and may be unacceptable when processing shear–sensitive materials. In microfiltration, crossflow rates of up to 8 m s^{-1} are used currently. An alternative to increasing the crossflow velocity is to rotate the membrane, or a surface close to the membrane. Such a filter is called a "dynamic membrane filter", or "high-shear crossflow filtration". The former term is also sometimes used to represent the secondary membrane, see Section 10.1, and such a use should be discouraged. The rotational surfaces have the same effect as increasing the crossflow velocity, i.e. increasing the fluid shear gradient near the membrane surface. A commercial rotary filter has been developed [Murkes and Carlsson, 1988], and there are several studies of rotary microfiltration [Rushton and Zhang, 1988]. An alternative to rotating the membrane, or a surface, is to use fluid energy to cause rotation on the membrane surface by a tangential inlet to the module, or by a forced helical flow path through a tubular filter [Holdich and Zhang, 1992]. In general, a high shear rate slows down the rate of flux decline and often results in a higher equilibrium value. However, any technique that increases the shear at the surface of the membrane also increases the likelihood of removing the coarser particles from the deposit. The remaining particles, of a finer diameter, tend to provide a greater resistance to filtration. The advantage in operating at higher shear rates may not be as great as anticipated because of the resulting variable deposit permeability. However, this effect is not important if the particles in the deposit are of uniform size.

There are few, if any, materials that provide sufficient flux during microfiltration for cleaning to be unnecessary. The type of cleaning depends on the type of fouling that has caused the flux rate to decline. Section 10.4 illustrated the types of resistances that occur in membrane filtration, and the necessary cleaning depends on which of these resistances dominate. In the most ideal filtration the flux decline is due solely to a loose deposit formed on top of the membrane. The deposit can, therefore, be dislodged from the membrane by a simple mechanical shock, such as momentarily lowering the transmembrane pressure caused by fully opening the retentate valve or by back–flushing with some clean permeate, air or other fluid. In many applications a combination of these mechanical techniques are used. If the retentate valve is used to increase the transmembrane pressure, opening it fully may increase the crossflow rate and reduce the transmembrane pressure at the same time, which can be an effective method of membrane cleaning. When reducing the transmembrane pressure, or back–flushing, filtration must be momentarily interrupted and, after resuming filtration, the permeate flux rate may increase substantially before falling off again, and another cleaning cycle. An illustration of the resulting "saw-tooth" flux curve is provided in Figure 10.27.

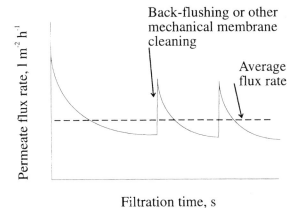

Figure 10.27 Permeate flux rate with cyclical membrane cleaning

This type of cleaning is effective with deposits of a mineral nature that are not bound together with organic or gelatinous matter, and are resting close to, and on, the surface of the membrane. If particle penetration of the membrane has occurred then back–flushing becomes less effective under most circumstances. However, it may still be sufficient to clean out the larger pores. In most microfiltrations, on membranes of appreciable pore size variation, the larger pores will take a very substantial part of the permeate flow so cleaning by this technique may still be effective. If, however, penetration is accompanied by deforming or gelatinous particles which may become strongly lodged within the membrane and glued by an organic content then chemical cleaning must be employed.

Back–flushing with a gas is not practical during water filtration if the membrane is so hydrophobic that it needs wetting to overcome the capillary pressure resisting water ingress, before filtration will commence or recommence, see Equation (6.12). Back–flushing with chemical cleaning fluid may be possible, but some of that fluid will inevitably report to the process fluid. If membrane penetration and internal blocking is found to be a major problem a secondary membrane, or precoat, could be employed in a similar fashion to conventional precoat filtration. Back–flushing could then be used to remove the precoat, which has become internally clogged, followed by a further precoat cycle on the membrane filter [Holdich et al, 1991].

Other physical or mechanical means of membrane cleaning, or minimisation of flux decline, include the following.:

1) Foam balls of a diameter slightly greater than that of the membrane tube, passed through at regular intervals, typically 10 min, have been employed. The ball deforms to enter the filter tube and scrapes off the deposit on the membrane surface. For this technique to be effective the membrane pores should be very much smaller than the deposited material, otherwise the deposit will be pushed further into the membrane. It is, therefore, more applicable to ultrafiltration membranes which may be performing a microfiltration duty.

2) Electric fields have been used to clean membranes and to try to prevent the deposition of charged particles onto a membrane surface by means of electrophoresis

[Bowen & Sabani, 1992; Wakeman and Tarleton, 1987]. Tubular metal microfiltration membranes are particularly suited for this application; a thin metal counter-electrode positioned centrally provides acceptable current distribution. A DC electric field is applied to provide a current density of 100–200 mA cm^{-2} near to the membrane surface. Use of a DC field creates gas bubbles at the membrane surface that leads to removal of the attached particles [Fairey Microfiltrex, undated]. The membrane is the cathode to prevent oxidation at the membrane surface causing chemical attack of the metal.

3) Combined ultrasonic and electric fields have been investigated, and a synergistic membrane cleaning effect has reported [Wakeman and Tarleton, 1991].

4) Oscillatory or pulsatile flow of the process fluid over the membrane surface has also been employed [Boonthanon et al, 1991]. Such a flow condition is known to increase the mass or heat transfer coefficient in various process equipment. Baffles have also been included into the oscillatory flow system; these lead to vortices with high shear at the membrane surface between the baffled sections [Finnigan and Howell, 1989; Li et al, 1998, Millward et al, 1995].

5) Turbulence promoters, air scour and abrasives have also been investigated as a means to reduce the boundary layer or deposit thickness during microfiltration [Dejmek, et al, 1974; Milisic and Bersillon, 1986]. In general, these are not so effective during microfiltration as they can be during ultrafiltration.

6) Surface alteration of the membrane to reduce the likelihood of deposition and adhesion by biologically treating membranes has been used [Hall and Protheroe, 1991]. In microfiltration the nature of the membrane has less significance than in ultrafiltration, however, the flux decay and membrane fouling may be due to species other than the major constituent of the process fluid. Under these circumstances the membrane chemistry may be significant even in microfiltration.

Chemical cleaning in order to dissolve a species adhering to the membrane surface, either entirely or just the part which is causing the particle to become fixed, depends on the nature of the foulant. Fouling by iron, or other metals, is common at high pH. Precipitation of common hydroxides and hydrated metal salts occurs at pH values over 2, in the absence of metal complexing agents. Hence, if there is a significant metal concentration and a pH close to neutrality, colloidal metal precipitates on the microfilter will need to be removed. Organic fouling is generally removed under alkaline conditions. Proprietary cleaners have a natural pH in water of either 5 or 11, and may contain metal complexing agents, enzymatic detergents and dispersing agents. Acid and alkaline chemical cleaning can be achieved by the following:

1) Citric acid, 5% w/v, used to remove colloidal iron deposits.
2) Sodium hydroxide, at pH 11–13 at 50^{0}C, often followed by 0.3 mol/l nitric acid at 45^{0}C.

Both cleaning regimes need to be followed by careful washing by clean water prior to reintroduction of process liquor. The time required to achieve adequate membrane cleanliness depends entirely on the nature of the fouling and must be established by experiment. Cleaning under the above conditions for 20–50 min is common, however.

10.7.1 Surface Microfilters

A clean or new microfilter often provides flux rates in excess of $1000 \, 1 \, m^{-2} \, h^{-1}$. However, during operation the flux rate drops to a value close to that given by an ultrafiltration membrane and may decay to an even lower value if material becomes lodged within the filter. Thus crossflow microfiltration has the potential to provide very high flux rates during solid–liquid separation but rarely achieves this because of fouling. If that fouling is limited to a deposit on the surface of the filter then mechanical means to remove, or minimise, it are possible as discussed in the previous section. It is often the internal penetration and deposition of material that fouls a filter irreversibly. Conventional microfilters available for process scale applications rely on depth filtration mechanisms to achieve their pore retention rating, see Figures 10.2, 10.9 and 10.10. By contrast, a surface microfilter is one which does not rely on a tortuous flow channel to achieve its pore size rating. One example of a surface filter used primarily in laboratory applications is a Nuclepore filter. These are formed by nuclear bombardment of a polymer sheet and subsequent chemical etching. This gives rise to a filter with highly uniform pores that pass straight from one side of the filter to the other without the need of a tortuous flow channel to achieve the required pore rating. They can, therefore, be described as sieves for filtering out the oversize particles, and other suspended material that may become trapped in the deposit retained on the filter surface.

There have been other developments to produce surface filters for commercial application of crossflow microfiltration in process scale applications. The objectives are to produce a microfilter with a very high open area to flow, very low pore opening diameter and to be as thin as possible whilst retaining adequate mechanical properties.

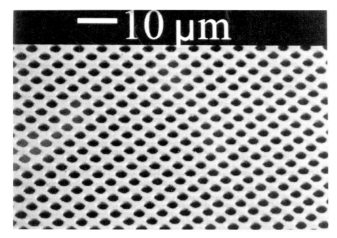

Figure 10.28 SEM of a surface microfilter [Courtesy of Cees van Rijn, Aquamarijn Microfiltration, B.V.]

Such a filter is likely to lead to wider scale adoption of crossflow microfiltration in large scale process applications. The current filters under development and trial are usually based on optical engineering and chemical etching to form the pore structure. One example is illustrated in the SEM picture reproduced in Figure 10.28 [Kuiper et al., 1998].

Surface microfilters have been used to filter suspensions containing yeast, blood cells and oil droplet dispersions and provide stable flux rates in the 100's and even 1000's of $1 \, m^{-2} \, h^{-1}$ [Kuiper et al, 1998, Holdich et al, 1998]. Conventional microfilters employing depth filtration mechanisms would provide pseudo–equilibrium fluxes considerably below $100 \, 1 \, m^{-2} \, h^{-1}$ when filtering these materials.

Crossflow microfiltration is the last of the three pressure driven membrane processes (RO, UF and MF) to be widely commercialised, but the developments in true surface microfilters should significantly increase the process scale applications of this technology.

10.8 Applications and Investigations

Crossflow microfiltration is being used and investigated for both clarification and thickening of concentrated suspensions in industrial applications. The following provides some details of these applications.

General. Clarification and thickening duties, recycling of pigment dispersions, latex product recovery and concentration.

Potable water. To remove suspended material that is harmful to human health.

Biotechnological. Cell by–product separation, desalting or buffer exchange (diafiltration), removal of bacteria and cellular debris, cell harvesting, beverage clarification.

Oil and effluents. Filtration of seawater prior to injection into oil reservoirs [Abdel-Ghani et al, 1988; Holdich et al, 1991], separation of oil from dispersions in water, separation of water from dispersions in oil, textile finishing effluents, biological effluent sludges [Al-Malack and Anderson, 1996]

Nuclear. Thickening of radioactive contaminated solids prior to encapsulation.

In the last example crossflow microfiltration has been chosen because a slurry of a specified concentration can be achieved, cement is added when this has been obtained to form an encapsulated effluent for disposal [Empsall and Hebditch, 1994]. The ability of microfiltration to provide particle-free permeate, above a certain threshold, has led to the adoption of skid-mounted crossflow filters for heavy metal effluent treatment, see Figure 10.29 [Metal Finishers Association, 1992].

The metal-bearing influent is treated chemically to precipitate metal salts and then filtered in multiple pass crossflow filtration. When the suspension is at a suitable solids content, approximately 5% w/w, the slurry is fed to a filter press for further dewatering. The crossflow microfilters are used for concentration and to guarantee effluent quality. The only discharge stream from the circuit is from the crossflow filters, and the heavy metal content is <0.1 ppm.

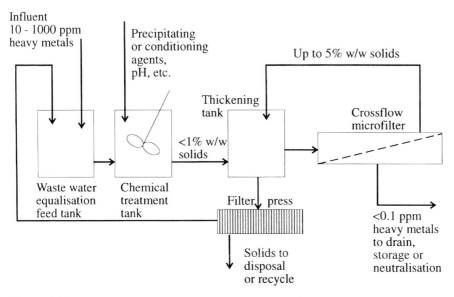

Figure 10.29 Flowsheet for heavy metal removal including crossflow microfiltration

Crossflow filtration of drinking water (potable water) is expected to become a major market for microfiltration [McIlvaine, 1999]. Microfiltration is capable of filtering out bacteria and other material that are detrimental to human health such as cryptosporidium oocysts and giardia protozoa. The cryptosporidium parasite may hatch from an oocyst after it has been ingested by a person, possibly resulting in illness. Filtration to remove oocysts is, therefore, important to the drinking water supply companies. The oocysts are spherical and between 3 to 6 μm in diameter, see Figure 10.30.

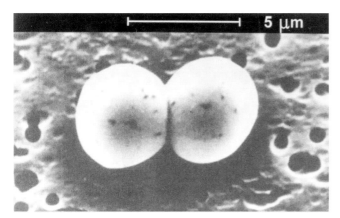

Figure 10.30 SEM of cryptosporidium oocysts filtered on a 0.8 μm surface microfilter

Thus microfiltration should be capable of completely removing this, and other, water borne parasites, and many investigations have been reported [MacCormick, 1999].

Microfiltration of viruses from water, providing significant removal efficiency has also been reported [Madaeni et al, 1995], and excellent virus rejection can be achieved with ultrafiltration membranes.

In general, microfiltration or ultrafiltration is used when a material is difficult to filter by conventional equipment, possibly because it is very fine, compressible or dilute. It is also used when the intention is to produce a slurry of higher concentration than can be achieved by gravity settling. The shear-thinning nature of most suspensions makes crossflow microfiltration particularly suited for this application. The process equipment required consists mainly of a pump and membrane module(s). Hence the cost of a microfilter system can be low compared with alternative solid/liquid separation systems, for small-scale operation. However, if the membranes need frequent replacement the costs may become high. The running costs are very dependent on the permeate flux rates that can be achieved, and the techniques described to enhance the flux rate are aimed at reducing this significant cost.

10.9 References

Abdel-Ghani, M.S., Jones, R.E. and Wilson, F.G., 1988, Filtrat. Separat., 25, pp 105–109.

Al-Malack, M.H. and Anderson, G.K., 1996, Wat. Res., 31, pp 3064–3072.

Anon, 1994, Filtrat. Separat., 31, pp 443–547.

Baker, R.J., Fane, A.G., Fell, C.J.D. and Yoo, B.H., 1985, Desalination, 53, pp 81–93.

Belfort, G. and Mark, B., 1979, Desalination, 28, pp 13–30.

Belfort, G., Mikulasek, P., Pimbley, J.M. and Chung, K.Y., 1993a, J. Membrane Sci., 77,   pp 23–39.

Belfort, G., Pimbley, J.M., Greiner, A. and Chung, K.Y., 1993b, J. Membrane Sci., 77, pp 1–22.

van den Berg, G.B. and Smolders, C.A., 1988, Filtrat. Separat., 25, pp 115–121.

Bhattacharya, D., Jumawan, A.B., Grieves, R.B. and Harris, L.R., 1979, Sep. Sci. Tech., 14, pp 529–549.

Blake, N., Cumming, I.W. and Streat, M., 1992, J. Membrane Science, 68, pp 205–216.

Blatt, W.F., Dravid, A., Michaels, A.S. and Nelsen, L., 1970, Solute polarization and cake formation in membrane ultrafiltration: causes, consequences and control techniques in Membrane Science and Technology, Flinn, J.E., (Ed.), Plenum, New York, pp 47–97.

Boonthanon, S., Hwan, L.S., Vigneswaran, S., Ben Aim, R. and Mora, J.C., 1991, Filtrat. Separat., 28, p 199.

Bowen, W.R. and Sabuni., H.A.M. 1992, Ind. Eng. Chem. Res., 31, p 512.

Cumming, I.W., Holdich, R.G. and Ismail, B., 1999, J. Membrane Sci., 154, pp 229–237.

Dejmek, P., Funeteg, B., Hallstrom, B. and Winge, L., 1974, J. Food Science, 39, p 1014.

Empsall, R.D. and Hebditch, D.J., 1994, The I.ChemE Research Event, UCL, London, Inst. Chem. Eng., Rugby, UK, pp 410–412.

Fairey Microfiltrex, undated, Precision Filtration FM 28, Fareham Industrial Park, Fareham, Hants.

Fane, A.G.,1986, in Progress in Filtration and Separation, vol. 4, R.J. Wakeman, Elsevier, (Ed.), Amsterdam.

Field, R.W., Wu, D., Howell, J.A. and Gupta, B.B., 1995, J. Membrane Sci., 100, pp 259–272.

Finnigan, S.M. and Howell, J.A., 1989, Chem. Eng. Res. Des., 67, p 278.

Granger, J., Dodds, J. and Leclerc, D., 1985, Filtrat. Separat. 22, pp 58–60.

Gutman, R.G., "Membrane Filtration", Adam Hilger, Bristol U.K., 1987.

Gutman, R.G., 1977, The Chem. Engr., pp 510–523.

Gutman, R.G. and Knibbs, R.H., 1989, in Future Industrial Prospects of Membrane Processes, L. Cecille and J.C. Toussaint, (Eds.) Elsevier, Amsterdam, pp 16–30.

Hall, G.M., and Protheroe, R.G., 1991, Filtrat. Separat., 28, pp 45– 48.

Holdich, R.G. and Zhang, G.M., 1992, Trans. I.Chem.E., 70, Part A, pp 527–536.

Holdich, R.G., Zhang, G.M. and Boston, J.S., 1991, Filtrat. Separat., 28, pp 117–122.

Holdich, R.G., Cumming, I.W. and Ismail, B., 1994, Trans. Inst. Mining and Metallurgy,   Section C, pp C188–C192.

Holdich, R.G., Cumming, I.W. and Smith, I.D., 1998, J. Membrane Sci., 143, pp 263–274.

Kuiper, S., van Rijn, C.J.M., Nijdam, W. and Elwenspoek, M.C., 1998, J. Membrane Sci., 150, pp 1–8.

Li, Hong-yu, Bertram, C.D. and Wiley, D.E., 1998, AIChE J., 44, pp 1950–1961.

Madaeni, S.S., Fane, A.G. and Grohmann, G.S., 1995, J. Membrane Sci., 102, pp 65–75.

McIlvaine, R.W., 1999, Filtrat. and Separat., 36, pp 24–25.

MacCormick, A., 1999, Filtrat. and Separat., 36, pp 16–19.

Metal Finishers Association, 1992, Effluent Treatment in Electroplating Plants in the U.S.A., 10 Vyse Street, Hockley, Birmingham, B18 6LT, U.K.

Milisic, V. and Bersillon, J.L., 1986, Filtrat. and Separat., 23, pp 347–349.

Millward, H.R., Bellhouse, B.J. and Walker, G., 1995, J. Membrane Sci., 106, pp 269–279.

Murkes, J. and Carlsson, C.G., 1988, Crossflow Filtration, Wiley, Chichester.

Raasch, J., 1987, The place of membrane techniques in the treatment and purification of industrial water, Conference, Institut de la Filtration et des Techniques Separatives, Agen, France, pp 43–56.

Rautenbach, R. and Albrecht, R., 1989, Membrane Processes, Wiley, Chichester.

Rushton, A. and Zhang, G.S., 1988, Desalination, pp 379–394.

Tarleton, E.S. and Wakeman, R.J., 1993, Chem. Eng. Res. Des., 71 pp 399–440.
   ibid, 1994a, Chem.Eng.Res.Des., 72, pp 431–440.
   ibid, 1994b, Chem.Eng.Res.Des., 72, pp 521–528.

Wakeman, R.J. and Tarleton, E.S., 1991, Chem. Eng. Res. Des., 69, pp 386–397.

Wakeman, R.J. and Tarleton, E.S., 1987, Chem. Eng. Sci., 42, p 829.

Yazhen Xu-Jiang, Dodds, J. and Leclerc, D., 1995, Filtrat. Separat., 32, pp 795–798.

10.10 Nomenclature

A	Membrane area	m^2
c	Solute concentration	$kg\ m^{-3}$
d	Channel diameter	m
d_p	Pore channel diameter	m
J	Permeate flux rate	$m^3\ m^{-2}\ s^{-1}$
k	Membrane deposit permeability	m^2
k_0	Empirical coefficient in Equation (10.23)	
k_1	Empirical coefficient in Equation (10.23)	
L	Membrane length	m
N	Suspended solid or solute concentration	$kg\ m^{-3}$
n	Rheological power law model "flow index"	
n_p	Number of pores in membrane	
ΔP	Axial Pressure Differential in filter tube	$N\ m^{-2}$
ΔP_c	Pressure Differential over membrane deposit	$N\ m^{-2}$
ΔP_m	Pressure Differential over membrane	$N\ m^{-2}$
ΔP_{TMP}	Transmembrane pressure differential	$N\ m^{-2}$
Q	Volume flow rate	$m^3\ s^{-1}$
Q_p	Permeate volume flow rate	$m^3\ s^{-1}$
R_c	Resistance of deposit on membrane	m^{-1}
R_m	Membrane resistance	m^{-1}
R_t	Membrane retention	–
t	Time	s
V	Permeate volume	m^3
V_o	Volume of suspension in system	m^3
$\bar{v}$	Mean suspension velocity of feed	$m\ s^{-1}$
$\bar{x}$	Mean particle diameter	m

Greek Symbols

δ	Deposit depth at membrane surface	m
γ_w	Shear rate at wall	s^{-1}
η	Rheological power law model "consistency coefficient"	$kg\ s^{n-2}\ m^{-1}$
μ	Liquid viscosity	$Pa\ s$
μ_s	Viscosity of feed slurry	$Pa\ s$
ρ	Fluid density	$kg\ m^{-3}$
ρ_m	Mean density of suspension	$kg\ m^{-3}$
ρ_s	Solid density	$kg\ m^{-3}$
τ_w	Shear stress at deposit surface	Pa

11 Filtration Process Equipment and Calculations

11.1 Introduction

The purpose of this chapter is to describe some of the large-scale vacuum and pressure filters used industrially for solid–liquid separation. In view of space limitations, only a representative few of the many available machines can be mentioned; comprehensive filtration machinery descriptions are available elsewhere in relevant literature [Purchas, 1981; Dickenson, 1992].

The important process of filter selection is covered in detail in recent filtration literature [Purchas,1984; Svarovsky, 1981; Shirato et al, 1987]. An expert system is available [Wakeman, 1995; Wakeman and Tarleton, 1999] which facilitates plant selection. In application, the more information on the filtration characteristics of the suspension in question, the more precise will be the selection result. In general, even when the economic aspects of plant capital and operating costs are considered, no single answer to a specific problem will be produced. Several types of filter may answer the problem, with varying degrees of efficiency. Complete failure is rare, except perhaps in filter media selection. The general trend, with both users and suppliers, is that when a particular filter is successfully applied to a certain process, extensions of the latter will usually include the same filter.

In Table 11.1, some of the slurry properties are listed, which influence plant selection [Tiller & Crump, 1977]. Figure 11.1 presents a broad classification of suitable filters against the particle size distribution parameters. Of course, such classifications cannot be viewed as being rigorous in all circumstances, but the data indicate that pressure filters will be found most commonly in those processes containing slow-settling, slow-filtering suspensions.

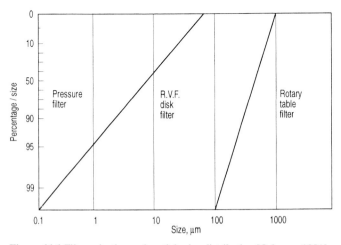

Figure 11.1 Filter selection and particle size distribution [Osborne, 1981]

Table 11.1 Slurry properties

Slurry type	Fast-filtering	Medium-filtering	Slow-filtering
Cake formation rate, cm/s	2–4	0.02 0.08	0.002–0.01
Normal concentration, wt%	>20	10–20	1–10
Settling rate	Rapid	Medium	Slow
Leaf test rate kg $m^{-2} h^{-1}$	2500	250–2500	25–250
Filtrate rate $m^3 m^{-2} h^{-1}$	10	0.5–10	0.0025–0.05
Typical slurry	Crystal solids	Salts	Pigments

From Figure 11.1, it may be inferred that suspensions of fine particles will require pressure filters whilst a coarse suspension will be suitably processed in pan or table filters. Where the size range is wide, the actual plant choice may be affected by the concentration of fines: a slurry with 10% greater than 10 μm may require pressure filtration; at the same average size an increase in proportion greater than 10 μm may make a rotary filter possible.

11.1.1 Filter Productivity Fundamentals

The productivity of all filters is related to the time required to complete a full separation cycle. In addition to the time required for filtration t_f, further periods may be necessary for dewatering t_d and cake washing t_w. Finally, time will be required for filling the filter, discharging the filter cake, etc. It is usual to lump the latter operations into a "down-time" t_{dw}. In certain pressure operations, the filtration period may be extended to allow for compression of the filter cake by inflatable diaphragms.

The overall cycle time t_c is thus given by:

$$t_c = t_f + t_d + t_w + t_{dw} \tag{11.1}$$

In large-scale, batch-operated pressure units, e.g. plate and frame filters a considerable proportion of the overall cycle may be taken up in frame filling, cake discharging, etc.

The productivity of the unit $P*$ (kg dry solids per second) may be estimated from:

$$P* = (cV)/t_c \tag{11.2}$$

where V is the volume of filtrate produced per cycle (m^3) and c is the mass of dry solids deposited per unit of filtrate volume. The value of c is given by (see Chapter 2):

$$c = \rho s / (1- ms) \tag{11.3}$$

in which ρ is the fluid density (kg/m^3), s is the mass of solids per unit mass of feed slurry and m is the ratio of mass wet/mass dry filter cake.

In process specifications, it is important to have estimates of the filter cake thickness produced per cycle. The mass of solids deposited and the associated cake thickness have an important effect on cake discharge mechanisms. As will be noted below, manufacturers of continuous filters provide information on the minimum cake thickness required to avoid cake discharge problems.

The mass of dry solids deposited from V volumes of filtrate and the associated cake thickness can be calculated from:

$$M \ = \ cV \ = \ \rho_s \ (1- \ \varepsilon) \ AL \tag{11.4}$$

where M is the mass of solids per cycle, (kg), ρ_s is the density of the solids (kg/m^3), $\bar{\varepsilon}$ is the average porosity of the cake, A is the filter area (m^2) and L is the cake thickness (m).

Chapter 2 contains information on the methods used for estimating the filtration period t_f. A basic filtration equation:

$$q \ = \ (dV \ / \ dt) \ = \ (A^2 \Delta P) \ / \mu(\bar{\alpha} \ cV + \ AR_m) \tag{11.5}$$

relates the flow rate q or (dV/dt) m^3/s through the filter cake of average specific resistance $\bar{\alpha}$ (m/kg) and the filter medium of resistance R_m (m^{-1}), after the production of V volume of filtrate in t process seconds.

The above expression can be integrated under appropriate process conditions of:

a) Constant pressure differential ΔP
b) Constant flow rate $(dV/dt) = q$
c) Variable pressure — variable rate

Condition (a) could apply to vacuum units, or pressure filters operated under constant ΔP.
Condition (b) could apply to pressure filters fed by constant-rate pump devices. Use of a centrifugal pump will produce condition (c), as discussed in Chapter 2.

Pressure filtration may ensue in conditions of constant filtration flow, in which case the Equation (2.28) can we written as:

$$\Delta P \ = \ K_1 \ q^2 \ t_f + \ K_2 \ q \tag{11.6}$$

where:

$$K_1 \ = \ (\bar{\alpha}\mu c / A^2) \tag{11.7}$$

$$K_2 \ = \ (\mu R_m \ / \ A) \tag{11.8}$$

The above equation indicates that in these operations, the pressure differential will rise with increase in filtration time, t_f. Arrangements are made to terminate the separation when the pressure differential reaches an upper, safe limit. In some cases, the filter is pre-coated with filter aid, prior to the separation. The time required for pre-coat must be added to the overall cycle time, in process productivity calculations.

Practical studies have been reported [Silverblatt et al, 1974], [Rushton and Matsis, 1994] which pointed to optimum flow rates, in this mode of separation. The corresponding minimum filter area requirements have been identified [Shirato et al, 1987].

Several of the continuous filtration units discussed below may be operated in pressure and vacuum conditions; thus the continuous rotary vacuum drum, disc or horizontal belt filter can be enclosed in a pressure vessel. Separation may then ensue by use of a gas pressure blanket in the vessel. This technique is often used in solvent filtrations. Some processes merit the simultaneous use of pressure upstream, with vacuum downstream of the filter; in this manner, the continuous filtration characteristic can be coupled with higher filtration differential pressures.

Finally, some process equipment, e.g. large-scale plate and frame units used in sludge separations, may operate in a mixed mode. Filling of the press and early build up of "cake" deposit may ensue at a controlled constant flow; final deposits, with dewatering, may be effected at a higher-level, constant pressure differential [Svarovsky, 1981].

These general comments and the theoretical developments provided in Chapter 2 point to the need for information on the effect of pressure differentials on filter cake resistance, porosity, moisture content, etc. Test work must be designed to quantify the compressibility characteristics of the suspension of interest. This information facilitates the estimation of filter size and productivity in various operating conditions. Quite often, the level of pressure used in plant-scale separations are quite different to those available in the test laboratory.

11.1.2 Filter Cake Dewatering and Washing

Methods for estimating dewatering and washing times are presented in Chapter 9, mainly based on information presented in the literature [Wakeman and Tarleton, 1990]. Other sources of information [Stahl and Nicolaou, 1990], highlight the various stages in rotary vacuum filtration processes, including calculation of solids throughput, discharged cake moisture, vacuum pump requirements, etc.

The graphical techniques reported in Chapter 9, Figure 9.7, relate the so-called reduced saturation S_R to a complex function θ which includes dewatering time t_d, to attain a saturation level S:

$$S_R = (S - S_\infty) / (1 - S_\infty) \tag{11.9}$$

$$\Theta = (k\,p_b\,t_d) / \mu\,L^2\,\varepsilon\,(1 - S_\infty) \tag{11.10}$$

These expressions include: k, the filter cake permeability (m^2); p_b, the threshold pressure below which air will not penetrate the filter cake pores (N/m^2); S_∞, the equilibrium, irreducible cake saturation, a function of the capillary number N_c, (defined below); ε the porosity of the filter cake.

Use of these expressions requires further relationships for p_b, k, S_∞ and N_c. These relationships are usually reported in terms of the average size of particles in the filter cake. Alternative relationships were developed [Rushton & Arab, 1986] in terms of the average specific cake resistance $\bar{\alpha}$; it was hypothesized that the latter, if measured immediately prior to dewatering, would serve as a useful basis for correlating dewatering kinetics. From Chapter 2, a relationship between $\overline{\alpha}$ and k, and the permeability may be written:

$$k \ = \ 1 / (\overline{\alpha} \ (1 - \ \varepsilon) \rho_s)$$

Further relationships between $\overline{\alpha}$ and particle size are developed in Chapter 2 and Appendix A. Thus for spheres:

$$\overline{\alpha} = \ 180 \ (1 - \ \varepsilon) / (\varepsilon^3 \rho_s \ x^2)$$

where x is the mean surface diameter of the particles; deviations from such relationships are related to particle shape.

For spheres, the various process relationships then take the form:

$$N_c = \ 180 \ [\rho g L + \Delta P] \ / (\overline{\alpha} \rho_s \ (1 - \ \varepsilon) \ L \gamma \qquad (11.11)$$

where γ is the surface tension of the fluid:

$$\Theta = \ (p_b t_d) / [\overline{\alpha} \rho_s (1 - \varepsilon) \ L^2 \ \mu \varepsilon (1 - S_\infty)] \qquad (11.12)$$

Actual measurements of S_∞ for filter cakes of calcium carbonate took the form:

$$S_\infty \ = 0.39 \ (1 + 0.0256 \sqrt{N_c}) \qquad (11.13)$$

in which N_c is calculated by Equation 11.11.

It was also suggested that, in dewatering processes, the graphical data contained in Figure 9.7, Chapter 9 could be correlated by:

$$S_r = \ 0.45 \ (\Delta P^* \Theta)^{-0.25} ; \ \Delta P^* > 2.5 \qquad (11.14)$$

$$S_r = \ 0.58 \ (\Delta P^* \Theta)^{-0.22} ; \ \Delta P^* < 2.5 \qquad (11.15)$$

where:

$$\Delta P^* = \ (P\text{air, inlet} \ / \ P_b) - \ (P\text{air, outlet} \ / \ P_b) \qquad (11.16)$$

and

$$P_b \ = \ (0.118 \varepsilon \gamma^2 \rho_s (1 - \varepsilon))^{0.5} \qquad (11.17)$$

The use of these expressions is demonstrated in Example 11.2. Since the above relationships are based on a limited amount of experimental data, widespread application to materials other than those used in producing the correlations is not advised. This is demonstrated by the differences between Equation 11.13, for calcium carbonate, and:

$$S_\infty \ = 0.155 \ (1 + 0.031 \ N_c^{-0.49}) \ for \ N_c > 10^{-4} \qquad (9.35)$$

reported in Chapter 9. The latter equation was obtained with sand-like solids, glass spheres, etc.

These differences in the equations recorded for S_∞ (hence S at drainage time t) may be compared with the industrial trials [Carleton & Mehta, 1983] on large-scale filters producing cakes of sand, gypsum, pigments, chalk and "impurities". A general interest here was in the differences between solids filtration rate, as measured by the leaf test and filters (drum, table and belt). It was found that, as $\bar{\alpha}$, the average specific resistance of the cake, increased, the productivity of large-scale units was somewhat less than that produced in the laboratory, on the leaf. This effect is shown in Figure 11.2A. Differences in filter performance with changes in scale were also noted in tests on 0.1, 0.3 and 1.0 m^2 RVFs [Rushton, 1981]. These differences were attributed to mixing and sedimentation effects.

On the other hand, excellent agreement was obtained [Carleton & Mehta, 1983] between the moisture contents from large-scale tests and those predicted by Wakeman's equations. Figure 11.2B shows the level of this agreement for cakes with properties in the ranges:

$$5.0 \times 10^6 \,(\text{sand}) < \alpha \,(\text{m/kg}) < 2 \times 10^{12} \,(\text{pigment})$$

In these studies, the filter cake thickness produced ranged from 70 mm for the free filtering material, down to 1 mm for some high-resistance pigments. These results may be compared with practical trials on filtering centrifuges, in Chapter 8, where the published correlations seem to over predict the drainage kinetics. Obviously, more information is needed in the latter subject.

Perhaps the principal area of ignorance in both the filtration and dewatering of precipitates is the effect of crystal morphology. Serious theoretical and practical problems arise in cases where the "crystal" in fact possesses internal porosity. Liquors bound inside a crystal will be impossible to remove by normal dewatering methods. Again, in process calculations and correlations of experimental results, uncertainties arise in the values of particle density for these materials.

Changes in crystal shape, with alterations in the process conditions, have been studied in relation to the filtration and washing of gypsum from phosphoric acid solutions [Van der Sluis, 1989]. Optimal filterability conditions were observed, in terms of acid concentrations, resistance times, etc. and differences between various filtration models were discussed, with reference to the appearance of needle-shaped crystal clusters.

It is to be expected that these materials would also exhibit deviation from relationships such as the Kozeny–Carman relationship, Chapter 2, for permeation of stationary packed beds. On the other hand, the methods available in Chapter 9 can be used with confidence when dealing with solid, granular precipitates. In other conditions, experimental values for $\bar{\alpha}$, S, S_∞, etc, need to be acquired.

The process of solute removal in cake washing is usually represented by curves of the type shown in Figure 11.3 A. Here the change in concentration is observed by measuring the outlet concentration c_s of solutes in solution. This may then be normalised by division

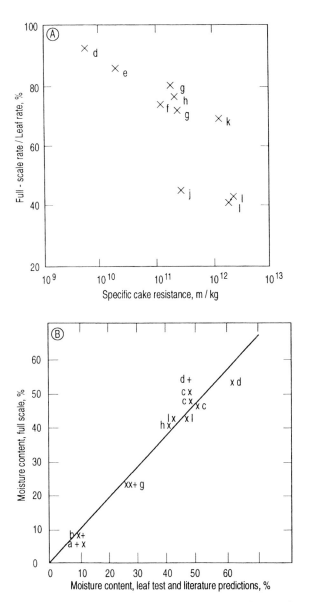

Figure 11.2 Productivity and dewatering by vacuum: large- and small-scale [Carleton and Metha; 1983]
Crosses, leaf test; plus signs, Wakeman predictions

	α, m/kg		α, m/kg		
a: sand 1	5.9×10^6	e: pigment 1	1.9×10^{10}	j: pigment 4	2.7×10^{11}
b: sand 2	1.4×10^7	f: pigment 2	1.2×10^{10}	k: pigment 5	1.3×10^{12}
c: gypsum	1.9×10^9	g: chalk	2.1×10^{11}	l: pigment 6	2×10^{12}
d: impurity	5.5×10^9	h: pigment 3	2.1×10^{11}		

by the initial solute concentration c_0. The ratio c_s/c_o is plotted against wash time t_w. Alternatively, the ratio of the wash delivered up to a certain time, divided by the liquid content in the filter cake at the start of washing may be used. These so called void volumes are clearly related to time, in view of the fact that the wash will proceed at a certain rate.

Despite the extensive studies on cake washing, as reported in Chapter 9, complete mathematical description of the processes involved in the removal of dissolved solutes is still awaited. Nevertheless, the generalised approach [Wakeman & Tarleton, 1990; Carleton & Taylour, 1991] involving the changes in cake solute with additional wash volumes and as a function of the "dispersion parameter": (vL/D), Figure 11.3 B, presents a reliable theoretical basis for the subject.

In this group of variables: v is the wash velocity (m/s), L is the cake thickness (m), D is the axial dispersion coefficient (m^2/s). High values of the latter parameter produce a "plug-flow" washing curve; lower values, as shown in Chapter 9, tend to produce long washing "tails", with increasing demand on wash time to produce acceptable solute removal.

The situation is made more complicated by the fact that the filter cake may be dewatered before washing. A dewatered cake may have channels in the cake through which the wash will flow preferentially. Removal of solute may then ensue only by diffusion from stagnant pockets near the points of contact between particles. This situation has been considered [Wakeman & Rushton, 1974] in a "side pore" model which finds application to the calculation of the recovery of solution from drained cakes.

In vacuum, or constant-pressure, process calculations, the wash is assumed to be conducted at the end of the filter formation period; if the same differential pressure is used as that during filtration, the wash rate will equal the filtration rate at the final stage of filtration.   Wash efficiencies are defined in terms of the fraction of solute removed after the passage of one void volume of wash. Theoretical calculations suggest that wash efficiency will vary with the dispersion parameter discussed above [Wakeman & Attwood, 1988]. Some approximations to these calculations are:

vL/D	900	50	10	6	2	1	0.5	0.1
Wash efficiency, i.e. % removed at $n = 1$	98	92	87	83	69	60	48	34

Washing efficiencies on large-scale filters vary from 35 to 90%; the main causes for such inefficiencies are uneven cakes, maldistribution of wash water, and cake cracking.

In RVF operation, the filtrate rate at the end of the cake formation period, in combination with the limited time available for washing on the drum, make it unlikely that more than three void volumes of wash will be delivered to the cake. For high dispersion numbers, this does not present a problem in that, despite practical inefficiencies, satisfactory cake washing may still be obtained. However, at lower values of the dispersion group, much larger void volumes will be required for adequate washing.

In other systems, where batch operation is practised, e.g. in large-scale vacuum filters or pressure units, extra time may be allowed, in order to achieve high solute removal. Generally, extra wash time means lower productivity per unit area; when t_w becomes

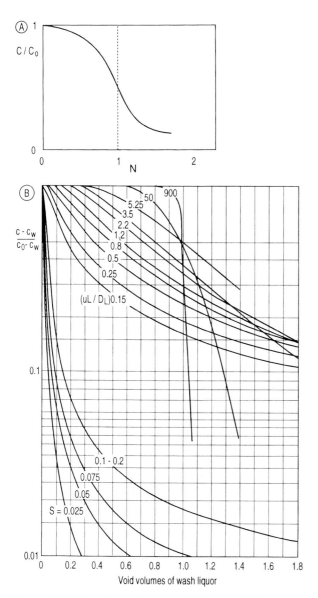

Figure 11.3 Filter cake washing curves [Wakeman, 1981]

unduly high, steps need to be taken to alter the flow conditions in the filter during the wash phase (e.g. changes in applied pressure, interrupting the flow of wash water).

Disturbance of the filter cake during washing can also be advantageous in continuous systems, e.g. horizontal belt filtration of pigments [Dresner & Bailey, 1973]. Here the wash water was delivered through staggered jets which give a re-slurrying effect in the wet cake.

11.2 Continuous, Large-Scale Filters

Examples of these units include:

	Figure
a) Rotary drum	11.4
b) Rotary disc	11.5
c) Horizontal tilting pan	11.6
d) Horizontal belt	11.7

Comprehensive descriptions, including mechanical and operational details are available in relevant literature [Dahlstrom & Silverblatt, 1986].

11.2.1 Rotating Drum

As depicted in Figure 11.4, the filtering surface is usually situated on the outer face of a cylindrical drum; plant does exist, however, where filtration proceeds inside the drum. For the second case considered, filtered solids are deposited on the outer surface of the vertical disc, Figure 11.5.

The axis of rotation for both filters is horizontal, and there is a partial immersion of the rotating surface in the slurry. In both cases, the active surface is divided into independent elementary portions which are connected to the vacuum source by means of a circular collector or rotary valve. The latter is made up with mating surfaces, one of which corresponds to the division of the filtering area and is rotating with the drum or disc, whilst the other has openings corresponding to the division of the process cycle. After cake formation, dewatering by vacuum displacement takes place, followed by cake discharge at the end of the rotational cycle. Washing may be effected, if required; since the geometry of these filters makes it difficult to maintain liquid coverage, washing usually proceeds by spraying. Again, the time available for washing is limited by obvious geometric constraints. Both systems require uniform slurry suspension to ensure even cake deposition. Homogeneous suspension can be obtained by integral mixers or by controlled pumping through the trough containing the slurry.

Drum design has developed in two ways: (a) units where the whole drum is under vacuum and (b) the open-ended variety, in which only an outer annulus compartment is evacuated, as in Figure 11.4. The latter is the most common since mechanical troubles can be created with type (a) by the large buoyancy force developed by drum immersion.

In the metallurgical field, or elsewhere for heavy solids, arrangements sometimes have to made to deposit the cake inside the drum or to use top-feed, rotary filters involving slurry troughs situated towards the top of the filter; advantage is thus taken of the fast settling characteristics of the particles. In these circumstances and where large filtering areas are required, the horizontal belt filter discussed below will probably be preferred.

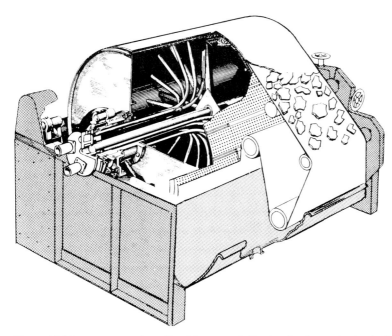

Figure 11.4 Rotary vacuum drum filter (Courtesy of Filtration Services Ltd, Macclesfield, U.K.)

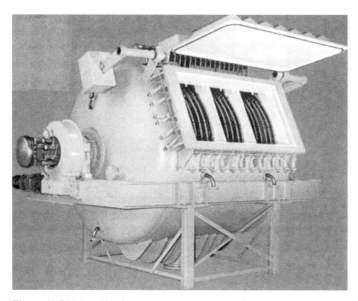

Figure 11.5 Rotary disc filter (Courtesy of Gaudfrin SA, St. Germaine-en-Laye, France)

Figure 11.6 Tilting pan filters (Courtesy of Baker Process, South Walpole, MA, USA)

Figure 11.7 Horizontal belt filters [Prinssen, 1979]

As may be noted in Figure 11.8, the positioning of the discharge knife relative to the point of entry into the slurry, produces a proportion of the drum surface not involved in the filtration process [Pierson, 1981].

11.2.1.1 Drum Filter Productivity

Elementary filtration theory in Chapter 2 may be used to produce a process equation related to the rotary filter. Thus from the parabolic equation for constant-pressure separation of incompressible particles:

$$(\alpha\mu c / 2)V^2 + \mu A R_m V = A^2 \Delta P \ t_f \qquad (11.18)$$

Multiplying across by $(c/\mu A^2)$ and using the material balance relationship $w = (c\,V)/A$ leads to the equation below for the dry solids yield per unit area produced by t_f seconds of filtration:

$$(\alpha / 2)\,w^2 + R_m w = (c\,\Delta P\ t_f)/\mu \qquad (11.19)$$

Using the algebraic solution to equations of the type $a\,x^2 + b\,x - c = 0$ produces:

$$w = [-R_m \pm (R_m^2 + 4\,k_1 k_2)^{0.5}]/2\ k_1 \qquad (11.20)$$

in which:

$$k_1 = \alpha/2; \qquad k_2 = c\Delta P t_f/\mu$$

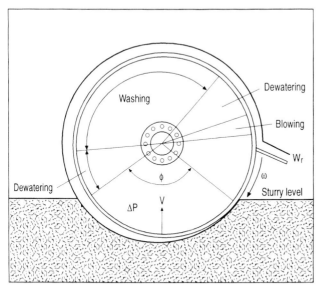

Figure 11.8 Rotary vacuum filter cycle

This is the dry solids discharge (kg/m^2) per cycle of the drum. The cycle time and the time for filtration are determined by the fraction of filter surface submerged in the slurry ψ. Thus $t_f = \psi t_C$. Dividing Equation 11.20 by t_c gives the production rate of the filter (kg m^{-2} s^{-1}):

$$w_r = [-R_m + (R_m^2 + 4k_1k_2)^{0.5}]/(2k_1t_c) \qquad (11.21)$$

An equivalent expression may be derived in terms of the volumetric filtrate rate, V_r $(m^3m^{-2}s^{-1})$

Example 11.1

Calculate the dry solids from a 10 m² RVF operating at 68 kN/m² filtering uranium-bearing gangue, in the conditions itemised below:

Cake resistance	α: 1×10^{10} m/kg
Medium/precoat resistance	R_m: 1×10^{10} m⁻¹
Drum speed	ω: 0.105 rad/s; (1 rpm)
Fraction submerged	ψ: 0.3
Solids concentration: kg solids/kg slurry	s: 0.1
Cake moisture: kg wet cake/kg dry cake	m: 3.5
Liquid density	ρ: 1000 kg/m³
Liquid viscosity	μ: 0.001N m⁻²s⁻¹

NB: Drum speed is sometimes reported in terms of ω rad/s.

1 rev = 2π radians (1 rpm = 0.105 rad/s)

Submergence may also be reported in terms of the angle ϕ rad. Fraction submerged ψ = $\phi/2\pi$.

Solution

1. Cycle and form time

Cycle time t_c = 60 s; form time t_f = 20 s; drum speed ω = 0.105 rad/s.

2. k_1 and k_2

$$k_1 = \alpha/2 = 0.5 \times 10^{10} \text{ m/kg}$$

$$\text{NB: } k_2 = \frac{\Delta Pct_f}{\mu} = \frac{68000 \times 153.8 \times 20}{1 \times 10^{-3}} = 2.092 \times 10^{11} \text{ kg/m}^3$$

$$c = \frac{\rho s}{(1-ms)} = \frac{1000 \times 0.1}{(1 - 3.5 \times 0.1)} = 153.8 \text{ kg/m}^3$$

3. Calculate dry solids yield

$$w_r = \{ -1\times10^{10} + [(1\times10^{10})^2 + 4\times0.5\times10^{10}\times2.092\times10^{11}]^{0.5} \} / 60\times10^{10}$$

$$w_r = 9.24 \times 10^{-2} \text{ kg} / \text{m}^2 \text{ s}$$

For a cycle time of 2 min, i.e. 0.5 rpm:

$t_c = 120$ s; $t_f = 40$ s; $\omega = 0.052$ rad/s.

$$k_2 = \frac{6800 \times 153.8 \times 40}{0.001} = 4.184 \times 10^{11}$$

$$w_r = \{ -1\times10^{10} + [(1\times10^{10})^2 + 4\times0.5\times10^{10}\times4.184\times10^{11}]^{0.5} \} / 120\times10^{10}$$
$$= 6.833\times10^{-2} \text{ kg m}^{-2}\text{s}^{-1}$$
$$= 2.460 \text{ tonnes/h for the filter}$$

Thus the slower drum speed has a lower productivity. If R_m is negligible, compared with the cake resistance, the productivities at different speeds may be estimated from:

(Productivity at speed A)/(Productivity at speed B) $= \sqrt{\omega_A / \omega_B}$

These theoretical estimates of cake productivity indicate an advantage at higher speeds, with concomitant thinner filter cakes and higher solids yield. A thin cake can be produced by increased speed of rotation; the final optimum condition is determined by mechanical limitations and the minimum thickness required for discharge.

Calculations of this type are embodied in recent rotary vacuum filter scale-up procedures, using computer spreadsheets [Holdich, 1990] included in Appendix C.

11.2.1.2 Cake Discharge and Thickness

Cake discharge can sometimes be a problem, and various means have been developed for this step, including:

1. Knife discharge following a previous reverse air blast. Here a reasonably thick cake is required, and sometimes it is necessary to protect the filter medium, which covers the drum, by using a rubber- or plastic-edged knife. The blow-back technique can create a re-wetting of the partially dewatered cake with filtrate which remains undrained in the conduits behind the filter cloth.

2. Rollers may be used when processing sticky filter cakes.

3. Strings or springs are available for the discharge of particularly difficult materials.

In some drum filters, blinding of the medium can be ameliorated by a continuous removal of the cloth during rotation, with back-washing. Again, this produces a loss of available filtration area on the drum, Figure 11.8; this effect is not present in horizontal belt filters.

Discharge may be particularly difficult in circumstances where thin, sticky cakes are produced. The theoretical result that thin cakes lead to high productivity has to be related to the difficulty of discharging these cakes, and, of course, to the mechanical problems attaching to the rotation of large machines. Practical experience leads to the various minimum cake thicknesses which are presented in Table 11.2; these are needed to avoid problems at the point of discharge, at least with the filters considered here.

Table 11.2. Recommended Cake Thickness

Filter	Discharge mechanism	Cake thickness Minimum, mm	Maximum, mm
Drum	Scraper	3–5	5–75
	Wire	2–3	6
	String	5–6	2
	Belt	3–5	25
	Roll	Thin–1	3
	Precoat	Thin	25–30
Disk below	Scraper or snap	6–9	25–75
Horizontal	Table (scroll discharge)	20–25	100
	Belt	4–6	100
	Pan (tilting)	20–25	100

The mass calculable by Equation 11.4 can be related to the above recommended minimum thicknesses to define operation conditions (drum speed, submergence, etc.) for optimum production of solids [Wakeman, 1983].

The thickness of the cake may be related to w_r by the equation:

$$w_r t_c = L (1 - \varepsilon) \rho_s \tag{11.22}$$

Substituting for $w_r t_c$ in Equation 11.21 yields the expression:

$$(2 \pi \psi c \Delta P) / \omega = \mu L (1 - \varepsilon) \rho_s (L (1 - \varepsilon) \alpha \rho_s + 2 R_m) \tag{11.23}$$

where ψ is the fraction submerged.

Given that data are available on the slurry properties, the above expression can be used to estimate the interaction of mechanical variables ΔP, ψ and drum speed ω to produce a recommended thickness *L*. The slurry concentration has an effect on cake pick–up and preconcentration should be considered when optimum conditions are being sought. The equation may also be used to estimate the effect of process changes on the effectiveness of dewatering and washing. A similar equation is available for horizontal filters.

11.2.1.3 Theory *vs* Practice: Rotary Vacuum Drum Filter

A series of practical pilot plant measurements using 0.1, 0.3 and 1.0.m^2 rotary vacuum filters indicated that, in many systems, serious deviations can occur between theoretical predictions using leaf tests obtained at bench scale, and practical results [Rushton & Wakeman, 1977].   Practical yields are affected by factors not considered in the theoretical model, which assumes that the mass of filter cake deposited by the passage of *V* volumes of filtrate may be calculated from $w = cV$, where *c* is is given by Equation 11.3. Furthermore, cake cracking may occur, in which case the effective vacuum for particle pick-up may be reduced.

Cake cracking has received limited attention [Rushton & Hameed, 1969; Wakeman, 1974] in the literature. Recent contributions [Yagishita et al, 1990] list the methods normally suggested to avoid cake cracking:

a) Formation of cakes from dilute feeds, rather than concentrated ones. Thus, in the formation of pre-coats from filter aids (diatomaceous earth, perlite, etc.), 1–2 mm layers are usual, from feed slurries of 3–6 wt%. These deposits of 0.5 kg/m^2 (fine grade) to 1.0 kg/m^2 (coarse grade) provide a matrix for the entrapment of fine solids.

b) Avoidance of uncontrolled filter cake dewatering.

c) Formation of thin cakes, rather than thick; cake thickness can be controlled by changes in drum speed.

d) Pre-coat formation under lower vacuum than that used in the filtration process.

Cake cracking is attributed to sudden changes in cake porosity, e.g. during cake washing operations. The porosity of a filter cake and its resistance is usually a strong function of solids concentration, as shown in Figure 11.9. During washing, the input solids content of the flowing liquor is zero; this may lead to changes in the structure of the deposited cake.

Further work [Gross et al, 1989] suggests that filter cake cracking may be controlled by suitable filter media which are divided, on the surface of the fabric, into sets of small permeable zones; the cake forms over the latter, spreading out in honeycomb formation. The use of filter media with fine pores and/or membranous qualities can produce advantages in vacuum filtration and in those processes by mechanical/gas blowing dewatering techniques [Anlauf, 1990].

Cake drop-off in sludge processing [Gale, 1971], is a further problem, and coefficients have been suggested to correct for the effect; drop-off may occur immediately after the new filter cake leaves the trough.

Table 11.3. Dry solids and filtrate flow: $0.1\,m^2$ RVF, calcium sulphate filtration

Dry solids $w_r \times 10^3$, kg m^{-2} s^{-1}	Filtrate $V_r \times 10^5$, m^3 m^{-2} s^{-1}	Theoretical $w_r = cV_r \times 10^3$, kg m^{-2} s^{-1}	Agitator speed, s^{-1}
1.91	4.90	8.87	0
2.93	4.99	9.04	0.42
3.42	5.11	9.26	0.83
2.71	5.23	9.40	1.67
2.67	5.21	9.43	1.67
1.79	5.37	9.72	2.50
1.28	5.22	9.46	3.34
1.01	5.19	9.38	4.17

$\Delta P = 54\,kN/m^2$; s $= 0.12$; $R_m = 2.4 \times 10^{11}$ m^{-1}; $t_c = 58\,s$

These data show the difference between actual solid yield w_r and that calculated from the filtrate flow. It may be noted that the actual dry solids yield is less than that predicted from the filtrate volume via: $w_r = cV_r$. These differences are attributed to the loss of solids via sedimentation, etc. It is also noticed that the initial improvement in cake pick-up created by increased agitation is offset by the stripping action at high speeds; a maximum is created in the yield curve, Figure 11.10.

Whilst cake cracking and drop-off are principal sources of differences in the yield of actual plant, compared with laboratory-scale tests, other considerations include: (1) equality of the pressure differential available for cake formation and (2) equivalence of media condition and resistance. The latter will depend on the age of the medium and the effectiveness of cloth washing operations. Filtration equipment is generally poorly instrumented. The location of the ΔP measuring device and the pressure (vacuum) losses present between the measurement and point of filtration will vary from station to station.

The use of probes can be used to measure the variation in ΔP across the cake and medium during passage through the slurry; these measurements demonstrate sinusoidal variations due to changes in submergence.

Despite these observations, other work [Carleton & Cousens, 1982] in the industrial evaluation of eight vacuum filters (one table filter, one tipping pan and six drum units), reported useful application of theoretical relationships for filtration and dewatering.

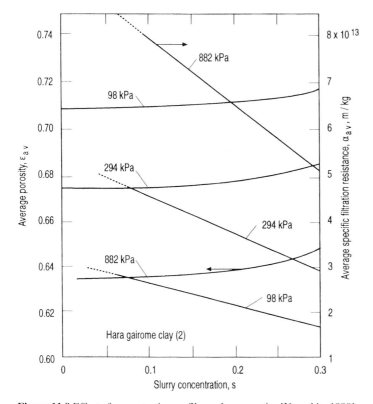

Figure 11.9 Effect of concentration on filter cake properties [Yagashita,1990]

Conversely, this work also listed many of the causes of poor performance by large-scale vacuum units. These studies were later [Carleton & Salway, 1993] extended to include cake dewatering by gas blowing and squeezing.

The difference in cake yield in upward and downward filtering systems has been studied extensively [Rushton & Rushton, 1973] and a model allowing for particle sedimentation away from the filtering zone has been proposed. In this way, the particle size characteristics are included in the analysis. The data above demonstrates that cake removal mechanisms created by vigorous agitation also need to be included.

Cake pick-up can be strongly influenced by the resistance of the medium which predetermines the initial filtrate velocity and the drag forces created by the latter, which stabilise the cake layers. Figure 11.11 shows the influence of medium resistance on the rate of cake filtration, to that expected from liquid flow, plotted against agitator speed.

An indication of these possible difficulties can be obtained by careful laboratory scale leaf or Buchner tests. Differences in solid yield, when using the leaf filter in upward filtering mode, compared with downward-directed Buchner tests, is a clear indication that filtration is accompanied by sedimentation effects. Severe sedimentation is manifested in a change in slope of the $(t/V, V)$ diagram attached to Buchner trials; cake formation may be completed by the combined filtration–sedimentation mechanism. In these circumstances, a

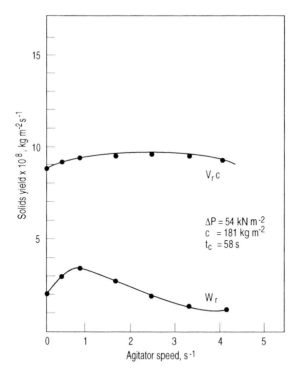

Figure 11.10 Practical and predicted solids yields in rotary vacuum filtration of calcium sulphate

quantity of the original liquor may remain on the surface of the cake. Passage of this supernatant, by permeation through the cake, augments V with time. Permeation of a fixed cake thickness leads to a horizontal $(t/V, V)$ relationship.

In upward filtering systems, where sedimentation is present, or where the resistance of the medium reduces the filtration velocity, deposition of the finer particles in a suspension may be promoted. This will result in a high-resistance, first-layer deposit on the filter.

Practical suggestions to avoid some of these difficulties include delaying the application of vacuum to the rotating "cells" until the latter are well immersed in the slurry. This is to avoid first deposition of fine particles unaffected by sedimentation [Tiller et al, 1986].

Obviously, this discussion suggests that where possible, pilot plant trials are recommended for vacuum systems, particularly in upward filtering systems and attention should be directed to cake loss effects in scale-up applications.

Industrially, sedimentation effects during filtration merely create conditions of filter thickening. If the latter takes place in the slurry trough attached to certain filters, eventually uneven cake pick-up will ensue.

In vacuum-operated plant the best possibilities for adequate scale-up are obtained with horizontal belt and table filters. The performance of these units can be simulated by Buchner trials, since loss of solids is absent, and the mixing or gravitational effects present in the rotary drum or disc filters does not interfere with cake formation. Slurry mixing in the filter is unnecessary in horizontal filters.

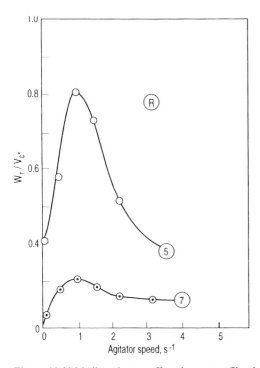

Figure 11.11 Media resistance effects in vacuum filtration

11.2.1.4 Dewatering and Washing on the RVF

In the design and optimisation of filters, it is often found that the time for dewatering and/or washing is a controlling factor. Provision of sufficient dewater/wash time is often difficult in RVF applications, in view of the cycle times normally involved.

Thus in RVF design, the rate of cake formation, taken along with the submergence time available, may be used to predict the solids yield. Further constraints, e.g. the necessity to deliver several void volumes of wash or the requirement to dewater down to an acceptable level, may cause considerable change in the filter size or operating conditions, compared with those based on cake formation only. It will be recalled that, at an "average" operating speed of 1 rpm and with 30% submergence, only 40 s are available for cake washing, dewatering and discharge.

Long wash times involve the use of a slow drum speed and, in order to maintain the required cake thickness, etc. a low submergence. The overall effect is to reduce the solids yield considerably. This comment demonstrates the importance of wash time in filter sizing and operation. Equivalent examples can be quoted for dewatering — a factor of great importance in solids handling and in legislation pertaining to the dumping of waste.

The relative monetary values of solid and liquid products have an effect on the optimum cake thickness to be used in RVF operations, involving filtration and washing [Wakeman, 1983]. This follows from the theoretical evidence that thin cakes lead to higher productivities, but better washing is obtained with thick cakes.

Figure 11.12 shows practical measurements of the variation in liquor content of the filter cake around the external surface of a 0.3 m^2 RVF. It will be noted that increases in applied vacuum from 17 to 54 kN/m^2 has little effect on the wetness of the filter cake and at the point of discharge. The porosity of some filter cakes is particularly sensitive to changes in pressure differential, in vacuum systems.

The applied differential has first to overcome the effect of capillary pressure, then to provide a suitable dewatering rate; changes in the cake properties towards a more resistant assembly will minimise the effect of increased ΔP. This is a manifestation of cake compressibility; relatively large changes in porosity are often created by small changes in ΔP in low-pressure systems.

It is also observed that filter cake washing efficiencies can be low in these filters. Such effects are attributed to maldistribution of wash water; some of the latter inevitably cascades over the cake surface, to be collected in the feed trough. As before, the limited time for washing is also a problem. In some cases, two stages (or more) of filtration may be considered, with interstage, re-slurry washing. These components also apply to rotary disc filters discussed in the next section.

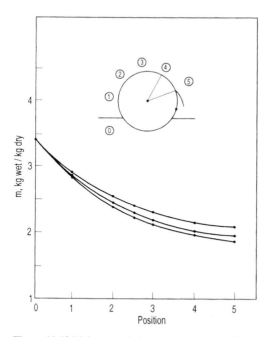

Figure 11.12 Moisture variations in rotary vacuum filtration

11.2.2 Rotary Disc Filters

Rotary disc filters find many applications industrially, particularly in the large-tonnage mineral industry. These units provide larger filtration areas, per unit of floor space, than rotary drum units and are, therefore, generally less expensive. Of course, as with drum filters, practical effects limit the available speed, vacuum, etc. of large disc units; the thickness of the filter cake has also to be controlled in order to avoid difficulties in cake discharge, high moisture etc. A particular disadvantage with disc filters is the difficulty of producing high washing efficiencies on vertical surfaces. These considerations have been taken into account in an interesting practical approach [Gaudfrin, 1981] to the relationships between F_k (the filterability constant measured at 1 bar), the operating pressure ΔP, the drum or disc speed ω and the filtrate flow rate. Figure 11.13 was produced by superimposing a graph of F_k versus rotational speed, for various filtrate flowrates (m^3/h) onto a graphical relationship between F_k and operating pressure. The latter data, for the crystalline materials dealt with in sugar refining operations, tends towards a limiting value at high pressure. The value of this operating approach to filter design relates to the defined operating limits on the graph, between thin and thick cakes, in the shaded areas of the graph.

It should be noted that F_k (used in European sugar industry calculations) is defined by:

$$F_k = (\mu \alpha c) / \Delta P \tag{11.24}$$

Hence:

$$F_k = (A^2 K_1) / \Delta P$$

where

$$K_1 = (\alpha \mu c) / A^2$$

Another version of the disc filter is shown in Figure 11.14. Here, the discs are made from porous ceramics [Ekberg & Rantala 1991]. The latter have the property of permitting water flow whilst, by capillary pressure control, preventing passage of air. This reduces vacuum losses and costs; better dewatering/washing performance is claimed.

In these units, as with drum filters, the feed suspension is supplied continuously into troughs, in which the liquor flow is arranged in the same direction as the rotating discs. The latter move at various speeds, in the range 0.5–3.5 rpm. Submergences up to 50% of the filtering surface can be arranged by level control. Disc filters are available in areas from 0.5–300 m^2.

Improved disc filter operation has been obtained with a disc design containing 30 segments instead of 12 and gave capacity increases of 10%. A higher operational speed was permitted with the new design leading to a 35% increase in productivity. These practical results have been predicted in a theoretical model of the disc filter [Rushton,

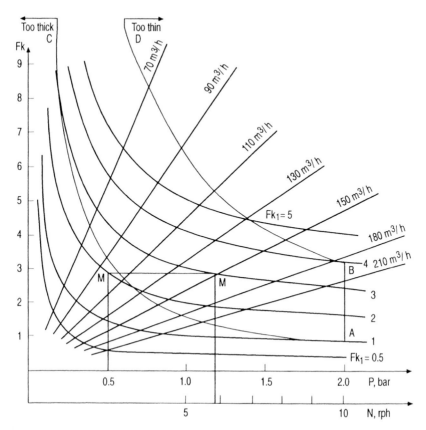

Figure 11.13 Combined graph for continuous filter operation control (Courtesy of Gaudfrin SA St. Germaine en laye France, 1981)

1974] which allows for the distribution in submergence time over the leaf, along with the variable filtration pressure over each segment of the disc. Further improvements were obtained by providing individual slurry compartments for each disc, and ensuring adequate slurry velocities to avoid sedimentation effects. Attention to the filtrate drainage from individual discs and the use of disc filters inside a pressure vessel, have facilitated the handling of thin cakes, produced at relatively high disc speeds, with high production capacities. Figure 11.14 shows disc filters composed of ceramic materials; the latter reduce vacuum losses by prevention of air flow through the small pores in the discs [Ekberg, 1990].

Technical brochure information rarely reports the "basis" of cake moisture. Laboratory tests on the latter [Bosley, 1986] involve careful sampling of the wet cake; weighed quantities of the latter are dried at 105^0C to constant dry weight. The amount of liquor evaporated is then expressed as a percentage of the wet (or dry) cake. In the table below, a wet basis is assumed.

Table 11.4 presents some typical productivities obtained with the use of drum and disc filters in industry. Other data are available [Purchas, 1981; Dickenson, 1992]. Such information demonstrates the wide variation in the production rates and moisture contents; these variations depend on concentration, particle size and range, temperature, pH and other variables.

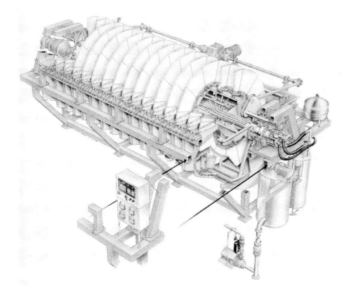

Figure 11.14 Ceramic disc filters; capillary pressure control (Outokumpu Mintech Oy, Espoo, Finland)

Table 11.4 Vacuum filter productivity[*]

Application	Feed Solids, wt%	Particle Size, %>325 mesh	Production, $kg\ m^{-2}\ h^{-1}$	Cake Moisture, wt%
Cement	65	70	85–120	75–80
Copper Concentrate	50	96	290–480	1–13
Flue Dust	50	95	145	22
Iron Magnetite	60	90	720–1440	10–13
Lead Concentrate	75	70	120–360	8
Pyrite Concentrate	70	95	190–480	9

*Svedala Pumps and Process AB, Sala, Sweden

11.2.3 Horizontal Filters

The geometric constraints inherent in the design of rotary drum (upward filtration) and vertical disc units, have been discussed in relation to poor cake pick-up and washing. In contrast, presentation of a wide particle size spectrum to a horizontal unit may prove advantageous, if the larger particles in the feed can be encouraged to deposit as a first layer on the filter medium. In effect, the suspension would then have a self-pre-coating action on the cloth.

Where such a spread in size exists, and where cloth blinding effects are prevalent, particle classification of the feed prior to filtration may prove beneficial. As outlined in Chapter 4, an alternative is to allow passage of the fine material in the early stages of a separation; these "fore-runnings" may then be recycled to the filter.

Three industrial units are available in this category; all these horizontal filters take advantage of the effect of gravity in cake formation. The units are:

11.2.3.1 Tilting Pan Filter

The operating features of these units may be imagined as a set of Buchner funnels, arranged in a circle. Upon rotation, each funnel is a) filled with slurry; b) vacuum filtered; c) washed counter-currently (maximum three washes); d) dewatered; e) cake discharged, as shown in Figure 11.6.

During the cake discharge stage, the funnel (or pan) is turned completely upside-down. After cake release, the pan returns to an upright position, and is refilled with slurry. This is the most expensive of this type of filter. It generally occupies a greater floor area per unit filter area, is mechanically complex and suffers processing difficulties (e.g. cloth blinding, poor dewatering levels. On the other hand, high washing efficiencies can be attained in the filter which, at present, is used mainly in the fertiliser industry for phosphoric acid filtration.

11.2.3.2 Rotating Table Filter

This is a horizontal, circular table constructed from several sectors. The latter are connected to a vacuum source, with filter cake formation proceeding on the upper porous surface. Each sector is covered with an appropriate filter medium; the liquor emanating from different sectors may be collected separately if necessary. The mode of cake discharge here is by a radial scroll which is situated slightly above the rotating table and cloth, Figure 11.15.

The action of the scroll is such as to leave a residual "heel" of filter cake (up to 1.5 cm thick) on the surface of the filter medium; the continued presence of the heel, under the washing conditions required by most processes, can lead to particle migration, or scouring

Figure 11.15 Rotation table filter (Courtesy of Ucego Filter Co., Dametal, Rouen France)

into the filter medium. The latter may then tend to blind. The unit is the cheapest of the three filters described and is usually applied to the filtration of free filtering crystalline materials.

11.2.3.3 Horizontal Belt Filter

This unit is the most popular version of the continuous horizontal filters discussed in this section. A schematic diagram of a typical unit is shown in Figure 11.16. Again, these are gravity-assisted units, with the feed slurry presented to the upper surface of the horizontal filter cloth. In turn, the latter is supported by an endless belt which is situated above the vacuum box filtrate receivers.

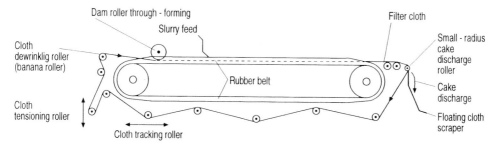

Figure 11.16 Horizontal belt filter

Filtrate passes through the filter medium, along cross-grooves in the belt, and down through centrally situated drainage pipes. Figure 11.16 shows the position of the belt in relation to two rollers; the discharge end roller acts as the drive roller.

The basic principles used here are not new, of course; horizontal units of this type have been used for many years in the paper-making industry. In the latter, however, the filtered material is highly porous and constitutes a small resistance to filtration and dewatering processes. It follows that the high-speed paper machine requires only a small pressure differential for the separation of fibres.

In contrast, much higher ΔP is required for the filtration and dewatering of minerals. This higher vacuum requirement leads to increased friction between the rubber/elastomer belt and the vacuum box filtrate receiver.

These units require careful "tracking" of the belt and filter cloth; facilities are also required for "de-wrinkling", the latter whilst on the machine. New techniques are available, which provide ± 0.5 cm lateral stability. Careful consideration also has to be given to cloth selection. As outlined in Chapter 4, expensive errors can be made in the use of easily blinded or mechanically weak cloths when applied to machines which can require large areas of materials. Horizontal belt units up to 200 m^2 in area have been reported; these are used in large-scale mining applications, where in past operation, mistakes in cloth specification led to frequent cloth-changing requirements.

The horizontal belt filter finds widespread application in the mining and chemical industries [Schonstein, 1991] with higher productivities being claimed, when compared with other vacuum devices.

Several equipment vendors supply the continuous, rubber belt type filter depicted in Figure 11.17. Variations in design are usually directed at reducing the friction between the belt (with the supported cake mass) and the vacuum box. Attention is given to the use of low-friction materials (PTFE) for the contact surfaces. Air or filtered water may be directed at the interface between the sliding surface area and the box.

Efficient cake discharge can be effected by separation of the belt from the cloth and directing the latter over a set of discharge rollers. Here a sharp turn in cloth travel, over a small diameter roller produces cake cracking and discharge. Another area which has received attention, is that of the control of the down-stream pressure drops which ensue with the two-phase flow of air and water through the vacuum system.

Measures to permit fast separation of the air phase from the filtration (as achieved in Buchner flask operations) can reduce the frictional pressure loss in the filtration pipework. This results in a higher ΔP being available for the filtration operation.

The air–water separating device is depicted in Figure 11.18. Further care must be given to the alignment of the connecting ports in the vacuum box and the belt. Belt velocity is not a free choice in a given application [Schwalbach,1990]. It is conditioned by the filtration characteristics of the material being separated. This point is demonstrated in Example 11.2.

Standard belt widths are 1.0, 1.5, 2.0, 3.0 and 4.0 m with a length/width ratio in the range 4–15. High values of the latter rates lead to higher belt speeds with associated wear and tear. Low speeds tend to have higher belt costs and process liquor distribution difficulties. Thyristor-controlled drive units may be used for belt speed control — usually in the range of 3–30 m/min.

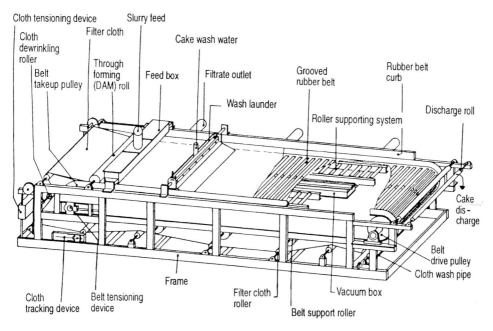

Figure 11.17 Rubber horizontal vacuum filter (Courtesy of Dorr–Oliver, Croydon, Surrey, U.K.)

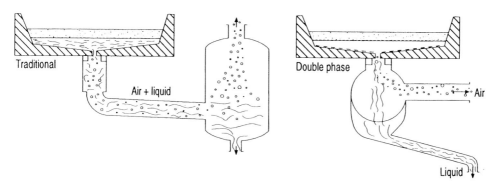

Figure 11.18 Air–filtrate separation in belt filtration (Courtesy of Gaudfrin SA, St. Germain-en-Laye, France)

Increases of the rubber belt width leads to increased belt cost, which can be 22–40% of the overall cost of the filter. Cost considerations, and the need to find a replacement for rubber in solvent and food processing, led to the development of module-type horizontal belt filters. In these cases, the rubber belt is replaced by stainless steel or plastic trays. Three designs of module filter are available, each having different tray arrangements.

a) Stationary Filter Tray or Pan (Fig. 11.19). Here the feed launder moves back and forth over the stationary tray/cloth, with vacuum applied beneath the latter [Pierson, 1981a,b]. After formation of a prescribed quantity of filter cake, the vacuum is released, and the cloth is moved forwards towards the discharge end of the filter. Several trays are situated beneath the cloth, so that after clearing the first tray, the filter cake can be connected to other trays for further processing i.e. dewatering/washing. The continual venting of the unit adds to the processing (vacuum) costs; these must be compared with the capital savings when the purchase cost of these units is compared with the rubber belt model. Filter cake cracking may be ameliorated by placing a plastic sheet over the surface of the moving cake, in the dewatering section.

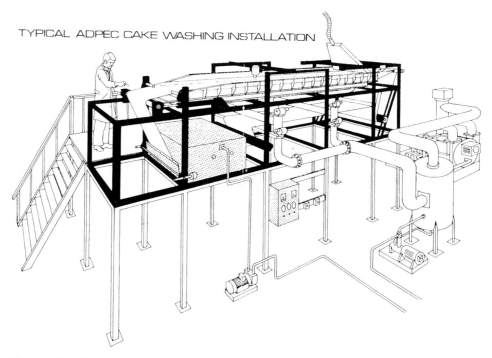

TYPICAL ADPEC CAKE WASHING INSTALLATION

Figure 11.19 Adpec stationary pan horizontal filter (Courtesy of B.W.F. GmbH, Sontheim, Germany)

b) Reversible Trays. In this design, the filter cloth moves continuously beneath a stationary feed pipe. The filter trays connect intermittently with the cloth, return towards the feed end of the filter after travelling with the cloth a certain distance and time. Application of the vacuum creates an adhesion between the cloth/cake and the tray. Venting is synchronised with tray return. As may be noted in Figure 11.20, the feed point is covered beneath, at all times, by the pans.

Successful industrial applications have been reported [Groh & Van der Werff, 1982], particularly in the pressing of solvents, which would preclude the use of elastomer belts. Noxious, corrosive vapours are handled by enclosing the entire filter in vapour-tight enclosures. A further important advantage over other SLS systems is the control of cake cracking by minimisation of the area between flooded filtration and wash zones on the belt.

c) Continuous Circulating Trays and Cloth. This design, Figure 11.21, operates on a fully continuous basis. Adjacent trays are designed to seal against each other during passage over the vacuum box. Replaceable rubber profile seals are fitted to the trailing edge of each tray to

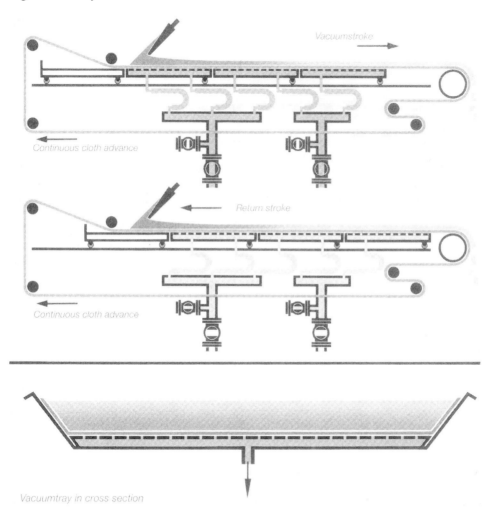

Figure 11.20 Reciprocation tray filter (Courtesy of Paunevis, Utrecht, The Netherlands)

prevent air leaks, between the trays, into the vacuum system. The trays are located on the filter by attachment to two rubber chains which are situated either side of the vacuum box. Other mechanical details are available [Smith & Beaton, 1989], along with process capacity data. The latter publication lists claimed advantages of this "rigid belt" filter over the rubber belt design:

i) Lower costs, particularly at larger sizes
ii) Reduced down-time in maintenance periods
ii) Reduced installation costs

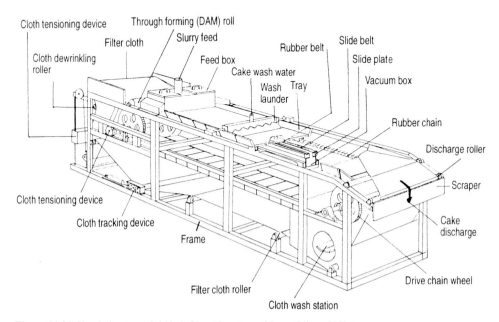

Figure 11.21 Circulating tray: rigid belt filter (Courtesy of Dorr–Oliver, U.K.)

Further advantages are claimed for reversible trays and continuous circulating tray/cloth modules, mainly linked to the continuous nature of the operation, and the absence of frequent vacuum release with such units.

Example 11.2

Calculate the dimensions and operational speed of a horizontal belt filter required to produce 5 m³/h of filtrate from a slurry with the properties given below. Wash is to be conducted co-currently with the delivery of two void volumes followed by dewatering to a saturation of 75%.

Data

$$\alpha = 5 \times 10^9 \text{ m/kg}$$
$$\Delta P = 60 \text{ kN/m}^2$$
$$c = 350 \text{ kg/m}^3$$
$$\rho_s = 2000 \text{ kg/m}^3$$

ρ = 1000 kg/m^3
μ = 10^{-3} Pa s
σ = 0.065 N. m.$^{-1}$
ε = 0.43

Washing characteristics

c_s/c_o	0.34	0.03	0.01
Void volume	1	2	3

Solution

The instantaneous filtration rate after V m^3 of filtrate is given by:

$$dV/dt = \Delta P\, A^2/\alpha\, \mu\, c\, V$$
$$= 3.428{\times}10^{-5}\, (A^2/V)\ \text{m}^3/\text{s}$$

Assume unit area and compute filtration time required to produce various values of V.

Calculate corresponding values of dry solids yield:

$$w = c\, V$$

Calculate corresponding filter cake thickness :

$$L = w/(1-\varepsilon\,)\rho_s$$

where w is the solids load per unit area

V (assumed) m^3	Instantaneous filtration rate m^3/m^{2}s	Average rate (from start) m^3/m^{2}s	w kg/m^2	L m	t_f s
0.01	0.00343	0.00686	3.5	0.0031	1.46
0.05	0.00069	0.00137	7.7	0.0154	36.46
0.10	0.00034	0.00068	35.0	0.0310	146.00

Assume 1.54 cm thick cake as a design basis, filtration time 36.46 s. (N.B. this could be changed in accordance with discharge requirements).

Required filtration rate $= 5 \text{ m}^3/\text{h} = 1.388 \times 10^{-3} \text{ m}^3/\text{s}$
Time $= 36.46 \text{ s}$
Average rate per unit area $= 0.00137$
Area required $= 1.012 \text{ m}^2$

Calculate belt speed: assume two belt widths 1 and 2 m:

Belt width	1 m	2 m
Length, m	1.012	0.506
Speed (length/t_f) m/s	0.028	0.014

Wash area requirements: wash with two void volumes co-currently to reduce the residual solute concentration to 3%.

Wash rate = Final filtration rate   = $0.00069 \text{ m}^3 / \text{m}^{-2} \text{ s}^{-1}$

Volume of cake liquor per second   =   Cake width×cake thickness ×
porosity×belt speed

=   0.000185 m³/s for both widths

Volume of wash required per second =   2×volume of cake liquor per second
=   0.00037 m³/s

Wash area   = $0.00037/0.00069 = 0.541 \text{ m}^2$

Length of wash zone

Belt width	1 m	2 m
Wash length	0.541	0.270
Belt speed, m/s	0.028	0.014
Wash times, s	19.32	9.32

Dewatering Requirements:

a) Calculate capillary number $N_c = 180 \, [\rho g L + \Delta P]/\alpha \, \rho_s (1 - \varepsilon) \, L\sigma$
b)

$$N_c = 180 \frac{[9.81 \times 1000 \times 0.0154 + 60\,000]}{5 \times 10^9 \times 2000 \times 0.57 \times 0.0154 \times 0.065} = 1.898 \times 10^{-3}$$

b) Calculate saturation at infinite drainage time: S_∞

$$S_\infty = 0.39 \, [1 + 0.0256/\sqrt{Nc}] = 0.619$$

c) Calculate threshold pressure

$P_b = [0.118 \, \varepsilon \, \sigma^2 \, \alpha \, \rho_s \, (1-\varepsilon)]^{0.5}$
$P_b = [0.118 \times 0.43 \times 0.065^2 \times (5 \times 10^9) \times 2000 \times 0.57]^{0.5}$
$\quad = 34\,956 \text{ N/m}^2$

d) Calculate available ΔP^* for drainage

Pressure of air above cake = Pair, inlet   = 101 300 N/m^2

Pressure of air beneath cake Pair, outlet = (101 300 – 60 000)

$$= 41\ 300\ \text{N/m}^2$$

$$\Delta P^* = (P\text{inlet}/P_b) - (P\text{outlet}/P_b) = 2.89 - 1.18$$
$$= 1.71$$

e) Calculate S_R value at required saturation:

Saturation required in dewatered cake $S = 0.075$

$$S_R = (S - S_\infty)/(1 - S_\infty) = (0.75 - 0.619)/(1 - 0.619) = 0.3438$$

f) Calculate θ:

From $S_R = 0.58\ (\Delta P^* \theta)^{-0.22}$

$$S_R = 0.3438$$

Hence:

$$(\Delta P^* \theta)^{-0.22} = 0.5928$$

Thus:

$$\Delta P^* \theta = 10.8$$

$$\Delta P^* = 1.71$$

Hence:

$$\theta = 6.29$$

g) Calculate t

$$\theta = (P_b\ t)/\ \overline{\alpha}\ \rho_s\ (1 - \varepsilon)\ L^2\ \mu\varepsilon(1 - S_\infty)$$

Hence:

$$t = (\theta \overline{\alpha} \rho_s (1 - \varepsilon) L^2 \mu\varepsilon(1 - S_\infty))/P_b$$

$$= 6.29 \times \frac{5 \times 10^9 \times 2000 \times 0.57 \times 0.01542^2 \times 10^{-3} \times 0.43 \times 0.381}{34\ 956}$$

$$t = 39.85\ \text{seconds}$$

Filter specifications

Belt width	1 m	2 m

Belt length, m

Filtration	1.013	0.507
Wash	0.541	0.270
Dewater	1.116	0.558

Total belt length	2.67	1.335
Length/width ratio	2.67	1.335

(Normal length/width range 4–5)

Time, s

Filtration	36.46	36.46
Wash	19.32	19.32
Dewater	39.85	39.85

Total cycle time	95.63	95.63

Belt speed, m/s	0.028	0.014
m/min	1.68	0.84

(Normal range 3–30 m/min)

These calculations suggest that a smaller belt width should be considered and demonstrate that the belt area, speed etc., follow on from the decision on cake thickness. Again, the latter should be related to recommended discharge values. Gas rates required for this separation are discussed in Chapter 9.

This elementary estimate of horizontal belt dimensions highlights the importance of having available information on the process of cake washing, in terms of time or delivered volume of wash. In the problem, it was decided to wash the cake co-currently with two void volumes.

In horizontal belt filtration applications, the various filtrates and washes can be collected in separate receivers, if required. Again, the overall effectiveness of the wash operation can be changed by the use of counter-current washing of the filter, with two or three wash stages. Such counter-current operation could be effected on rotary vacuum filters, with intermediate re-slurrying of the filter cake with wash water, between filters. Horizontal belt units are economically more attractive in that extra equipment is not needed for re-slurrying, etc. This type of process, shown in Figure 11.22 has received much attention [Wakeman, 1981; Hermia, 1981]. Various options are available, in such arrangements, on the division and placement of the wash water. Rigorous analysis suggests that it is probably always a good option to place all the clean wash water at the final stage (cake exit position) in counter-current operation.

11.3 Batch-Operated, Large-scale Vacuum Filters

Examples of these units include:

a) Large-scale Nutsche filters
b) Horizontal table filters
c) Vertical leaf filters

These units can also usually be operated under pressure-filtration conditions, as required, either by operation of a gas blanket in a), and b), or by use of a pumped feed in c).

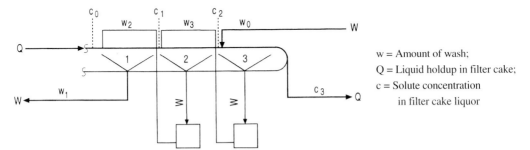

w = Amount of wash;
Q = Liquid holdup in filter cake;
c = Solute concentration
     in filter cake liquor

Figure 11.22 Counter–current washing principles [Tomiak, 1979]

11.3.1 Nutsche Filters

This unit is a scaled-up version of the simple Buchner funnel, used in laboratory separations and tests (Chapter 2), which finds use in quite large, batch separations, particularly where rigorous washing is required [Bosley, 1974]. Modern versions of these machines have been developed to allow for reaction (crystallisation, precipitation, extraction, etc.), filtration (separation of solids, washing, re-slurrying, dewatering) and drying (vacuum and convection) to be conducted in the same vessel. Figure 11.23 shows the various operations available on the Nutrex, Rosenmund machine which is extensively used in pharmaceutical separation. It will be noted that the reactor/filter can be tilted, to facilitate cake discharge. In some separations, the filter medium used is of woven metallic fibre; the latter can be produced to provide surface or depth clarification (Chapter 4) and is stable, even in the most stringent sterilisation conditions. These units are available from 0.32 to 11.0 m^3 vessel volume, providing corresponding filter areas of 0.25 –6.3 m^2 and filter cake volumes 0.08–3.15 m^3.

The agitators are fitted to facilitate slurry agitation, cake smoothing prior to washing and, in the filtered condition, cake removal. Crack elimination is provided via cake detectors which monitor wash flow through the system [Schafer, 1993].

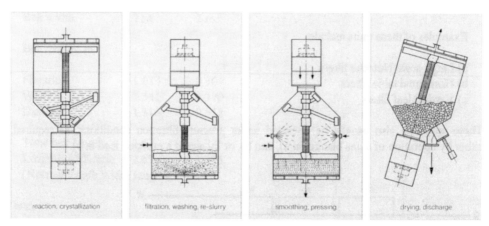

Figure 11.23 Nutrex Nutsche–type filter (Courtesy of Rosenmund AG, Liestal, Switzerland): A) Reaction, crystallization; B) Filtration, washing, re-slurry; C) Smoothing, pressing D) Drying discharge

Horizontal Table Filter (Fig. 11.24). This type of unit is typified by the Triune filter [Nield, 1994] an automated Nutsche-type unit designed to overcome some of the washing difficulties experienced with other systems (plate and frame, rotary vacuum, etc.). The overall features of the unit are similar to the other horizontal vacuum filters discussed earlier, except that during filtration, washing and dewatering, the filter element is stationary. This facilitates optimisation of filter cake thickness, wash times, etc. The totally enclosed system is opened to allow band movement for cake discharge. Process advantages have been claimed for sticky materials, such as antibiotics and filtrations involving solvents, e.g. methanol, isopropanol, dimethylformamide, etc. Again, in view of the enclosure, fume extraction is not necessary here, as it would be with the open Nutsche or belt filter.

Vertical Leaf Vacuum Filters. Batch-operated vacuum units are, of course, a special case of pressure filtration operation. Early large-scale use of simple vacuum leaves, immersed in open top tanks, was practised in sugar beet processing [Purchas, 1981]. Vacuum leaf filters for use in china clay processes have also been reported [Brociner, 1972]. Where washing of the filter cake was required, the whole set of filter leaves were lifted from the slurry tank and immersed in vessels containing fresh water. The ensuing low productivities of such devices prompted the development of automated units; an example of the latter is shown in Figure 11.25, where, in the Universal filter [Gaudfrin, 1978], the filter leaves are contained in a single vessel, with separate compartments for slurry feed and wash water. These units find widespread application in systems containing slow-filtering solids, e.g. first-carbonation juice filtration in beet sugar processing, titanium dioxide separations. It may be noticed that in Figure 11.25, the filter can be operated as a pressure vessel. This trend, with attention to rapid cake release, has produced the developments reported below on pressure leaf filters [Bosley, 1974].

As discussed in Chapter 9, optimisation of the batch operation of vacuum leaf filters has been discussed in the literature [Maracek and Novotny, 1980] in which a mathematical

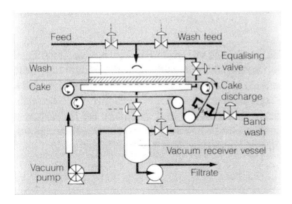

Figure 11.24 Triune horizontal belt filter (Courtesy of Cumberland Process Equipment, Cumbria, U.K.)

model for the functioning of a vacuum leaf filter assembly was developed. This model was verified by comparison with 27 full-scale, industrial experiments. The model included an interesting approach to the problem of including wash time in the optimisation model. Here, washing was described by three stages: (1), constant solute removal rate (plug flow displacement); (2) exponential decrease of solute concentration with time, down to low solute levels; (3) slower exponential removal by diffusional processes.

The exponential stages were described in the manner adopted by Rhodes (Eq. 9.1), i.e. by graphical plots of log (c_s/c_o) versus wash time. The slopes of these graphs produced "washing constants" for the second and third stages of washing. A parametric sensitivity analysis of the static leaf process was reported. It was shown that the principal parameters affecting the final concentration of solids in discharged cakes was the viscosity of the liquid phase — both during filtration and washing. Temperature control of the wash was indicated.

11.4 Pressure Filters

Many suspensions of solids in liquids are difficult to separate by gravitational, centrifugal, or vacuum techniques by reason of slow settling characteristics, poor filterability, high solids content and other factors. In these circumstances, where large separating areas may prove necessary, the use of pressure filtration equipment is indicated.

Slow-filtering slurries, containing finely divided solids tend to require the application of elevated pressures, in order to maintain adequate flow. Stringent washing requirements may lead to the consideration of the batch-operated, pressure filters described in this section.

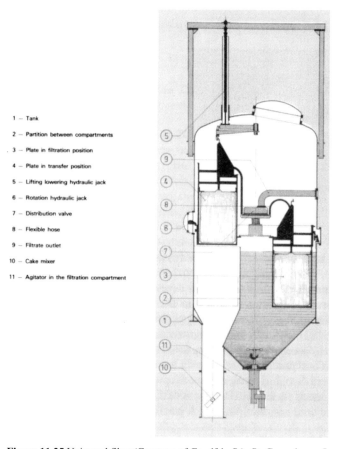

1 — Tank

2 — Partition between compartments

3 — Plate in filtration position

4 — Plate in transfer position

5 — Lifting lowering hydraulic jack

6 — Rotation hydraulic jack

7 — Distribution valve

8 — Flexible hose

9 — Filtrate outlet

10 — Cake mixer

11 — Agitator in the filtration compartment

Figure 11.25 Universal filter (Courtesy of Gaudfrin SA, St. Germaine en Laye, France)

11.4.1 Filter Presses

In the units described below, fluid pressure necessary for filtration is generated by pumping. Pressures are usually below 600 kN/m^2 but in some machines, higher values are used, e.g. up to 140 bar in those machines fitted with facilities for cake squeezing after filtration.

An extremely large variety of presses are available in which three groups may be noted:

Group A: Filter press containing plate and frames (flush plates) or recessed plates.

Group B: Pressure vessel containing tubular or flat filter elements, in vertical or horizontal positions.

Group C: Variable-chamber presses containing means of squeezing the solids after deposition.

In each group, arrangements may be made for filter cake washing. Examples of group A presses are shown in Figures 11.26 and 11.27.

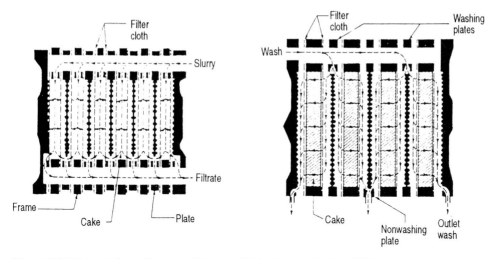

Figure 11.26 Plate and frame filter press (Courtesy of Baker Process, Rugby, U.K.)

Usually an increase in fluid pressure creates an increase in filtrate flow, although strict proportionality is seldom gained since the resistance to flow of most filter cakes increases with pressure. Plugging of the filter medium can also reduce the advantage of high-pressure use, and in order to avoid this phenomenon, plant is usually operated at low pressure in the initial stages of the process. In other applications, the filter may be filled at an initial constant feed rate followed by continued separation at constant pressure. Plant which can be operated in this manner includes a positive displacement pump with incorporated pressure-reducing control.

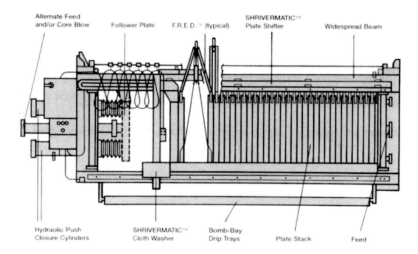

Figure 11.27 Shriver recessed plate filter press: plate shifter and washer (Courtesy of Baker Process, Rugby, U.K.)

The filter press continues to be the most common filter machine in the chemical industry, since it may be used in processing materials of widely varying properties. Because of its versatility, the press has undergone continuous development. Most units are batch operated and after filling with solids, the machine must be disassembled for cake discharge. The time taken in discharging the cake and re-fitting the filter is of great importance in deciding the overall economics of the process; where the cake discharge is conducted manually, high labour costs will be incurred, particularly with larger units.

Modern trends are to increase mechanisation of the press so that full economic advantage may be gained with the unit. For a fully mechanised unit, with one operator, cake discharge rates may be as high as 0.25 m^3 of cake per minute; this approximates to ten times the rate for a similarly sized unmechanised press requiring two operators [Moss,1970].

A fully automated press [Kurita & Suwa, 1978] is depicted in Figure 11.28. As shown, the unit is available in large scale, and is often used in the handling of difficult compressible materials such as sewage sludge. This unit has been developed as a membrane (diaphragm) press [Kurita, 1993] which gives lower cake moisture and higher solids output rate for many materials. The cloth-lifting mechanism, in Figure 11.28, accompanied by vibration and cloth washing serves to facilitate continued high productivity and extend cloth life.

In plate and frame assemblies, square plates are often used, although rectangular shapes are also available. The plates vary in size (250–1450 mm); two plates are brought together, separated by a frame to create an enclosure into which the slurry is pumped. Thus there are two filtering surfaces per frame. The plates are covered by a suitable filter medium; filtrate flows through the medium and out of the filter via channels cut into the surface of the plate. In some systems, consideration is given to providing a robust "underlay" between the filter cloth and plate. This is to ensure a clear passage for the filtrate; such effects can have a serious positive influence on filter productivity.

Figure 11.28 Automatic filter press (Courtesy of Kurita Machinery Mfg. Co., Osaka, Japan)

In those cases where cake washing is required, the assembly is made up of two types of plate; a filtering plate P and a washing plate W. The latter are designated by one-dot and three-dot identifiers in Figure 11.29. The frame F is designated as two-dot. These components are arranged in the sequence P(.), F(:), W(:), F(:), P(.). These designs allow for frame filling, during the filtration stage of the process followed by the "through" washing-of the cake, as shown below. The frame can be supplied in different depths, which provide various cake holding capacities for the solids.

A first step in the selection of a suitable filter of this type is to decide on the area required for filtration. Provision of this area will be decided by the size of plate selected and the corresponding number; the filter area provided in each chamber (approximately twice the area available per plate) will vary from design to design as shown in Table 11.5.

Obviously, it is possible to select various combinations of plate size and number. Generally, it is good practice to reduce the number of plates by using a larger size, given that an alternative exists.

Practical data are available from filter manufacturers which indicates that costs can be minimised by careful consideration of the plate size/number combination. Such data has been used in preparing Table 11.6 which relates to the area required to the "best option" in terms of plate size, etc. for three designs: metal recessed plate, metal plate and frame, and wooden plate and frame.

It will be realised that the above information is approximate, and manufacturers' literature should be consulted for accurate up-to-date information. This comment also applies to

Table 11.5 Approximate plate and frame dimensions

Plate size, m	Area per chamber, m^2		Capacity per chamber, m^3 (2.5 cm frame)	
	Metal	Wood	Metal	Wood
1.06 (42")	2.02	1.39	0.024	0.016
0.91 (36")	1.45	1.00	0.018	0.012
0.76 (30")	0.97	0.71	0.013	0.009
0.61 (24")	0.65	0.39	0.008	0.006
0.46 (18")	0.35	0.18	0.005	0.003
0.30 (12")	0.14	0.09	0.002	0.001

Table 11.6 Filter area and economic plate selection

Area required, m^2	Metal recessed plate		Metal plate and frame		Wooden plate and frame	
	Size, m	Approx. number	Size, m	Approx. number	Size, m	Approx. number
0.93	0.30	6	0.30	6	0.30	10
2.32	0.30	15	0.30	15	0.46	12
4.65	0.30	30	0.30	25	0.61	14
9.29	0.46	25	0.46	25	0.61	20
18.6	0.61	30	0.61	30	0.76	28
27.9	0.76	30	0.76	30	0.81	34
37.2	0.76	40	0.76	40	0.81	45
46.5	0.76	50	0.81	40	0.81	50
55.7	0.76	60	0.91	40	0.81	65
65.0	0.81	60	0.91	46	0.91	65
74.3	0.81	66	0.91	52	1.06	50
83.6	0.81	75	0.91	60	1.06	55
92.9	0.81	84	0.91	65	1.06	64
111.5	0.91	75	0.91	80		
130	1.06	65	1.06	65		
139.4	1.06	70	1.06	70		

Table 11.6 . An important design aspect of such filters is the mechanism used to close the press and maintain a sealed unit during filtration. Large forces are transmitted to the seals; such forces increase with the large units which are usually specified with lower maximum operating processes, e.g. 6 bar at 0.61 m plates and above.

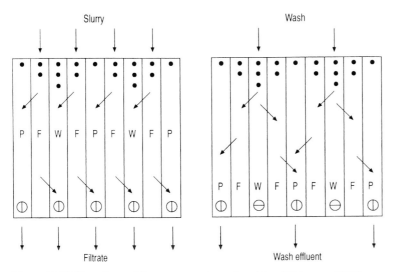

Figure 11.29 Filter plate and frames: separation and washing [Wakeman, 1981]

11.4.1.1 Optimum Filtration Time Cycle

As discussed above, with continuous vacuum filters, production of thin filter cakes leads to high overall productivity. This follows since the discharge time of the cake is a negligible proportion of the overall time cycle. In a batch process involving pressure (or vacuum) filters, it is necessary to allow time for cake discharge in each cycle. The more frequent discharging of thin cakes contrasts with the relatively shorter times spent in the discharge of thicker cakes; dismantling and discharging times are assumed to be independent of cake volume.

In these circumstances, an optimum cake thickness is suggested; the latter will depend on the process times involved in filtration, washing/dewatering and discharge. Analysis for operation at constant pressure involves the following expression for filtration time, derived from Equation 11.5:

$$t_f = \frac{(\overline{\alpha}\mu c)V_f^2}{2A^2\Delta P} + \frac{(\mu R_m)V_f}{A\Delta P} \tag{11.25}$$

If washing proceeds at the same pressure differential as that used in the filtration stage, the washing rate can be estimated from the filtration rate ensuing at the end of the filtration period.

In plate and frame operations, wash water passes through twice the cake thickness produced during filtration. Also, only one-half of the area per frame is involved. It is thus usual to use one- quarter of the final filtration rate as an estimate of the washing rate.

t_w = (volume of wash)/wash rate

Wash rate: $(0.25 \, \Delta P)/(2 \, K_1 \, V_f + K_2)$: from Equation 11.5.

where:

$K_1 = (\overline{\alpha}\mu c)/2A^2$

$K_2 = (\mu \, R_m)/A$

The periods t_f and t_w can thus be expressed in terms of the volume filtered V_f ; productivity per cycle $P*$ may then be estimated from:

$P* = V_f / f_1 \, (V_f) + f_2 \, (V_f) + t_{dw}$

where: $f_1 \, (V_f)$ and $f_2 \, (V_f)$ refer to the filtration and wash periods respectively.

A maximum productivity may be found by differentiating the above expression to find $dP*/dV$. The differential coefficient may then be equated to zero to find the optimum filtered volume per cycle V_f^*. The second differential is negative, at the maximum. Optimum values of V_f^* may then be related to maximum dry solids yield and optimum cake or frame thicknesses. These steps are illustrated in Example 11.2.

Generally, the optimum filtered volume is given by Equation 11.26, where washing is not required:

$$V_f^* = (t_{dw} \Delta P / K_1)^{0.5} \tag{11.26}$$

It follows that, if the medium resistance term is negligible, the optimum time for the filtration stage approaches the down-time, i.e. $t_f^* = t_{dw}$. In normal circumstances, $t_f^* > t_{dw}$, because of the presence of the used filter medium.

For washed cakes, optimum values of V^*, dry solids W^* and cake thickness l^* can also be related to other process variables by the equations involving the number of void volumes of wash n:

$$V* = [A^2 \Delta P t_{dw} / \overline{\alpha} c \, (0.5 + K_w)]^{0.5} \tag{11.27}$$

$$W* = [A^2 \Delta P t_{dw} c / \overline{\alpha}\mu \, (0.5 + K_w)]^{0.5} \tag{11.28}$$

$$l* = [4\Delta P t_{dw} c / \overline{\alpha}\mu \, (1-\varepsilon)^2 \rho_s^2 \, (0.5 + K_w)]^{0.5} \tag{11.29}$$

where $K_w = [(4\ (m-1))\ n\ c]/\rho$. The derivations of these expressions are presented in Example 11.4. The washing factor K_w is related to the filter cake wetness factor m and the liquid density ρ. For simple filtration, where no washing is required, $K_w = 0$.

These equations have been used [Rushton and Daneshpoor, 1983] to demonstrate the effect of washing requirements on filtration process productivities, e.g., as illustrated in Figure 11.30, which records the theoretical reductions in productivity with increased downtime and/or number of wash void volumes.

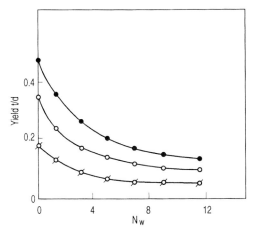

Figure 11.30 Theoretical effect of wash time on filter productivity
$t_d = 450s$ (full circles); 900s (open circles); 3600s (circles with checkmark); $\rho_s = 2400$ kg/m³ $\alpha = 10^{12}$ m/kg; $\Delta P = 100$ kN/m²; C = 330 kg/m³: $\varepsilon = 0.42$; $n = 0.4$

Example 11.3

Calculate the maximum plant capacities and most economic frame thickness for the filter cake and press specified below. Assume a wash pressure equal to the filtration pressure and a down-time of 1800 s. Take the volume of wash to be 0.15 times the volume of filtrate.

Data

Filter area	A	$= 37.2$ m²
Specific resistance	$\bar{\alpha}$	$= 5\times10^{10}$ m/kg
Viscosity	μ	$= 1\times10^{-3}$ N s m⁻²
Pressure differential	ΔP	$= 100$ kN/m²
Medium resistance	R_m	$= 4\times10^{11}$ m⁻¹
Cake moisture	m	$= 1.4$
Mass fraction of solids in slurry feed	s	$= 0.2307$
Solids density	ρ_s	$= 2400$ kg/m³
Liquid density	ρ	$= 1000$ kg/m³

Calculate dry cake mass per unit volume filtrate:

$$c = \frac{\rho s}{(1 - ms)} = \frac{230.7}{1 - 1.4 \times 0.2307} = \frac{230.7}{0.677} = 340.8 \ kg \ / \ m^3$$

Calculate filtration parameters K_1 and K_2:

$$K_1 = \frac{\bar{\alpha}\mu c}{2A^2} = \frac{5 \times 10^{10} \times 10^{-3} \times 340.8}{2 \times 37.2^2} = 6.16 \times 10^6$$

$$K_2 = \frac{\mu R_m}{A} = \frac{10^{-3} \times 4 \times 10^{11}}{37.2} = 1.08 \times 10^7$$

Calculate filtration times:

$$t_f = \frac{\bar{\alpha}\mu c}{2A^2 \Delta P} V_f^2 + \frac{\mu R_m V_f}{A \Delta P}$$

$$= \frac{6.16 \times 10^6}{10^5} V_f^2 + \frac{1.0^8 \times 10^7 V_f}{10^5}$$

Hence:

$$t_f = 61.6 \ V_f^2 + 108 V_f$$

Calculate wash time:

$$t_w = \frac{\text{Volume of wash}}{\text{Wash rate}}$$

Volume of wash: $0.15 \ V_f$

$$\text{Rate of wash} = \frac{0.25 \Delta P}{2 K_1 \ V_f + K_2}$$

Hence:

$$\text{Wash rate} = \frac{0.25 \times 10^5}{2 \times 6.16 \times 10^6 \ V_f + 1.08 \times 10^7}$$

$$t_w = \frac{0.15 \ V_f \ [12.32 \times 10^6 \ V_f + 1.08 \times 10^7]}{0.25 \times 10^5}$$

$$t_w = 73.92 \ V_f^2 \ + 64.8 \ V_f$$

Downtime $t_{dw} = 1800$ s

Total cycle time $(t_f + t_w + t_{dw}) = 135.52 \ V_f^2 + 172.8 \ V_f + 1800$

Production rate $P_r = \dfrac{Total \ filtrate \ collected \ per \ cycle}{Total \ cycle \ time}$

$$P_r = \frac{V_f}{135.52 \ V_f^2 + 172.8 V_f + 1800}$$

Calculate optimum production rate:

Optimum production rate is found by differentiation of this (P_r, V_f) relationship and setting $(dP_r / dV_f) = 0$

$$P_r = V_f / (132.52 \ V_f^2 + 172.8 \ V_f + 1800)$$

Hence:

$$(dP_r / dV) = \frac{1}{(135.52 V_f^2 + 172.8 \ V_f + 1800)} - \frac{V_f \ (2 \times 135.52 \ V_f + 172.8)}{(135.52 \ V_f^2 + 172.8 \ V_f + 1800)^2}$$

Set $(dP_r / dV_f) = 0$

Hence:

$$(135.52 V_f^2 + 172.8 V_f + 1800) - V_f \ (271.04 V_f + 172.8) = 0$$

i.e. $1800 - 135.52 \ V_f^2 = 0$

or

$$V_f^* = (1800/135.52)^{0.5} = 3.64 \ \text{m}^3$$

Cycle time:

$$135.52 V_f^{*2} + 172.8 \ V_f^* + 1800 = 42255 = 1.173 \ \text{h}$$

Calculate average porosity from moisture and densities:

$$\bar{\varepsilon} = \rho_s (m - 1) / (\rho + \rho_s (m - 1))$$
$$= 0.4898$$

Calculate optimum cake and frame thickness:

The optimum cake thickness is given by:

$l^* = W^*/A\,(1-\varepsilon)\,\rho_s$. Thus $l^* = (340.8\times3.64)/(37.2\times0.5102\times2400) = 0.0272$ m.

In the press, filtration occurs on two surfaces per frame. It follows that the optimum frame thickness is 0.0545 m. Information is required on the capacity per chamber in order to estimate the number of frames.

Cycle Time and Productivity

Cycle time = 1.173 h; Cycle/day = 20.45
Productivity = $(c\,V_f^*)\times$(cycles/day) = 25.37 t dry solids/day

The example allows for only a limited amount of wash water:

Mass of dry cake per batch:	$c\,V_f^*$	= 1240.51 kg
Mass of wet cake:	$m\,c\,V_f^*$	= 1736.72 kg
Mass of liquid in cake:	$(m-1)\,c\,V_f^*$	= 496.20 kg
Volume liquid in cake:	$(m-1)\,c\,V_f^*/\rho$	= 0.4962 m^3
Wash volume (V_w):	$0.15\,V_f^*$	= 0.546 m^3
Number of void volumes wash:	n	= 1.1

In flood wash conditions, theory suggests that a high value of the dispersion number (vL/D) would be necessary for effective removal of solute, at the above value of n.

This optimisation calculation can be solved graphically by plotting the filtrate volume time data with the origin displaced on the time axis to the point required for cake discharge. A tangent from zero to the filtrate curve then identifies the optimum point, as shown below in Figure 11.31. This method has the advantage that the filtration data need not conform to the parabolic expression in Equation (11.25).

In the plate and frame press calculation, Example 11.3, particle movement inside the frame was assumed to take place on two, opposite horizontal paths. Hence, twice the area of the filter plate is available in each frame for cake deposition. This assumption is discussed below.

11.4.1.2 Practical Deviations from Theoretical Predictions

The model of horizontal particle movement was checked experimentally [Rushton and Metcalfe, 1973] by injecting quantities of a carbon-black dispersion into the feed to a filter processing slurries of: calcium carbonate, zinc oxide and magnesium carbonate suspensions. In all separations, after a short, initial even deposition, nonlinearities began to appear. Particle sedimentation into the base of the frame resulted in a gradual reduction in the available area for filtration.

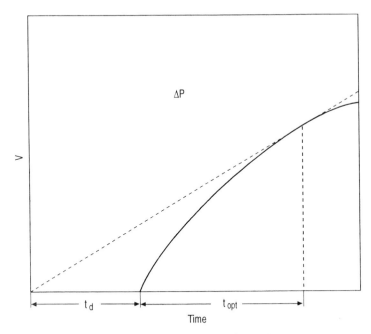

Figure 11.31 Optimisation of filter cycle: graphical analysis

The calcium carbonate cakes produced the nearest approach to the theoretical assumption. In other cases, particularly zinc oxide, filter cake bridging across the frame was observed towards the end of the separation. Further tests on the filter cakes produced in this study showed that the moisture content varied from place to place within the frame. These effects were probably related to the small area pilot filter (0.4 m^2) used in the study; industrial-scale equipment which may be operated at higher pressures (up to 15 bar) generally produce better control of moisture content.

Filter cake handling difficulties associated with thixotropic wet solids may be ameliorated by use of increased filtration pressures [Bosley et al, 1986]. Thus a sewage sludge cake produced at 15 bar may be handled more easily than one produced at lower pressure, say 7 bar. The associated cheaper cake handling costs must be compared with the higher capital investment involved with the high-pressure plant. The question of dewatering at higher filtration pressure is taken up below in the section dealing with variable chamber filters.

Maldistribution of flow in these units can also produce poor washing performance; this may lead to the necessity of operating with thin cakes, despite the theoretical conclusion on optimum cake thickness, large dispersion parameters, etc. Alternative designs (leaf filters) are available which are aimed at improving productivity, cake washing and release, etc. whilst retaining the large surface areas available in filter presses. Reduced downtime in these alternatives provides an opportunity of working with thin filter cakes.

Example 11.4

Calculate the effect of specific cake resistance, in the range $10^{10} > \bar{\alpha} > 10^{13}\,\text{m}/\text{kg}$, on the area of plate and frame press required to produce 100 m^3/d of water from the slurry described below:

Assume a plant downtime $t_{dw} = 1800$ s for cake discharge and re-charging. The filter is to be operated under essentially constant pressure differential of 1.5 bar. The filter cake is to be washed with 2.5 void volumes of wash water, delivered at the same pressure differential used in the filtration phase.

Data

Resistance of filter medium	R_m	$= 1 \times 10^{11}\,\text{m}^{-1}$
Concentration of solids	c	$= 200\,\text{kg/m}^3$ filtrate
Cake (wet/dry) mass ratio	m	$= 2.0$
Liquid viscosity	μ	$= 0.001$ Pa.s
Liquid density	ρ	$= 1000\,\text{kg m}^{-3}$
Solids density	ρ_s	$= 2800\,\text{kg m}^{-3}$

This problem can be solved using Equation 11.27 which is derived as follows:

Mass of dry cake: $= c\,V_f$
Mass of wet cake: $= m\,c\,V_f$
Mass of liquid in cake: $= (m-1)\,c\,V_f$

Hence:

Volume liquid in cake: $= [(m-1)\,c\,V_f]/\rho$
Volume of wash $= [n\,(m-1)\,c\,V_f]/\rho$

Filtration time: t_f

$$t_f = \left(\frac{\bar{\alpha}\mu c}{2\,A^2\,\Delta P} \right) V_f^2 + \left(\frac{\mu R_m}{A\Delta P} \right) V_f$$

Wash time: t_w

$$t_w = (\text{Volume of wash/Wash rate})$$

$$\text{Wash rate} = 0.25\,\Delta P\,/(2\,K_1\,V_f + K_2)$$

where:

$$K_1 = \bar{\alpha}\,\mu c/2A^2$$
$$K_2 = \mu R_m/A$$

Hence:

$$t_w = (4\,n\,(m-1)\,cV_f\,(2\,K_1 V_f\,+\,K_2))\,/\,\rho\Delta P$$

The total cycle time $t_c = t_f + t_w + t_{dw}$ can be used to estimate the productivity $P_r = V_f/t_c$, where V_f is the filtrate volume produced per cycle.

These expressions may be rearranged to give:

$$t_c = a\,V_f^2 + b\,V_f + t_{dw}$$

where:

$$a = (K_1/\Delta P)\,\{1\,+\,[8\,n\,(m-\,1]\,c)\,/\,\rho\}$$

$$b = (K_2/\Delta P)\,\{1\,+\,[4n\,(m-\,1)c]\,c\,/\,\rho\}$$

Optimum values of V_f can be found from the differentiation of the productivity expression, to give : (dP_r/dV_f), which is then equated to zero.

Thus:

$$P_r = (V_f\,/\,t_c) = V_f\,/\,(a\,V_f^2 + b\,V_f + t_{dw})$$

$$(dP_r\,/\,dV_f) = \frac{1}{(a V_f^2\,+\,bV_f\,+t_{dw})} - \frac{V_f\,(2\,a\,V_f\,+\,b)}{(a\,V_f^2\,+\,bV_f\,+\,t_{dw})^2}$$

For an optimum:

$$dP_r/dV = 0$$

Hence:

$$(a\,V_f^2\,+\,bV_f\,+\,t_{dw}) = 2\,aV_f^2\,+\,bV_f$$

Thus:

$$t_{dw} = aV_f^2$$

$$V_f^* = (t_{dw}\,/\,a)^{0.5}$$

Substituting for a and K_1 gives:

$$V_f^* = \left(\frac{2\,t_{dw}\,A^2\,\Delta P}{\overline{\alpha}\mu c\,(1\,+\,8\,n\,(m-\,1)\,c)\,/\,\rho)}\right)^{0.5}$$

These expressions may be used to relate the optimum cycle time t_c^* to other variables:

$$t_c^* = 2 \, t_{dw} + R_m \left[\frac{2 t_{dw} \, \mu \left[1 + (K_w)\right]^2}{\bar{\alpha} c \, \Delta P \left[1 + 2 \, K_w\right]} \right]^{0.5} \tag{11.30}$$

where $K_w = (4 \, n \, (m-1) \, c)/\rho$

Equation 11.30 shows that in circumstances where R_m is negligible, t_c^* is twice the downtime t_{dw}, or

$t_{dw} = (t_f + t_w)$.

Substitution of the design data yields:

$$t_c^* = 3600 + (4.647 \times 10^7) / \sqrt{\bar{\alpha}}$$

This information can be used to calculate the optimum cycle time and filtration cycles per day for various assumed levels of the average specific resistance $\bar{\alpha}$. The corresponding V_f^* per cycle and the filter area requirements are listed in the table below.

The frame thickness l is estimated from:

$$l = (2 \, c \, V) / (1 - \bar{\varepsilon}) \rho_s \, A$$

in which the average porosity is given by:

$$\bar{\varepsilon} = \rho_s \, (m-1) / [\rho + \rho_s \, (m-1)]$$

The optimum cycle time relationship in Equation 11.30 takes the form:

$$t_c^* = 2 \, t_{dw} + R \, C_l \, (t_{dw} / \bar{\alpha})^{0.5} \tag{11.31}$$

in which:

$$C_l = [2 \, \mu \, (1 + K_w^2) / c \, \Delta P \, (1 + 2 \, K_w)]^{0.5}$$

This may be used to produce analytical expressions for the corresponding optimum filter area and frame thickness:

$$A^* = V^* \left[2 \left(\frac{\mu c (1 + 2 \, K_w) t_{dw} \, \alpha}{2 \Delta P} \right)^{0.5} + \frac{\mu}{\Delta P} \, (1 + K_w) \right] \tag{11.32}$$

where V^* is the overall productivity in m^3/s.

$$L_f^* = \frac{2\sqrt{2}}{(1-\varepsilon)\rho_s}\left[\frac{\Delta P\, c\, t_{dw}}{\mu\,(1\,+2\,\,K_w)\overline{\alpha}}\right]^{0.5} \qquad (11.33)$$

Filter Dimensions *vs* $\overline{\alpha}$ for 100 m^3 filtrate per day

Specific resistance α, m/kg	Optimum batch volume, m^3	Cycles per day	Filter area, m^2	Frame thickness, m
10^{10}	4.70	21.26	20.22	0.13
10^{11}	4.34	23.06	59.06	0.04
10^{12}	4.22	23.70	181.60	0.013
10^{13}	4.18	23.90	569.00	0.004

These results indicate that as the filterability of the material reduces, so does the optimum cycle time. Practical experience [Bosley, 1986] suggests that difficult-to-filter materials, e.g. alum sludge or high protein containing materials require thin cake filtration conditions, with cake thicknesses 1.5–2.0 cm. Readily filtered materials are generally produced optimally as thicker cakes, up to 7.0 cm in thickness. This experience points to the usefulness of the models discussed here, in process calculations and plant specifications.

Another feature of analysis, is the reduction in optimal batch volume and corresponding cake thickness with increase in the specific resistance. In multiproduct processes, or where large changes in the filterability of batches may be expected, optimal operation using a filter of fixed dimensions is impossible. This situation is ameliorated by use of variable-chamber presses, discussed below; in these units, the filtration stage can be stopped at any time, followed by a squeezing action on the cake, as discussed in Section 11.4.2.

11.4.1.3 Filtration of Highly Compressible Materials

The separation of highly compressible "solids" from dilute suspensions constitutes an important technological application for filter presses. As an example, the latter find widespread use in sludge filtration and dewatering in waste water treatment processes.

Filter presses are effective in producing high solids concentrations (> 30%) in the filter cake formed from dilute (2–3%) suspensions. As a technique, pressing competes mainly with solid-bowl centrifuges, and, to a lesser extent, continuous vacuum filter. Within the filter press area, extensions include variable-chamber presses and belt presses; both these systems are discussed below.

In all these cases of sludge separation, filtration and dewatering is preceded by chemical treatment (coagulants and flocculants) to improve the filtration characteristics of the suspension. Physical treatments, e.g. thermal conditioning (200^0C for 30–90 min) are also included in some cases. Freezing and melting, although not an economic option, is known to speed up subsequent filtration stages.

The separation of sludges has been described in terms of three phases [Hoyland & Day, 1983].

Filling Phase. Here the sludge is pumped at low pressure into the filter chambers. The filling time is related to the total volume of the press and the pump capacity.

Cake Growth Phase. Here the experimental (*V*, *t*) curve exhibits a parabolic shape, as expected from theory. The specific resistance–pressure relationship for newly conditioned sludge takes a value of unity, for the compressibility coefficient. This suggests that $\bar{\alpha}$ is directly proportional to pressure. If this phase is conducted at constant pressure, elementary theory leads to the following relationship for filtration volume:

$$V^2 = 2\Delta P\,(t - t_o)/\bar{\alpha}\mu c$$

in which the cloth resistance has been neglected; t_o is the filling time.

Exponential Compression Phase. Here further increases in the solids concentration inside the filter chambers are attained by compression of the contents. The rate of filtration now deviates from classical filtration theory with the (*V*, *t*) curve taking on an exponential shape, rather than a parabolic one.

The transition point from growth to compression depends on the compressibility and mobility of the cake. Much of the process time (70–80%) can be taken up in exponential pressing; Figure 11.32 shows the extended compression phases obtained with cakes that are resistant to movement in the frame.

11.4.2 Variable Chamber Filters

In those cases where variable filterability between batches of the same product occur, or where multiproduct operation on the same filter results in a wide range in filtration times, optimised operation is impossible with a fixed frame depth. In these circumstances, the principal variable is the volume of filter cake per batch. Where the wet volume of the cake changes from batch to batch, in plant using a fixed-volume press, the latter can be oversized for many products, and blanking or dummy plates are required. By use of the dummy, the multichamber press can be reduced to any required smaller number. Care must be taken to identify clearly the blank plate; confusion in fitting the latter may result in loss of batch by incorrect sealing of the unit.

Although it is known that a drier, more easily handled cake will be obtained by the use of higher pumping pressures, these are not always available, and cake solids contents can be quite low. Quite often filtration is followed by drying and the economics of that operation can be affected considerably by prior production of dry, handleable cakes. Some

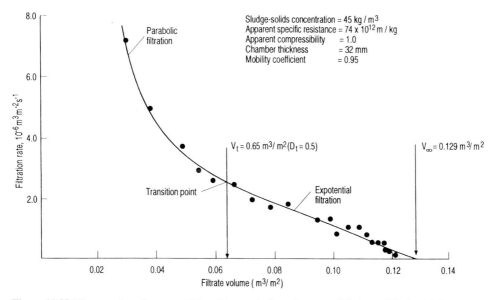

Figure 11.32 Filter pressing of compressible solids: parabolic and exponential phases [Hoyland & Day, 1982]

of these problems may be resolved by the use of variable-chamber presses; as shown in the process flowsheet in Figure 11.33.

As may be noted in the diagram, variable-chamber presses contain inflatable diaphragms which are expanded at the end of the normal filling phase. The diaphragm is inflated by hydraulic fluid or compressed air. Liquid is expressed out of the chamber, leaving behind a drier, harder filter cake. Several filters of this type are available.

It is often found that washing of the cake is improved by compression, possibly because at lower porosity the probability of channelling is reduced. A principal factor which has made possible the success of the variable-chamber press is the simultaneous development of closely woven cloth, of polyester and polypropylene which, whilst retaining the fine particles, does not tear under the imposed pressures and, more importantly, does not blind or present cake release problems.

Perhaps the most important feature of the unit is that filtration may be terminated at any time and, by compression, a dry cake can still be produced. Most machines are automatic so filtration time and compression time can be varied quite easily.

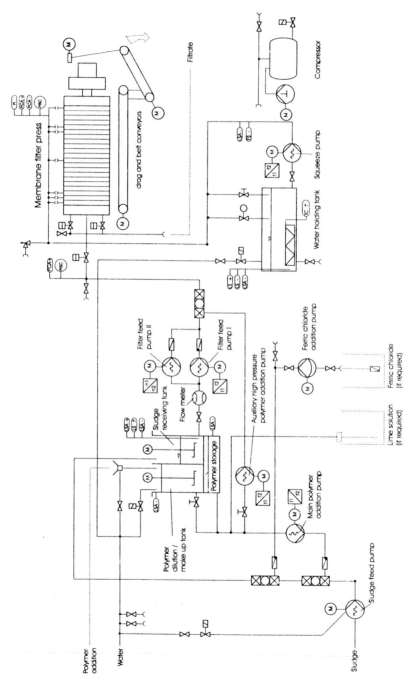

Figure 11.33 Variable chamber press flowsheet (Courtesy of MSE GmbH, Nöttingen, Germany)

11.4.2.1 Filter Cake Compression

Changes in the solid content of a filter cake with the application of compression may be calculated from the relationship:

$$(1-\varepsilon) = (1-\varepsilon_0)\ P_s^{\beta} \tag{11.34}$$

where P_s (kPa) is the compressive pressure and $(1-\varepsilon)$ represents the fractional volume of solids in the filter cake.

Values of $(1-\varepsilon_0)$ and β for a limited number of materials are given in Table 11.7. These values were measured in a compression cell (Section 2.9.2, Chapter 2). Figure 11.34 relates to a typical filtration–compression cycle, compared with a conventional separation.

Table 11.7 Compression characteristics

Material	Pressure range, k Pa	$(1-\varepsilon_0)$	β
Calcium carbonate	7–550	0.209	0.06
Kaolin	7–700	0.374	0.06
Titanium dioxide	7–700	0.203	0.13
Clay	10–100	0.177	0.10

Information on compressibility constants for other materials is available in the literature [Shirato et al, 1987]. It is important to check the pressure units used in calculating the $(1-\varepsilon_0)$ and β "constants" and to realise that this type of information will vary, for any particular substance, with changes in particulate properties. Also, the absolute values recorded will depend on the equipment used to measure compression effects. Thus information from compression cells may differ from data collecting using pilot-scale variable-chamber pressure.

Figure 11.35 shows the effect of compression on materials where $(1-\varepsilon_0)$ remains constant and the exponent β takes values of 0.05, 0.1 and 0.2. The results refer to "equilibrium" porosity conditions after prolonged squeezing.

11.4.2.2 Compression Kinetics

When the variable-chamber concept was first introduced as an alternative, or development, of the conventional press, controversy persisted on the relative merits of the two units [Cherry, 1978]. The essential problem related to the possible economic gain in adopting diaphragm plates. This, in turn, depends on the rate at which a filter cake solidity

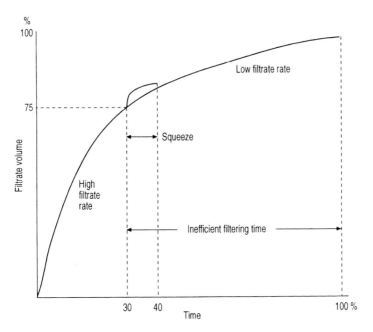

Figure 11.34 Filtrate volume–time curves for conventional and diaphragm filters [Moon, 1981]

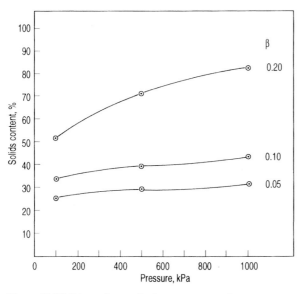

Figure 11.35 Cake moisture–diaphragm pressure plot

can be attained by either a prolonged, "exponential" filtration phase, referred to above; or administration of an expression effect, via the diaphragm, to expel liquor from the wet cake.

As discussed in the literature [Shirato et al, 1987] dehydration by expression follows consolidation of the filter cake in different hydraulic conditions from those obtained during filtration. When the solids concentration in the chamber reaches a certain level, the particles in contact are stressed by the applied diaphragm pressure, which is in excess of the hydraulic pressure.

The changes in filter cake volume, concomitant with the application of diaphragm pressure have been studied extensively [Shirato, 1981; Yim & Ben Aim, 1986].

Figure 11.36 depicts a typical dewatering by compression of a filter cake of original thickness L_1. The changes in cake thickness L as a function of compression time θ_c have been described by equations of the type below for U_c the average consolidation rate.

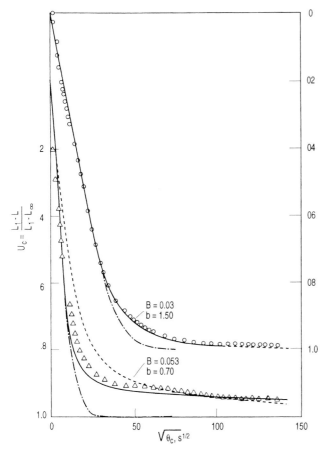

Figure 11.36 Expression kinetics for semisolids [Shirato,1981]
Circles, Korean kaolin; triangles, 50 wt% Mitsukuri-Gairome clay–50 wt% standard Supercel Mixture

$$U_c = \frac{(L_1 - L)}{(L_1 - L_\infty)} = \left\{ \frac{4}{\pi} \frac{(i^2 \, C_e \, \theta_c)}{w_o^2} \left[1 + \left(\frac{4}{\pi} \frac{i \, C_e \theta_c}{w_o^2} \right)^v \right]^{1/v} \right\}^{0.5} \qquad (11.35)$$

In which:

L_∞ is the cake thickness at infinite squeeze time, m
i is the number of drainage surfaces in the filter
C_e is a "consolidation coefficient", m²/s.
w_o is the total solids volume in the cake per unit area, m³/m²
v is a "consolidation behaviour index".

The complexity of such equations follows the attempts to describe the primary consolidation where the total voidage ε dependent on local compressive pressure *and* a secondary consolidation, the so-called creep effect.

These models are based on original work in soil mechanics [Terzaghi & Peck, 1948]. Attempts to use Equation 11.35 to describe practical expressions point to values of $v <$ 2.85. Practical applications to primary expressions may be followed by use of the simple relationships:

$$U_c = 1 - B \exp(-\eta\, \theta_c)$$

Where B and η are "creep" constants as reported in Table 11.8.

Table 11.8 Typical expression constants

Material	Creep constants	
	B	η, s^{-1}
Kaolin	0.03	5.16×10^{-5}
Solka floc	0.075	3.47×10^{-5}
Clay/Supercel mix	0.053	3.30×10^{-5}

Experimental curves, Figures 11.36 and 11.37 for the expression of filter cakes and homogenous semisolid material (concentrated deposit) exhibit differences.

The linear $U_c - \sqrt{\theta_c}$ relationship, at low values of θ_c obtained with concentrates, contrasts with the inverted S-shaped curves produced for wet filter cakes. These differences, which disappear at high θ_c values, are attributed to variations in ε which persist during the filtration of compressible materials. In the latter, layers near the filter medium are compressed to a higher degree than the "soupy" outer layers. Thus expression rates for filter cakes will be initially less than for the semisolid material.

Consideration has been given to the optimisation of the diaphragm plate cycle [Shirato, 1981]. This involves a graphical procedure, as described in Figure 11.31.

Fundamental studies of the expression process [Wakeman & Tarleton, 1989] have shown that the rate of deliquoring depends on the conditions at the surface of the particles which promote mutual repulsion of the latter. Surface forces can be altered by changes in

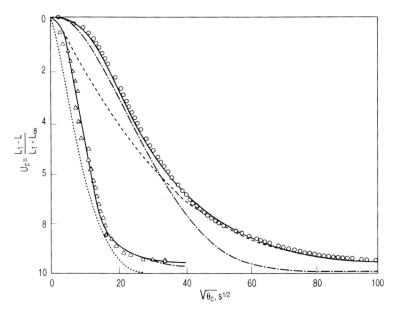

Figure 11.37 Expression kinetics for wet filter cakes [Shirato, 1981]
Circles, Japanese Mirin mash; triangles, 50 wt% Hara - Gairome clay–50 wt% Solka-Floc mix

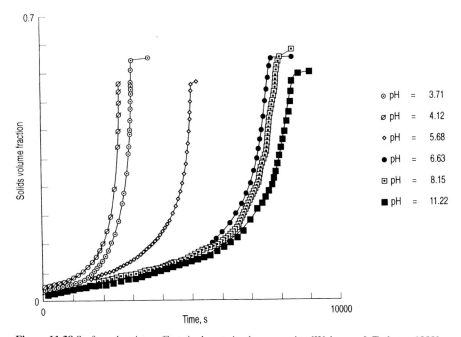

Figure 11.38 Surface chemistry effects in dewatering by expression [Wakeman & Tarleton, 1989]

the pH of the surrounding fluid. Figure 11.38 shows the effect of such changes in the dewatering of 0.5 μm titanium dioxide crystals. In these studies, which included calcite (calcium carbonate), kaolinite and hydromagnesite (magnesium carbonate), expression pressures up to 45 bar were used with varying results. Pressure level had a major effect in some systems, but minimal effect in others. Again, whilst higher pressure accelerated the formation of pressed cake, above a certain pressure, little advantage was gained, neither in growth rate nor in moisture reduction. These differences in behaviour were attributed to the range over which repulsive surface forces extended.

Another contribution [Stahl & Pfuff, 1993] suggests that increased dewatering potential can be realised by the combined action of mechanical pressure and gas displacement, particularly at low process pressure.

11.4.2.3 Variable-Chamber Filter Types

Various types of filter are available which are designed for cake formation and squeezing. The broadest classification is between the batch and continuous units. Batch variable-chamber filters are all characterised by having a diaphragm fitted inside the filter frame; at a predetermined point in the separation, the feed to the system is interrupted and the diaphragm is inflated. A typical filtrate volume–time curve is depicted in Figure 11.34. Typical examples of this type of unit are:

Diaphragm Plate Filter Presses. These are recessed plate-type units; the diaphragm plate shown in Figure 11.39 consists of a solid body plate onto which the flexible diaphragms are heat laminated. Thus each diaphragm section acts as a single element in assembly. The surface of the diaphragm is provided with drainage pips which lend support to the filter medium and facilitate movement of filtration and washings during processing. The action of the diaphragm plate in filtration and squeezing periods is shown in Figure 11.33.

Early claims [Heaton, 1978] that installation of these plates could more than double production, in certain cases, were supported by plant data of the type shown in Table 11.9.

Table 11.9 Productivity of conventional and diaphragm presses

Plate data	Conventional press	Diaphragm press
Size	1.3×1.3 m	1.3×1.3 m
No. of chambers	70	66
Chamber thickness	3 cm	3 cm
Press type	Hydraulic overhead	Hydraulic overhead
Filter cloth	Saran 092	Saran 092

Table 11.9 Continued

Sludge data

Type of sludge	Mixed primary	Humus
Dry solids	8.8%	8%
Conditioning	15% Copperas	28 wt%

Cyle time

Filtration at 6 bar	270 min	30 min
Filtration at 8 bar		15 min
Discharge time	30 min	5 min

Total	300 min	60 min
Filtration data		

Cake thickness	3 cm	2 cm
Wet cake weight	2895 kg	1703 kg
Dry solids content	33%	45%
Dry solids weight	956	762

Output rate data
(including discharge time)

Dry solids/h/press	193	762
Slurry processed		
m^3/h/press	174	6.86
Output ratio	1.0	3.94

Capacity comparisons

Pressing to handle	2 presses	1 press
170 t of slurry per week	2 pressing/day	5 pressings/day
	16 h overtime/man	no overtime

* The dryness of the cake after squeezing may be unnecessarily high; the squeezing time could be
  reduced here, with an increase in productivity

These advantages are offset, of course, by machine capital costs. In the field of waste treatment (sludge handling) particularly, a competitive market continues to lead to new designs aimed at cost reduction. This is typified by the so-called plateless filter membrane press, Figure 11.40 in which the relatively thick plates of conventional presses are replaced by thin frames, [Kurita et al, 1993].

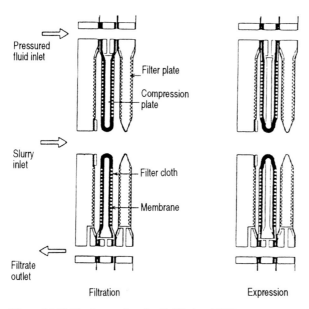

Figure 11.39 Membrane plate details [Kurita, 1993]

Diaphragm plate filters of this type are available in areas up to 1800 m^2, with operating pressures up to 60 bar. Automation of these machines has produced significant differences in the economics attached to diaphragm plate usage. This is particularly true in the problem of filter medium washing; during the filter cycle, particles are pressed into the surface of the cloth. An efficient automatic washing mechanism is a prerequisite to a successful application. [Kurita, 1978]. Modified versions of this type of filter continued to find new applications, e.g. in the variable-chamber mesh filter developed for brewery applications [Hermia & Rahier, 1990]. Improved methods involving automatic cloth lifting and shaking have been reported by Mayer 1995.

Attention to automation [Keinänen et al, 1994] and cloth cleaning is incorporated into the design of horizontal diaphragm plate, presented in Figure 11.41. Here, the endless, moving cloth is treated to spray washing, as part of the overall cycle. The cloth movement also promotes discharge of the filtered, washed and dewatered cake. Figure 11.41 depicts such a discharge of a copper residue filter cake. Typical claimed productivities of this unit are presented in Table 11.10 . The use of the word "membrane" to describe diaphragm filters can cause confusion with ME, UF and RO membrane filters.

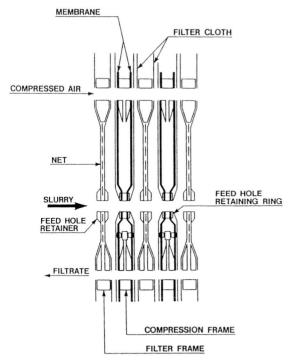

Figure 11.40 "Plateless" filter press [Kurita, 1993]

Table 11.10 Performance data: Larox filter

Material	Capacity, kg dry solid m^{-2} h^{-1}	Cycle time, min	Cake moisture, wt%
Aluminium hydroxide	20	23	60
Clay	140	16	27
Copper Concentrate	430	8	8.2
Iron Oxide	50	20	30
Gypsum	250	9	14
Magnesium Hydroxide	294	40	26
Protein Ferment Mycelium	28.5	14	62
Red Mud	69	13	31

Cylindrical Variable-Chamber Filters. Cylindrical filtering surfaces are inherent in the tube press, Figure 11.42, and the VC filter, Figure 11.43. In the latter, the filter is mounted on a horizontal axis. The perforated filter tube is covered by a suitable filter medium, and this assembly is inserted into a larger diameter cylinder. The inside surface of the latter is covered with an inflatable diaphragm; filtration ensues in the annulus between the two cylinders. After deposition of a predetermined cake thickness, the latter can be partially squeezed and wash water directed radially through the filter cake [Bailey, 1979].

Figure 11.41 Horizontal belt pressure filtration and discharge (Courtesy of Larox Oy, Lappenranta, Finland)

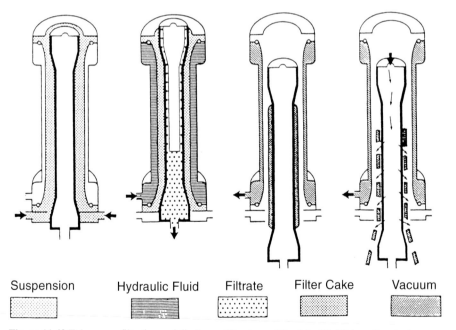

Figure 11.42 Tube press: filtration and discharge (Courtesy of Svedala AB, Malmo, Sweden)

Figure 11.43 VC filter (Courtesy of VC Filters International, Huddersfield U.K.)

This radial process produces extremely high washing efficiencies. The latter, coupled with the enhanced handling characteristics of the 15 bar pressed cake, can produce up to 20-fold increases in productivity for some materials (e.g. pigments) when compared with normal plate and frame separations. These filters are available in areas up to 6 m^2.

The tube press dewaters cakes at pressures up to 140 bar producing easily handled, dry filter cakes. In its first applications in the china clay industry, [Gwilliam, 1973], considerable savings were made in separation and drying costs over established processes involving ordinary pressure filters. This point is stressed again in later reports [Brown, 1979] in extensions to the filtration of sugar muds, magnesia from sea-water, steel industry effluents, etc. In the case of magnesia, filter cakes with moisture contents of 25 and 20 wt% were produced at pressure levels of 100 and 140 bar respectively. These figures compared with moisture levels of 40 wt% from conventional plate and frame presses. Limitations of this unit with respect to fibrous materials, plastic cakes and low particulate concentrations have also been reported in the latter publication.

Applications of the tube press continue to extend, e.g. to gold, vegetable oil and coal suspensions [Johns, 1991]. High-pressure filtration costs versus evaporation costs have been studied using a tube press module in the filtration of baker's ycast, colour pigments, harbour sediments and *E.coli* bacteria dispersions [Hess, 1990]. In the case of baker's yeast, in the pressure range 25–150 bar, minimum filtration costs were found in the concentration range 23–26% dry solids. Filtration costs were between 36 and 44% of the evaporative costs. These gains will increase as fuel costs rise.

Further studies on the economic optimisation of a variable chamber press have been reported. The overall costs of filtering and thermal drying to an equilibrium moisture content were contrasted with the comparable costs of filtering on a RVF followed by drying. The results for two materials are shown in Figure 11.44. The paper mill effluent was tested at two squeeze pressures on a pilot scale VC filter and the results used in an economic model for a full scale unit [Ward et al,1996]. For comparison, the data for the baker's yeast was obtained on a 46 cm^2 piston press and then used in a similar model for an ECC press [Hess, 1990]. The existence of a minimum in the relative cost is evident for all three cases. In the figure this a shown at an optimum final cake solids concentration which corresponds to an optimum squeeze time.

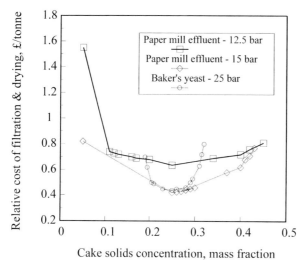

Figure 11.44 Optimum cake concentration in variable chamber filter types [Ward, 1996]

It may be concluded that high pressure-filtration with diaphragms can produce real savings, particularly in processes involving subsequent drying stages. On the other hand, care has to be taken in considering the feeding and discharging arrangements.

11.4.3 Continuous Filtration and Expression

Continuous belt presses, Figure 11.45, compete with solid-bowl centrifuges and variable-chamber presses in sludge dewatering applications. Table 11.11 presents typical productivities of such units; it can be seen that each of these units has advantages and disadvantages. Usually, a clear optimum choice of filter is difficult to identify.

Thus, whilst the moisture content of filter cakes emanating from belt presses is higher than in those produced in batch processes, in general the continuous nature of the filter, its power
requirements, etc. are positive selection features.

Modern centrifuges are designed, with larger residence times, and modified "beach" geometry, to enhance the dewatering process.

A recent technical publication [Young, 1991] lists the operating costs per tonne of dry solids as:

Machine	Cost (1991) £ per tonne dry solids.
Modern centrifuge	30.0
Diaphragm press	17.5
Old-style centrifuge	15.5
Belt press	13.5

Table 11.11 Typical productivity in sewage handling

Unit	Pretreatment	Capacity kg DS/m^2 h	Dryness wt%	Energy kW/T
Belt press	Flocculants 1–5 kg/TDS.	50–500*	11–36	2–20
Vacuum filter	FeCl$_3$ 2–12% Lime 3–37%	15–40	18–32	50–150
Pressure filter Filter	FeCl$_3$ 3–12% Lime	1.5–5	33–60	15–40
Centrifugal decanter	Flocculants 2–7 kg/TDS	100–500**	9–26	30–60

TDS = tonnes dry solids   * Capacity, kg DS/h (metre belt width) (belts 1–3 m)

** Centrifuge 40–45cm bowl diameter (kg dry solids/h)

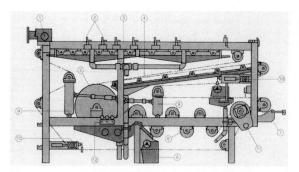

1. Feed distribution box
2. Multiple rows of plows
3. Long gravity drainage zone
4. Large converging inclined wedge zone
5 Large diameter stainless steel perforated roll
6. Seven decreasing diameter compression rolls

7. Counter weight discharge scrapper
8. Totally enclosed belt wash assemblies
9. Belt tensioning box frame swing arm assembly
10. Belt tracking
11. Hydraulic motor and gear reducer
12. Hydraulic control manifold

Figure 11.45 Continuous belt press (Courtesy of Baker Prozess, South Walpole, MA, USA)

Despite these figures, other factors, e.g. cake moisture content may lead to the adoption of higher processing cost levels.

Belt presses have been studied, both theoretically and practically, providing useful information on the absolute level and positional aspects of pressure as it develops in the moving filter cake [Badgjugar & Chiang, 1989; Austin, 1978].

The relative motion of the two belts in Figure 11.46 is considered to promote a shearing action on the filter cake. During passage through the various rolls, the two cloths continue to change position on the upper and lower parts of the filter cake; changes in cloth speed induce the shearing action. These effects and compression by sliding are recognised in similar machinery such as the rotary press [Barbulescu, 1993] and screw presses [Shirato et al, 1983].

As shown in Figure 11.47, considerable investment has been put into the development of combined vacuum and expression devices [Smith & Beaton, 1989]. The continuous belt in this application is attached to the upper part of the rotating drum filter. Compression rolls deliver a dewatering squeeze to the cake, on its progress towards the discharge point.

Considerable effort has been put into the development of continuous pressure units for the filtration, thickening and washing of materials which may exhibit rheological changes during processes. Such units involve the generation of an intensive shearing action, which is caused to move through an annular gap adjacent to a moving filter membrane [Tobler, 1979]. Quite often, such materials need washing, and here, attempts are made to take advantage of thin cake conditions [Tiller et al. 1982, 1986] as depicted in Figure 11.48. Methods have been proposed for determining the number of stages required in operations in processing such materials as metallic hydroxides, pigments and other materials with difficult filtration characteristics [Toda, 1981]. The development of large-scale, continuous-pressure filters continues to receive attention [Bott, 1990].

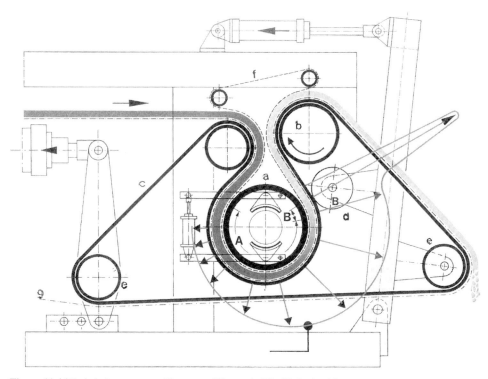

Figure 11.46 Twin belt movement (Courtesy of Paunevis, The Netherlands)

a) Press drum with brake mechanism;

b) Top roller (driving roller);

c) Press belts (25 mm wide, 2 mm apart);

d) Squeeze roller;

e) Guiding rollers for filter cloth and press belts;

f) Auxiliary cloth;

g) Filter cloth (as used for filtration on integrated machines)

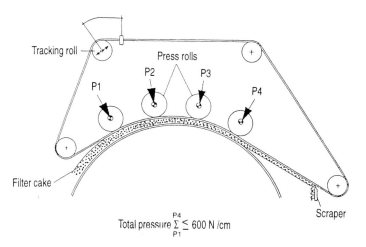

Figure 11.47 Arrangement of rolls on press belt drum filter (Courtesy of Dorr–Oliver, U.K.)

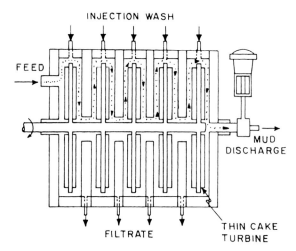

Figure 11.48 Rotary disc filter [Tiller, 1982, 1986]

In these designs, horizontal belt, disc or drum filters are installed inside pressure vessels. Sufficient room is provided in the latter for access by personnel, cake discharge, etc. Such plants are operated under compressed air blankets, alone or in conjunction with vacuum downstream of the filter.

11.4.4. Pressure Leaf and Candle Filters

These units are sometimes described as pressure vessel filters; the vessels used to contain the filter elements are cylindrical designs, positioned vertically or horizontally. These vessels operate in batch mode, with filter cycles similar to plate and frame filters. The variations available in the type of elements (leaf or candle) and its position in the pressure vessel, lead to a large number of available designs [Purchas, 1981], General operating features of pressure vessel filters have been described [Bosley, 1986]; these units are considered to be cleaner and more reliable than other pressure filters, provided adequate attention is given to filter cake properties.

In this respect, a principal effect is the stability of the cake on the filter element. If the latter is in the vertical position, loss of pressure during the filter cycle can result in portions, or all, of the filter cake falling away from the element. On the other hand, sticking of cakes on filter media can produce difficulties in cake discharge.

Such effects are often more important in dewatering filter cycle time than cake formation and washing; practical information on the cake discharge problem [Carleton & Heywood, 1983] is discussed in Chapter 4.

A typical rectangular leaf unit, with vertical elements contained in a vertical pressure vessel, is shown in Figure 11.49. In these units, gravitational forces will compete with the stabilising forces caused by fluid flow and friction. Any change in fluid velocity through the cake may create instability. Thus, care has to be taken during changeover from filtration to dewatering or washing. Problems associated with cake loss can be eliminated by use of horizontal elements, Figure 11.50.

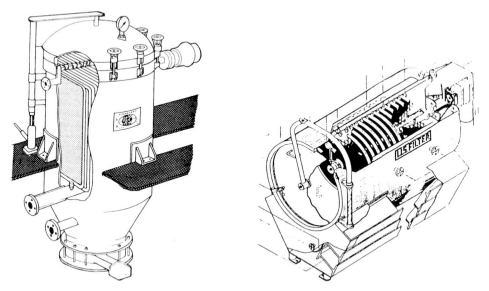

Figure 11.49 Vertical leaf filters: back–flushing and sluicing discharge (Courtesy of Goodtech, Vanpipe Co. Ltd, Burstem, Stoke, U.K.)

Figure 11.50 Calmic horizontal plate press filter (Courtesy of Euro-Vent, Ltd., Stoke, U.K.)

Here, filtration proceeds only on the upper surface of the leaf. The latter may be provided in the form of a tray, to ensure through washing of the cake. A typical arrangement here would be the provision of a stack of leaves or trays (up to 25 m^2 filter area). After filtration, the pressure vessel has to be opened for removal of the filled stack. Normally, the latter would be replaced immediately with a clean stack. Horizontal leaves are also useful in the processing of fast sedimenting solids.

The problem of discharge from horizontal leaf filters for larger sizes than indicated above has been resolved in filters which spin the stack after completion of the cycle, Figure 11.51; the economics of these filters, their design and operation have been reported [Howard, 1979].

Fully automatic operation is essential to take complete advantage of this mode of cake discharge. Design changes involving a reduction in vessel size, even for dry cake discharge, reduced overall costs. These units are also useful in liquor clarifications involving the use of filter aids, as a pre-coat and/or as a body feed.

In these units, only half the available leaf area is available for filtration. Also, the horizontal units are limited to sizes up to 65 m^2 so that, despite the inherent cake loss problem, vertical leaf filters are usually specified for larger tonnages.

Vertical vessels can accommodate up to 80 m^2 of vertical plate; larger areas, up to 300 m^2, are attainable with vertical leaves in horizontal vessels. Plate spacing must be considered, in relation to the volume of cake filtered per batch. Adequate space must be allowed between the leaves for cake deposition and subsequent discharge. Bridging of solids between plates can lead to lengthy discharge times and, in some circumstances, damage to the filter medium and plate.

Filter leaves are generally of rigid construction; a metal frame covered with a metallic mesh is common. In some applications, the leaf will be covered by a suitable filter cloth. Open-pore metallic mesh (~130 μm) can be used in pre-coating applications in clarification processes.

In the latter, it is usual to deposit a thin layer of filter aid (~2 mm) on the surface of the medium. Pre-coating slurry will be supplied from a separate vessel; the feed will be recycled at a sufficient velocity to promote even deposition of filter aid on the surface of the filter. Sedimentation of solids is to be avoided in such operations. The size of filter specified for a particular duty and the quantity of filter aid used per batch, will depend on the cycle time assumed in the specification of plant size. The capital investment and the consumable cost of filter aid will vary with the cycle time. The possibility of an optimum choice of cycle time and filter area exists, as demonstrated by elementary analysis in Example 11.5.

Example 11.5 Pressure Leaf Filter Optimisation

A pressure leaf filter is required for the clarification of a vegetable oil, operating at a constant flowrate of 3 m^3/h. After pre-coating the filter medium with a suitable filter aid, filtration ensues up to a maximum pressure differential of 4.5 bar.

The filtration cycle time t_c is related to the filtration time t_f and the downtime t_{dw} by:

$$t_c = t_f + t_{dw}$$

Figure 11.51 Centrifugal discharge pressure filter (Courtesy of Schenk Filterbau GmbH, Waldstetten, Germany)

Downtime includes filling, precoating and emptying the filter. In order to simplify the problem, it may be assumed that $t_{dw} = 0.3\ t_c$, irrespective of the size of the filter. Calculate the filter area requirements for cycle times of 1, 2, 7, and 14 h. If the pre-coating requirement involves the deposition of 2 mm of filter aid per cycle on the filter area A m^2, calculate the effect of cycle time on the sum of the purchase cost of the filter and the annual cost of filter aid. Assume an operating period of 14 h/d and 340 working days per annum.

Data

Specific resistance of cake α	2×10^{11} m/kg
Resistance of pre-coated medium, R_m	1×10^{11} m^{-1}
Concentration of solids, c	50 kg/m^3 of filtrate
Liquid viscosity, μ	0.01 Pa.s
Bulk density of precoat layer	350 kg/m^3

Cost of filter aid   £ 350 per tonne
Estimated cost of filter of area A m^2 £ $(1000 \times A^{0.9})$

The calculations below are, of course, subject to the various assumptions made in the statement of the problem.

Solution

Constant rate design equation:

$$\Delta P = (K_1 V + K_2)\, q \qquad (11.6)$$

where:

$$K_1 = (\alpha \mu c / A^2) \qquad (11.7)$$

and:

$$K_2 = (\mu R_m / A) \qquad (11.7)$$

also:

$q = 3.0$ m^3/h $= 8.33 \times 10^{-4}$ m^3/s

Allowable $\Delta P = 4.5$ bar $= 450\,000$ N/m^2

Cycle time $t_c =$ Filtration time and downtime

$t_{dw} = 0.3\, t_c$

Hence:

$t_f = 0.7\, t_c$

Filtrate volume per cycle:

$V = q\, t_f = 0.7\, q\, t_c$

Substituting design data in the constant-rate equation:

$$\Delta P = [K_1 q^2 t_f + K_2 q]$$

$$450\,000 = \left[\frac{(2 \times 10^{11} \times 0.01 \times 50) \times (8.333 \times 10^{-4})^2}{A^2} \times 0.7\,t_C + \frac{(0.01 \times 1 \times 10^{11}) \times 8.333 \times 8.33 \times 10^{-4}}{A} \right]$$

Hence:

$$A^2 - 1.852\,A - 0.108\,t_c = 0$$

Solve for A in terms of cycle time t_c:

t_c, h	1	2	7	14
A, m^2	20.7	28.8	53.1	74.7

Cost Analyses:

Volume of pre-coat	A×0.002 m^2
Mass of pre-coat	A×0.002×350 = 0.7 A kg
Number of pre-coats per annum	(14×340)/t_c
Pre-coat coat cost per annum	(14×340×0.7 A×350)/(t_c ×1000)
	1166.2 (A/t_c)

t_c h	A m^2	Filter cost £	A/t_c	Pre-coat cost, £	Total cost, £
1	0.7	15 288	20.7	24 140	39 428
2	28.8	20 580	14.4	16 793	37 373
7	53.1	35 693	7.59	8847	44 544
14	74.7	48 527	5.34	6223	54 754

This simple analysis suggests that a cycle time of 2 h and a filter area of 28.8 m^2 will produce the minimum initial cost. A computer spreadsheet based on the above example is given in Appendix C. Of course, the capital charge would be distributed, according to some fiscal policy, in a more rigorous analysis.

Recognition of the importance of filter cake discharge in its effect on overall cycle time has produced recent design changes in pressure leaf and candle filters. Adequate control of cake discharge time leads to the possibility of operation in "thin cake" filtration conditions. As noted in the section dealing with rotary vacuum and belt filters, discharge of thin cakes can present unexpected difficulties and it is usual to recommend minimum thicknesses for discharge.

The unit shown in Figure 11.52 is a stationary filter leaf model in which the elements are automatically back-flushed at the end of the filtration period. Cake stability is ensured by maintenance of a positive gas pressure in the vessel, during changeover from filtration to washing. The filter leaves vary in design, depending on the particular application of conventional filtration, thickening, dry cake discharge, washing or pre-coat operation in clarification duties [Simonart and Gaudfrin, 1993]. Areas up to 600 m² can be arranged in one vessel, by adoption of a star-shaped arrangement of the filter leaves. Despite these large areas, individual leaves can be isolated, if circumstances require; the filtrate from each leaf can be inspected for clarity.

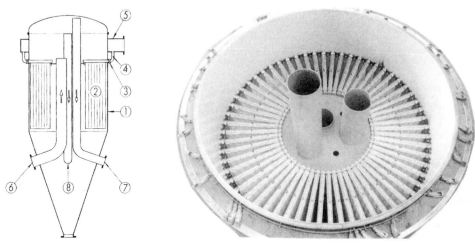

Figure 11.52 Stationary leaf Diastar pressure filter (Courtesy of Gaudfrin, SA, France)

1) Filter/tank body;	5) General collector;
2) Drained cloth;	6) Supply/feed pipe;
3) Collector;	7) Air decompression pipe
4) Isolating valve and inspection hole;	8) Levelling pipe;

Many other designs are available [Purchas, 1981] which are aimed at a resolution of the problem of cake discharge. Thus, cake discharge may be achieved by:

i) Blow-back of gas (air); the latter provided by compressor or from a trapped volume in the filter
ii) Cake vibration or shaking
iii)Leaf sluicing by filtrate; the discharge here will obviously be wet or slurried, Figure 11.49B

In tubular applications, Figure 11.53, successful applications include those designs which ensure adequate cake discharge facilities. The filter candle shown allows expansion of the filter medium during back-flushing [Mueller, 1984]. In another application, [Rushton & Malamis, 1985], the filter medium is supported on metallic springs which contract with the rise of pressure level inherent in the development of the filter cake. At a pre-set ΔP level, the pressures upstream and downstream of the medium are equalised; the spring returns to its original position, discharging the solids. Some liquid back-flush is also induced in the latter action; this produces an effective media cleaning and long process life.

Control of the back-pressure level required for effective back-flushing has been claimed with the use of flat perforated plate elements in the "Cricket" filter [Filtration and Separation, Sep/Oct., 1991]. Automatic, semicontinuous cleaning of tubular candle filters can also be realised by momentary reversal of the filtrate flow; a resulting bump is delivered to the cake, which is dislodged and falls to the cone-shaped bottom of the unit, Figure 11.54. The effectiveness of these filters is improved by the use of filter media which can handle very dilute suspensions of fine (< 0.5 μm) particles, e.g. PTFE membranes.

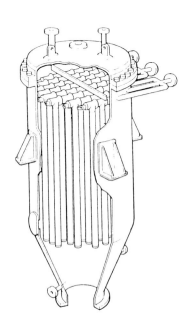
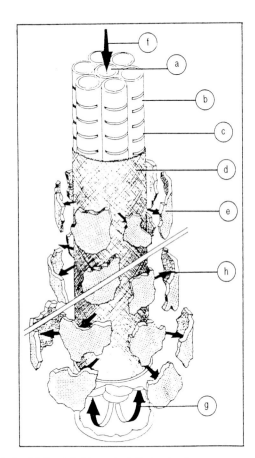

Figure 11.53 Candle filter (Courtesy of DrM, Dr. Müller AG, Männedorf, Switzerland)

a) Central tube;

b) Concentric tubes;

c) Horizontal slots for blow–back;

d) Filter medium lifted off during blow–back;

e) Filter cake thrown off during blow-back ;

f) Blow–back air down central tube;

g) Blow–back rising into concentric tubes;

h) Blow–back air throwing off cake For slurry discharge, neither blow–back nor drying is required; discharge is by back–wash with clear water or other suitable fluid; back–wash and blow–back path are identical

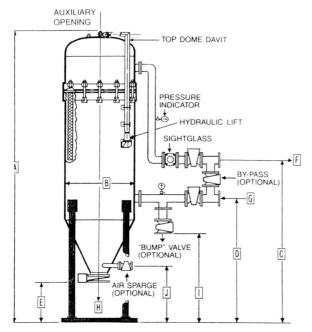

Figure 11.54 Autopulse candle filter (Goodtech, Vanpipe, Co. Ltd, Bursten, Stoke, U.K.)

Serious attention has to be given to candle spacing in these designs, in order to avoid filter cake bridging between the elements. Such effects can lead to poor overall performance of these filters.

Theoretically, deposition on the outer surface of cylindrical shapes is advantageous in filtration systems in view of the increase in area, with cake growth.

Equations are available [Shirato et al, 1981] for calculating flow rates on two-dimensional
surfaces, such as cylinders. The increase in filter area can be allowed for by using an "augmentation factor" J_n :

$$J_n = 0.5 \frac{\left[(d_1/d_2)^2 - 1\right]}{\ln(d_1/d_2)} \tag{11.36}$$

The pressure differential over the filter cake can then be estimated from the constant-rate equation:

$$\Delta P_c = (\alpha \mu c V q) / J_n A^2 \tag{11.37}$$

A relationship between the outer and inner cake diameters d_1, d_2 and filtration time is required for the calculation of the pressure–drop–time variation:

$$(d_1 / d_2) = \left(\frac{4\,c\,q\,t_f}{N\,\pi\,l\,\rho_s\,(1-\varepsilon)} + d_2^2 \right)^{0.5} / d_2 \qquad\qquad (11.38)$$

where N is the number of candles. These equations predict a downward trend, or longer filtration periods, in the $(\Delta P, t)$ profile when filtering on candles. The effect, of course, has real significance for thicker deposits, e.g. pre-coated candles. Whether or not such trends are of economic interest will depend on the circumstances attaching to cake discharge, etc.

Pressure leaf and candle filters are used on a wide scale in process industries. A noticeable trend, perhaps linked to a "technological empathy" and based on long, practical experience, is towards simplicity of design and operation. The desirable features of versatile filter machinery will include [Mueller, 1984], inter alia: (a) large surfaces in the smallest possible space, (b) cake discharge without assistance (mechanical, rotational, etc.) in dry and wet conditions, (c) easily installed and replaceable filter media, (d) effective cake washing with minimal wash fluid, (e) microprocessor control. Improved media for candle filters, which respond more easily to cleaning by back washing have been reported [Egas, 1998]. The principle here is the control of the size and shape of the surface pores in the medium. The use of ultra-sonic cleaning to augment the back flushing effect has also been incorporated into new designs [Rantala, 1998].

11.5 References

Anlauf, H.,1900, 5th World Filtration Congress, Nice, 2, pp 211–283.
Austin, E.P., 1978, Filtrat.Separat., 15, p 649.
Badgjugar, M.N. and Chiang, S.H., 1989, ibid, 26, p 364.
Bailey, P.C.C.H., 1979, ibid, 16, p 649.
Barbulescu, A., 1993, 6th World Filtration Congress, Nagoya, Japan. p 392.
Bosley, R., 1974, Filtrat. Separat. 11, p 138.
Bosley, R., 1986, Chap.10, in Solid–Liquid Equipment Scale-Up, 2nd edn., D. Purchas and R. Wakeman, (Eds.) Uplands Press.
Bott, R., 1990, 5th World Filtration Congress, Nice, 2, p 416.
Brociner, R.E., 1972, Filtrat. Separat., 9, p 562.
Brown, A., 1979, ibid, 16, 468.
Carleton, A.J. and Cousens, T.W., 1982, ibid, 19, p 136.
Carleton, A.J. and Heywood, N.I., 1983, Filtrat. Separat., 20, p 357.
Carleton, A.J. and Mehta, K.B., 1983, Filtech Conference, London, p 120.
Carleton, A.J. and Salway, A.G., 1993, Filtrat. Separat., 30, p 641.
Carleton, A.J. and Taylour, J.M., 1991, Filt. Conference, Karlsruhe, p 281.
Cherry, G.B., 1978, ibid, 15, p 313.

Dahlstrom, D.A. and Silverblatt, C.E., 1986, Chap.11 in Solid-Liquid Equipment Scale-Up,   2nd edn., D. Purchas and R. Wakeman, (Eds.), Uplands Press.

Dickenson, C., 1992, Filters and Filtration Handbook, 3rd edn., Elsevier.

Dresner, J.R. and Bailey, M.G.D., 1973, I.Chem.E, Yorkshire Branch, Filtration Symposium,   pp 601–614.

Ekberg, B., 1990, 5th World Filtration Congress, Nice, 2, p 422.

Ekberg, B. and Rantala, P., 1991, Filtech Conference, Karlsruhe, 1, p 297.

Egas, F., 1998, Scandinavian Filtration Society meeting, Stavanger, Norway.

Gale, R., 1971, ibid., 8, p 531.

Gaudfrin, G., 1978, International Symposium KVIV, Belgian Filtration Society, Antwerp.

Gaudfrin. G., 1981, International. Symposium, Belgian Filtration Society Louvain..

Groh, R.E. and Van der Werff, H., 1982, 3rd World Filtration Congress, Downingtown, American Filtration Society, 1, p 333.

Gross, H. et al., 1989, Filtech Conference, Karlsruhe, 2, p 502.

Gwilliam, R.D., 1973, Filtrat. Separat., 10, p 711.

Heaton, H.M., 1978, ibid, 15, p 232.

Hermia, J., 1981, International Symposium, Belgian Filtration Society, p 177.

Hermia, J. and Rahier, G., 1990, 5th World Filtration Congress, Nice, 2, p 441.

Hess, W.F., 1990, ibid, 2, p 393.

Holdich, R.G., 1990, Filtrat. Separat., 27, p 435.

Howard, M.G., 1979, ibid., 16, p 150.

Hoyland, G. and Day, M., 1983, Filtrat. Separat., 20, p 302.

Johns, F.E., 1991, Filtech Conference, Karlsruhe, 1, p 183.

Keinänen, T. et al., 1994, Filtrat. Separat, 31, p 67.

Kurita, T. and Suwa, S., 1978, ibid, 15, p 109.

Kurita, T. et al., 1993, 6th World Filtration Congress, Nagoya, p 309.

Maracek, J. and Novotny, P., 1990, Filtrat. Separat., 17, p 34.

Mayer, L., 1995, Filtech Conference, Karlsruhe, Germany, p 161 (Filtration Society, U.K.)

Moon, P.J., 1981, International Symposium, Belgian Filtration Society p 315.

Moss, A., 1970, The Chemical Engineer, 237, CE70.

Mueller, H.K., 1984, Filtrat. Separat., 21, p 259.

Nield, P., 1994, ibid, 31, p 691.

Osborne, D.G., 1981, Chap. 13, in Solid–Liquid Separation, 2nd edn., L. Svarovsky, (Ed.).

Osborne, D.G. and Robinson, H.Y., 1972, ibid., 9, p 397.

Pierson, W., 1981a, Chap.20 in Solid–Liquid Separation, 2nd edn., L. Svarovsky, (Ed.).

Pierson, W., 1981b, ibid, Chap.13, Part II: Vacuum Filtration.

Prinssen, A., 1979, Filtrat. Separat., 16, p 176.

Prinssen, A., 1982, ibid, p 396.

Purchas, D.B., 1981, Solid–Liquid Separation Technology, Uplands Press.

Purchas, D.B., 1994, U.K. Filtration Society 30th Anniversary Symposium, p 1.

Rantala, P., 1998, CST Workshop in Filtration, Lappeenranta Univ. Tech. Finland.

Reid, D.A., 1974, Filtrat. Separat., 11, p 293.

Rushton, Alan, 1974, C.S.I.R. Dept., C.Eng M-225, C.S.I.R. Pretoria.

Rushton, A., 1981, Hydrometallurgy 81, Soc. Chem. Ind., London, C1/19.

Rushton, A. and Arab, M.A.A., 1986, 4th World Filtration Congress, Antwerp.

Rushton, A. and Daneshpoor, S., 1983, Filtech Symposium, London, p 68.

Rushton, A. and Hamead, M.S., 1969, Filtrat. Separat., 6, p 136.

Rushton, A. and Malamis, J., 1985, ibid., 22, p 368.

Rushton, A. and Matsis, V.M., 1994, ibid., 31, p 643

Rushton, A. and Metcalfe, M., 1973, ibid, 10, p 398

Rushton, A. and Rushton, Alan., 1973, ibid, 10, p 267.

Rushton, A. and Wakeman, R.J., J., 1977, Powder a Bulk Solids Tec. 1, p 58.

Schafer, C., 1993, Filtech Conference, Karlsruhe, 1, p 369.

Schonstein, P.J., 1991, Filtrat. Separat., 28, p 121.

Schwalbach, W., 1990, 5th World Filtration Congress, Nice, 2, p 400.

Shirato, M. et al., 1981, 2nd World Congress, Chemical Engineering, Montreal, 4, p 107.

Shirato, M. et al., 1983, Pachec '83, 4, p 377.

Shirato, M. et al., Chap.6, in Filtration Principles and Practices, 2nd edn, M.J. Matteson, (Ed.) Dekker, New York, 1987.

Silverblatt, C.E. et al., 1974, Chem. Eng., 4, p 127.

Simonart, H. and Gaudfrin, G., 1993, Filtech Conference, Karlsruhe, 1, p 363.

Smith, D. and Beaton, N.C., 1989, Filtech Conference, Karlsruhe, 2, p 478.

Stahl, W. and Nicolaou, I., 1990, 5th World Filtration Congress, Nice, 2, p 37.

Stahl, W., and Pfuff, T., 1993, 6th World Filtration Congress, Nagoya, p 377.

Svarovsky, L., 1981, Solid-Liquid Separation, 2nd edn, Butterworths.

Terzaghi, K. and Peck, R.B., 1948, Soil Mechanics in Engineering Practice, p 51. Wiley, New York.

Tiller, F.M. and Crump, J.R., 1977, Chem. Eng. Progr., 73, p 65.

Tiller, F.M. and Risbud, H., 1974, A.I.Ch.E.J., 20, p 36.

Tiller, F.M. et al., 1982, Filtrat. Separat., 19, p 119.

Tiller, F.M. et al., 1986, Chap.11, in Solid–Liquid Equipment Scale-Up. 2nd edn., D. Purchas and R. Wakeman, (Eds.) Uplands Press.

Tobler, W., 1979, Filtrat. Separat., 16, p 630.

Toda, T., 1981, ibid., 18, p 118.

Tomiak, A., 1979, Filtrat. Separat., 16, p 354.

Van der Sluis, S. et al., 1988, ibid, 26, p 105.

Wakeman, R.J., 1974, ibid, 11, p 357.

Wakeman, R.J., 1981, International Symposium, Belgian Filtration Society Louvain, p 159.

Wakeman, R.J., 1983, Filtech Conference, London, p 78.

Wakeman, R.J., 1995, Filtrat. Separat., 32, p 337.

Wakeman, R.J. and Attwood, J., 1988, ibid., 25, p 272.

Wakeman, R.J. and Rushton, A., 1974, Chem. Eng. Sci., 29, p 1857.

Wakeman, R.J. and Tarleton, S., 1990, 5th World Filtration Congress, Nice, 2, p 21.

Wakeman, R.J. and Tarleton, E.S., 1999, Filtration: equipment selection modelling and process simulation, Elsevier, Oxford, UK.

Ward, A.S., Laflin, S and Wood, J., 1996, 7th World Filtration Congress, Hungarian Chemical Society, Budapest, pp 37-41.

Yagishita, A. et al., 1990, ibid, Nice, 2, p 122.

Yim, S.S. and Ben Aim, R., 1986, 4th World Filtration Congress, Ostend, 4, p 1.

Young, I., 1991, Filtrat. Separat., 28, p 146.

11.6 Nomenclature

A	Area	m^2
B	Creep Constant: ratio of secondary to total consolidation	–
c	Dry mass of solids per unit filtrate volume	$kg\,m^{-3}$
c_o	Initial concentration of solids or solutes	$kg\,m^3$
c_e	Consolidation Coefficient	$m^2\,s^{-1}$
c_s	Concentration of solutes	$kg\,m^{-3}$
d_1, d_2	Inner and outer diameter of cylindrical cake deposit	m
D	Axial dispersion coefficient	$m^2\,s^{-1}$
F_k	Filtration Constant: $F_k = \alpha\mu c/\Delta P$	$s\,m^{-2}$
g	Gravitational constant	$m\,s^{-2}$
i	Number of drainage surfaces, Equation 11.35	–
K	Filter cake permeability	m^2
K_w	Washing factor, Equation 11.27	–
l	Filter Candle length	m
L	Cake thickness	m
L_1	Cake thickness at end of filtration period, $\theta_c = 0$	m
L_∞	Final thickness of compressed cake, $\theta_c = \infty$	m
L^*	Optimum cake thickness in batch filtration	m
m	Ratio of wet to dry cake mass	–
M	Mass of solids	kg
N	Number of candles	–
N_c	Capillary number	–
P	Pressure	$N\,m^{-2}$
P_b	Capillary or threshold pressure	$N\,m^{-2}$
P_s	Solids compressive pressure	$N\,m^{-2}$
ΔP	Pressure Differential	$N\,m^{-2}$
P^*	Productivity	$kg\,s^{-1}$
q	Filtrate flowrate	$m^3\,s^{-1}$
R_m	Medium Resistance	m^{-1}
s	Mass fraction of solids in feed mixture	–
S	Saturation: volume of liquid in cake/volume of cake pores	–
S_R	Reduced Saturation: $S_R = (S{-}S\infty)/(1{-}S\infty)$	–
S_∞	Residual Equilibrium (irreducible) saturation, $t = \infty$	–
t	Time	s
t_c, t_c^*	Cycle Time, optimum cycle time respectively	s
t_d	Dewatering time	s
t_f	Filtration time	s
t_o	Press filling time	s
t_w	Washing time	s
t_{dw}	Down time for filling and discharging	s
v	Wash velocity	$m\,s^{-1}$
U_c	Average consolidation ratio: Equation 11.35	–

V	Volume of Filtrate	m^3
V^*_f	Optimum volume of filtrate in batch operation	m^3
V^*	Overall volumetric productivity	$m^3\,s^{-1}$
w	Mass of solids deposited per unit area	$kg\,m^{-2}$
x	Particle size	m

Greek Symbols

$\alpha,\ \bar{\alpha}$	Local and average specific cake resistance respectively	$m\,kg^{-1}$
$\varepsilon,\ \bar{\varepsilon}$	Local and average filter cake porosity	–
ε_o	Empirical constant in Equation 11.34	–
η	Creep constant	s^{-1}
θ	Time function in dewatering, Equation 11.10	–
θ_c	Compression time	s
μ	Filtrate viscosity	$Pa\,s$
υ	Consolidation behaviour index	–
$\rho,\ \rho_s$	Fluid and solid density, respectively	$kg\,m^{-3}$
ϕ	Submergence angle of rotary vacuum filter	–
ω	Angular velocity	s^{-1}
γ	Surface tension	$N\,m^{-1}$

Appendix A

Particle Size, Shape and Size Distributions

A particle may be defined as an aggregation of matter without limit to upper size and a system containing particles which are small in relation to their surroundings is a "particulate system". Such systems have a continuous phase, which is usually a fluid, and a discrete phase — the particles — which may be liquid, gas or solid. Examples are powders, where the continuous phase is air or other gas, suspensions, dispersions and emulsions. The relative proportion of the discrete and continuous phases and the size of the particles in the discrete phase are of fundamental importance in describing, understanding and quantifying the behaviour of particulate systems in process machinery.

Particles have shape and in some circumstances this may be regular. Liquids and gases form drops and bubbles which are spherical. Some solid particulate materials occur as spheres, cubes or other regular shapes such as tetrahedra. Obviously if the typical particle is regular in its shape then it can be described easily by a simple linear dimension such as the diameter of a sphere or the side length of a cube. For irregular particles the matter of describing their size is more complicated because no simple linear dimension is evident. In this case resort is made to the concept of "the equivalent sphere".

A comparison is made between the irregular particle and a sphere of the same density on a basis which has some relevance to the process being examined. So for example the basis for comparison could be equivalent volume or surface area, or sieve diameter, or projected area, or settling rate, etc. Most of the possible bases for comparison are associated with a particular particle size analysis technique and it will be noted that for the irregular particle the equivalent sphere diameters will differ depending on the analytical methods used in the measurement. It is implicit in particle size analysis results that they are calculated on an "equivalent sphere basis". If the irregularity of the particles is important in the study of the process, the particle size analysis data should be modified to account for the shape (see below).

Few particulate systems are comprised of particles that are all the same size; most systems have a range of sizes present. The most complete description of a powder or particulate system is obtained from its particle size distribution curve; the latter may be presented on a cumulative basis or in its derivative form as a frequency distribution curve as illustrated in Example 1.

Particle Size Analysis

Several methods are available for the measurement of the size of particles, e.g. microscopic inspection, sieving, elutriation, sedimentation. Generally, the results obtained will depend on the method of measurement since different fundamental dimensions are involved in the different methods. For irregularly shaped particles it is usual to use a particle size technique which duplicates the process of interest, e.g. to use projected diameter (microscopic) or area in paint or pigment studies; diameters based on surface area determinations would have relevance in chemical reactions involving solids, such as catalysis and adsorption.

Some of the more common methods are listed in Table 1 and are categorised as either "direct" or "indirect". The former are methods where each particle is put to a gauge and the latter are methods where a comparison is on the basis of a particular property of the particle.

Table 1 Methods of particle size analysis

	Technique	Approximate size range, μm	Diameter measured
Direct			
Field scanning	Microscope	1–100	Length, projected area statistical diameters
	Electron microscope	0.001–5	
	Scanning electron microscope	0. 1–5	
	Sieves	45–1000+	Mesh size
Indirect			
Sedimentation	Andreasen pipette	2–60	Stokes' diameter
	Sedigraph		
	Photosedimentometer		
	Centrifuges	0.5–5	
Stream scanning	Electro sensing (Coulter)®	1–100	Volume
	Light obscuration (Hiac)®	2–150	Projected area
	Light scattering	0.3–10	Projected area
	Laser diffraction	0.5–800	Volume or projected area*
	Scanning laser microscopy	0.5-1000	Random distribution of scanned chords
Surface area	Gas adsorption		Volume-surface or
	Permeability		Sauter mean diameter

*The use of laser diffraction size analysers has become widespread because of their convenience, but there is doubt about their comparison basis in terms of the equivalent spherical diameter. Reproducible and repeatable results are obtained so the method is fine for quality control work, but users of the technique in technological studies should be cautious in interpreting data.

Every laboratory dealing with particulate systems should possess a microscope. A simple one is adequate for most purposes and should have the following features.

1) An eyepiece lens of magnification 10×, into which a graticule may be placed so that a size comparison may be made with the particle images.
2) Two objective lenses, of 10× and 20× magnification, with as large a numerical aperture as possible.
3) A stage micrometer to calibrate the eyepiece.
4) The capability of moving the stage in two perpendicular directions in the horizontal plane.
4) An external light source.

The particles are sized by comparison of their images with the scale in the eyepiece, which is calibrated from the scale of the stage micrometer. It is unlikely that the process engineer would wish to conduct microscope count analyses routinely, but familiarity with the instrument will provide a general appreciation of the range of particle sizes and shapes that are present in the system. Such knowledge can prove invaluable in developing solutions to processing problems.

Statistical Diameters

There are exceptions to the "equivalent sphere" principle. Examples of these are statistical diameters which are related to the microscope counting method. When viewed in the microscope, the particle appears two-dimensional and from this image, an estimate of particle size may be made in several ways (Figure A.1).

Martin's diameter M is the length of line which bisects the image of the particle. The line may be drawn in any direction but once chosen the direction must remain constant for all measurements in the distribution.

Feret's diameter F is the perpendicular distance between two parallel tangents on opposite sides of the particle, parallel to a fixed direction.

The image shearing diameter IS refers to an optical microscope technique in which two images of the same particle are obtained and displaced so that they are just touching. The length of the displacement is the image shearing diameter of the particle.

It has been found empirically that $M < IS < F$ and $M < x_a < F$, where x_a is the equivalent spherical diameter based on projected area.

Diameters such as these are meaningless when applied to a single particle, but are of value if large numbers (>500) are measured. The measurements are useful in automated optical microscopy when very large numbers of particles can be examined

For some materials the quotient F/M approaches a constant, e.g. for Portland cement F/M = 1.2 and for ground glass F/M = 1.3. As the number of observations increases $x_a \rightarrow F$

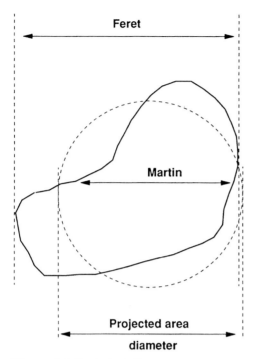

Figure A.1 Statistical diameters

Particle Size Distributions

In Figure A2 when size is plotted against frequency a skewed distribution (positive, right-hand skew in figure) usually results. Frequency distributions are characterised by the parameters which measure the central tendency in terms of the mean, median or mode and the dispersion about the central tendency which is the standard deviation.

Mean or arithmetic average $\bar{x}$: the centre of gravity or balancing point of the distribution, affected by all the information, particularly extreme values

Median x_{50}: the value that divides the distribution into two equal areas with the same amount of material above or below that size.

Mode: the most frequently occurring size in the distribution.

A cumulative distribution is defined by ogives; 25 or 75% of the total amount of material is above or below these sizes which are denoted as x_{75} or x_{25} sizes. The shape of the curve depends on whether the distribution is by number, length, surface, mass or volume. Since the larger particles contribute more to surface and volume, the curves for those bases will be skewed to the right-hand side, as shown in Example 2.

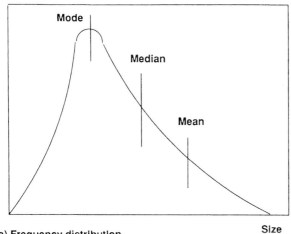

(a) Frequency distribution

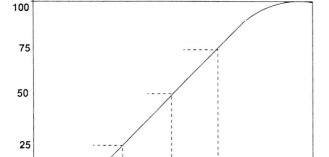

(b) Cumulative distribution

Figure A.2 Frequency and cumulative distibutions. A) Frequency distribution; B) Cumulative distribution

Calculation of means for a distribution of particles, where n is the number of particles of size x

Arithmetic or number–length mean:

$$\bar{x} = \frac{\Sigma nx}{\Sigma n}$$

or

$$\bar{x} = \frac{\int nxdx}{\int ndx}$$

Length–surface mean:

$$\frac{\Sigma n x^2}{\Sigma n x} = \frac{\int n^2 \mathrm{d}x}{\int n x \mathrm{d}x}$$

Surface–volume mean:

$$\frac{\Sigma n x^3}{\Sigma n x^2} = \frac{\int n x^3 \mathrm{d}x}{\int n x^2 \mathrm{d}x}$$

Standard deviation:

$$\sigma = \left\{ \frac{\Sigma n (x - \bar{x})^2}{\Sigma n} \right\}^{1/2}$$

when n is large

The coefficient of variation, which is useful when comparing distributions e.g. in mixing studies, is $s / \bar{x}$. In manipulating the data for presentation it is usual to normalise the frequency values so that they are on a fractional basis. This is defined as the condition when:

$$\sum_{x_{min}}^{x_{max}} \frac{n}{\Delta x} = 1$$

or

$$\int_{x_{min}}^{x_{max}} n \mathrm{d}x = 1$$

However the fraction of a cumulative distribution is defined as the proportion of the size distribution above or below a certain size:

$$f = \sum_{x_{min}}^{x} \frac{n}{\Delta x} \left/ \sum_{x_{min}}^{x_{max}} \frac{n}{\Delta x} \right. = \sum_{x_{min}}^{x} \frac{n}{\Delta x}$$

when normalised, or

$$f = \frac{\int_{x_{min}}^{x} n \mathrm{d}x}{\int_{x_{min}}^{x_{max}} n \mathrm{d}x} = \int_{x_{min}}^{x} n \mathrm{d}x$$

when normalised.

It is important to observe these rules when plotting size distribution data because in this way the information is presented on a truly comparable basis. Frequency data may be graphed as a histogram where the relative number per micrometre range is shown against each size interval, which will take account of the varying width of size intervals found in many analytical techniques such as sieving and microscope counting.

The normal or gaussian distribution is a function closely followed by many biological systems such as pollen grains, blood cells, starch grains. It takes the form:

$$\frac{n}{\Sigma n} = \frac{1}{\sigma \pi^{1/2}} \exp\left\{ -\frac{1}{2}\left(\frac{x - \bar{x}}{\sigma} \right)^2 \right\}$$

When a skewed distribution is transformed to the normal shape (Fig. A2(a)) the mean, median and mode coincide. In this form the distribution is amenable to all the statistical procedures developed for these distributions which are symmetrical about a mean value. Thus the mean particle size $\bar{x}$ and the standard deviation completely define the distribution.

Many fine-particle systems formed by reduction in comminution processes or by growth in crystallisation have particle size distributions obeying the log-normal distribution function.

$$\frac{n}{\Sigma n} = \frac{1}{\log \sigma \pi^{1/2}} \exp\left\{ -\frac{1}{2}\left(\frac{\log x - \log \bar{x}}{\log \sigma} \right)^2 \right\}$$

where:

$$\log \bar{x} = \frac{\Sigma n \log x}{\Sigma n}$$

is the geometric mean diameter and:

$$\log \sigma = \left[\frac{\Sigma n (\log x - \log \bar{x})^2}{\Sigma n} \right]^{1/2}$$

is the geometric standard deviation.

Both these functions may be linearised by the probit transformation which uses a substitution. Hence for the normal curve the substitution is:

$$\frac{x - \bar{x}}{\sigma} = u$$

so that u, the normal equivalent deviate, is zero at the mean size, negative at sizes smaller than mean size and positive at larger sizes. To the values of u calculated, 5 is added to make all values positive and the resulting values called probits are used as the proportion axis of a cumulative plot. Special probability paper is available which allows direct plotting of the data. The "normal" version has a linear scale for particle size and the "log-normal" has a log scale. Data which follows the distribution will give a straight line plot on the appropriate paper.

Several empirical equations have been developed to describe distributions that may arise with certain materials or within the technology of a particular industry. The Rosin–Rammler equation was devised for use with mineral breaking processes, especially coal, and takes the form:

$$P = 100 \exp(-bx^m)$$

where P is the percentage of particles less than size x; b and m are constants. Taking logarithms twice gives:

$$\log\left(\log\frac{100}{P}\right) = \log b + m \log x$$

Hence a graph of P as log-log reciprocal vs. log x gives a line of slope m and intercept log b. The Gaudin–Schumann equation was originally used to treat the results of sieving tests by using a $\sqrt{2}$ increment in size. It employs two empirical constants in the expression:

$$W = \left(\frac{x}{K}\right)^m$$

where W is the weight of particles of size x with K and m as constants. A log-log plot of W vs. x gives a straight line of slope m and intercept K.

These treatments may linearise the data, but they render the information insensitive and make it difficult to use interpretively.

Calculus of Size Distributions

Much modern analytical equipment can produce a great deal of data automatically which may either be dealt with numerically or converted to give a mathematical function for the distribution. In that case it is useful to be able to manipulate the function analytically, which can be done with the calculus of size distributions.

The cumulative distribution is the fractional number above (or below) a given size x and is represented as $N_o(x)$, and the frequency distribution is the fractional number in the size interval x to $x + dx$ shown as $n_o(x)$. The relationship between these two distributions is:

$$n_o(x) = \frac{dN_o(x)}{dx}$$

or

$$\int_{x_{min}}^{x} n_o(x)\mathrm{d}x = N_o(x)$$

Also

$$N_o(x_{max}) = \int_{x_{min}}^{x_{max}} n_o(x)\mathrm{d}x = 1$$

i.e the function is normalised.

The notation is extended to other distributions (mass, volume, surface area, etc.) by the use of subscripts; 0 for number, 1 for length, 2 for area, 3 for mass or volume. So if mass is measured as a function of size, the frequency distribution is $n_3(x)$ vs. x and the cumulative $N_3(x)$ and so on.

Specific Surface

The specific surface, or surface area to volume ratio, of the particles is an important parameter in solid–liquid separation technology and can be calculated readily. For a number distribution of spheres, the total surface area is:

$$\int_{x_{min}}^{x_{max}} \pi x^2 n_o(x)\mathrm{d}x$$

and the total volume is:

$$\int_{x_{min}}^{x_{max}} \frac{\pi x^3}{6} n_o(x)\mathrm{d}x$$

From a volume distribution of spheres, the total surface area is:

$$\int_{x_{min}}^{x_{max}} \pi x^2 \frac{n_3(x)}{\pi \frac{x^3}{6}} \mathrm{d}x = 6\int_{x_{min}}^{x_{max}} \frac{n_3(x)}{x} \mathrm{d}x$$

and the volume is:

$$\int_{x_{min}}^{x_{max}} n_3(x)\mathrm{d}x$$

which equals unity if the $n_3(x)$ function is normalised. In that case the total surface area is also the specific surface.

The diameter of the sphere having the same equivalent specific surface as the particle is sometimes termed the surface volume mean or the **Sauter mean diameter**.

Specific surface may be expressed on a mass basis using the material density to modify the volume.

Particle Shape

The shape of particles can have important effects e.g. on the specific surface. The concept of the equivalent sphere may be extended by shape factors to take account of the real surface or volume when compared with that of the equivalent sphere. Surface area is proportional to x^2, i.e. surface area $= fx^2$ (for a sphere πx^2) and volume is proportional to x^3, i.e., volume $= kx^3$, (for a sphere $\pi x^3/6$). Hence f is a surface factor or coefficient and k is a volume factor or coefficient. The coefficients f and k are functions of the geometrical shape and the relative proportions of the particle; their values depend on the equivalent sphere diameter used. Sphericity is defined as:

$$\Psi = \frac{\text{surface area of sphere of same volume as particle}}{\text{surface area of particle}}$$

which has the advantage of a range of values from 0 to 1, with the more rounded particles tending to a sphericity of 1.

Values of f_a, k_a, (subscript a refers to projected area diameter) and Ψ for various shapes are given in Table 2.

Table 2 Surface and volume factors and sphericity

	k_a	f_a	Ψ
Rounded particles: water-worn sands, atomised metals, fused flue dust	0.32–0.41	2.7–3.4	0.82
Angular particles of crushed minerals	0.2–0.28	2.5–3.2	0.66
Flaky particles: talc, gypsum	0.12–0.16	2.0–2.8	0.54
Thin flaky particles: mica, graphite, aluminium	0.01–0.03	1.6–1.7	0.22

The value of the shape coefficients can be calculated for various equivalent sphere diameter bases. Let subscript a = projected area diameter; v = volume diameter; s = surface area diameter; S_t = Stokes' diameter; m = mesh size. The volume of particles may be expressed as $k_a x_a^3 = k_v x_v^3 = k_s x_s^3 = k_{St} x_{St}^3 = k_m x_m^3$. Hence $k_v = k_a (x_a/x_v)^3$ and so on.

Example 1 Sieve analysis

Columns A and B show the raw data produced by the analytical laboratory which is in the form of cumulative wt% undersize *vs* size in micrometres.

A Sieve size μm	B wt% under size	C size range width	D mean size of range $\bar{x}$	E wt% range	F wt% per μm range	G Add oversize + undersize to col E	H $m\bar{x}$
1400	99.8						
1200	98.7	200	1300	1.1	0.0055	1.3	1690
1000	86.4	200	1100	12.3	0.6105	12.3	13530
850	77.1	150	925	9.3	0.062	9.3	8602.5
600	49.2	250	725	27.9	0.1116	27.9	20228
500	39	100	550	10.2	0.102	10.2	5610
425	30.3	75	462.5	8.7	0.116	8.7	4023.8
300	17.3	125	362.5	13	0.104	13	4712.5
212	9.3	88	256	8	0.0909	8	2048
150	5.7	62	181	3.6	0.0581	3.6	651.6
90	1.7	60	120	4	0.0667	4	480
75	1.1	15	82.5	0.6	0.04	0.6	49.5
45	0.2	30	60	0.9	0.03	1.1	66
		Total		99.6		100 (Σm)	61691.4 ($\Sigma m\bar{x}$)

Mean size by mass = $\Sigma m \bar{x} / \Sigma m$ = 616.9 μm.

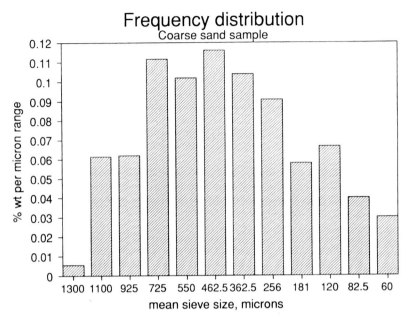

Figure A.3 Frequency plot: Column F *vs* Column D

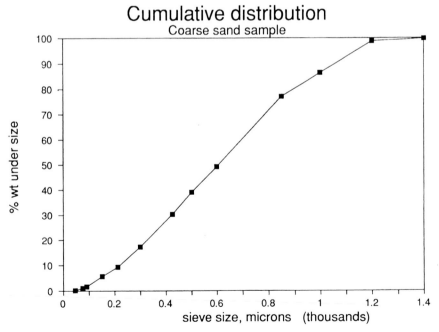

Figure A.4 Cumulative undersize plot: Column B *vs* Column A

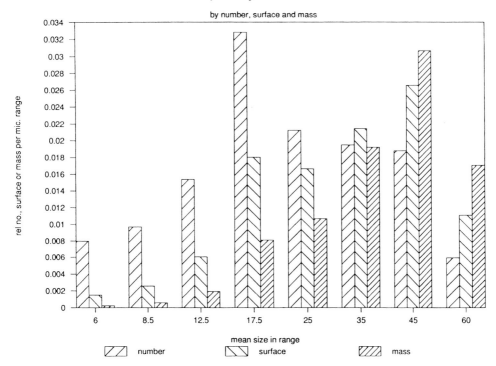

Figure A.5 Frequency distributions by number, surface, and mass

Example 2 Microscope count data

The first two columns on the left-hand side show the results from the laboratory and are for projected area diameter. The calculation shows the respective means and the specific surface.

Size range	No. n	Mean size $\bar{x}$	$n\bar{x}$	relative number $\dfrac{n\bar{x}}{\Sigma n\bar{x}}$	$n\bar{x}^2$	relative surface $s = \dfrac{n\bar{x}^2}{\Sigma n\bar{x}^2}$	$s\bar{x}$	$n\bar{x}^3$	relative mass $m = \dfrac{n\bar{x}^3}{\Sigma n\bar{x}^3}$	$m\bar{x}$
5– 7	42	6	252	0.015976	1512	0.003007	0.018046	9072	0.000462	0.002776
7–10	54	8.5	459	0.029099	3901.5	0.007761	0.065969	3162.75	0.001691	0.014380
10–15	97	12.5	12135	0.076869	15156.25	0.030150	0.376875	189453.1	0.009665	0.120813
15–20	148	17.5	2590	0.164199	45325	0.090164	1.577871	793187.5	0.040464	0.708135
20–30	134	25	3350	0.212381	83750	0.166602	4.165052	2093750	0.106813	2.670343
30–40	88	35	3080	0.195264	107800	0.214444	7.505548	3773000	0.192481	6.736855
40–50	66	45	2970	0.188290	133650	0.265867	11.96401	6014250	0.306820	13.80690
50–70	31	60	1860	0.117919	111600	0.222003	13.32021	6696000	0.341599	20.49599
Total	660		15774	1.0	502694.7	1.0	38.99359	19601875	1.0	44.55620

Mean size by number $= \dfrac{15774}{660} = 23.9\ \mu m$

Mean size by surface $= 38.99\ \mu m$

Mean size by mass $= 44.56\ \mu m$

Specific surface $= 6\ \dfrac{\Sigma n\bar{x}^2}{\Sigma n\bar{x}^3} = 6 \times \dfrac{502694.7}{19601875} = 0.154\ \mu m^{-1}$ or $1.54 \times 10^5\ m^2 / m^3$

Sauter mean diameter $= \dfrac{\Sigma n\bar{x}^3}{\Sigma n\bar{x}^2} = \dfrac{19601975}{502694.7} = 39.0\ \mu m$

Reference

Allen, T., Particle Size Measurement, 1997, Vol. 1 Powder sampling and Particle Size Measurement, Vol.2 Surface Area and Pore Size Determination, Chapman and Hall, London.

Appendix B

Slurry Rheology

A basic appreciation of slurry rheology, or flow behaviour, is important in many solid–liquid separations, e.g. when feeding pressure filters, pumping thickener underflow, hydrocyclone feed and exit streams and during cross-flow filtration. This Appendix is designed to introduce some of the terminology and basic concepts. A more thorough text such as Wilkinson [1960] should be referred to for further details, if necessary.

Definitions

A slurry or suspension is characterised by the behaviour of its shear stress as a function of the shear rate applied. There are many different types of behaviour, as illustrated in the two figures below.

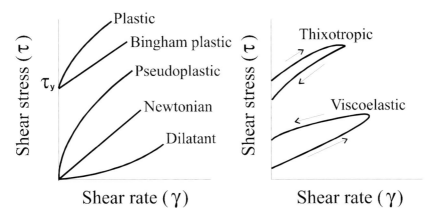

Rheograms: no time dependency   Rheograms: with time dependency

The simplest rheogram is that of a Newtonian liquid, such as water, where the following equation is obeyed:

$$\tau = \mu\gamma \tag{B.1}$$

the constant of proportionality between shear stress and rate is known as the dynamic viscosity. All materials which do not obey Equation (B.1) are known as non-Newtonian fluids. One of the most common equations applied to non-Newtonian liquids is the power law

$$\tau = \eta\gamma^{n} \tag{B.2}$$

where η is the consistency coefficient and n is the flow index. When the flow index is less than 1 Equation (B.2) describes the behaviour of pseudoplastic material, a flow index greater than 1 is characteristic of dilatant material. When the flow index is unity Equation (B.2) reduces to Equation (B.1), i.e. Newtonian flow. Plastic and Bingham plastic flows can be modelled in a similar way with the addition of yield stress (τ_y).

The apparent viscosity of a material is the viscosity of a Newtonian liquid that would flow under the same condition (shear rate and stress) as the fluid observed. Combining Equations (B.1) and (B.2) provides:

$$\mu_a = \eta \gamma^{n-1}$$

where μ_a is the apparent viscosity. If the apparent viscosity is dependent on the shear history, i.e. shear rate used previously or time over which the shear has been applied, the flow is time dependent, as illustrated in the right-hand figure above. Time dependent materials which exhibit reducing apparent viscosity are thixotropic, materials exhibiting increasing apparent viscosity are viscoelastic.

Slurries may also be characterised with reference to their ability to segregate: homogeneous suspensions have a uniform distribution of solids throughout the pipe or channel in which they are flowing, heterogeneous suspensions have a distinct solid concentration gradient within the channel, due to sedimentation of the suspended material. Most slurry flow problems involve finely divided material at a solid concentration sufficient to hinder solid settlement, with little time dependency. Homogeneous suspension flow equations are, therefore, generally applicable, and the mixture may be treated as if it was a pure liquid of similar rheological properties to the suspension.

Equations of Flow

In many instances the engineer needs to know the pressure drop required to pump a suspension at a design flow rate, or the maximum flow, given a value of pressure head. For laminar flow the following equations are widely used.
For Newtonian flow:

$$Q = \frac{\pi d^2}{4} \frac{d^2}{32\mu} \frac{\Delta P}{L} \tag{B.3}$$

where Q is the volume flow rate, d is the pipe diameter and L is the pipe length. Equation (B.3) is known as the Hagen–Poiseuille equation.
For non-Newtonian flow (based on a power law rheological model):

$$Q = \frac{\pi d^2}{4} \frac{nd}{2(3n+1)} \left(\frac{d\Delta P}{4\eta L} \right)^{1/n} \tag{B.4}$$

Equation (B.4) is known as Wilkinson's equation, which reduces to Equation (B.3) when the flow index is unity and the consistency coefficient becomes the dynamic viscosity.

In the case of turbulent flow, correlations between the friction factor and the Reynolds number must be employed. For Newtonian liquids, with $2500 < Re < 1 \times 10^7$:

$$\left(\frac{f}{2} \right)^{-1/2} = 2.5 \ln \left[Re \left(\frac{f}{2} \right)^{1/2} \right] + 0.3 \tag{B.5}$$

where the friction factor is defined as:

$$\frac{f}{2} = \frac{\tau_w}{\rho_m v^2} \tag{B.6}$$

v is the bulk velocity of the slurry and ρ_m is the density of the liquid or suspension being pumped. A force balance at the wall of a pipe of circular cross-section gives the following relation between pressure drop and wall shear stress:

$$\Delta P = \frac{4 L \tau_w}{d} \tag{B.7}$$

Hence, for a given flow rate or pressure drop an iterative solution to Equations (B.5) to (B.7) may be obtained.

For non-Newtonian liquids, or suspensions, over a similar Reynolds number range:

$$\left(f \right)^{-1/2} = \frac{4}{n^{0.75}} \log \left[Re^* \left(f \right)^{(1-n/2)} \right] - \frac{0.4}{n^{1.2}} \tag{B.8}$$

However, a modified Reynolds number is employed in Equation (B.8), defined for power law fluids as [Metzner & Reed, 1955]:

$$Re^* = \frac{\rho_m v^{2-n} d^n}{\eta \left(\dfrac{6n + 2}{n} \right)^n} \tag{B.9}$$

An iterative solution to Equations (B.6) to (B.9) will provide the required pressure in order to maintain a given flow rate, or vice versa. The critical modified Reynolds number, marking the transition between laminar and turbulent flow, is approximately 2300, i.e. the same value as the conventional critical Reynolds number under Newtonian flow conditions.

Worked Example: A pilot study has provided thickened sludge with the following rheological characteristics:

Shear rate s^{-1}	5.5	11	173	346	692	1390	2770
Shear stress Pa	3.0	4.9	35	58	95	160	250

It is intended that the sludge will be continuously discharged as thickener underflow. Fit the data to a power law model and investigate the pressure drop flow rate relation up to a maximum pressure drop of 0.8 bar, given that the suspension density is 1100 kg m^{-3} and the pipe diameter and length are 0.128 and 30 m respectively.

A linear regression conducted on the above data in a logarithmic form provides the following

$$\tau = 0.886\gamma^{0.715}$$

Up to a flow rate of 1900 l min^{-1} Equation (B.4) can be used to investigate the flow-pressure drop relation, as the flow is laminar. Above this flow rate, Equations (B.8) was iteratively solved as follows:

i)   the pressure drop was fixed for an increment (in 5000 Pa increments),
ii) the velocity was calculated by a rearranged and expanded form of Equation (B.4),
iii) the generalised Reynolds number was calculated from Equation (B.9),
iv) followed by the friction factor from Equations (B.6) and (B.7) combined.

The left- and right-hand sides of Equation (B.8) were then calculated; if the values had not converged the velocity was reduced by an increment (0.01 m s^{-1}) and the above process repeated from the calculation of the generalised Reynolds number.

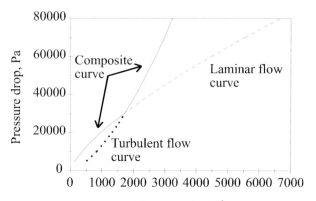

The composite curve on the accompanying figure is the result of the analysis; the flow –pressure relation follows the laminar curve up to 1900 l min^{-1}, and the turbulent curve at greater flow rates. The complete analysis was performed on a computer spreadsheet.

References

Metzner, A.B. and Reed, J.C., 1955, AIChE J., 1, pp 434 - 440.
Wilkinson, W.L., 1960, Non-Newtonian Fluids, Pergamon, Oxford.

Appendix C

Computer Spreadsheet Files

The following spreadsheet files can be used to analyse laboratory, or process, data and assist in the optimisation of filter operation by means of process simulation undertaken on a standard computer spreadsheet package. The minimum amount of information needed to program the spreadsheet has been provided, and the cell formulae will need to be copied into appropriate new cell locations after entry. Brief notes accompany each spreadsheet to enable the reader to understand the origin of the equations used in the cells. Advanced spreadsheet programming, such as the use of Macros, has been avoided so that the cell formulae are readily understandable and applicable to any of the currently available and popular spreadsheet packages. An example printout of each of the spreadsheets has been provided so that the reader can check a typed-in spreadsheet against the example illustrated.

It is hoped that the inclusion of these spreadsheets will help to reinforce and illustrate some of the mathematical developments provided in the main text of this book. Many of the worked examples can be investigated using the following files, and some of the figures used in the main text were derived using the graphical capability of the spreadsheet package on these examples. The graphs have not, therefore, been reproduced in this Appendix.

The spreadsheet "notebook" files were originally written in Quattro Pro for Windows v5. However, they are now also available in Microsoft Excel Office'97 format and they can be downloaded without cost from the World Wide Web site:

http://www-staff.lboro.ac.uk/~cgrgh/

and other locations. These are obtainable by sending an e-mail to:

richard_holdich@bigfoot.com

Please address all correspondence about these files to the same author. Where appropriate, all trademarks are acknowledged as belonging to their rightful owners.

SPECIFIC.WB1

A spreadsheet to calculate the specific surface area per unit volume from a particle size distribution in cumulative mass, or volume, percent less than particle diameter. Data inputs are contained in columns A and B, and cells F6 and F7 must also contain data if the specific surface area per unit volume is to be used in the calculation of cake permeability (by the Kozeny–Carman equation), and then to calculate a specific resistance to filtration. All other cells contain calculated values or text, as illustrated on the accompanying printout. The cell formulae which will need copying into other areas of the table, and accompanying notes are given in the following table.

Cell	Cell formulae	Equations and comments
D37	+A36	Top size in grade comes from previous increment
E37	+A37	Bottom size in grade comes from the same increment
F37	(A36+A37)/2	Average size in grade
G37	(B36–B37)/100	Amount in grade expressed as a fraction: percentage/100
H37	+G37/F37	
H39	@SUM(H12..H37)	
E42	6*H39*1000000	Specific surface area per unit volume is the surface area over the volume shape factors (6 for spheres), multiplied by the sum of the mass fraction in each increment over the average particle diameter in that increment (see Appendix A)
D43	1/H39	
D44	(1-F7)^3/(5*F7^2*E42^2)	This is Equation (2.4) for cake permeability: $k = \dfrac{\varepsilon^3}{5(1-\varepsilon)^2 S_v^2}$
D45	1/(D44*F7*F6)	This is Eequation (2.11) linking cake permeability and specific resistance: $\alpha = \dfrac{1}{kC\rho_s}$
		In practice, measured resistance may be up to two orders of magnitude greater than this value owing to the migration of fines within the filter cake

	A	B	C	D	E	F	G	H	I
1	SPECIFIC.WB1								
2			Specific surface area per unit volume from a size distribution						
3			*with cake resistance during filtration via Kozeny-Carman*						
4									
5	INPUTS:	**	SECTION FOR SPECIFIC RESISTANCE:					**	
6		**	Solid density:			2650	kg/m^3	**	
7		**	Cake concentration:			0.38	v/v	**	
8	Particle	Cumul've	**	Topsize	Bottom	Average	Mass	Mass fract'n	
9	size	mass	**	of	of	size	fraction	over	
10		undersize	**	grade	grade			mid point	
11	(um)	(%)	**	(um)	(um)	(um)	(-)	(1/um)	
12	88.1	100	**	102.1	88.1	95.1	0	0	
13	76	99.7	**	88.1	76	82.05	0.003	3.7E-05	
14	65.6	98.8	**	76	65.6	70.8	0.009	0.00013	
15	56.6	97.2	**	65.6	56.6	61.1	0.016	0.00026	
16	48.8	94.5	**	56.6	48.8	52.7	0.027	0.00051	
17	42.1	90.7	**	48.8	42.1	45.45	0.038	0.00084	
18	36.3	85.6	**	42.1	36.3	39.2	0.051	0.0013	
19	31.3	79.1	**	36.3	31.3	33.8	0.065	0.00192	
20	27	71.5	**	31.3	27	29.15	0.076	0.00261	
21	23.3	63.6	**	27	23.3	25.15	0.079	0.00314	
22	20.1	55.9	**	23.3	20.1	21.7	0.077	0.00355	
23	17.4	49.2	**	20.1	17.4	18.75	0.067	0.00357	
24	15	43	**	17.4	15	16.2	0.062	0.00383	
25	12.9	36.6	**	15	12.9	13.95	0.064	0.00459	
26	11.1	29.9	**	12.9	11.1	12	0.067	0.00558	
27	9.6	24.3	**	11.1	9.6	10.35	0.056	0.00541	
28	8.3	19.7	**	9.6	8.3	8.95	0.046	0.00514	
29	7.2	15.9	**	8.3	7.2	7.75	0.038	0.0049	
30	6.2	12.6	**	7.2	6.2	6.7	0.033	0.00493	
31	5.3	9.2	**	6.2	5.3	5.75	0.034	0.00591	
32	4.6	6.3	**	5.3	4.6	4.95	0.029	0.00586	
33	4	3.9	**	4.6	4	4.3	0.024	0.00558	
34	3.4	2	**	4	3.4	3.7	0.019	0.00514	
35	3	1	**	3.4	3	3.2	0.01	0.00313	
36	2.6	0.2	**	3	2.6	2.8	0.008	0.00286	
37	2.2	0	**	2.6	2.2	2.4	0.002	0.00083	
38	**	**	**					---------	
39								0.08155	
40	CALCULATED VALUES:								
41	SPECIFIC SURFACE IS:			6 times	0.08155	1/microns			
42	(ie assuming spheres)			=		489294	1/metres		
43	Sauter mean diameter:				12.3	microns			
44	Permeability (by K & C):			1.4E-12	square metres				
45	Specific resistance:			7.2E+08	m/kg				

CONPRESS.WB1

A spreadsheet to calculate the specific cake and medium resistances from constant-pressure filtration data if filtration time and volume filtrate are measured. Other required inputs are the total filtration pressure used, physical data on the slurry and the masses of wet and dry cake–or a sample of the cake. Cake moisture content is calculated from the last two inputs, and is needed for the calculation of mass of dry cake per unit volume filtrate and, hence the specific cake resistance from the gradient of the line of best fit between the time over filtrate volume plotted against filtrate volume. The example data were used in Figure 2.7, and the worked example following that figure. Reference should be made to Equations (2.24) and (2.25).

The spreadsheet is used to analyse laboratory data, but could also be used to investigate data obtained from a process-scale constant-pressure filter, such as a plate and frame filter after achieving the maximum working operating pressure.

Data inputs are contained in columns A, B and C, and cells F7–F12. All other cells contain calculated values or text as illustrated on the accompanying printout. This spreadsheet uses the facility for linear regression for the time over filtrate volume plot. The cell defined as containing the regression output is cell H14. The cell formulae which will need copying into other areas of the spreadsheet, and accompanying notes are given in the following table.

Cell	Cell formulae	Equations and comments
E19	+A19/C19	Filtration time/cumulative filtrate volume
F19	+$M$10*$M$8*C19/$F$7/($M$10*$M$8	
	* C19/$F$7÷$M$11)*$M$9	The fraction of the total pressure drop going to form the cake is: $\dfrac{R_c}{R_c + R_m}$
		where the filter cake resistance at any instant in time is: $R_c = \alpha c V / A$
		The cake forming pressure is not required in the calculation of cake or medium resistances, but is needed when determining the relation between pressure and either resistance or concentration. These are covered on the next spreadsheet
B26	@AVG(B18..B25)	The average pressure allows for some fluctuation in the total pressure over the cake and medium.
M7	+F11/F12	Moisture ratio: mass of sample of wet cake over mass of same sample when dry
M8	+F10*F9/(1-F10*M7)	Dry mass of solids per unit filtrate volume (Eq. 2.18): $c = \left[\dfrac{sp}{(1-sm)} \right]$

CONPRESS.WB1 Continued

Cell	Cell formulae	Equations and comments
M9	+B26*100000	
M10	+J21*2*F7^2*M9/(F8*M8)	Equation (2.24) rearranged for $\alpha = \dfrac{gradient 2A^2 \, \Delta P}{\mu c}$
M11	−K15*F7*M9/F8	Equation (2.25) rearranged for medium resistance $R_m = \dfrac{intercept A \Delta P}{\mu}$
M12	@AVG(F23..F25)	

	A	B	C	D	E	F	G	H	I	J	K	L	M	N
1	CONPRESS.WB1													
2			Constant pressure filtration readings from vacuum or pressure test											
3						cake and cloth resistance follows from								
4						equation (2.23)								
5														
6	INPUTS **********************************							CALCULATED VALUES: ********************************						
7	Filter area:					2.72	m^2	Moisture ratio of cake:			2.01575			
8	Liquid viscosity:					0.001	Pa s	Dry cake per filtrate volume (equn 2.18):			125.247	kg/m^3		
9	Liquid density:					1000	kg/m^3	Filtration pressure:			300000	Pa		
10	Solids in slurry weight fraction:					0.1	-	Cake specific resistance:			5.2E+11	m/kg		
11	Mass of wet cake sample:					1280	g	Cloth resistance:			2.9E+12	m^-1		
12	Mass of dry cake sample:					635	g	Cake forming pressure:			134177	Pa		
13	******************************						*********	***						
14	Filtration	Filtration	Filtrate	*********	Time	Pressure	Fitted				Regression Output:			
15	time	pressure	Volume	*********	over	forming	t/V				Constant	3506.64		
16				*********	volume	cake					Std Err of Y Est	37.1443		
17	(s)	(bar)	(m^3)	*********	(s/m^3)	(Pa)	(s/m^3)				R Squared	0.99435		
18	0	3	0	*********							No. of Observations	7		
19	92	3	0.024	*********	3833.33	50295	3859.7852	Degrees of Freedom			5			
20	160	3	0.039	*********	4102.56	73977	4080.5015							
21	232	3	0.054	*********	4296.3	93557	4301.2177	X Coefficient(s)	14714.4					
22	327	3	0.071	*********	4605.63	112013	4551.3628	Std Err of Coef.	495.945					
23	418	3	0.088	*********	4750	127440	4801.5079							
24	472	3	0.096	*********	4916.67	133856	4919.2233							
25	538	3	0.106	*********	5075.47	141235	5066.3674							
26	Average:	3	bar											

EXPONENT.WB1

A spreadsheet to calculate the exponents and constants in the constitutive equations linking specific resistance and concentration with cake forming pressure, Equations (2.38) and (2.40).

Data inputs are contained in columns A, B and C, and cells F6 and F7. The data used here could be the result of tests analysed using the previous spreadsheet. Two equations are being investigated, hence two linear regressions are conducted: the regression for the concentration equation is based on cell I6, and the regression for the specific resistance equation is based on cell M6. Both regressions use a natural logarithm plot of the data entered in columns A and B, and data calculated in column E.

Cell	Cell formulae	Equations and comments
E14	1/(1+(C14-1)*$F$6/$F$7)	Moisture ratio converted into a cake concentration by volume fraction from a knowledge of the solid and liquid densities: $C = 1 / \left(1 + \dfrac{(m-1)\rho_s}{\rho} \right)$
F14	@LN(E14)	
G14	@LN(B14)	
H14	@LN(A14)	
K16	@EXP(L7)/(1-K17)	After linear regression the intercept provides a value for C_o, in accordance with the log form of Equation (2.40), i.e. $C_o = \exp(intercept)/(1-m)$
K17	+K13	The exponent m comes from the gradient of the regression
K18	@EXP(P7)/(1-K19)	After linear regression the intercept provides a value for α_o, in accordance with the log form of Equation (2.38), i.e. $\alpha_o = \exp(intercept)/(1-n)$
K19	+O13	The exponent n comes from the gradient of the regression

EXPONENT.WB1

Exponents and constants in resistance and concentration equations

follows from laboratory constant pressure tests

	A	B	C	D	E	F	G	H	I	J	K	L	M	N	O
1	EXPONENT.WB1														
2			Exponents and constants in resistance and concentration equations												
3				*follows from laboratory constant pressure tests*											
5	INPUTS *********	*********	*********	*********	*********	*********	*********	********Concentration:	*********			Alpha:		Regression Output:	
6	Solid density:					2650	kg/m^3	*********	********	Regression Output:					
7	Liquid density:					1000	kg/m^3	*********	Constant			-2.068	Constant		
8					*********	*********	*********	*********	Std Err of Y Est			0.01557	Std Err of Y Est		
9					CALCULATED VALUES:				R Squared			0.96015	R Squared		
10	Cake	Specific	Moisture	***:	Cake	ln(C)	ln(a)	ln(DP)	No. of Observations			4	No. of Observations		
11	forming	resistance	ratio	***:	concn				Degrees of Freedom			2	Degrees of Freedom		
12	pressure			**											
13	(Pa)	(m/kg)	(-)	***:	(v/v)				X Coefficient(s)		0.08594		X Coefficient(s)		0.3944
14	55000	1.9E+11	1.79	***:	0.32326	-1.129303	2.597E+01	10.9151	Std Err of Coef.		0.01238		Std Err of Coef.		0.11811
15	100000	2.4E+11	1.74	***:	0.33772	-1.085527	2.622E+01	11.5129	GIVES: *********	*********	*********	*********	*********	*********	*********
16	200000	3.8E+11	1.65	***:	0.36731	-1.001551	2.667E+01	12.2061	Constant Co:		0.138	*********	*********	*********	*********
17	280000	3.3E+11	1.65	***:	0.36731	-1.001551	2.651E+01	12.5425	Exponent m:		0.086	*********	*********	*********	*********
18	*********	*********	*********	*********	*********	*********	*********	*********	Constant (alpha)o:		4.3E+09	*********	*********	*********	*********
19									Exponent n:		0.394	*********	*********	*********	*********

RVF.WB1

A spreadsheet to investigate the effect of the feed concentration of the slurry on throughput, assuming the medium and cake resistances are not a function of this, i.e. true incompressible behaviour. Cake and medium properties have to be evaluated separately. This spreadsheet provides a simulation of the filter, where the effect of the following on throughput can also be investigated: rotational speed, drum submergence, drum area and total filtration pressure.

The solid density is needed only if volume slurry throughput or cake depth are also required; the latter is often important for the assessment of cake discharge.

Cell	Cell formulae	Equations and comments
H6	+D6/60	
H7	+D7/H6	Calculates the amount of time during each cycle that any portion of the drum is in the tank and forming new cake
H21	+A21/100	
I21	+H21*$D$11/(1−$D$14*H21)	Dry cake mass per unit filtrate volume: Equation (2.18): $c = \left[\dfrac{s\rho}{(1 - sm)} \right]$
J21	($D$10*I21*$D$12)/(2*$D$8^2* $D$7^2*$D$9)	Equation (2.23) rearranged as a quadratic: $aV^2 + bV - t = 0$ this is coefficient a
K21	($D$10*$D$13)/($D$8*$D$7* $D$9)	Coefficient b in the above equation
L21	(−K21+(K21^2−4*J21* (−$H$7)^(0.5))/(2*J21)	Solution to the above quadratic equation, giving volume of filtrate produced per cycle
C21	+I21*L21/($D$8*$D$7)*(1/$D$16+ ($D$14−1)/$D$11)*1000	Calculates cake depth from solid mass balance: $\dfrac{cV}{FA}\left[\dfrac{1}{\rho_s} + \dfrac{(m-1)}{\rho} \right]1000$ where F is the fractional submergence

RVF.WB1 Continued

Cell	Cell formulae	Equations and comments
D21	+I21*L21*3600/$H$7	Product of dry solids per unit filtrate volume and filtrate volume, converted from per cycle to per hour
E21	($D$14*I21*L21+L21*$D$11)* 3600/$H$7	Mass of wet cake and filtrate per hour, i.e. $cV'm + V'\rho$
F21	(I21*L21*(1/$D$16+($D$14-1)/$D$11)	Volume slurry throughput using given densities:

$$\left[cV'\left(\frac{1}{\rho_s} + \frac{(m-1)}{\rho} \right) + V' \right] \frac{3600}{form - time} 1000$$

+L21)*3600/$H$7*1000

Cell addresses in bold italics indicate results which are out of alphabetic order; this has been done to improve the appearance of the spreadsheet.

	A	B	C	D	E	F	G	H	I	J	K	L
1	RVF.WB1											
2			Prediction of Throughput of a RVF - effect of feed preconcentration									
3			*based on leaf test data for moisture, cake and cloth resistance*									
4												
5	INPUTS:	********	********	********	********	********	CALCULATED VALUES:					
6	Drum speed in rpm:			0.333	r p m		rps:	0.00555				
7	Fractional submergence:			0.3	(-)		form time	54.0541	seconds			
8	Drum area:			1	m^2							
9	Total filtration pressure:			64386.6	Pa							
10	Liquid viscosity:			0.001	Pa s							
11	Liquid density:			1000	kg/m^3							
12	Specific resistance:			2.2E+11	m/kg							
13	Cloth resistance:			5.5E+09	m^-1							
14	Moisture ratio:			1.8	(-)							
15	IF CAKE DEPTH IS NEEDED:					********	********	Intermediate values used in calculations:				
16	Solid density:			2650	kg/m^3							
17	********	********	********	********	********	********	********	*	*	*	*	
18	Slurry		Cake	Dry sol.	Slurry	Slurry	*	Slurry	Dry	First	Second	Volume
19	conc		depth	thru'	thru'	thru'	*	conc.	weight	coeff. of	coeff. of	filtrate
20	(% w/w)		(mm)	(kg/h)	(kg/h)	(l/h)	*	as mass	per unit	equation	equation	produced
								fraction	volume	(2.23)	(2.23)	(m^3)
21	20		4	62	312	273	*	0.2	312.5	5932052	286.292	0.00299
22	22		4	67	306	264	*	0.22	364.238	6914180	286.292	0.00278
23	24		4	73	302	257	*	0.24	422.535	8020803	286.292	0.00258
24	26		5	78	300	252	*	0.26	488.722	9277194	286.292	0.0024
25	28		5	84	300	248	*	0.28	564.516	1.1E+07	286.292	0.00223
26	30		5	90	301	245	*	0.3	652.174	1.2E+07	286.292	0.00208
27	32		6	97	304	243	*	0.32	754.717	1.4E+07	286.292	0.00193
28	34		6	105	308	243	*	0.34	876.289	1.7E+07	286.292	0.00179
29	36		7	113	314	244	*	0.36	1022.73	1.9E+07	286.292	0.00166
30	38		7	123	323	247	*	0.38	1202.53	2.3E+07	286.292	0.00153

PLFCOST.WB1

A worked example showing how basic filtration data can be used to assess the optimum cycle time, and hence filter area required to complete a clarification of a vegetable oil using precoat filtration on a pressure leaf filter. The filtration is assumed to be constant rate, after having formed a precoat of 2 mm on the pressure leaves prior to clarification. The cleaning and reforming of the precoat takes approximately 30% of the total cycle time, hence active filtration time is only 70% of any given cycle time. The analysis is based on a 14 h working day, and 350 days per year. These are easily altered in cells D15 and D16, respectively.

Various cycle times are used in column A from 1 to 8 h, the optimum, i.e. lowest cost per unit volume of filtrate, is 2 h. Alternatively, the lowest total annual cost is given by a 3 hour cycle, but with a significantly lower yield of cleaned oil.

Cell	Cell formulae	Equations and comments
J17	+D6/D7	Constant rate fixed by feed pump
J18	+D8*D10*J17	$\mu R_m V / t$
J19	+D8*D9*D11*J17	$\mu c \alpha V / t$ — at this concentration, and under these operating conditions, the slurry concentration will be approximately equal to the dry solids mass per unit volume filtrate
C23	(1-D$13/100)*A23	Filtration time, which is also the time taken to obtain the maximum pressure differential
D23	3600*C23*J$17	Cumulative filtrate volume produced when the maximum pressure differential has been reached
E23	(J$18+(J$18^2+4*D$12*100000*J$19*D2 3)^(0.5))/(2*D$12*100000)	Equation (2.28) rearranged into a quadratic equation in terms of filter area, and solved for the positive root, at the end of the filtration, i.e. $\Delta P A^2 - \left(\mu R_m \dfrac{V}{t}\right) A - \mu c \alpha \dfrac{V}{t} V = 0$
F23	1000*E23^(0.9)	Cost of filter area empirically fixed by the equation (£) = $1000\,A^{0.9}$
G23	@INT(D$15/A23)*D$16	Only an integer number of precoats per year may exist
H23	+D$14/1000*E23*D$17*G23*D$18/1000	Product of mass of precoat and cost of precoat per unit mass giving total precoat cost in a year
I23	+H23+F23	
J23	+G23*D23	
K23	+I23/J23	

	A	B	C	D	E	F	G	H	I	J	K
1	PLFCOST.WB1										
2		Pressure leaf filter sizing and costing - effect of cycle time									
3			*capital and precoat costs considered*								
4											
5		INPUTS: *********	*********	*********	*********	*********	*********				
6		Volume filtered:		3	m^3	*********					
7		in unit time:		3600	seconds	*********					
8		Viscosity:		0.01	Pa.s	*********					
9		Cake specific resistance:		2E+11	m/kg	*********					
10		Cloth resistance:		1E+11	m^-1	*********					
11		Slurry concentration:		50	kg/m^3	*********					
12		Max. pressure differential:		4.5	Bar	*********					
13		Downtime as % of cycle:		30	%	*********					
14		Precoat depth:		2	mm	*********					
15		Day length:		14	hours	*********					
16		Days per year:		350	days	*********					
17		Precoat bulk density:		350	kg/m^3	*********	Filtrate rate:			0.00083	m^3/s
18		Cost of filter aid:		350	£/tonne	*********	mu.Rm.q:			833333	
19		*********	*********	*********	*********	*********	mu.alpha.w.q:			8.3E+07	
20	Cycle	*********		Volume	Filter	Filter	Number	Precoat	Annual	Oil	Cost per
21	time	*********	Form time	filtrate	area	cost	precoats	cost	cost	cleared	volume
22	(hours)	*********	(hours)	(m^3)	(m^2)	(£)	(annual)	(£)	(£)	(m^3)	(£/m^3)
23	1.0	*********	0.7	2.1	20.7	15267	4900	24812	40079	10290	3.89
24	2.0	*********	1.4	4.2	28.8	20600	2450	17305	37905	10290	3.68
25	3.0	*********	2.1	6.3	35.1	24588	1400	12038	36625	8820	4.15
26	4.0	*********	2.8	8.4	40.4	27895	1050	10387	38282	8820	4.34
27	5.0	*********	3.5	10.5	45.0	30773	700	7723	38496	7350	5.24
28	6.0	*********	4.2	12.6	49.2	33349	700	8445	41793	8820	4.74
29	7.0	*********	4.9	14.7	53.1	35699	700	9108	44807	10290	4.35
30	8.0	*********	5.6	16.8	56.7	37870	350	4863	42733	5880	7.27

CONRATE.WB1

If the material being filtered is incompressible a spreadsheet to simulate the performance of constant-rate filtration based on Equation (2.28) is straightforward. When filtering compressible materials the cake properties will be constantly changing, however, as the cake forming pressure changes. Also, the feed rate may remain constant but the filtrate rate will vary with cake compression. A simulation can be undertaken based on an incremental calculation of cake properties at each stage: i.e. average cake concentration and resistance. This simulation also takes account of the medium resistance, which is assumed to remain constant. The pressure drop over the cloth will vary, and is calculated for each increment. It is assumed that this pressure drop remains constant for the following increment. An iterative solution would overcome this simplifying assumption, but the errors introduced are small and can be minimised by reducing the incremental pressure interval contained in cell E18.

Cell	Cell formulae	Equations and comments
H19	+D6/60000	
K19	1/(1+(1-E7)/E7*E8/E9)	
A25	+$E$18*A24	
C25	(A25-B24)	Current total pressure minus pressure across the cloth in the previous increment - minimise pressure increment to reduce any error here
D25	+$E$15*(1-$E$13)*(C25)^$E$13	Equation (2.40): $C_{av} = C_o (1 - m)\Delta P_c^{\,m}$
E25	+$E$16*(1-$E$14)*(C25)^$E$14	Equation (2.38): $\alpha_{av} = \alpha_o (1 - n)\Delta P_c^{\,n}$
F25	+$H$19*(1-$K$19/D25)	Feed rate (q) minus volume solids and liquid retained in cake, gives instantaneous filtrate rate
H25	+H24+F25*(G25-G24)	
I25	1/((1-$E$7)/($E$7*$E$9)-(1-D25)/(D25*$E$8))	Equations (2.13) and (2.18) combined and rearranged: $1 \Big/ \left(\dfrac{(1 - s)}{s\rho} - \dfrac{(1 - C_{av})}{C_{av}\rho_s} \right)$

CONRATE.WB1 Continued

Cell	Cell formulae	Equations and comments
J25	+I25*H25	Height of cake deposited in increment is: $\delta q s_{v/v} / AC_{av}$ which is added to the previous height for total height.
K25	(G25-G24)*$H$19*$K$19/($E$11*D25)+K24	Pressure forming new cake is total pressure over cake minus that due to filtrate flow through previous cake layers, which is: $\mu\alpha_{av}C_{av}\rho_s \dfrac{dV}{dt} L / A$
L25	+C25-E$10*E25*D25*$E$8*F25*K24/$E$11	Pressure over cake minus pressure lost due to filtrate flow through previous layers of cake
M25	+$E$12-2*K25	
B25	+$E$10*$E$17/$E$11*F25	Uses Darcy's law for pressure drop over cloth
G25	+G24+L25/($E$10*E25*$E$8*F25*$H$19/$K$19/($E$11^2)	Total time is previous time plus new time increment required to form new cake, which is obtained by combining Darcy's law and a mass balance on a newly formed layer of cake

CONRATE.WB1

Prediction of throughput of a filter under constant rate conditions
based on test data for concentration, cake and cloth resistances

INPUTS: *********		*********
Constant feed rate:	70 litres per minute	*********
Slurry concentration:	0.05 by mass	*********
Solid density:	2800 kg/m^3	*********
Liquid density:	1000 kg/m^3	*********
Liquid viscosity:	0.001 Pa.s	*********
Area of filter cell:	9.4 m^2	*********
Vessel height or clearance:	0.05 m	*********
Pressure exponent in concentration:	0.08	*********
Pressure exponent in alpha equation:	0.5	*********
Constant in concentration equation:	0.15	*********
Constant in alpha equation:	4.5E+08	
Medium resistance:	1E+11 1/m	
Pressure increment multiplier:	1.32	*********

CALCULATED VALUES:
Constant feed rate: 0.001167 m^3/s
Slurry concentration: 0.01845 by volume

Total Filtration pressure (Pa)	Pressure drop cloth (Pa)	Pressure drop cake (Pa)	Average cake concn. (v/v)	Average alpha (m/kg)	Filtrate rate (m^3/s)	Filtration time (s)	Volume filtrate (m^3)	Average dry cake per filtrate (kg/m^3)	Mass of dry cake (kg)	Height of cake (m)	Pressure forming new cake (Pa)	Clearance between plates (m)
12411	12411	0	0	0		0	0.000	0	0	0.000	0.000	
16383	11556.24	3971.631	0.27	1.4E+10	0.001086	378	0.411	55.48317	22.78313	0.003	3971.631	0.044
21626	11617.57	10069.3	0.29	2.3E+10	0.001092	570	0.620	55.19027	34.20464	0.005	3220.642	0.040
28546	11649.88	16928.14	0.30	2.9E+10	0.001095	720	0.784	55.03719	43.15151	0.006	3280.445	0.038
37680	11675.65	26030.45	0.31	3.6E+10	0.001098	876	0.956	54.91573	52.49443	0.007	4255.624	0.036
49738	11697.67	38062.39	0.32	4.4E+10	0.0011	1045	1.141	54.81233	62.56105	0.008	5553.019	0.033
65654	11717.32	53956.53	0.33	5.2E+10	0.001101	1230	1.346	54.72042	73.62662	0.010	7277.407	0.031
86664	11735.33	74946.23	0.34	6.2E+10	0.001103	1436	1.573	54.63646	85.94661	0.011	9560.909	0.028
114396	11752.13	102660.6	0.35	7.2E+10	0.001105	1668	1.829	54.55833	99.78135	0.012	12580.12	0.025
151003	11768.01	139250.4	0.36	8.4E+10	0.001106	1930	2.118	54.4847	115.4099	0.014	16569.22	0.022
199323	11783.16	187555.4	0.36	9.7E+10	0.001108	2226	2.447	54.41466	133.1401	0.016	21837.92	0.018
263107	11797.7	251323.7	0.37	1.1E+11	0.001109	2564	2.821	54.34761	153.3162	0.018	28795.4	0.014
347301	11811.72	335503.4	0.38	1.3E+11	0.00111	2948	3.248	54.28311	176.3275	0.020	37982.09	0.009

Chart — y-axis: Filtration pressure, Pa (0; 50000; 100000; 150000; 200000; 250000; 300000; 350000); x-axis: Filtration time, s (0; 500; 1000; 1500; 2000; 2500; 3000).

FILTER.WB1

Constant-pressure compressible cake filtration, solving the basic flow and filtration equations on an incremental time basis. Cloth resistance is included, hence the pressure drop forming the filter cake changes according to the total applied pressure minus that due to the flow rate of filtrate through the filter cloth. This is solved by means of a fixed number of iterations.

Graphs have been inserted onto the spreadsheet to provide an instant assessment of altering a process variable during the simulation. Two printouts from the spreadsheet are provided, as the tabulated results are deduced after the sixth iteration. The table below contains the minimum information required to program the spreadsheet. It should be copied into other cells in the usual way. Further details are provided in [Holdich, 1994].

Cell	Cell formulae	Equation and comments
E15	1/(1+(1-E4)/E4*E5/E6)	Conversion of concentration by mass into concentration by volume fraction
E16	+$E$11*(1-$E$9)*E3^$E$9	First iteration assumes all pressure acts across cake, i.e. negligible cloth resistance to start solution
E17	1/((1-$E$4)/($E$4*$E$6)-(1-E16)/(E16*$E$5))	First calculation of dry solids per unit volume filtrate using above concentration
E18	+$E$12*(1-$E$10)*E3^$E$10	Specific resistance calculated using full pressure drop over cake, i.e. neglecting cloth resistance in first instance $m/(1-m-n)$
E19	+$E$9/(1-$E$9-$E$10)	
E20	0.009*E8	Arbitrary multiplyer of filter area used to fix time increment during solution
E21	+$E$7*$E$18*$E$17/(2*$E$8^2*$E$3)	First coefficient of quadratic equation in terms of filtrate volume, when filtering at constant pressure
E22	+$E$7*$E$13/($E$8*$E$3)	Second coefficient - which does not change during iterations, i.e. $\dfrac{\mu R_m}{A \Delta P}$
E24	+$E$20/D27+D24	Cumulative filtrate time
E26	(@SQRT($E$22^2+4*$E$21*E24)-$E$22)/(2*$E$21)	Equation (2.41) solved for filtrate volume

FILTER.WB1 Continued

Cell	Cell formulae	Equation and comments
E27	1/($E$21*2*E26+$E$22)	The instantaneous filtrate rate comes from the following equation $$\frac{dV}{dt} = \left[\frac{\mu\alpha_{av}c_{av}V}{A^2\Delta P} - \frac{\mu R_m}{A\Delta P} \right]^{-1}$$
E28	+$E$3*(1-$E$22*E27)	Total pressure minus pressure drop over filter cloth
E29	+$E$11*(1-$E$9)*E28^$E$9	New concentration from Equation (2.40) taking into account pressure drop that acts over the cake, i.e. after removing pressure loss over cloth
E30	1/((1-$E$4)/($E$4*$E$6)-(1-E29)/(E29*$E$5))	$$c = \cfrac{1}{\left(\cfrac{1-s}{s\rho} - \cfrac{1-C_{av}}{C_{av}\rho_s} \right)}$$
E31	+$E$12*(1-$E$10)*E28^$E$10	New value of specific resistance using cake pressure drop only, i.e. having removed pressure drop over the cloth
E32	+$E$7*E30*E31/(2*$E$8^2*$E$3)	New value of first coefficient of Equation (2.41) rearranged into a quadratic
E70	(@SQRT($E$22^2+4*E68*E$24)-$E$22)/(2*E68)	See above cells for explanation
E71	1/(E68*2*E70+$E$22)	
E72	(E71-E62)*100/E62	
E73	+$E$3*(1-$E$22*E71)	
E74	+$E$11*(1-$E$9)*E73^$E$9	
E75	1/((1-$E$4)/($E$4*$E$6)-(1-E74)/(E74*$E$5))	
E76	+$E$12*(1-$E$10)*E73^$E$10	

FILTER.WB1 Continued

Cell	Cell formulae	Equation and comments
E77	+$E$7*E75*E76/(2*$E$8^2*$E$3)	
E79	(E73)^(1-$E$9-$E$10)/($E$5*$E$11*$E$12*$E$7*(1-$E$9-$E$10))*1000*($E$8/E71)	Equation (2.50) for cake height, using values obtained after final iteration
E80	+E70*E75	Dry mass of cake produced: cV
E81	+E80+(1-E74)*E79/1000*$E$8*$E$6	
E82	+$E$3	
E83	100*E73/E82	
E84	+E24/E70	

Reference: Holdich, R.G., 1994, Simulation of compressible cake filtration, Filtration and Separation, 31, pp 825–829.

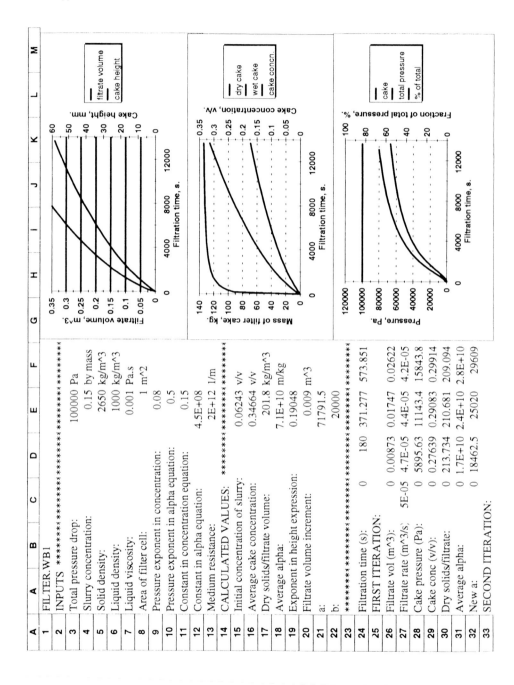

	A	B	C	D	E	F
1	FILTER.WB1					
2	INPUTS ******************************					
3	Total pressure drop:				100000	Pa
4	Slurry concentration:				0.15	by mass
5	Solid density:				2650	kg/m^3
6	Liquid density:				1000	kg/m^3
7	Liquid viscosity:				0.001	Pa.s
8	Area of filter cell:				1	m^2
9	Pressure exponent in concentration:				0.08	
10	Pressure exponent in alpha equation:				0.5	
11	Constant in concentration equation:				0.15	
12	Constant in alpha equation:				4.5E+08	
13	Medium resistance:				2E+12	1/m
14	CALCULATED VALUES: **********************					
15	Initial concentration of slurry:				0.06243	v/v
16	Average cake concentration:				0.34664	v/v
17	Dry solids/filtrate volume:				201.8	kg/m^3
18	Average alpha:				7.1E+10	m/kg
19	Exponent in height expression:				0.19048	
20	Filtrate volume increment:				0.009	m^3
21	a:				71791.5	
22	b:				20000	
23	**					
24	Filtration time (s):		0	180	371.277	573.851
25	FIRST ITERATION:					
26	Filtrate vol (m^3):		0	0.00873	0.01747	0.02622
27	Filtrate rate (m^3/s;		5E-05	4.7E-05	4.4E-05	4.2E-05
28	Cake pressure (Pa):		0	5895.63	11143.4	15843.8
29	Cake conc (v/v):		0	0.27639	0.29083	0.29914
30	Dry solids/filtrate:		0	213.734	210.681	209.094
31	Average alpha:		0	1.7E+10	2.4E+10	2.8E+10
32	New a:		0	18462.5	25020	29609
33	SECOND ITERATION:					

A	B	C	D	E	F	G	H	I	J	K	L	M
69	SIXTH ITERATION:											
70	Filtrate vol (m^3):	0	0.00898	0.01837	0.02806	0.03791	0.04785	0.05781	0.06775	0.07765	0.0875	0.0973
71	Filtrate rate (m^3/s):	5E-05	5E-05	4.9E-05	4.8E-05	4.6E-05	4.5E-05	4.3E-05	4.1E-05	4E-05	3.8E-05	3.6E-05
72	dt/dV change (%):	0.00	0.04	0.09	0.13	0.15	0.15	0.14	0.13	0.11	0.09	0.07
73	Cake pressure (Pa):	0	553.079	2050.56	4339.57	7217.2	10467.3	13900.5	17372.1	20782.8	24070.9	27201.7
74	Cake conc (v/v):	0	0.22872	0.254	0.2697	0.2809	0.28938	0.29602	0.30135	0.3057	0.30931	0.31235
75	Dry solids/filtrate:	0	227.575	219.378	215.293	212.738	210.971	209.676	208.69	207.917	207.295	206.786
76	Average alpha:	0	5.3E+09	1E+10	1.5E+10	1.9E+10	2.3E+10	2.7E+10	3E+10	3.2E+10	3.5E+10	3.7E+10
77	New a:	0	6021.04	11175.9	15955.4	20332.1	24282.4	27811	30944.3	33720.5	36181.6	38368.3
78												
79	Cake height (mm):	0	3.79848	6.68676	9.38051	11.9751	14.5071	16.9943	19.4465	21.87	24.2689	26.6465
80	Dry cake (kg):	0	2.0425	4.03031	6.0403	8.0654	10.0951	12.1213	14.1388	16.1448	18.1384	20.1193
81	Wet cake (kg):	0	4.9722	9.01865	12.8909	16.6767	20.4041	24.0849	27.7251	31.3292	34.9006	38.4427
82	Total pressure (Pa):	100000	100000	100000	100000	100000	100000	100000	100000	100000	100000	100000
83	Fraction of total (%)	0	0.55308	2.05056	4.33957	7.2172	10.4673	13.9005	17.3721	20.7828	24.0709	27.2017
84	t/V data (s/m^3):	0	20055.6	20209.3	20453.6	20777.9	21169.1	21614.5	22102.4	22623.5	23170.2	23736.6

PROFILE.WB1

The solid concentration profile, i.e. the height from the filter medium as a function of solid volume fraction concentration, is calculated in this spreadsheet. Equations (2.46) to (2.50) are used in the analysis. The spreadsheet uses the same method of solution as that described in the previous file, except that the original slurry height, or clearance, is included in cell E9; this provides the initial concentration profile which is referenced in cells E85 and E86.

Only the cell formulae for the table of results, row 81 and beyond, are provided because of the similarity of the earlier cell formulae with the previous spreadsheet. Use of the spreadsheet commands BLOCK-INSERT-ROW when the cursor is positioned on row 9 of the previous spreadsheet will ensure that all the appropriate row renumbering, to accommodate the extra line, is completed successfully.

Finally, the pressure forming the filter cake increases during the filtration, as the flow rate of filtrate through the cloth reduces and hence that pressure drop also diminishes. Hence the cake concentration increases during the apparently constant pressure filtration. This can be seen on the accompanying figure. If the cloth resistance is negligible an iterative solution is not necessary, as the pressure drop forming the cake is a constant, and the concentration profile will also be constant when plotted in dimensionless terms in accordance with Equation (2.46).

Cell	Cell formulae	Equation and comments
E81	(E75)^(1-$E$10-$E$11)/($E$5*(E73/$E$8)*$E$12*$E$13*$E$7*(1-$E$10-$E$11))	Equation (2.50) for cake height, using values given by final iteration
E85	1000*$E$9	Height of slurry
E86	1000*$E$9	Height of slurry again, cells E85 and E86 are used to provide the rectangular box shape, at filtration time zero, on the graph shown on the PROFILE.WB1 worksheet reproduced in the following figure
E87	(E$81-$A$87*E$81)*1000	Equations (2.46) and (2.50) combined to give solid concentration by volume fraction within the cake in terms of actual distance from the filter cloth, rather than dimensionless distance used in Equation (2.46)

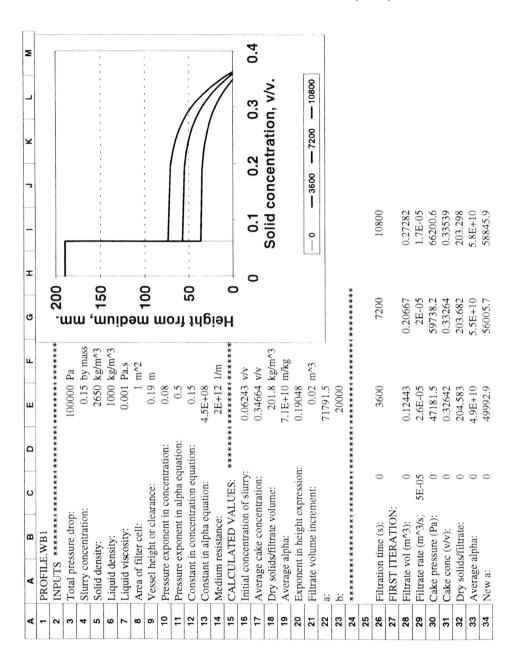

PROFILE.WB1

INPUTS ***

Label	Value	Units
Total pressure drop:	100000	Pa
Slurry concentration:	0.15	by mass
Solid density:	2650	kg/m^3
Liquid density:	1000	kg/m^3
Liquid viscosity:	0.001	Pa.s
Area of filter cell:	1	m^2
Vessel height or clearance:	0.19	m
Pressure exponent in concentration:	0.08	
Pressure exponent in alpha equation:	0.5	
Constant in concentration equation:	0.15	
Constant in alpha equation:	4.5E+08	
Medium resistance:	2E+12	1/m

CALCULATED VALUES: ***

Label	Value	Units
Initial concentration of slurry:	0.06243	v/v
Average cake concentration:	0.34664	v/v
Dry solids/filtrate volume:	201.8	kg/m^3
Average alpha:	7.1E+10	m/kg
Exponent in height expression:	0.19048	
Filtrate volume increment:	0.02	m^3
a:	71791.5	
b:	20000	

****

Filtration time (s):	0	3600	7200	10800
FIRST ITERATION:				
Filtrate vol (m^3):	0	0.12443	0.20667	0.27282
Filtrate rate (m^3/s):	5E-05	2.6E-05	2E-05	1.7E-05
Cake pressure (Pa):		47181.5	59738.2	66200.6
Cake conc (v/v):	0	0.32642	0.33264	0.33539
Dry solids/filtrate:	0	204.583	203.682	203.298
Average alpha:	0	4.9E+10	5.5E+10	5.8E+10
New a:	0	49992.9	56005.7	58445.9

Chart: Height from medium, mm (0, 50, 100, 150, 200) vs Solid concentration, v/v. (0, 0.1, 0.2, 0.3, 0.4). Legend: 0 — 3600 — 7200 — 10800

A	A	B	C	D	E	F	G	H	I
71	SIXTH ITERATION:								
72	Filtrate vol (m^3):		0		0.13733		0.22463		0.29334
73	Filtrate rate (m^3/s)	5E-05			3.1E-05		2.3E-05		1.9E-05
74	dt/dV change (%):		0.00		0.03		0.01		0.00
75	Cake pressure (Pa):		0		38324.5		54655.5		62710.9
76	Cake conc (v/v):		0		0.32104		0.33029		0.33394
77	Dry solids/filtrate:		0		205.398		204.019		203.5
78	Average alpha:		0		4.4E+10		5.3E+10		5.6E+10
79	New a:		0		45236.3		53658.7		57330.9
80									
81	Cake height (m):		0		0.03632		0.05735		0.07388
82	Dimen'ss	Cake	Cake	Cake	Cake	Cake	Cake	Cake	Cake
83	height	concn.	height	concn.	height	concn.	height	concn.	height
84	(-)	(v.f.)	(mm)	(v.f.)	(mm)	(v.f.)	(mm)	(v.f.)	(mm)
85		0	190	0	190	0	190	0	190
86		0.06243	190	0.06243	190	0.06243	190	0.06243	190
87	0	0.06243	0.00	0.06243	36.32	0.06243	57.35	0.06243	73.88
88	0.04	0	0.00	0.18901	34.87	0.19446	55.05	0.19661	70.93
89	0.08	0	0.00	0.21569	33.42	0.2219	52.76	0.22436	67.97
90	0.12	0	0.00	0.23301	31.96	0.23972	50.47	0.24237	65.02
91	0.16	0	0.00	0.24613	30.51	0.25322	48.17	0.25603	62.06
92	0.2	0	0.00	0.25682	29.06	0.26422	45.88	0.26714	59.10
93	0.24	0	0.00	0.2659	27.61	0.27356	43.58	0.27658	56.15
94	0.28	0	0.00	0.27382	26.15	0.28171	41.29	0.28482	53.19
95	0.32	0	0.00	0.28087	24.70	0.28896	39.00	0.29216	50.24
96	0.36	0	0.00	0.28725	23.25	0.29552	36.70	0.29879	47.28
97	0.4	0	0.00	0.29307	21.79	0.30151	34.41	0.30485	44.33
98	0.44	0	0.00	0.29844	20.34	0.30704	32.11	0.31043	41.37
99	0.48	0	0.00	0.30343	18.89	0.31217	29.82	0.31562	38.42
100	0.52	0	0.00	0.30809	17.43	0.31696	27.53	0.32047	35.46
101	0.56	0	0.00	0.31247	15.98	0.32147	25.23	0.32502	32.51
102	0.6	0	0.00	0.3166	14.53	0.32572	22.94	0.32932	29.55
103	0.64	0	0.00	0.32052	13.08	0.32975	20.64	0.3334	26.60
104	0.68	0	0.00	0.32424	11.62	0.33358	18.35	0.33727	23.64
105	0.72	0	0.00	0.32779	10.17	0.33723	16.06	0.34096	20.69
106	0.76	0	0.00	0.33118	8.72	0.34072	13.76	0.34449	17.73
107	0.8	0	0.00	0.33443	7.26	0.34407	11.47	0.34787	14.78
108	0.84	0	0.00	0.33756	5.81	0.34728	9.18	0.35112	11.82
109	0.88	0	0.00	0.34056	4.36	0.35037	6.88	0.35425	8.87
110	0.92	0	0.00	0.34346	2.91	0.35335	4.59	0.35726	5.91
111	0.96	0	0.00	0.34625	1.45	0.35623	2.29	0.36017	2.96
112	1	0	0.00	0.34895	-0.00	0.35901	-0.00	0.36298	-0.00

BATCH SEDIMENTATION.XLS

The batch sedimentation spreadsheet employing a finite difference solution to Equation (3.37) requires nine pages in the workbook. It is, therefore, too complicated to describe in the same way as the previous spreadsheets. However, the file can be downloaded from the World Wide Web site provided earlier. The accompanying figures illustrate the first page of the spreadsheet and show the required inputs. These are: initial suspension conditions of height and concentration, values determining the compressibility of the solids forming the sediment, values determining the permeability of the sediment and physical properties such as densities and liquid viscosity. Further notes on how to use the spreadsheet to investigate the batch sedimentation of compressible compacts are provided on the spreadsheet itself, starting in cell P18.

A	B	C	D–F	G	H–I	J
BATCH SEDIMENTATION.XLS			Three time-level method - with initiation of calculation			
richard_holdich@bigfoot.com			NB. REQUIRES ITERATION			
Initial concentration:		0.2 v/v		Solid density:		2578 kg/m^3
Initial suspension height:		0.329 m		Liquid density:		1000 kg/m^3
Specific surface of particles:		800000 m^2/m^3		Liquid viscosity:		0.001 Pa s
Coefficient in C v Ps power law:		0.2265739		Kozeny gradient:		63.703
Exponent in C v Ps power law:		0.0442918 i.e.	C = Co Ps^n	Kozeny intercept:		-1.4341
dw =	h =	0.1				
dt =	k =	0.0035 NOTE BOUNDARY CONDITION FORMULA NOT SAME AS ELSEWHERE ON SPREADSHEET				
***********	***********	***********	***********	***********	***********	
2r =	2k/h^2	0.7 This must be less than 1 for explicit solution		Initial Kozeny constant:		11.3065
Hindered settling velocity:		5.477E-06 m/s		Multiplier in PI*:		0.0009817 1/Pa
to:		60074.213 s		Multiplier in a(PI*):		22.083008
wo		169.6324 kg/m^2		Multiplier in b(PI*):		45.115421
Po:		1018.5958 Pa		(1-p/ps)g:		6.0047246
0.0598069 Pa, i.e. solids pressure at initial C				Initiator cell (0 or 1):	1 CHECK THIS CELL:	0.2775318
*********** *********** DIMENSIONLESS EXCESS HYDRAULIC PRESSURE ***********						

Real time (s)	i =	Time*	0	1	2	3	4	5	6	7	8	9
(w =)			0	0.1	0.2	0.3	0.4	0.5	0.6	0.7	0.8	0.9
j =			0.9999413	0.8999413	0.7999413	0.6999413	0.5999413	0.4999413	0.3999413	0.2999413	0.1999413	0.0999413
0	0	0	0.9999	0.8999	0.7999	0.6999	0.5999	0.4999	0.3999	0.2999	0.1999	0.0999
210.25974	1	0.0035	0.9499	0.8999	0.7999	0.6999	0.5999	0.4999	0.3999	0.2999	0.1999	0.0999
420.51949	2	0.007	0.9249	0.8999	0.7999	0.6999	0.5999	0.4999	0.3999	0.2999	0.1999	0.0999
630.77923	3	0.0105	0.9124	0.8997	0.7999	0.6999	0.5999	0.4999	0.3999	0.2999	0.1999	0.0999
841.03898	4	0.014	0.9061	0.8995	0.7999	0.6999	0.5999	0.4999	0.3999	0.2999	0.1999	0.0999
1051.2987	5	0.0175	0.9028	0.8990	0.7999	0.6999	0.5999	0.4999	0.3999	0.2999	0.1999	0.0999
1261.5585	6	0.021	0.9009	0.8982	0.7999	0.6999	0.5999	0.4999	0.3999	0.2999	0.1999	0.0999
1471.8182	7	0.0245	0.8996	0.8969	0.7999	0.6999	0.5999	0.4999	0.3999	0.2999	0.1999	0.0999
1682.078	8	0.028	0.8982	0.8946	0.7998	0.6999	0.5999	0.4999	0.3999	0.2999	0.1999	0.0999
1892.3377	9	0.0315	0.8964	0.8911	0.7998	0.6998	0.5999	0.4999	0.3999	0.2999	0.1999	0.0999

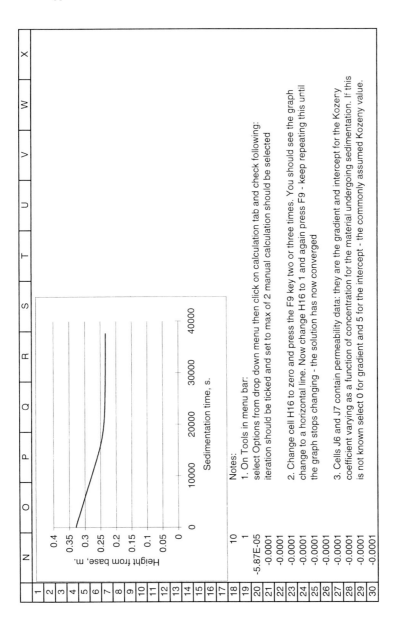

	N	O	P	Q	R	S	T	U	V	W	X
1											
2											
3											
4											
5											
6											
7											
8											
9											
10											
11											
12											
13											
14											
15											
16											
17											
18	10										
19	1										
20	-5.87E-05										
21	-0.0001										
22	-0.0001										
23	-0.0001										
24	-0.0001										
25	-0.0001										
26	-0.0001										
27	-0.0001										
28	-0.0001										
29	-0.0001										
30	-0.0001										

Notes:

1. On Tools in menu bar:
select Options from drop down menu then click on calculation tab and check following:
iteration should be ticked and set to max of 2 manual calculation should be selected

2. Change cell H16 to zero and press the F9 key two or three times. You should see the graph change to a horizontal line. Now change H16 to 1 and again press F9 - keep repeating this until the graph stops changing - the solution has now converged

3. Cells J6 and J7 contain permeability data: they are the gradient and intercept for the Kozery coefficient varying as a function of concentration for the material undergoing sedimentation. If this is not known select 0 for gradient and 5 for the intercept - the commonly assumed Kozeny value.

Index